Preparation, Physico-Chemical Properties and Biomedical Applications of Nanoparticles

Preparation, Physico-Chemical Properties and Biomedical Applications of Nanoparticles

Editor

Simona Cavalu

MDPI • Basel • Beijing • Wuhan • Barcelona • Belgrade • Manchester • Tokyo • Cluj • Tianjin

Editor
Simona Cavalu
Faculty of Medicine and
Pharmacy
University of Oradea
Oradea
Romania

Editorial Office
MDPI
St. Alban-Anlage 66
4052 Basel, Switzerland

This is a reprint of articles from the Special Issue published online in the open access journal *Materials* (ISSN 1996-1944) (available at: www.mdpi.com/journal/materials/special_issues/Preparation_Physico-Chemical_Biomedical_Nanoparticles).

For citation purposes, cite each article independently as indicated on the article page online and as indicated below:

LastName, A.A.; LastName, B.B.; LastName, C.C. Article Title. *Journal Name* **Year**, *Volume Number*, Page Range.

ISBN 978-3-0365-4534-9 (Hbk)
ISBN 978-3-0365-4533-2 (PDF)

Contents

About the Editor

Simona Cavalu

Professor Simona Cavalu serves as Full Professor in the Department of Preclinical Sciences, Faculty of Medicine and Pharmacy, University of Oradea, Romania and Head of Cells Culture Laboratory. She obtained an International Ph.D in Sciences at Babes-Bolyai University, Cluj – Napoca, Romania, and she has worked as an invited professor in several international institutions including Istanbul Technical University, Debrecen University and ISM-CNR Rome, Italy. She is also member of several Academic Societies: International Society for Ceramic in Medicine, Romanian and European Society of Biomaterials, Romanian Society of Pure and Applied Biophysics, Romanian Society of Medical Physics. Professor Simona Cavalu's research focuses on bio-nano-materials for orthopaedic, dental and tissue regeneration applications, natural polymeric composites for controlled and targeted drug release, titanium cranioplasty, nanoparticles production and characterization, nanomedicine, animal model (in vivo biocompatibility tests), in vitro biocompatibility tests, and natural compounds for different therapeutic strategies. She also served as an invited lecturer at prestigious conferences and co-organizer of international meetings, international reviewer, editorial board member for top-ranked international journals in MDPI, Frontiers, Springer, Elsevier. She has been awarded with various research prizes from international organizations: "Daniel Bunea" Award of Romanian Society of Biomaterials 2014, Excellence Award SRB 2015, Best Conference Chairperson EMN 2016, Gold Medal – National Research Council of Thailand (NRCT) 2018, Special Honor of Invention – Toronto International Society of Innovation & Advanced Skills 2018, Special Award & Gold Medal – Malaysian Research & Innovation Society, and Special Award & Medal – Association of Polish Inventors 2018, as a group member of the invention "Cranial implant with osteointegrating structures and functional coatings".

Preface to "Preparation, Physico-Chemical Properties and Biomedical Applications of Nanoparticles"

The capability to produce nanoparticles in the same size domain as proteins has led to a wide range of applications in the biomedical field, for example, nanoparticle-based strategies for optical diagnosis, cancer therapies, nanomedicine, contrast agents, nano-carriers as a vehicle for nucleic acid delivery, gene and drug delivery systems, and nano-biosensors, to name but a few. The various applications require precisely defined nanoparticle characteristics related to reaction conditions, particle morphology, chemical composition and crystallinity, which can be tailored by fabrication strategy, either "top-down" or "bottom-up". Special attention is paid to "green synthesis" techniques and eco-friendly protocols. To manufacture nanoparticles of appropriate size and form, organisms ranging from primary bacteria to very sophisticated eukaryotes may all be employed. On the other hand, fungi and plant extracts for manufacturing nanoparticles are affordable, readily scaled up, and environment-friendly. Moreover, plant extracts possess the ability to generate nanoparticles with a specified size, shape and content, with high potential to be employed in current medical processes such as fluorescent labelling in immunoassays, targeted administration of therapeutic medications, hyperthermia, and as antibacterial agents.

Even if rapidly developing nanotechnologies offer many benefits, possible toxicity, as a side effect, has to be considered not only to prevent nanoparticle intoxication, but to prepare custom-made health treatments. Nanotechnology has penetrated various fields such as drug delivery and other biomedical applications. Moving these things out of the lab and into the real world is the only way to bring them to life.

Nowadays, the impact of nanotechnology on applications in medicine and biomedical sciences has broader societal and economic effects, enhancing awareness of the business, regulatory, and administrative aspects of medical applications. The selected papers included in the present Special Issue gives readers a critical, balanced and realistic evaluation of existing nanomedicine developments and future prospects, allowing practitioners to plan and make decisions.

The topics covers the use of nanoparticles and nanotechnology in medical applications including biomaterials for tissue regeneration, diagnosis and monitoring, surgery, prosthetics, drug delivery systems, nanocarriers, and wound dressing.

I would like to express my gratitude to all contributors to this issue, who have given so much of their time and effort to help create this collection of high quality papers.

Simona Cavalu
Editor

materials

MDPI

Review

Nanoparticles and Nanostructured Surface Fabrication for Innovative Cranial and Maxillofacial Surgery

Simona Cavalu [1], Iulian Vasile Antoniac [2,3], Aurel Mohan [1,*], Florian Bodog [1,*], Cristian Doicin [2], Ileana Mates [2,4], Mihaela Ulmeanu [2], Roman Murzac [2] and Augustin Semenescu [2,3,*]

1 Faculty of Medicine and Pharmacy, University of Oradea, 10 P-ta 1 Decembrie, 410073 Oradea, Romania; scavalu@uoradea.ro
2 Faculty of Materials Sciences Engineering, University Politechnica Bucharest, 313 Splaiul Independentei, 060042 Bucharest, Romania; iulian.antoniac@upb.ro (I.V.A.); cristian.doicin@upb.ro (C.D.); ileana-mariana.mates@upb.ro (I.M.); mihaela.ulmeanu@upb.ro (M.U.); roman.murzac@upb.ro (R.M.)
3 Romanian Academy of Scientists, 3 Ilfov, 077190 Bucharest, Romania
4 Carol Davila University, Emergency Military Hospital, 134 Calea Plevnei, 010825 Bucharest, Romania
* Correspondence: amohan@uoradea.ro (A.M.); fbodog@uoradea.ro (F.B.); augustin.semenescu@upb.ro (A.S.); Tel.: +40-721-129-386 (A.S.)

Received: 2 October 2020; Accepted: 23 November 2020; Published: 27 November 2020

Abstract: A novel strategy to improve the success of soft and hard tissue integration of titanium implants is the use of nanoparticles coatings made from basically any type of biocompatible substance, which can advantageously enhance the properties of the material, as compared to its similar bulk material. So, most of the physical methods approaches involve the compaction of nanoparticles versus micron-level particles to yield surfaces with nanoscale grain boundaries, simultaneously preserving the chemistry of the surface among different topographies. At the same time, nanoparticles have been known as one of the most effective antibacterial agents and can be used as effective growth inhibitors of various microorganisms as an alternative to antibiotics. In this paper, based on literature research, we present a comprehensive review of the mechanical, physical, and chemical methods for creating nano-structured titanium surfaces along with the main nanoparticles used for the surface modification of titanium implants, the fabrication methods, their main features, and the purpose of use. We also present two patented solutions which involve nanoparticles to be used in cranioplasty, i.e., a cranial endoprosthesis with a sliding system to repair the traumatic defects of the skull, and a cranial implant based on titanium mesh with osteointegrating structures and functional nanoparticles. The main outcomes of the patented solutions are: (a) a novel geometry of the implant that allow both flexible adaptation of the implant to the specific anatomy of the patient and the promotion of regeneration of the bone tissue; (b) porous structure and favorable geometry for the absorption of impregnated active substances and cells proliferation; (c) the new implant model fit 100% on the structure of the cranial defect without inducing mechanical stress; (d) allows all kinds of radiological examinations and rapid osteointegration, along with the patient recover in a shorter time.

Keywords: titanium cranioplasty; endoprosthesis; patented solutions

1. Introduction into Cranioplasty and Craniofacial Reconstruction Using Titanium Meshes and Plates for Major Cranial Defects

As a result of traumatic injuries, congenital deformities, decompressive craniectomies, or bone flap loss due to infections, cranioplasty is performed as a neurosurgery procedure to reshape the imperfections of the skull, with both cosmetic and functional outcomes. This surgical intervention has a long history, dating back to 3000 B.C., according to archeological findings [1]. In 19th century, the use of bone from different donor sites (allographs), such as the ribs or tibia, gained wide population, following the successful case of reimplantation of cranial bone reported by Sir William Macewen in 1885, promoting autografts for cranioplasty [1]. In autograph or autologous cranioplasty, the patient's own bone flap is stored and reused in during the waiting period, in order to perform bonny closure [2,3]. This technique evolved at the same time with the beginning of modern general anesthesia, but has been abandoned due to high infection and absorption rates. Military conflicts all over the world have proved the necessity for advanced techniques in cranioplasty, requiring immediate treatment with immediate decompressive craniectomy, careful brain debridement, followed by watertight dural repair [2].

The reconstruction of the large cranial vault defects thus continues to be a challenge to craniofacial surgeons and neurosurgeons. Solutions to the challenging problem of repairing skull defects continue to evolve in order to improve patient outcomes. Although autologous bone is still considered the gold standard in cranioplasty, via the replacement of previously removed bone flap, alloplastic material is indicated in difficult situations when autologous bone in unavailable, such as graft infection, resorption, or insufficient donor site availability [4]. Cranial reconstruction using bone grafting techniques, although ideal from the perspective of immunological compatibility, has many disadvantages, such as complex and intense surgical work, time consumption, and high resorption rate [5]. It has been reported that, in the pediatric population, a much higher bone absorption rate was noticed compared to adults when autologous bone was used [6]. Although many different methods have been described in literature, using autologous or alloplastic material, there is still no consensus on which method is better, and continuous multidisciplinary researches are conducted aiming to develop the ideal reconstruction materials. However, there is a general consensus about the features of an ideal material to be used in cranioplasty [7–9], including:

- perfect fit into the cranial defect in order to get fast and complete closure
- radiolucent
- resistance to infections
- heat non-deformable
- strengths comparable to that of the surrounding bone
- biomechanical resistance
- easy to manipulate and contour
- inexpensive
- fast application

PMMA (Poly-methylmethacrylate), introduced in the 1960s [10], is a polymerized organic compound of acrylic acid, which can be used either prefabricated, hand-formed, or templated. Being easy to shape, lightweight, and radiolucent, it gained large popularity in recent decades as the most frequently used material for skull reconstruction [10]. Even if the long-term results of PMMA cranioplasty have generally been acceptable, there are some limitations of PMMA, including its exothermic setting reaction (between 70 and 120 °C, which may cause thermal necrosis of healthy bone, and hence additional surgical procedures are required for such saturation with saline solution for protection) and the osteolysis induced by unpolymerized MMA (methylmetacrylate) monomer. Moreover, in vivo fibrous encapsulation may occur as PMMA does not chemically bond to surrounding tissue [11]. However, the possibility of using prefabricate or template patient specific PMMA implants clearly demonstrates their benefits, avoiding the risk of thermal or chemical necrosis [4].

Titanium cranioplasty using meshes and plates gained large popularity for the reconstruction of moderate to large-sized defects, being readily available and providing strong and malleable mechanical features to be shaped intraoperatively [12]. Unalloyed titanium for implant applications is available in grades 1 to 4, according to a concentration of impurities: 0.2–0.5% Fe and 0.18–0.4 % O_2 [13]. Titanium is also compatible with magnetic resonance imaging scans [6]. Other favorable properties of titanium mesh for cranioplasty are the stimulation of the bone ingrowth, based on the concept of osteointegration and osteoinduction [14] and its resistance to corrosion from bodily fluids. The standard mesh is a square of 90/90 mm and a thickness of 2.5 mm, which can be easily manipulated by cutting and shaping according to size of the defect. The fixation system is comprised of helically threaded micro-screws, with a screw head of 3 mm in diameter, with at least four points of anchorage [15], but in the case of larger defects, additional fixation points for anchorage are made for optimum geometry. In Figure 1 are presented the surgical details of classical titanium cranioplasty, with the placement and fixation of titanium mesh.

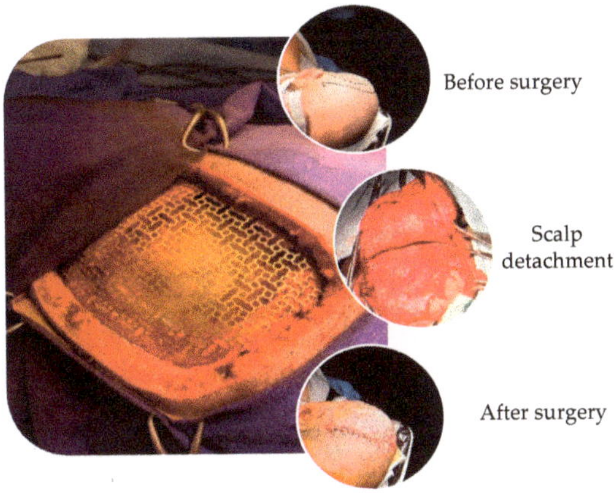

Figure 1. Intraoperatively, surgical details of titanium cranioplasty procedure in the case of a large defect (from private collection of Assoc. Prof. Aurel Mohan).

However, adjacent titanium screws for fixation, which is achieved through self-anchoring screws, may cause excessive tension in the titanium mesh, producing titanium mesh edge tilt, temporal muscle compression, and pain [16]. Moreover, titanium cranioplasty (using classical meshes or plates) is associated with a significant risk of complications. There are some large studies in the literature reviewing the complications of titanium cranioplasty which report complication rates as high as 34% [17] in terms of infection, poor cosmesis, haematoma, headache, and seizures, often necessitating re-operation and plate or mesh removal [17,18]. For example, the study performed by Hill et al. [18] aimed to determine the risk factors of complications following TC (titanium cranioplasty), in a heterogeneous population with various demographic features, taking into account the sizes of cranioplasty. They concluded that infection, seizures, and haemorrhage were the commonest reported complication, while skull defect size was considered to be a determining factor. Instead, the size-related cranioplasty was directly correlated to the hospitalization time. In a similar study conducted by Mukherjee et al. [17], the authors concluded that, besides the size of defect and the traumatic aetiology, the timing of cranioplasty may be important, with late (>12 months) TC associated with a higher rate of complications.

Titanium mesh or plates can be used in cranioplasty either alone or in conjunction with other synthetic materials, such as hydroxyapatite, calcium phosphate, and polyethylene. With commercially available materials such as Medpor Titan [19], a thin sheet of titanium mesh was embedded in a high-density porous polyethylene implant, with better malleability being easily shaped with scissors. This combination provided a fast and effective method for pterional reconstruction after frontotemporal and orbitozygomatic craniotomy, with excellent cosmetic results and patient satisfaction [19]. However, the use of this material in other cranio-anatomical sites or in large defects is not reported. The combination hydroxyapatite-titanium mesh has the advantages of high capacity for osteoconduction and osseointegration, but a high rate of infection was reported when placed in contact with the frontal sinus [20].

Nowadays, one of the strategies adopted in order to overcome the drawbacks and limitations in classical TC is the choice of PEEK (polyetheretherketone) custom implants, prefabricated according to 3D computed tomography and digital modeling of both the cranial defect and the surrounding skeleton, also named "patient-specific implant". Besides the aesthetic outcomes, there are several advantages to using prefabricated implants for cranioplasty, such as decreased surgical complexity and operative times, which consequently minimize the exposure and risk of contamination [21]. This approach offer a precise fit, along with strength, stiffness, durability, temperature resistance, and radiolucency. However, their high costs might be prohibitive, especially in the case of large defects. Moreover, according to a very recent study performed by Rosinsky et al. [22] infection rates are higher among patients receiving custom implants compared to those receiving titanium meshes, but their long-term complication profile remains to be studied. Therefore, the ideal method for alloplastic cranioplasty with large defects would enable a cost-effective implant choice with the possibility for intraoperative modification, including the cases of tumor resection in which the size of the skull defect is unknown preoperatively, and hence prefabricated implants cannot be used.

This aim of this review is to emphasize the main reasons why titanium mesh is preferred for large skull reconstructions, along with the importance of developing innovative surface structures with a dual benefit: improved osteointegration and enhanced antibacterial activity to reduce the risk of post-surgical infection. Based on literature research, a comprehensive review was conducted on the mechanical, physical, and chemical methods for creating nano-structured titanium surfaces along with the identification of the main nanoparticles used for the surface modification of titanium implants, the fabrication methods, their main features, and the purpose of use.

2. The Importance of Nano-Structured Surface on Titanium Implants

Titanium is the most employed implant material for cranial and maxillofacial applications owing to a specific combination of strength and biocompatibility. Its excellent corrosion resistance in body fluids and mechanical properties that closely match human cortical bone in terms of Young's modulus (E = 100–110 GPa for pure titanium grade II), resulting in smaller stress shielding compared to other biometals, are the main features taken into account in developing new and versatile implantable devices [23]. Being a bioinert material, it hampers tissue integration, leading to a short lifespan. In Figure 2, metallographic aspects related to the surface analysis of titanium mesh samples are presented using different microscopic techniques.

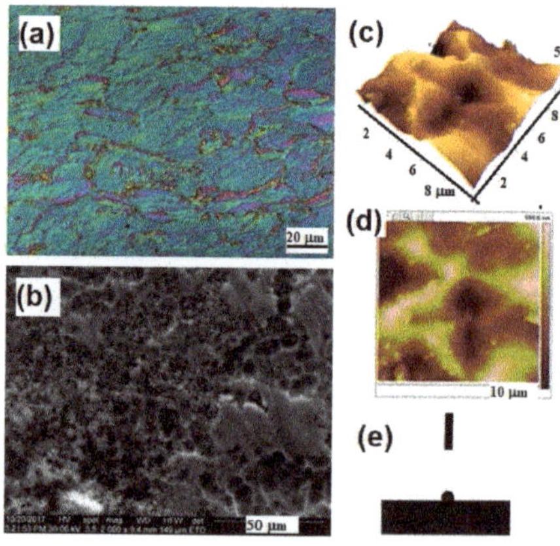

Figure 2. Surface properties of Ti mesh for cranioplasty evidenced by different microscopic techniques: (**a**) light microscopy image in phase contrast, longitudinal section, 500×, Kroll reagent; (**b**) Scanning Electron Microscopy 2000×; (**c**,**d**) 3D and 2D Atomic Force Microscopy images; (**e**) contact angle investigation on the surface of the titanium mesh.

So, strong efforts have been made to develop innovative, suitable implant surface structures, with the aim to promote osseointegration and to prevent periprosthetic infection. Following the rule of "smaller, faster, cheaper", nanopatterning with dimensions smaller than 100 nm has encountered clinical applications in orthopedics and traumatology [24,25]. Many studies in the literature have reported that nanometer-controlled titanium surfaces can be fabricated by a variety of surface modifications techniques, including mechanical, chemical, and physical methods, or a combination of these [25–32].

2.1. Mechanical, Physical and Chemical a Methods for Creating Nano-Structured Titanium Surfaces

Nowadays, there are commercially available cranial titanium implant whose surface is modified by mechanical methods (either machined or grit blasting), or attrition for obtaining nanophase materials), chemical methods (acid or alkali treatment, electrochemical processes, sol-gel, biochemical treatment), and physical methods (thermal or plasma treatment, coatings), or by their combination. Table 1 contains an overview of the commonly applied methods to modify titanium surface structures into micro- or nano- scale for applications in the field of cranio and maxillofacial surgery.

Table 1. Mechanical, physical and chemical methods used for surface modification of Ti implants for cranial and maxilofacial surgery.

	Mechanical Methods	
Technologies	**Texture/Roughness Size**	**Outcome/Reference**
Machining Grinding Blasting	~1 μm, rough surface formed by subtraction process	Specific surface topographies. Improved adhesion and bonding. Auxiliary method to remove contamination. Rarely solely used [25–32]
Shoot peening	20–80 nm grains on the surface	Improved fatigue resistance, hardness and wear [33,34]
Friction stir processing (FSP)	<1 μm, ultrafine grained surface	Improved sliding friction and wear resistance. Incorporation of AgNPs, Zn with antibacterial effect [34,35]
Attrition	<100 nm grains on the surface	Improved tensile properties and surface hardness, higher hydrophilicity, better biological affinity [24,36,37]
Hydrothermal & pressure (HPT)	flake-like titanate layer on Ti substrate, pore size of 300–600 nm	Minimize the time-consumption and the manufacturing cost. Enhance the in vitro cell-material interactions [38]
	Physical Methods	
Technologies	**Texture/Roughness Size**	**Outcome/Reference**
Thermal (flame or plasma) spraying	~30 to ~200 μm of coatings, such as TiO_2, HA, CaP, Al_2O_3, ZrO_2, TiO_2	Improved wear/corrosion resistance and biocompatibility [24,39–41]
Physical vapor deposition: evaporation, sputtering, ion plating	<1 μm, TiN, TiC, TiCN, TiO_2, amorphous carbon films, full density	Improved wear/corrosion resistance and blood compatibility [34,36,42,43]
Ion implantation and deposition	~10 nm of surface modified layer and/or thin film such	Improved hardness, wear, fatigue/corrosion resistance
	as Ti–O, Ti–N films	and blood compatibility [44,45]
Plasma treatment	<100 nm, TiO_2, TiN, TiOH, TiCN layers, full density	Clean and sterilize surface, remove native oxide layer. Improved hardness, wear and corrosion resistances, fatigue limit and biocompatibility [46,47]
Plasma polymerization	Not reported	Bioactive surface. Improved cell adhesion [48]

Table 1. *Cont.*

Chemical Methods		
Technologies	**Texture/Roughness Size**	**Outcome/Reference**
Acidic treatment (HF, HCl, H_2SO_4)	~10 nm oxide layer on the surface	Remove oxide scales and contamination. Used in combination with other treatments (blasting), higher roughness promoting osteoblasts attachment [49,50]
Alkali treatment (NaOH, KOH)	~1 µm sodium titanate gel on the surface	Improved biocompatibility, bioactivity or bone conductivity [32,51]
Hydrogen peroxide treatment	Inner oxide layer <10 nm; outer porous oxide layer up to 40 nm	Improved biocompatibility or bioactivity [34,52]
Passivation treatment (nitric acid, phosphoric acid)	~2–30 nm oxide layer dominated by TiO_2, uniform, full density	Enhanced corrosion, resistance and wear resistance, better bioactivity compared to mechanical treatment [34,53]
Electrochemical methods (anodization, electrodeposition)	~10 nm–10 µm uniform, controllable thickness of TiO_2 layer; adsorption and incorporation of electrolyte anions	Improved adhesive bonding, corrosion resistance, bioactivity, specific surface topographies [54,55]
Chemical vapor deposition	~1 µm of TiN, TiC, TiCN, diamond and diamond-like carbon thin film, nearly full density	Extremely high hardness and wear resistance compared with Ti substrate. Improved corrosion resistance and blood compatibility [56,57]
Sol-gel	<10 µm of thin ceramic coatings, such as $Ca_3(PO_4)_2$, TiO_2, SiO_2	Highly homogeneity and improvement in bioactivity [58,59]
Biochemical methods (by soaking-peptide, proteins immobilization, functional molecules, drug loaded)	self-assembled monolayers, does not ensure controlled deposition	Improved bioactivity, biocompatibility, and/or antibacterial functions [60,61]

The mechanical techniques, like grinding, polishing, machining, or blasting, have the main objective to obtain specific surface topographies designed to improve the adhesion of bioactive molecules and to accelerate the biomineralization as a result of increased surface area. There are some disadvantages in the case of metallic materials, as machining usually produces deformations, while blasting and grinding procedures can induce abrasive pollution on the implant surface [31]. While the classical machining technologies, grinding and polishing are used to remove the contaminated surface layers and to obtain a low surface roughness, and sandblasting is used with the aim to obtain a rough surface with functional topographies, depending on the pressure of the air jet and the particles morphology. For this reason, these classical methods are rarely solely used, due to the limited improvements in terms of surface properties.

The attrition technique is an alternative to fabricate nanophase surface layers on titanium of commercial purity which improve the tensile properties and surface hardness, rough morphology, and higher hydrophilicity [24,31]. As an evolutionary technology from sandblasting, shot peening can be used for surface-modified layers with refined grains at about 25–80 nm, with improved corrosion resistance [34]. Friction stir processing and attrition treatment are also advanced mechanical technologies used in order to obtain ultra-fine Ti grains with better biological affinity compared with coarse Ti grains, in a simple and economic way. On the other hand, conventional physical methods require expensive equipment that can produce a flame or plasma arc with high-speed gas flow and temperature between 1700 and 2700 °C (flame) in order to produce uniform coatings of hydroxyapatite, CaO–SiO_2, CaO–MgO–2SiO_2, TiO_2, and Al_2O_3 [24,34].

The plasma polymerization technique is an alternative method used to produce a bioactive surface on Ti substrates with the aim of the better immobilization of bioactive molecules, such as DNA, heparin, and glucose oxidase [48]. NH_2 and COOH based plasma polymers are most commonly used since these groups are known for their good chemical reactivity, or alternatively used for bacterial adhesion and biofilm prevention by coating the substrates with a suitable antibacterial agent (AgNPs). However, their aging and stability remain major issues for biological applications [48] along with the possibility of cracks occurrence during the remelting and solidification on the surface and low bond strength of the bioceramics coatings with substrates [34]. In order to produce a nano-scale texture on a titanium surface, physical vapor deposition techniques (including evaporation plating, ion plating, sputtering) alone or in combination may offer a variety of textures suitable for medical applications, but the best results were reported for plasma/ion implantation and deposition [24,34]. Even if sputtering is preferred for medical applications, due to its flexibility and ease of use, the physical methods and techniques are generally considered expensive, requiring complex systems and very high qualified technicians. The advanced physical techniques are continuously improved, reaching 1~10 nm of the surface modified layer, for improved wear resistance, corrosion resistance, and blood compatibility. In comparison with physical methods, chemical and electrochemical treatments are more affordable, being produced based on redox reactions.

Chemical methods include acidic or alkaline treatments [32,62], hydrogen peroxide treatment [63], chemical vapour deposition [64], electrochemical techniques (anodic oxidation) [65–67], and biochemical techniques for biomimetic coatings [68–72]. These techniques are able to produce specific surface topographies for improved corrosion resistance, improved biocompatibility, bioactivity, or bone conductivity. By tailoring the anodization conditions, different nanostructures can be obtained in terms of nanotubes, nanopores with a hole morphology, or nanochannels with different diameters from 15 nm up to 300 nm and different lengths, as Roy et al. [70] reviewed, pointing to the mechanisms involved in anodization and recent advances in the formation of nanostructured titanium surfaces. Oxide films prepared on Ti and Ti alloys by chemical and electrochemical treatments show complicated surface morphologies with nanostructures, among which the composite coatings proved their versatility due to their multiple functions or better properties [34]. For example, sol-gel deposition ensures the control of the purity and chemical composition of highly homogeneous films using simpler equipment and at lower cost compared to physical techniques. For biomedical applications, calcium and phosphate ions can be incorporated into nano sized anodic oxide films using an electrochemical setup. Moreover, by ailoring the anodization conditions, TiO_2 nanotubes with a diameter ranging from 15 to 300 nm can be obtained on the surface, with favorable properties, such as better adhesion, proliferation, ALP (alkaline phosphatase) activity, and bone matrix deposition [54,55].

The available biochemical techniques for treatment of titanium implant surface are: (i) physico-chemical adsorption, (ii) covalent binding of a specific biomolecule, and (iii) peptide inclusion into a carrier material. The specific cells response can be tailored by the means of surface immobilized peptides, proteins, or growth factors, but the main drawback remains in terms of not ensuring a controlled deposition. Moreover, with the Ti substrates, being bioinert in nature, a supplementary treatment is necessary before immobilization of peptides or proteins, such as plasma polymerization, in order to activate the titanium surface [34]. Also, antibiotics and or/growth factors can be incorporated into the surface layer by applying a combination of surface treatments to obtain multifunctional coatings for biomedical applications.

2.2. *Improved Biological Properties of Nano-Structured Surfaces*

In terms of mechanical properties, the surface modifications summarized in Table 1 will result in important achievements, such as improved wear, fatigue, and corrosion resistance. It is well known that the wear resistance of coatings materials is basically determined by the microhardness and elastic modulus. On the other hand, corrosion is one of the critical issues in biomaterials science. When in contact with biological fluids, titanium metallic atoms are transformed into cations, and develop a passive film by reacting with OH^- and, hence, act as natural corrosion protection of the underlying substrate. An appropriate surface modification is aimed for to retain the natural properties of underlying bulk Ti substrates as well as to improve their anti-corrosion potential.

Besides the wear and corrosion resistance improvement upon different surface treatments, the expected bioactivity should be taken into account when selecting an appropriate surface modification technique. The prediction of the bone response to a nanostructured titanium surface is very difficult and many factors have to be taken into consideration, as the osseous cellular interactions at the implant interface are complex process. Bones have multiple and various functions, and hence the mechanism underlying bone biomaterial-mediated osteogenesis is conducted with the cooperative participation of the host immune cells, the host bone cells, and the material itself, being directly correlated to the biochemical and biomechanical characteristics.

It has been demonstrated that topographical features of the implant surface are the determinant factor with respect to the cellular response [73,74]. Bone is a hierarchically structured material and highly anisotropic. The bone tissue itself consist of nanoscale architecture: osteocytes (bone-forming cells), collagen (type I) fibrils (the organic matrix), and hydroxyapatite (the inorganic bone component) are all nano-scaled structures in the range of 50–300 nm length and 0.5–5 nm width [75]. The network of collagen-crystal nano-arrangement represents the main factor influencing the mechanical properties [75]. There is a physiological, natural phenomenon, called bone remodeling, in which a continuous process of bone formation and degradation is balanced, in a very complex process regulated by growth factors and cytokines [76]. Although the detailed osteogenesis mechanisms at the bone–implant interface are still unknown, the general consensus is that the first response upon the contact with titanium implant is wetting followed by rapid protein adsorption to its surface, depending on physicochemical characteristics (pH, temperature, hydrophobicity, protein structure). Five distinct stages are assumed to occur in vivo, during the interaction between Ti surface and biological milieu [24], as represented Figure 3.

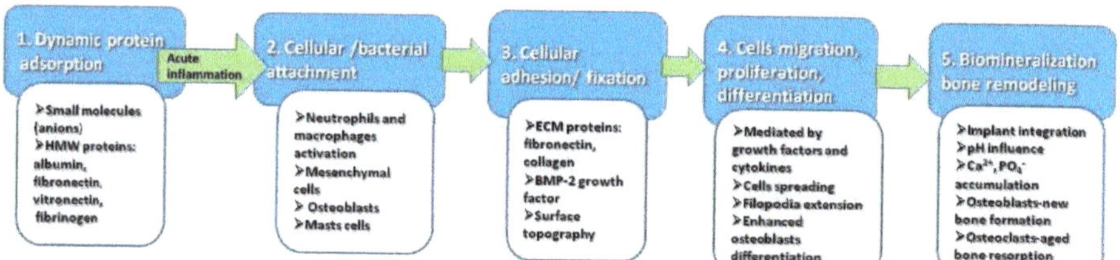

Figure 3. The main sequence of events occurring in vivo, during interaction between Ti surface and biological environment.

In the initial stage, the negative charged small molecules are attracted by the positive charge of titanium, followed by high molecular weight molecules with functional groups. This is a dynamic stage, involving the physical adsorption of fibronectin, albumin, fibrinogen, and immunoglobulin IgG. However, the in vivo competitive adsorption processes and their kinetics remain unclear. Initial acute inflammatory reaction may occur at the final of this stage, by activation of neutrophils. In the second stage, the nanostructured surface created by competitive protein adsorption, will promote both cellular and bacterial attachment. The crucial factor for the initiation of osseointegration is the involvement of healing macrophages in this stage. In the third stage, the attached cells are further fixed by extracellular matrix anchoring proteins. Integrins acts as a primary receptor in this stage. Mesenchymal cells, neutrophils, and macrophages are collectively involved releasing cytokines and growth factors, which subsequently interact with fibroblasts and osteoblasts. Proliferation and cell differentiation mediated by cytokines and specific growth factors occurs in the fourth stage. Enhanced osteoblasts spreading and filopodial extension are characteristics on nanorough titanium surface [38,49]. Finally, in the fifth stage, the biomineralization and osteointegration of the titanium implant occurs as a result of local changes of the pH and accumulation of Ca^{2+} and PO^{4-}. At this final level, the balance between osteogenesis and osteoclastogenesis is shifted towards the last one, the osteoclasts being involved in aged bone resorption [77]. Based upon these findings, it is largely accepted that tailoring the surface chemistry of implants opens the possibility to transform osteoconductive materials into osteoproductive ones. Therefore, the new trend in titanium cranioplasty is oriented and designed to modulate the immunological response while promoting the osseointegration.

3. Types of Nanoparticles Used for Surface Modification of Titanium Implants

Nowadays, it remains a challenge on the topic addressed to titanium surface modification, and the debates are related to the fabrication methods for a nanostructured, porous titanium surface with effective biological activity, using facile, eco-friendly, and cost-effective approaches. A novel strategy to improve the success of soft and hard tissue integration of titanium implants is the use of nanoparticles coatings made from basically any type of biocompatible substance, which can advantageously enhance the properties of the material, as compared to its similar bulk material [78]. Nanotechnology has enabled the addition of metals into their nanosize, leading to extreme changes in chemical, physical, and optical properties of metals, as the metallic nanoparticles are the most promising agents known for enhanced antibacterial activity. By these techniques, implant surfaces should be designed not only to support the attachment of target tissue cells, but also to prevent bacterial adhesion. One main strategy to decrease bacterial infiltration is to limit biofilm formation due to the initial contamination of the implant surface. Biofilms consist of microorganism aggregates adherent to the implant surface, comprising multiple species with high cell densities, which critically influence the implants function and may initiate local inflammation [79]. So, most of the physical methods approaches involve compaction of nanoparticles versus micron-level particles to yield surfaces with nanoscale grain boundaries, preserving in the same time the chemistry of the surface among different topographies. Based on literature research, in Table 2 are summarized the types of nanoparticles used for surface modifications of titanium implants, the fabrication methods, the main features, and the purpose of use.

As an alternative to antibiotics, nanoparticles are known as one of the most effective antibacterial agents, and nowadays, surface modification of titanium using antibacterial properties of metal nanoparticles is a popular approach in clinical treatments. Antimicrobial nanoparticles used for decorating the titanium surfaces mainly include Ag, Zn, Au, and Se [89–93], but also less investigated antimicrobial agents such as copper, fluorine, and calcium [83,95,96]. In a recent study, Esfandiari et al. [96] prepared Ag-decorated TiO_2 nanotubes by a combined electrochemical and UV-assisted reduction method, showing a synergetic bactericidal effect of ~100 nm TiO_2 nanotubes and 20 nm Ag NPs. Besides the antimicrobial effect, nanostructured titanium surfaces (created with different nanoparticles) demonstrated better implant fixation. Moreover, many efforts have been made to create nano-structured roughness to simultaneously increase healthy bone growth while inhibiting cancerous bone growth [93–95,97]. As selenium is a well-known anti-cancer chemical, nano-rough selenium was developed for orthopedic applications, including cranioplasty, involving bone cancer treatment [92,93]. In a comparative study performed by Tran and col. [94], the authors demonstrated a superior adhesion of healthy osteoblast on nano-rough selenium compacts versus the micro-rough ones. They proposed a mechanism in which the initial absorption of proteins (such as fibronectin and vitronectin) was favored by the nano-topography, followed by subsequent cell adhesion. Selenium nanoclusters were shown to promote healthy osteoblast functions after one day of culture, and more importantly, to inhibit cancerous osteoblast cell functions after three days of culture [94]. Recently, our research group proposed an improvement of the surfaces [92,93], in the case of titanium mesh for cranioplasty by the in-situ hydrothermal deposition of selenium nanoparticles obtained using an original method (starch, glucose, and galactose were selected as reducing agents), as summarized in Figure 4a. The surface modification (Figure 4b) may offer important benefits in terms of osteointegration, without using additional screws for fixation and closure procedure. However, considering the toxicity of selenium at high levels [98], a careful dosage of selenium in compacts is necessary in order to avoid unnecessary exposure and possible toxic effects.

Table 2. Nanoparticles used for surface modification of Ti implants.

Type of NPs	Fabrication Method	Main Features/Purpose	Reference
TiO_2	direct oxidation	Nanofibers, nanoneedles/better hydrophilicity, biocompatibility and antimicrobial activity, compared to Ti6Al4V	[80]
	Pulsed laser deposition	Nanorods/Nontoxicity, ability to increase the density of osteoblast cells on the implant, enhanced osseointegration, anticandidal effect, bone formation ability	[81]
	Anodic oxidation	Nanotubes incorporating Ca, P and Ag/cells migration on the Ti-based implants due to super hydrophilic properties of crystalline TiO_2 nanotubes; apatite formation in simulated body fluid, enhanced MC3T3-E1 cell adhesion and proliferation, antibacterial effect against S. aureus.	[82,83]
	plasma electrolytic oxidation	Nanostructured Zn-incorporated TiO_2 coatings, grains 20–100 nm/inhibition of *S. aureus* and *E. coli*	[84]
Al_3O_3	micro-arc oxidation	Nanostructured surface for improved microhardness and wear resistance	[81,85]
	dipping	Nanostructured surface/promote MSC commitment to	[86]
	-	the osteoblast phenotype, increase in bone-implant contact area and torque removal	-
Nano Hydroxyapatite	discrete crystalline deposition	Complex surface morphology via the bonded HA nanoparticles/progressive osseointegration profiles	[78,87]
Nano-crystalline diamond	Plasma spray	Nanosized crystallites/proteins immobilization on nanocrystalline diamond/osteoinductive effect in irradiated bone	[78,88]
Ag NPs	silanization method	Spherical morphology, 100 nm diameter, antibacterial and anti-adhesive activities towards *S. aureus* and *E. coli.*	[24,83,89]
ZnO NPs	EHDA spraying.	rod-shaped structure ~100 nm/significant antimicrobial activity against Staphylococcus aureus/early bone formation	[90]
Au Nps	Magnetron sputtering	40–80 nm thin layer of pure gold/early mechanical fixation	[91]
Se Nps	Hydrothermal deposition	Spherical, rods, wire nanostructure, using different saccharides as reducing agent/favorable results on RBC osmotic fragility and fibroblasts adhesion to accelerate osseointegration, bone cancer treatment.	[92–95]

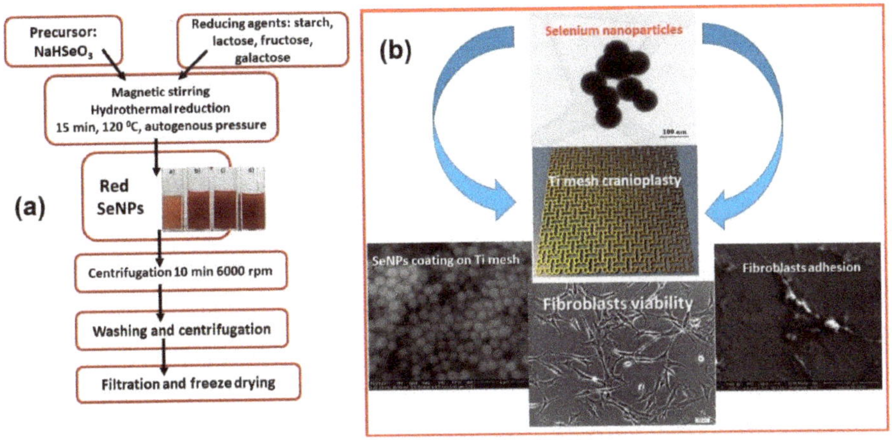

Figure 4. (**a**) The flow chart of SeNPs production via hydrothermal reaction using different saccharides as reducing agent; (**b**) TEM image of SeNPs used for the surface modifications of Ti mesh for cranioplasty, along with the surface morphology of the coating upon in situ SeNPs deposition and details of fibroblasts adhesion on the nanostructured Ti surface.

Osteoblast adhesion on Ti surfaces is a prerequisite for successful implant integration, as demonstrated by in vitro studies devoted to interaction between osteoblasts and nano-structured Ti surface. Webster et al. provided the first evidence of increased osteoblasts adhesion on nanophase Ti compared with conventional one [99]. The explanation was related to the increased surface area and nanoroughness which generated a better wettability and, consequently, promoted the adsorption of hydrophilic proteins (e.g., vitronectin, fibronectin) involved in bone cell attachment. Later studies demonstrated that osteoblast adhesion significantly increased by 33% on anodized Ti compared with conventional Ti, accompanied by an increase in fibronectin and vitronectin adsorption [100]. With respect to the size and morphology of the surface, Oh et al. [101] demonstrated that nanotube diameter regulates stem cell differentiation. Smaller nanotubes (~30 nm) are involved in mesenchymal stem cell adhesion, while larger nanotubes (~70–100 nm) are involved in cellular elongation and differentiation into osteoblast-like cells. Taking into account the scale-range of bone matrix represented by osteocytes, hydroxyapatite crystals, and collagen fibrils, respectively, 50–300 nm in length and 0.5–5 nm in width, the designed implant surfaces must fit the physiological milieu and mimic the growing conditions for osteogenic cells [24].

There are many controversies in the literature with respect to the effect of nanoscale topographic details of titanium implants. It was suggested that a minimal 100 μm pore size is required for bone tissue regeneration, while a smaller pore sized remains prone to fibrous tissue formation [84,102]. Clinical and in vivo information about these nano-size coatings and their beneficial applications is still rare. In 1983, Branemark et al. [103] described the osseointegration as a direct structural and functional bone to implant contact under functional load. Various animal models were performed in order to study the enhanced osseointegration by surface treatments at the nanometer scale. Kubo et al. observed a significant increase in bone-titanium interfacial strength after two weeks of implantation in femur rats using Ti nanotube (300 nm) surface modification [104], while in a similar in vivo experiment, Ogawa et al. [105] demonstrated that an increased surface area up to 40% was correlated to higher osseointegration of nanostructured Ti compared to an acid-etched surface. A very recent in vivo study of plasma-sprayed carbon nanotube (CNT)-reinforced hydroxyapatite (HA) coating on titanium implants embedded in rodents' bone demonstrated that CNT addition induces higher osseointegration as compared to HA [106]. In this case, another important evidence of implant mechanical integrity was proven by the similar value of elastic modulus of the newly grown bone compared with the distant bone. In another study, a 40–80-nm thin layer of gold nanoparticles was deposited by the magnetron sputtering technique on a porous Ti surface [91] and inserted in the proximal region of the humerus in canines. This study suggests that early biomechanical fixation was influenced by the gold nanostructured coating, pointing out that released gold ions are able to reduce the inflammatory process. However, the authors concluded that a long-term study is necessary to highlight the anticipated anti-inflammatory effects of the coating. A multi-database systematic literature review was conducted by Bral and Mommaerts [106] with the objective to provide a structural review of the fabrication techniques to produce a CaP coating with good in vivo results compared to micro-CaP–coated and uncoated titanium implants. This study was elaborated based on twenty-eight papers consisting of both animal models and human studies. Once again, it was demonstrated that titanium implants coated with nano CaP and nano-hydroxyapatite improves implant fixation, based on histomorphometrical healing properties. They also concluded that not all coating techniques have beneficial effects on biofunctionalization. Several factors might have a crucial importance in successful biofunctionalization: implant nano-roughness, coating thickness, calcium phosphate solubility, and nanotopography. However, the nano-HA pro-mimic coating has exhibited good results in vivo, in an animal model and in human studies involving early and intermediate healing periods [106]. Either chemical or physical deposition techniques proved to be effective methods for producing versatile and biocompatible nanostructured coatings with different chemical and morphological properties to promote a better bone tissue response. By tailoring the surface of titanium implants for cranioplasty and related fields, the results may contribute to the general efforts dedicated to continuous improvement of nano-biomaterials, opening new possibilities for long-term development and strategies in nanomedicine [93,107].

4. Patented Solutions to Improve Osteointegration of Cranial Implants

4.1. Cranial Endoprosthesis with Sliding System

The invention relates to a cranial stent with a sliding system, used for the repair of traumatic defects of the skull, through the cranioplasty surgical procedure.

Currently, the most used medical devices for cranioplasty are: plaques, nets, distractors, autografts and allografts [107,108], fastened by screw and mini-plates. The fixation is made in relation to the contour of the uninjured skull. The disadvantages of the above-mentioned solutions, known from the prior art, are related to the need for customization with low cost and in the same time avoiding additional drilling of the skull. The additive manufacturing systems evolved in parallel to the novel medical devices and nanotechnologies in order to develop new constructive forms as close as possible to the patient's anatomy. Due to its advantages, the manufacture of custom implants is currently considered to bring the highest level of benefits in terms of compliance with patient needs [106]. Also, a personalized implant significantly reduces the duration of surgery and, implicitly, postoperative complications. However, a cranial implant made by single production has high costs compared to standardized devices on the market [108–110].

We propose one innovative solution that supposes the adaptation to the customized dimensions of a patient without the need to process specific anatomical data and the manufacture of a unique cranial prosthesis. The cranial prosthesis, with a sliding system that consists of an upper sliding layer, a lower sliding layer, and a fixing system, eliminates the mentioned disadvantages, as the sliding system can be manufactured in several types of dimensions for the repair of cranial defects of different sizes and shapes (Figure 5). Also, the newly proposed solution solves the problem of complications arising from the perforation of the cranial box when fixing classic implants. The fixation in the case of cranial endoprosthesis with a sliding system is achieved by the constructive form of the stent, through a fixing system with clamps.

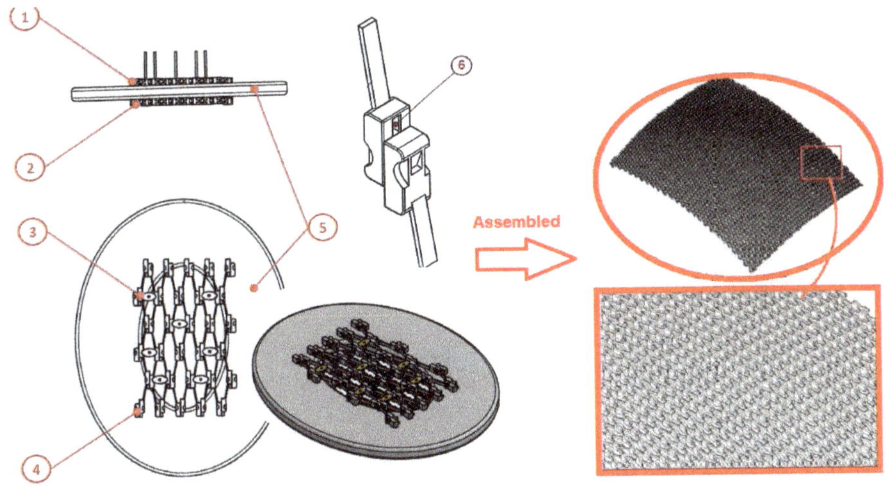

Figure 5. Left panel- Components of the cranial endoprosthesis with sliding system: Upper sliding layer ①; lower sliding layer ②; fixing system ③. The sliding layers ① and ② are composed of multiple mobile cells with sliding system ④; Positioning of the stent in relation to a schematic model of the cranial box ⑤; Conjugate sliding system ⑥. Right panel-The assembly of the mobile cells to obtain constructive anatomical curvature in order to facilitate customization.

The main advantage of the newly proposed cranial prosthesis with a sliding system consists in the adaptation of the prosthesis to various dimensions of the cranial defect, hence facilitating the serial production while maintaining the character of customization according to the patient's anatomy. The fixation system does not require drilling the patient's skull. In this way, post-traumatic complications are avoided, the duration of fixation is minimized, and consequently, the duration of entire intraoperatively procedure is also reduced, promoting a faster patient recovery.

The cranial prosthesis with sliding system consists of an upper sliding layer, a lower sliding layer and a fixing system, the sliding layers being composed of mobile cells (they have a general parallelepiped shape with connected edges, but can have any other shape that allows for the execution of the sliding movement, such as: spherical, cylindrical, prismatic, pyramidal, etc.) with the sliding system. For assembly, the lower sliding layer is positioned in a non-slided state, tangential to the lower surface of the skull, while for sliding and engaging the moving cells in the lower layer, an actuating key is required. On the side of the movable cells, the conjugate gear surfaces are provided, which come into contact with the actuating key for engaging the sliding movement. The upper and lower layers of the mobile cells are provided with slots for the orientation and positioning of the fastening system. The upper and lower layer has a conjugate surface for laying and orienting. The sliding movement is performed by means of the sliding and conjugate sliding surface, the movement being transmitted from one cell to another, propagating the sliding movement.

The assembly method of the mobile cells is shown in Figure 5 for a cranial prosthesis with overall dimensions between 80 mm and 120 mm. The sliding layers of the stent have a constructive anatomical curvature in three planes to facilitate customization according to the patient's anatomy and the shape of the traumatic hole.

4.2. *Titanium Mesh with Novel Geometry and Functional Nano-Coating Based on SeNPs*

According to the literature, complications of cranioplasties can be classified into three main categories: (1) septic complications; (2) neurological complications; (3) cerebrospinal fluid complications [110–116]. Infections and related complications may become dramatic, requiring special management with additional economic costs and prolonged hospitalization. It has been demonstrated that the timing of cranioplasty, the size of the defect, and the biomaterial choice may also influence the subsequent planning and post-operative management to reduce all types of complications [117–120].

In this context, the proposed solution relates to a structure surface and method for fixing a mesh-type cranial prosthesis, executable with biocompatible titanium alloys, having functional coatings based on selenium nanoparticles with osseointegration properties, used for cranioplasty and reconfigurations of major cranial defects caused by accidents, birth defects, and surgery due to tumor extractions or other cranial diseases. The disadvantages of titanium implants for cranioplasty are represented by the fixing system with titanium screws, as well as the absence of osseointegration, due to the inert character of titanium and its alloys in contact with human hard tissues.

The advantages of the cranial implant with osteointegration structures and functional coatings are based on mimicry, related to the basal cells with a spider web geometric structure, and geometric elements characterized by sharp angles, which favors the capillarity process for long-term maintenance [121–123]. Moreover, the particles deposited on the surface will promote osseointegration allowing for the fast restoration of human ordeal. Se nanoparticles are produced by eco-friendly processes, avoiding the use of harsh, toxic, and expensive chemicals [124]. Also, the system of alternating rigid and flexible bridges allows for the adaptation of the cranial implant with osseointegration structures to the patient's anatomy, maintaining the necessary mechanical strength characteristics.

As presented in Figure 6, the first constructive form of the osseointegration module consists of stratified basal cells and two rigid bridges arranged symmetrically on the diagonal of the osteointegration module. Basal cells are connected by a resistance structure. The same resistance structure connects all four basal cells, over the entire height of the osseointegration module. The rigid bridge consists of two connecting units and two resistance pillars with a semicircular contour. The resistance pillar is arranged symmetrically on the diagonal of the module, on the entire height of the osseointegration module, connecting all four basal cells. The connecting units, which together form the rigid bridge, are arranged alternately on the first and third layers of the basal cells in the osseointegration module.

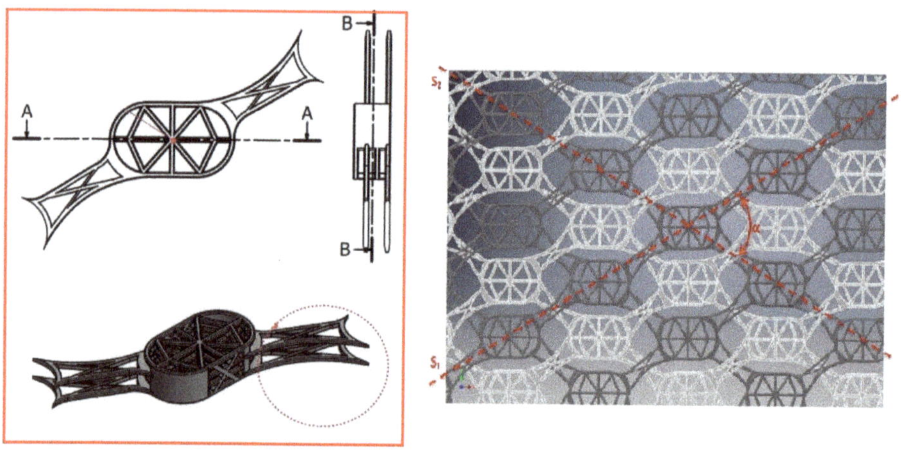

Figure 6. The components of the modular system. Left panel-stratified basal cells, the rigid and flexible connectors. Right panel-the arrangement of modular cells in alternately configuration, so that all rigid connecting bridges align in the S_1 direction, while the flexible ones align in the S_2 direction.

The second constructive form of the osseointegration module consists of the stratified arranged basal cells and two flexible connecting bridges arranged symmetrically on the diagonal opposite to the one containing the rigid bridges of the first constructive form. The two constructive forms of the osseointegration module have similar overall dimensions. The basal cells are connected by the resistance structure provided over the entire height of the osseointegration module. The flexible connecting bridge consists of two connecting units connected to the basic units in the module construction. The first connecting unit is connected to the basal cell arranged on the second layer, and the second unit to the basal cell arranged on the fourth layer, alternately to the connecting units in the rigid connecting bridge. The two constructive forms of the osseointegration module are alternately arranged so that all rigid connecting bridges align in the S_1 direction and the flexible connecting bridges align in the S_2 direction. The two directions intersect on the longitudinal axis of the implant forming the angle α.

The cranial prosthesis with osseointegration structures, according to Figure 7, consists of the osseointegration module ③, the rigid connecting bridge ④, and the flexible connecting bridge ⑤. The osseointegration module ③ consists of four basal cells and is defined in two constructive forms.

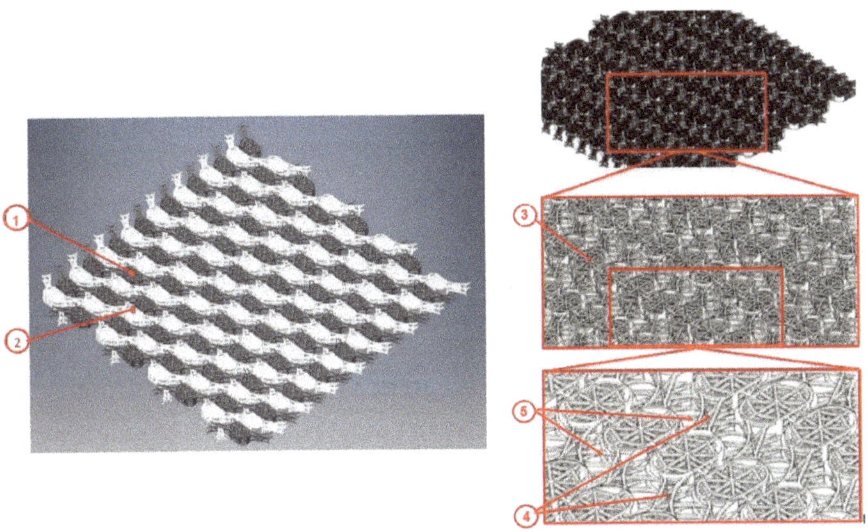

Figure 7. Interconnected layers of the modular system. Left panel-The fixed layer ① is assembled with the movable layer ② so that translations can be performed in the Ox and Oy directions, while maintaining the position of the layer ①. Right panel-The component elements of the cranial prosthesis presented in isometric view, consisting of four basal cells defined in two constructive forms ③, the rigid connecting bridge ④ and the flexible connecting bridge ⑤.

In height, the osseointegration module has dimensions between 1 mm and 2 mm, its width being between 2.5 mm and 3.5 mm and its length between 4 mm and 5 mm. The basal cells have the same values for length and width as the osseointegration module, and the layer thickness is between 0.1 mm and 0.2 mm.

As a manufacturing technique, the prosthesis is made in one piece, in an already assembled form. Primary post-processing involves detaching from the workpiece table and deburring the support structures. The upper and lower surfaces of each osseointegration module are post-processed with superfinishing processes and cleaned by pickling and in an ultrasonic bath. Subsequently, it is impregnated with Se nanoparticles, which accelerates the regeneration process of bone tissue, improving the degree of osseointegration of the cranial prosthesis with the surrounding anatomical bone structures, as well as the antibacterial effect.

The main clinical outcome of this novel approach is represented by providing solutions in particular cases in which the size of the cranial defect is unknown preoperatively or if the alteration of an existing skull defect is required, such as craniectomy performed to remove bony tumors or osteomyelitic bone [21,113]. In these cases, prefabricated implants cannot be used. The overall clinical benefit is reflected in saving clinical time, as a decreased surgical timeframe also means a decrease of blood loss and anesthesia exposure time. On the other hand, tailoring the titanium surface with functional nanoparticles offers important clinical applications due to simultaneous antibacterial and osteoconductive activities in patients with bone healing disorders associated with rheumatoid arthritis, osteoporosis, diabetes, or aging [25,94,96,106].

5. Conclusions and Future Perspective

Although autologous bone seems to be the standard material with the best biocompatibility properties, being also osteoconductive and osteoinductive, it has several limitations: only certain parts of the human body can provide an ideal bone graft, being difficult to model intraoperatively because it does not allow perfect coverage of the intracranial bone defect. For this reason, advance research and development of alloplastic biomaterials has become an important goal of materials science.

One can observe that, over time, there has been no ideal biomaterial developed for cranial prosthesis used in cranioplasty. However, biomaterials with high mechanical strength, resistant to infections, radiolucent, cheap, easy to handle and able to reintegrate into the skull, are the most used materials in cranioplasty. Each alloplastic biomaterial used to manufacture cranial prostheses has both advantages and disadvantages. It is important that the craniofacial surgeon, together with the medical engineer, can make the most appropriate choice, taking into account the anatomical and physiological characteristics of each patient and the type of trauma. Surgical replacement of bone tissue with a synthetic material has both a protective and an aesthetic role, which is why it is so difficult to choose a particular material for reconstruction. Although many different methods have been described in the literature, using autologous or alloplastic material, there is still no consensus on which method is better, and continuous multidisciplinary researches are conducted aiming to develop the ideal reconstruction materials.

This review emphasized the main reasons why titanium mesh is preferred for skull reconstructions along with the importance of developing innovative surface structures with a dual benefit in terms of improved osteointegration and enhanced antibacterial activity to reduce the risk of post-surgical infection, knowing that infections are the main complication in cranioplasty surgeries. Based on literature research, the identification of the nanoparticle types designed for the surface modification of titanium implants was conducted, along with the fabrication methods, their main features, advantages, and disadvantages with respect to their purpose of use.

It has been found that performing cranioplasty with titanium mesh prostheses has decreased infection rates in high-risk patients, such as military personnel who have suffered extensive scalp injuries and/or craniectomy surgeries during the war in Iraq and Afghanistan. In addition, computer-aided 3D modeling was used successfully to design titanium mesh prostheses which provide an excellent cosmetic effect, even in the case of large cranial defects.

Using the specialized software Autodesk Inventor Professional 2016, a novel intracranial prosthesis was designed. The design is a mesh type which does not require fixing to the skull with screws. The aim was to create a sliding system which allows for the adaptation of the cranial prosthesis to variable dimensions of bone deficiency, respecting the patient's anatomy, without the need for other fastening systems (screws, wires, etc.) that would cause additional trauma or prolonged surgery time. Different types of coatings (calcium phosphate, hydroxyapatite) can be used for this intracranial prosthesis, with the role of facilitating and accelerating the osseointegration process.

To promote the osseointegration process, the second type of intracranial prosthesis has been designed, whose constructive elements have a porous microstructure that promotes the absorption of osseointegration particles (Se nanoparticles) and promotes bone tissue regeneration. Selenium nanoparticles were proposed for developing the coatings with an osseointegration role, being fabricated by an eco-friendly method. Thus, the $NaHSeO_3$ salt (sodium hydrogen selenite) was used as a precursor, and molecules from the category of carbohydrates were used as reducing agents: starch (polysaccharide), lactose (disaccharide), galactose and fructose (monosaccharides) [124–130].

The future in cranioplasty is related to molecular biology, which involves the application of bone growth factors to help and stimulate bone formation in prosthesis. So, currently research related to cranioplasty is pursuing a new approach. The mechanism by which undifferentiated mesenchymal cells can be transformed into osteoprogenitor cells is being investigated, in order to favor the osseointegration of intracranial prostheses. Making an intracranial prosthesis with osseointegration properties is a complicated process that involves revascularization and bone formation in the prosthesis. The ideal prosthesis should function as a scaffold for the progressive proliferation of blood vessels and osteoprogenitor cells.

Through the osteoinduction process, osteoprogenitor cells should not migrate from the surrounding tissue. Instead, they will be produced "in situ" with the help of bone morphogenetic proteins, and then infused into a prosthesis structure. These bone growth factors could be integrated into sustained-release polymers to stimulate cellular responses to bone regeneration, providing immediate protection of the skull, with aesthetic benefits along the osteoconductive and osteoinductive advantages.

Also, the future belongs to the nanotechnology design, obtaining nanometric particles that can be used in medicine to cover different types of prostheses, so that the process of osseointegration is favored. Our work is aligned to the general efforts dedicated to continuous improvement in the field on nano-bio-materials, opening new possibilities for long-term development and strategies in nanomedicine, including antibacterial, osteoconductive, and osteoinductive nanomaterials, which may offer the desired biological response.

6. Patent

OSIM nr. 132417, release date 30/10/2019. Title: Cranial implant with osteointegrating structures and functional surface.

Author Contributions: Conceptualization: S.C. and A.S.; methodology: A.M., I.V.A.; software: C.D.; formal analysis: C.D. and F.B.; investigation: S.C., I.M., and F.B.; resources: A.M. and F.B.; data curation: S.C., A.S., M.U. and R.M.; writing, original draft preparation: S.C. and A.S.; writing, review and editing: I.V.A. and I.M.; visualization: I.M.; supervision: I.V.A.; project administration: S.C. and A.M.; funding acquisition: A.M. and F.B. All authors have read and agreed to the published version of the manuscript.

Funding: This research received no external funding.

Acknowledgments: The work was developed under the project "Work-based learning systems through entrepreneurial scholarships for doctoral and post-doctoral students (SIMBA)—MySMIS 124705" within Priority Axis 6—Education and skills—POCU 2014-2020.

Conflicts of Interest: The authors declare no conflict of interest.

References

1. Sanan, A.; Haines, S.J. Repairing holes in the head: A history of cranioplasty. *Neurosurgery* **1997**, *40*, 588–603. [PubMed]
2. Bonfield, C.M.; Kumar, A.R.; Gerszten, P.C. The history of military cranioplasty. *Neurosurg. Focus* **2014**, *36*, 18. [CrossRef] [PubMed]
3. Prolo, D.J.; Oklund, S.A. Composite autogeneic human cranioplasty: Frozen skull supplemented with fresh iliac corticocancellous bone. *Neurosurgery* **1984**, *15*, 846–851.
4. Kwarcinski, J.; Boughton, P.; Ruys, A.; Doolan, A.; Van Gelder, J. Cranioplasty and Craniofacial Reconstruction: A Review of Implant Material, Manufacturing Method and Infection Risk. *Appl. Sci* **2017**, *7*, 276. [CrossRef]
5. Grant, G.A.; Jolley, M.; Ellenbogen, R.G.; Roberts, T.S.; Gruss, J.R.; Loeser, J.D. Failure of autologous bone-assisted cranioplasty following decompressive craniectomy in children and adolescents. *J. Neurosurg.* **2004**, *100*, 163–168. [CrossRef]
6. Sheng, H.S.; Shen, F.; Lin, J.; Lin, F.C.; Yin, B.; Zhang, N. Titanium mesh implants exposure after cranioplasty in two children: Involvement of osteogenesis? *Chin. Neurosurg. J.* **2017**, *3*, 1–4. [CrossRef]

7. Zins, J.E.; Langevin, C.J.; Nasir, S. Controversies in skull reconstruction. *J. Craniofac. Surg.* **2010**, *21*, 1755–1760. [CrossRef]
8. Aydin, S.; Kucukyuruk, B.; Abuzayed, B.; Aydin, S.; Sanus, G.Z. Cranioplasty: Review of materials and techniques. *J. Neurosci. Rural. Pract.* **2011**, *2*, 162–167. [CrossRef] [PubMed]
9. Cavalu, S.; Banica, F.; Simon, V.; Akin, I.; Goller, G. Surface Modification of Alumina/Zirconia Ceramics upon Different Fluoride-Based Treatments. *Int. J. Appl. Cer. Technol.* **2014**, *11*, 402–411. [CrossRef]
10. Khader, B.A.; Towler, M.R. Materials and techniques used in cranioplasty fixation: A review. *Mater. Sci. Eng. C Mater. Biol. Appl.* **2016**, *66*, 315–322. [CrossRef] [PubMed]
11. Marchac, D.; Greensmith, A. Long-term experience with methylmethacrylate cranioplasty in craniofacial surgery. *J. Plast. Reconstr. Aesthetic Surg.* **2008**, *61*, 744–752. [CrossRef] [PubMed]
12. Rosinski, C.L.; Chaker, A.N.; Zakrzewski, J.; Geever, B.; Patel, S.; Chiu, R.; Rosenberg, D.; Parola, R.; Shah, K.; Behbahani, M.; et al. Autologous Bone Cranioplasty: A Retrospective Comparative Analysis of Frozen and Subcutaneous Bone Flap Storage Methods. *World Neurosurg.* **2019**, *131*, e312–e320. [CrossRef] [PubMed]
13. Malis, L.I. Titanium Mesh and Acrylic Cranioplasty. *Neurosurgersy* **1989**, *25*, 351–355. [CrossRef]
14. Brånemark, R.; I Brånemark, P.; Rydevik, B.; Myers, R.R. Osseointegration in skeletal reconstruction and rehabilitation: A review. *J. Rehabil. Res. Dev.* **2001**, *38*, 175–181.
15. Chiriac, A.; Stan, G.E.; Iliescu, B.; Poeata, I. The influence of host bone substitute in titanium mesh cranioplasty. *Digest J. Nanomat. Biostruct.* **2013**, *8*, 729–735.
16. Luo, J.; Liu, B.; Xie, Z.; Ding, S.; Zhuang, Z.; Lin, L.; Guo, Y.; Chen, H.; Yu, X. Comparison of manually shaped and computer-shaped titanium mesh for repairing large frontotemporoparietal skull defects after traumatic brain injury. *Neurosurg. Focus* **2012**, *33*, E13. [CrossRef]
17. Mukherjee, S.; Thakur, B.; Haq, I.; Hettige, S.; Martin, A.J. Complications of titanium cranioplasty—A retrospective analysis of 174 patients. *Acta Neurochir.* **2014**, *156*, 989–998. [CrossRef]
18. Hill, C.S.; Luoma, A.M.V.; Wilson, S.R.; Kitchen, N. Titanium cranioplasty and the prediction of complications. *Br. J. Neurosurg.* **2012**, *26*, 832–837. [CrossRef]
19. Kim, J.-K.; Lee, S.-B.; Yang, S.-Y. Cranioplasty Using Autologous Bone versus Porous Polyethylene versus Custom-Made Titanium Mesh: A Retrospective Review of 108 Patients. *J. Korean Neurosurg. Soc.* **2018**, *61*, 737–746. [CrossRef]
20. Verret, D.; Ducic, Y.; Oxford, L.; Smith, J. Hydroxyapatite Cement in Craniofacial Reconstruction. *Otolaryngol. Neck Surg.* **2005**, *133*, 897–899. [CrossRef]
21. Sunderland, I.R.; Edwards, G.; Mainprize, J. A technique for intraoperative creation of patient-specific titanium mesh implants. *Plast Surg.* **2015**, *23*, 95–99. [CrossRef]
22. Rosinski, C.L.; Patel, S.; Chaker, A.N.; Behbahani, M.; I Mehta, A. In Reply: A Retrospective Comparative Analysis of Titanium Mesh and Custom Implants for Cranioplasty. *Neurosurgery* **2020**, *87*, E268. [CrossRef] [PubMed]
23. Mazaheri, M.; Eslahi, N.; Ordikhani, F.; Tanjid, E.; Simchi, A. Nanomedicine applications in orthopedic medicine: State of the art. *Int. J. Nanomed.* **2015**, *10*, 6039–6054.
24. Jäger, M.; Jennissen, H.P.; Dittrich, F.; Fischer, A.; Köhling, H.L. Antimicrobial and Osseointegration Properties of Nanostructured Titanium Orthopaedic Implants. *Materials* **2017**, *10*, 1302. [CrossRef] [PubMed]
25. Durmus, N.G.; Webster, T.J. Nanostructured titanium: The ideal material for improving orthopedic implant efficacy? *Nanomedicine* **2012**, *7*, 791–793. [CrossRef]
26. Brunette, D.M.; Tengvall, P.; Textor, M.; Thomsen, P. *Titanium in Medicine: Material Science, Surface Science, Engineering, Biological Responses and Medical Applications*; Springer: Berlin/Heidelberg, Germany, 2001.
27. Goldman, M.; Juodzbalys, G.; Vilkinis, V. Titanium Surfaces with Nanostructures Influence on Osteoblasts Proliferation: A Systematic Review. *J. Oral Maxillofac. Res.* **2014**, *5*, e1. [CrossRef]
28. Lausmaa, J.; Kasemo, B.; Mattsson, H. Surface spectroscopic characterization of titanium implant materials. *Appl. Surf. Sci.* **1990**, *44*, 133–146. [CrossRef]
29. Wang, Y.; Yu, Z.; Li, K.; Hu, J. Effects of surface properties of titanium alloys modified by grinding, sandblasting and acidizing and nanosecond laser on cell proliferation and cytoskeleton. *Appl. Surf. Sci.* **2020**, *501*, 144279. [CrossRef]
30. Liu, X.; Chu, P.K.; Ding, C. Surface modification of titanium, titanium alloys, and related materials for biomedical applications. *Mater. Sci. Eng. R Rep.* **2004**, *47*, 49–121. [CrossRef]

31. Kulkarni, M.; Mazare, A.; Schmuki, P.; Iglič, A. Biomaterial surface modification of titanium and titanium alloys for medical applications. In *Nanomedicine; Chapter 5: Biomaterial Surface Modification of Titanium and Titanium Alloys for Medical Applications*; Alexander, S., de Mel, A., Deepak, M.K., Eds.; One Central Press: Altrincham, UK, 2014.
32. Kim, K.-H.; Narayanan, R.; Rautray, R.T. Surface Modification of Titanium for Biomaterial Applications. *Nova Publ. Sci. Inc.* **2010**, *42*, 201–228.
33. Unal, O.; Karaoglanli, A.C.; Varol, R.; Kobayashi, A. Microstructure evolution and mechanical behavior of severe shot peened commercially pure titanium. *Vacuum* **2014**, *110*, 202–206. [CrossRef]
34. Zhang, L.-C.; Chen, L.-Y.; Wang, L. Surface Modification of Titanium and Titanium Alloys: Technologies, Developments and Future Interests. *Adv. Eng. Mater.* **2020**, *22*, 1901258. [CrossRef]
35. Wang, L.; Xie, L.; Shen, P.; Fan, Q.; Wang, W.; Wang, K.; Lu, W.; Hua, L.; Zhang, L.-C. Surface microstructure and mechanical properties of Ti-6Al-4V/Ag nanocomposite prepared by FSP. *Mater. Charact.* **2019**, *153*, 175–183. [CrossRef]
36. Bagno, A.; Di Bello, C. Surface treatments and roughness properties of Ti-based biomaterials. *J. Mater. Sci. Mater. Med.* **2004**, *15*, 935–949. [CrossRef]
37. Jelliti, S.; Richard, C.; Retraint, D.; Roland, T.; Chemkhi, M.; Demangel, C. Effect of surface nanocrystallization on the corrosion behavior of Ti–6Al–4V titanium alloy. *Surf. Coat. Technol.* **2013**, *224*, 82–87. [CrossRef]
38. Guo, Z.; Jiang, N.; Chen, C.; Zhu, S.; Zhang, L.; Li, Y. Surface bioactivation through the nanostructured layer on titanium modified by facile HPT treatment. *Sci. Rep.* **2017**, *7*, 1–11. [CrossRef]
39. Zhou, H.; Li, F.; Wang, J.; Sun, B.-D. Microstructure analyses and thermophysical properties of nanostructured thermal barrier coatings. *J. Coat. Technol. Res.* **2008**, *6*, 383–390. [CrossRef]
40. Chen, H.; Zeng, Y.; Ding, C. Microstructural characterization of plasma-sprayed nanostructured zirconia powders and coatings. *J. Eur. Ceram. Soc.* **2003**, *23*, 491–497. [CrossRef]
41. Yılmaz, Ş. An evaluation of plasma-sprayed coatings based on Al_2O_3 and Al_2O_3–13 wt.% TiO_2 with bond coat on pure titanium substrate. *Ceram. Int.* **2009**, *35*, 2017–2022. [CrossRef]
42. LeClair, P.; Berera, G.P.; Moodera, J.S. Titanium nitride thin films obtained by a modified physical vapor deposition proces. *Thin Solid Film* **2000**, *1*, 376–379.
43. Jung, M.J.; Nam, K.H.; Shaginyan, L.R.; Han, J.G. Deposition of Ti thin film using the magnetron sputtering method. *Thin Solid Film* **2003**, *435*, 145–149. [CrossRef]
44. Wang, Z.G.; Zu, X.; Xiang, X.; Zhu, S.; Wang, L.M. Surface modification of Ti-4Al-2V alloy by nitrogen implantation. *J. Mater. Sci.* **2006**, *41*, 3363–3367. [CrossRef]
45. Rautray, T.R.; Narayanan, R.; Kwon, T.-Y.; Kim, K.-H. Surface modification of titanium and titanium alloys by ion implantation. *J. Biomed. Mater. Res. Part B Appl. Biomater.* **2010**, *93*, 581–591. [CrossRef] [PubMed]
46. Shibata, Y.; Miyazaki, T. Anode glow discharge plasma treatment enhances calcium phosphate adsorption onto titanium plates. *J. Dent. Res.* **2002**, *81*, 841–844. [CrossRef]
47. Aronsson, B.-O.; Lausmaa, J.; Kasemo, B. Glow discharge plasma treatment for surface cleaning and modification of metallic biomaterials. *Biomed. Mater. Res.* **1997**, *35*, 49. [CrossRef]
48. Aziz, G.; Ghobeira, R.; Morent, R.; De Geyter, N. Plasma Polymerization for Tissue Engineering Purposes. In *Recent Research in Polymerization*; Cankaya, N., Ed.; IntechOpen: London, UK, 2017. [CrossRef]
49. Lamolle, S.F.; Monjo, M.; Rubert, M.; Haugen, H.J.; Lyngstadaas, S.P.; Ellingsen, J.E. The effect of hydrofluoric acid treatment of titanium surface on nanostructural and chemical changes and the growth of MC3T3-E1 cells. *Biomaterials* **2009**, *30*, 736–742. [CrossRef]
50. Wen, H.B.; Liu, Q.; De Wijn, J.R.; De Groot, K.; Cui, F.Z. Preparation of bioactive microporous titanium surface by a new two-step chemical treatment. *J. Mater. Sci. Mater. Electron.* **1998**, *9*, 121–128. [CrossRef]
51. Lee, B.-H.; Kim, Y.D.; Shin, J.H.; Lee, K.H. Surface modification by alkali and heat treatments in titanium alloys. *J. Biomed. Mater. Res.* **2002**, *61*, 466–473. [CrossRef]
52. Pan, J.; Thierry, D.; Leygraf, C. Hydrogen peroxide toward enhanced oxide growth on titanium in PBS solution: Blue coloration and clinical relevance. *J. Biomed. Mater. Res.* **1996**, *30*, 393. [CrossRef]
53. Sul, Y.-T.; Johansson, C.B.; Kang, Y.; Jeon, D.-G.; Albrektsson, T. Bone Reactions to Oxidized Titanium Implants with Electrochemical Anion Sulphuric Acid and Phosphoric Acid Incorporation. *Clin. Implant. Dent. Relat. Res.* **2002**, *4*, 78–87. [CrossRef]

54. Li, B.; Hao, J.; Min, Y.; Xin, S.; Guo, L.; He, F.; Liang, C.; Wang, H.; Li, H. Biological properties of nanostructured Ti incorporated with Ca, P and Ag by electrochemical method. *Mater. Sci. Eng. C* **2015**, *51*, 80–86. [CrossRef] [PubMed]
55. Minagar, S.; Wang, J.; Berndt, C.C.; Ivanova, E.P.; Wen, C. Cell response of anodized nanotubes on titanium and titanium alloys—A review. *J. Biomed. Mater. Res. Part A* **2013**, *101*, 2726–2739. [CrossRef] [PubMed]
56. Rats, D.; Vandenbulcke, L.; Herbin, R.; Benoit, R.; Erre, R.; Serin, V.; Sevely, J. Characterization of diamond films deposited on titanium and its alloys. *Thin Solid Films* **1995**, *270*, 177–183. [CrossRef]
57. Baek, S.H.; Mihec, D.F.; Metson, J.B. The Deposition of Diamond Films by Combustion Assisted CVD on Ti and Ti-6Al-4V. *Chem. Vap. Depos.* **2002**, *8*, 29–34. [CrossRef]
58. Brinker, C.; Scherer, G. *Sol–Gel Science: The Physics and Chemistry of Sol–Gel Processing*; American Press: San Diego, CA, USA, 1990.
59. Piveteau, L.-D.; Gasser, B.; Schlapbach, L. Evaluating mechanical adhesion of sol–gel titanium dioxide coatings containing calcium phosphate for metal implant application. *Biomaterials* **2000**, *21*, 2193–2201. [CrossRef]
60. Ao, H.; Xie, Y.; Tan, H.; Wu, X.; Liu, G.; Qin, A.; Zheng, X.; Tang, T. Improved hMSC functions on titanium coatings by type I collagen immobilization. *J. Biomed. Mater. Res. Part A* **2013**, *102*, 204–214. [CrossRef]
61. Dettin, M.; Bagno, A.; Gambaretto, R.; Iucci, G.; Conconi, M.T.; Tuccitto, N.; Menti, A.M.; Grandi, C.; Di Bello, C.; Licciardello, A.; et al. Covalent surface modification of titanium oxide with different adhesive peptides: Surface characterization and osteoblast-like cell adhesion. *J. Biomed. Mater. Res. Part A* **2009**, *90*, 35–45. [CrossRef]
62. Kanemura, Y.; Sakai, D.; Komasa, S.; Sekino, T.; Okazaki, J. In vitro behavior of surface-modified implants after chemical processing at room temperature. *J. Osaka Dent. Univ.* **2014**, *48*, 29–35.
63. Karthega, M.; Nallaiyan, R. Hydrogen peroxide treatment on Ti–6Al–4V alloy: A promising surface modification technique for orthopaedic application. *Appl. Surf. Sci.* **2010**, *256*, 2176–2183. [CrossRef]
64. Hampden-Smith, M.J.; Kodas, T.T. Chemical vapor deposition of metals: Part 2. Overview of selective CVD of Metals. *Chem. Vap. Depos.* **1995**, *1*, 39–48. [CrossRef]
65. Frauchiger, V.; Schlottig, F.; Gasser, B.; Textor, M. Anodic plasma-chemical treatment of CP titanium surfaces for biomedical applications. *Biomaterials* **2004**, *25*, 593–606. [CrossRef]
66. Indira, K.; Ningshen, S.; Mudali, K.; Rajendran, N. Effect of Anodization Temperature on the Surface Morphology of Anodized Titanium. In *Thin Film and Nanomaterials*; Jayakumar, S., Kannan, M.D., Balasundaraprabhu, R., Prassana, S., Eds.; Macmillan Publishers: New York, NY, USA, 2011.
67. Gulati, K.; Santos, A.; Findlay, D.; Losic, D. Optimizing Anodization Conditions for the Growth of Titania Nanotubes on Curved Surfaces. *J. Phys. Chem. C* **2015**, *119*, 16033–16045. [CrossRef]
68. Bauer, S.; Schmuki, P.; Von Der Mark, K.; Park, J. Engineering biocompatible implant surfaces: Part I: Materials and surfaces. *Prog. Mater. Sci.* **2013**, *58*, 261–326. [CrossRef]
69. Mas-Moruna, C.E.M.; Montufar, E.; Mestres, G.; Aparicio, D.; Javier, G.F.; Ginebra., M. *Biomaterials Surface Science*; Wiley-VCH: Weinheim, Germany, 2013.
70. Roy, P.; Berger, S.; Schmuki, P. TiO_2 Nanotubes: Synthesis and Applications. *Angew. Chem. Int. Ed.* **2011**, *50*, 2904–2939. [CrossRef] [PubMed]
71. Wang, G.; Wan, Y.; Liu, Z. Construction of Complex Structures Containing Micro-Pits and Nano-Pits on the Surface of Titanium for Cytocompatibility Improvement. *Materials* **2019**, *12*, 2820. [CrossRef] [PubMed]
72. Liu, X.; Chu, P.K.; Ding, C. Surface nano-functionalization of biomaterials. *Mater. Sci. Eng. R Rep.* **2010**, *70*, 275–302. [CrossRef]
73. Beauvais, S.; Drevelle, O.; Jann, J.; Lauzon, M.A.; Foruzanmehr, M.; Grenier, G.; Roux, S.; Faucheux, N. Interactions between bone cells and biomaterials: An update. *Front. Biosci.* **2016**, *8*, 227–263.
74. Williams, D.F. On the mechanisms of biocompatibility. *Biomaterials* **2008**, *29*, 2941–2953. [CrossRef]
75. Schwarcz, H.P.; Abueidda, D.; Jasiuk, I. The Ultrastructure of Bone and Its Relevance to Mechanical Properties. *Front. Phys.* **2017**. [CrossRef]
76. Ratiu, C.; Brocks, M.; Costea, T.; Moldovan, L.; Cavalu, S. PRGF-Modified Collagen Membranes for Guided Bone Regeneration: Spectroscopic, Microscopic and Nano-Mechanical Investigations. *Appl. Sci.* **2019**, *9*, 1035. [CrossRef]
77. Ormanci, O.; Akin, I.; Sahin, F.; Yucel, O.; Simon, V.; Cavalu, S.; Goller, G. Spark plasma sintered Al_2O_3-YSZ-TiO_2 composites: Processing, characterization and in vivo evaluation. *Mater. Sci. Eng. C-Mater. Biol. Appl.* **2014**, *40*, 16–23. [CrossRef] [PubMed]

78. Parnia, F.; Yazdani, J.; Javaherzadeh, V.; Dizaj, S.M. Overview of Nanoparticle Coating of Dental Implants for Enhanced Osseointegration and Antimicrobial Purposes. *Pharm. Pharm. Sci.* **2017**, *20*, 148–160. [CrossRef] [PubMed]
79. Flemming, H.C.; Wingender, J.; Szewzyk, U.; Steinberg, P.; Rice, S.A.; Kjelleberg, S. Biofilms: An emergent form of bacterial life. *Nat. Rev. Microbiol.* **2016**, *14*, 563–575. [CrossRef] [PubMed]
80. Radtke, A.; Topolski, A.; Jedrzejewski, T.; Kozak, W.; Sadowska, B.; Wieckowska-Szakiel, M.; Piszczek, P. Bioactivity Studies on Titania Coatings and the Estimation of Their Usefulness in the Modification of Implant Surfaces. *Nanomaterials* **2017**, *7*, 90. [CrossRef]
81. Nathan, A.S.C.; Mohandoss, K.; Balasubramaniam, M.K. Comparative Evaluation of Antifungal Effect of Titanium, Zirconium and Aluminium Nanoparticles Coated Titanium Plates against C. albicans. *J. Clin. Diagn. Res.* **2016**, *10*, ZC56–ZC59.
82. Shokuhfar, T.; Hamlekhan, A.; Chang, J.-Y.; Choi, C.K.; Sukotjo, C.; Friedrich, C. Biophysical evaluation of cells on nanotubular surfaces: The effects of atomic ordering and chemistry. *Int. J. Nanomed.* **2014**, *9*, 3737.
83. Li, B.; Zhang, X.; Ma, J.; Zhou, L.; Li, H.; Liang, C.; Wang, H. Hydrophilicity of bioactive titanium surface with different structure, composition, crystal form and grain size. *Mater. Lett.* **2018**, *182*, 43–46. [CrossRef]
84. Hu, H.; Zhang, W.; Qiao, Y.; Jiang, X.; Liu, X.; Ding, C. Antibacterial activity and increased bone marrow stem cell functions of Zn-incorporated TiO_2 coatings on titanium. *Acta Biomater.* **2012**, *8*, 904–915. [CrossRef]
85. Demirbaşa, Ç.; Aydaya, A. The influence of Nano-TiO_2 and Nano-Al_2O_3 Particles in Silicate Based Electrolytes on Microstructure and Mechanical Properties of Micro Arc Coated Ti6Al4V Alloy. *Mater. Res.* **2018**, *21*, e20180092. [CrossRef]
86. Mendonça, G.; Mendonça, D.B.; Simões, L.G.; Araújo, A.L.; Leite, E.R.; Duarte, W.R.; Cooper, L.F.; Aragão, F.J. Nanostructured alumina-coated implant surface: Effect on osteoblast-related gene expression and bone-to-implant contact in vivo. *Int. J. Oral Maxillofac. Implants* **2009**, *24*, 205–215.
87. Lin, A.; Wang, C.J.; Kelly, J.; Gubbi, P.; Nishimura, I. The role of titanium implant surface modification with hydroxyapatite nanoparticles in progressive early bone-implant fixation in vivo. *Int. J. Oral Maxillofac. Implants* **2009**, *24*, 808–816. [PubMed]
88. Kloss, F.R.; Singh, S.; Hächl, O.; Rentenberger, J.; Auberger, T.; Kraft, A.; Klima, G.; Mitterlechner, T.; Steinmüller-Nethl, D.; Lethaus, B. BMP-2 immobilized on nanocrystalline diamond–coated titanium screws; demonstration of osteoinductive properties in irradiated bone. *Head Neck* **2013**, *35*, 235–241. [CrossRef] [PubMed]
89. Liao, J.; Zhu, Z.; Mo, A.; Li, L.; Zhang, J. Deposition of silver nanoparticles on titanium surface for antibacterial effect. *Int. J. Nanomed.* **2010**, *5*, 261–267.
90. Memarzadeh, K.; Sharili, A.S.; Huang, J.; Rawlinson, S.C.; Allaker, R.P. Nanoparticulate zinc oxide as a coating material for orthopedic and dental implants. *J. Biomed. Mat. Res. Part A* **2015**, *103*, 981–989. [CrossRef] [PubMed]
91. Zainali, K.; Danscher, G.; Jakobsen, T.; Baas, J.; Moller, P.; Bechtold, J.E.; Soballe, K. Assessment of modified gold surfaced titanium implants on skeletal fixation. *J. Biomed. Mater. Res. A* **2013**, *101*, 195–202. [CrossRef]
92. Antoniac, V.I.; Cavalu, S.; Costoiu, M.C.; Doicin, C.V.; Doicin, I.E.; Mates, I.M.; Mohan, A.G.; Murzac, R.; Semenescu, A.; Saceleanu, V.; et al. Cranial implant with osteointegration structures and functional coatings. Patent RO132417B1, 30 October 2019. Available online: https://worldwide.espacenet.com/patent/search/family/061724843/publication/RO132417A0?q=pn%3DRO132417A0 (accessed on 16 October 2020).
93. Cavalu, S.; Antoniac, I.V.; Fritea, L.; Mates, I.M.; Milea, C.; Laslo, V.; Vicas, S.; Mohan, A. Surface modifications of the titanium mesh for cranioplasty using selenium nanoparticles coating. *J. Adhes. Sci. Technol.* **2018**, *22*, 2509–2522. [CrossRef]
94. Tran, P.; Webster, T.J. Enhanced osteoblast adhesion on nanostructured selenium compacts for anti-cancer orthopaedic applications. *Int. J. Nanomed.* **2008**, *3*, 391–396.
95. Miller, D.; Webster, T.J. Anticancer orthopedic implants. In *Cancer Nanotechnology American Scientific*; Nalwa, H., Webster, T.J., Eds.; Springer: Berlin, Germany, 2007; pp. 307–316.
96. Esfandiari, N.; Simchi, A.; Bagheri, R. Size tuning of Ag-decorated TiO_2 nanotube arrays for improved bactericidal capacity of orthopedic implants. *J. Biomed. Mater. Res. A* **2014**, *102*, 2625–2635. [CrossRef]
97. Park, J.-W.; Han, S.-H.; Hanawa, T. Effects of Surface Nanotopography and Calcium Chemistry of Titanium Bone Implants on Early Blood Platelet and Macrophage Cell Function. *BioMed Res. Internat.* **2018**. [CrossRef]

98. Whanger, P.; Vendeland, S.; Park, Y.C.; Xia, Y. Metabolism of subtoxic levels of selenium in animals and humans. *Ann. Clin. Lab. Sci.* **1996**, *26*, 99–113.
99. Webster, T.J.; Ejiofor, J.U. Increased osteoblast adhesion on nanophase metals: Ti, Ti_6Al_4V, and CoCrMo. *Biomaterials* **2004**, *25*, 4731–4739. [CrossRef]
100. Ercan, B.; Webster, T.J. The effect of biphasic electrical stimulation on osteoblast function at anodized nanotubular titanium surfaces. *Biomaterials* **2010**, *31*, 3684–3693. [CrossRef] [PubMed]
101. Oh, S.; Brammer, K.S.; Li, Y.S.J.; Teng, D.; Engler, A.J.; Chien, S.; Jin, S. Stem cell fate dictated solely by altered nanotube dimension. *Proc. Natl. Acad. Sci. USA* **2009**, *106*, 2130–2135. [CrossRef] [PubMed]
102. Gongadze, E.; Kabaso, D.; Bauer, S.; Slivnik, T.; Schmuki, P.; van Rienen, U.; Iglic, A. Adhesion of osteoblasts to a nanorough titanium implant surface. *Int. J. Nanomed.* **2011**, *6*, 1801–1816.
103. Branemark, P.I.; Adell, R.; Albrektsson, T.; Lekholm, U.; Lundkvist, S.; Rockler, B. Osseointegrated titanium fixtures in the treatment of edentulousness. *Biomaterials* **1983**, *4*, 25–28. [CrossRef]
104. Kubo, K.; Tsukimura, N.; Iwasa, F.; Ueno, T.; Saruwatari, L.; Aita, H.; Chiou, W.-A.; Ogawa, T. Cellular behavior on TiO_2 nanonodular structures in a micro-to-nanoscale hierarchy model. *Biomaterials* **2009**, *30*, 5319–5329. [CrossRef] [PubMed]
105. Ogawa, T.; Saruwatari, L.; Takeuchi, K.; Aita, H.; Ohno, N. Ti nano-nodular structuring for bone integration and regeneration. *J. Dent. Res.* **2008**, *87*, 751–756. [CrossRef]
106. Bral, A.; Mommaerts, M.Y. In vivo biofunctionalization of titanium patient-specific implants with nano hydroxyapatite and other nano calcium phosphate coatings: A systematic review. *J. Cranio-Maxillofac. Surg.* **2016**. [CrossRef]
107. Chen, Z.; Klein, T.; Murray, R.Z.; Crawford, R.; Chang, J.; Wu, C.; Xiao, Y. Osteoimmunomodulation for the development of advanced bone biomaterials. *Mater. Today* **2016**, *19*, 304–321. [CrossRef]
108. Goh, R.C.W.; Chang, C.N.; Lin, C.L.; Lo, L.J. Customised fabricated implants after previous failed cranioplasty. *J. Plast. Reconstr. Aesthet. Surg.* **2010**, *63*, 1479–1484. [CrossRef]
109. Wohlers, T.; Caffrey, T. *Wohlers Report 2015: 3D Printing and Additive Manufacturing State of the Industry Annual Worldwide Progress Report*; Wohlers Associates: Fort Collins, CO, USA, 2015.
110. Li, A.; Azad, T.D.; Veeravagu, A.; Bhatti, I.; Long, C.; Ratliff, J.K.; Li, G. Cranioplasty Complications and Costs: A National Population-Level Analysis Using the MarketScan Longitudinal Database. *World Neurosurg.* **2017**, *102*, 209–220. [CrossRef] [PubMed]
111. Mrad, M.A.; Murrad, K.; Antonyshyn, O. Analyzing the Cost of Autogenous Cranioplasty versus Custom-Made Patient-Specific Alloplastic Cranioplasty. *J. Craniofac. Surg.* **2017**, *28*, 1260–1263. [CrossRef] [PubMed]
112. Zanaty, M.; Chalouhi, N.; Starke, R.M.; Clark, S.W.; Bovenzi, C.D.; Saigh, M.; Schwartz, E.; Kunkel, E.S.I.; Efthimiadis-Budike, A.S.; Jabbour, P.; et al. Complications following cranioplasty: Incidence and predictors in 348 cases. *J. Neurosurg.* **2015**, *123*, 182–188. [CrossRef] [PubMed]
113. Júnior, A.C.; Filho, P.T.; Gonçalves, M.P.; Neto, A.A.; Zanini, M. Cranioplasty: An Institutional Experience. *J. Craniofacial Surg.* **2018**, *29*, 1402–1405. [CrossRef]
114. Liang, E.S.; Tipper, G.; Hunt, L.; Chiung Gan, P.Y. Cranioplasty outcomes and associated complications: A single-centre observational study. *Br. J. Neurosurg.* **2016**, *30*, 122–127. [CrossRef] [PubMed]
115. Blake, D.P. The use of synthetics in cranioplasty: A clinical review. *Mil Med.* **1994**, *159*, 466–469. [CrossRef]
116. Kuttenberger, J.J.; Hardt, N. Long-term results following reconstruction of craniofacial defects with titanium micro-mesh systems. *J. Craniomaxillofac. Surg.* **2001**, *29*, 75–81. [CrossRef]
117. Antoniac, I. *Handbook of Bioceramics and Biocomposites*; Springer International Publishing: New York, NY, USA, 2016; pp. 1275–1307.
118. Dumitru, M.; Nica, D.; Mohan, A.; Feder, B.; Ciurea, A.V. New concept in cranioplasty based on biomaterials performance in various clinical cases. *Key Eng. Mater.* **2016**, *695*, 185–188.
119. Mohan, A.; Saceleanu, V.; Mohan, D.; Ciurea, A.V. Value of Dural Patches in Obliteration of Cranio-Ethmoidal Traumatic CSF Fistulas. *Key Eng. Mater.* **2016**, *695*, 196–199. [CrossRef]
120. Shuo-Tsung, C.; Cheng-Jen, C.; Wei-Chin, S.; Lin-Wan, C.; Hsuan, C.I.; Muh-Shi, L. 3-D titanium mesh reconstruction of defective skull after frontal craniectomy in traumatic brain injury. *Injury* **2015**, *46*, 80–85.
121. Stoodley, M.A.; Abbott, J.R.; Simpson, D.A. Titanium cranioplasty using 3-D computer modelling of skull defects. *J. Clin. Neurosci.* **1996**, *3*, 149–155. [CrossRef]

122. Patel, M.F.; Langdon, J.D. Titanium mesh (TiMesh) osteosynthesis: A fast and adaptable method of semi-rigid fixation. *Br. J. Oral Maxillofac. Surg.* **1991**, *29*, 316–324. [CrossRef]
123. Rosen, N.; Brady, B.R.; Buonanno, T.J.; Kistler, E.S.; Hinshaw, J.P.; Waldo, S. Potential hazards and artifacts of ferromagnetic and nonferromagnetic surgical devices in nuclear magnetic resonance imaging. *Radiology* **1982**, *147*, 139–148.
124. Cavalu, S.; Prokisch, J.; Laslo, V.; Vicas, S. Preparation, structural characterization and release study of novel hybrid microspheres entrapping nanoselenium, produced by green synthesis. *IET Nanobiotechnol.* **2016**, *11*, 426–432. [CrossRef] [PubMed]
125. Shah, C.P.; Dwivedi, C.; Singh, K.K.; Kumar, M.; Bajaj, P.N. Riley oxidation: A forgotten name reaction for synthesis of selenium nanoparticles. *Mat. Res. Bull.* **2010**, *45*, 1213–1217. [CrossRef]
126. Lin, Z.H.; Wang, C.R. Evidence on the size-dependent absorption spectral evolution of selenium nanoparticles. *Mater. Chem. Phys.* **2005**, *92*, 591–594. [CrossRef]
127. Zhang, J.; Taylor, E.W.; Wan, X.; Peng, D. Impact of heat treatment on size, structure, and bioactivity of elemental selenium nanoparticles. *Int. J. Nanomed.* **2012**, *7*, 815–825. [CrossRef]
128. Chen, Y.; Li, L.; D' Ulivo, A.; Belzile, N. Extraction and determination of elemental selenium in sediments—A comparative study. *Anal. Chim. Acta* **2006**, *577*, 126–133. [CrossRef]
129. Refat, M.; Elsabawy, S. Infrared spectra, Raman laser, XRD, DSC/TGA and SEM investigations on the preparations of selenium metal, (Sb_2O_3, Ga_2O_3, SnO and HgO) oxides and lead carbonate with pure grade using acetamide precursors. *Bull. Mater. Sci.* **2011**, *34*, 873. [CrossRef]
130. Kannan, S.; Mohanraj, K.; Prabhu, K.; Barathan, S.; Sivakumar, G. Synthesis of selenium nanorods with assistance of biomolecule. *Bull. Mater. Sci.* **2014**, *37*, 1631–1635. [CrossRef]

materials

MDPI

Review

Nanotechnology in the Diagnostic and Therapy of Hepatocellular Carcinoma

Florin Graur [1,2], Aida Puia [3,*], Emil Ioan Mois [1,2,*], Septimiu Moldovan [2], Alexandra Pusta [4], Cecilia Cristea [4], Simona Cavalu [5], Cosmin Puia [1,2] and Nadim Al Hajjar [1,2]

1 Department of Surgery, University of Medicine and Pharmacy "Iuliu Hatieganu", 400012 Cluj-Napoca, Romania; florin.graur@umfcluj.ro (F.G.); drpuia@yahoo.fr (C.P.); na_hajjar@yahoo.com (N.A.H.)
2 Regional Institute of Gastroenterology and Hepatology "Octavian Fodor", 400394 Cluj-Napoca, Romania; septimiu1995@yahoo.com
3 Department of General Practitioner, University of Medicine and Pharmacy "Iuliu Hatieganu", 400347 Cluj-Napoca, Romania
4 Department of Analytical Chemistry, University of Medicine and Pharmacy "Iuliu Hatieganu", 400347 Cluj-Napoca, Romania; alexandrapusta@gmail.com (A.P.); ccristea@umfcluj.ro (C.C.)
5 Department of Medical Biophysics, Faculty of Medicine and Pharmacy, University of Oradea, 410087 Oradea, Romania; simona.cavalu@gmail.com
* Correspondence: draidapuia@gmail.com (A.P.); drmoisemil@gmail.com (E.I.M.)

Abstract: Hepatocellular carcinoma is the most common liver malignancy and is among the top five most common cancers. Despite the progress of surgery and chemotherapy, the results are often disappointing, in part due to chemoresistance. This type of tumor has special characteristics that allow the improvement of diagnostic and treatment techniques used in clinical practice, by combining nanotechnology. This article presents a brief review of the literature focused on nano-conditioned diagnostic methods, targeted therapy, and therapeutic implications for the pathology of hepatocellular carcinoma. Within each subdomain, several modern technologies with significant impact were highlighted: serological, imaging, or histopathological diagnosis; intraoperative detection; carrier-type nano-conditioned therapy, thermal ablation, and gene therapy. The prospects offered by nanomedicine will strengthen the hope of more efficient diagnoses and therapies in the future.

Keywords: hepatocellular carcinoma; nanomedicine; cancer diagnosis; targeting; smart therapy

Citation: Graur, F.; Puia, A.; Mois, E.I.; Moldovan, S.; Pusta, A.; Cristea, C.; Cavalu, S.; Puia, C.; Al Hajjar, N. Nanotechnology in the Diagnostic and Therapy of Hepatocellular Carcinoma. *Materials* **2022**, *15*, 3893. https://doi.org/10.3390/ma15113893

Academic Editor: Hanna Dams-Kozłowska

Received: 15 April 2022
Accepted: 24 May 2022
Published: 30 May 2022

1. Introduction

Hepatocellular carcinoma (HCC) is ranked fifth among the most common neoplasms and ranked third among the causes of cancer-related deaths. The incidence of HCC is on the rise in Europe and the USA, due to new lifestyle-related risk factors, such as metabolic syndrome and nonalcoholic fatty liver disease [1,2]. The estimated five-year survival rate of HCC is only 18%, making it one of the most lethal cancers [3]. Thus, it is vital that this disease is diagnosed early and treated adequately, to ensure the best outcomes for the patient. Current diagnostic and treatment methods present certain limitations. The most commonly employed methods for the diagnosis of HCC are imaging techniques (ultrasound, computed tomography, or magnetic resonance imaging) and the quantification of biomarkers that are overexpressed in this pathology, such as alpha-fetoprotein (AFP) [1]. However, imaging equipment might not be available in all clinical facilities, and the results obtained are dependent on the user's expertise, while AFP poses certain sensitivity and selectivity issues [4]. Treatment methods for HCC include resection surgery, liver transplantation, chemotherapy, radiofrequency ablation (RFA), or transarterial chemoembolization (TACE) [3]. Resection surgery or liver transplantation are indicated in a very small number of patients, and systemic chemotherapy has moderate results (with high recurrence and multidrug resistance [MDR]) and significant side effects. Local therapies, such as RFA or TACE, have only a palliative role [5,6].

The marked increase in the number of patients diagnosed with liver cancer has led to the development of innovative diagnostic and therapeutic technologies, including nanomedicine. Its development was made possible by a more advanced understanding of the mechanisms of liver cancer, and by the collaboration of teams of doctors, chemists, materials specialists, and physicists, who have focused their research on this field. Although surgery and chemotherapy are still the standard treatments for HCC, the outlook suggests that new nano-conditioned therapies will be useful in clinical practice.

Owing to their intrinsic or artificially designed properties, nanomaterials find applications both in the diagnosis and treatment of hepatocellular carcinoma. Their unique properties and functionalities are due in part to their size (between 1 and 100 nm), and among them are increased biocompatibility, ability to detect early HCC, the possibility of use in imaging and diagnosis, and their use as genic or cytostatic transporters. In the context of diagnosis, nanomaterials can be employed for the functionalization of sensor surfaces, increasing the sensitivity of detection for certain biomarkers, or they can be used as tracers for imaging diagnosis. In the context of therapy, nanomaterials can be functionalized with ligands such as antibodies or aptamers that are specific for tumor cells, thus facilitating targeted transport of drugs directly to the tumor site, reducing non-specific toxicity and chemotherapy resistance. Moreover, nanomaterials can be employed for theragnostic approaches, a promising strategy that allows simultaneous therapy and diagnosis [7].

Given the burden of cancer on public health, many nanotechnology-based strategies have been developed for the treatment of cancer. The evolution of nanotechnology for medical applications in cancer is shown in Figure 1.

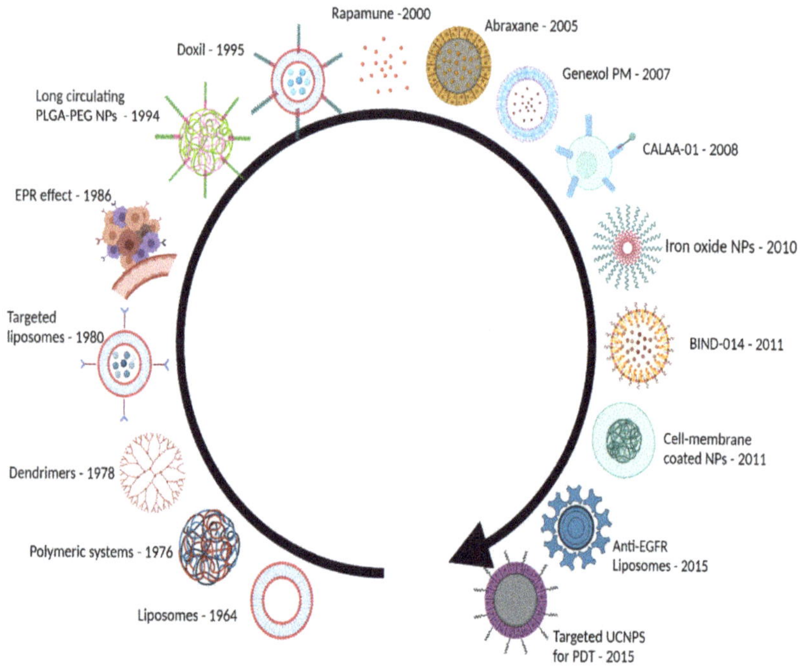

Figure 1. History of nanomedicine. EPR—enhanced permeability and retention; PLGA—poly (lactic-co-glycolic acid); PEG—polyethyleneglycol; UCNPs—upconverting nanoprticles; PDT—photodynamic therapy (created with BioRender.com).

The main types of nanomaterials used in the treatment of cancer are carbon-based nanomaterials (nanotubes, fullerene, and graphene), lipid-based nanoparticles (liposomes), metallic nanoparticles (gold nanoparticles), superparamagnetic iron oxide, silica-based nanoparticles, polymers, quantum dots, dendrimers, nanoshells, etc.

In this review, the most recent advances in the use of nanomaterials in the diagnosis and treatment of HCC will be presented. Firstly, the characteristics of the main types of nanomaterials will be, briefly, discussed, followed by the presentation of relevant examples of nano-mediated diagnostic (imaging, serologic, histopathologic, and intra-operatory) and therapeutic approaches. Finally, relevant examples of theragnostic approaches for HCC will be provided, and the future perspectives of nanomaterial applications in HCC management will be presented.

2. Types of Nanoparticles Used for Diagnosis and Treatment

In recent years, numerous types of nanoparticles have been employed for the diagnosis and therapy of cancers in general and HCC in particular. These nanoparticles can be classified into carbon-based nanomaterials (nanotubes, fullerene, and graphene), lipid-based nanoparticles (liposomes), metallic nanoparticles (gold nanoparticles), superparamagnetic iron oxide, silica-based nanoparticles, polymers, micelles, dendrimers, and virus-like particles (Figure 2). Each class presents unique properties as well as characteristic advantages and disadvantages and can be obtained using different synthesis methods (Table 1). Since the aim of this study was to highlight the medical applications of nanoparticles for HCC diagnosis and treatment, the following section only, briefly, details characteristics of the most common types of nanoparticles used for the targeted delivery or diagnosis of cancer. For a more comprehensive view on their synthesis and characteristics, the reader is referred to other works of the literature [8–13].

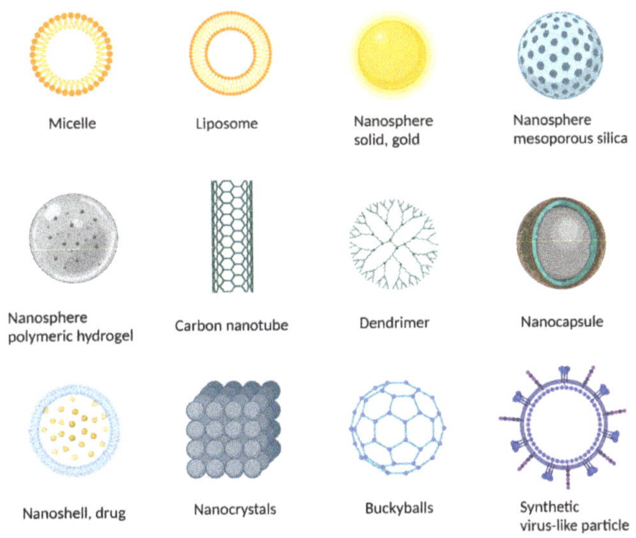

Figure 2. Different types of nanomaterials used in nanomedicine (created with BioRender.com).

2.1. Carbon-Based Nanomaterials

Numerous forms of carbon, such as carbon nanotubes (CNT), graphene (GRP), fullerenes (FUL), or quantum dots have, recently, gained increasing attention from the scientific community. These nanomaterials have been employed for the targeted delivery of drugs or for cancer diagnosis. However, there is still concern about their toxicity for both humans and the environment [14], so most of their cancer-related applications are limited to their use in the development of sensors for cancer biomarkers. Their high surface area and excellent electrical conductivity [15] make them suitable candidates for the development of the electrochemical sensors used for the detection of biomarkers that are expressed in HCC [16–18].

2.2. Liposomes

Liposomes are phospholipid vesicles that are comprised of two distinct regions: a hydrophillic core and a lipophillic lipid bilayer, thus being able to incorporate both hydrophillic and hydrophobic drugs [11,19,20]. Liposomes represent a promising carrier for targeted delivery, with many liposome-based drug formulations already on the market. Some of the shortcomings of liposomes, such as their increased uptake by immune cells, can be overcome by modifying their surface with polymers, such as polyethyleneglycol (PEG), or with targeting agents, such as antibodies or aptamers [19]. Liposomes are among the most commonly used carriers in the targeted treatment of cancers, with many examples of liposomes being used in the treatment of HCC. Doxorubicin was most commonly encapsulated into different types of liposomes [21–23] as well as sorafenib and short interfering RNA (siRNA) [24].

2.3. Metallic Nanoparticles

Metallic nanoparticles, such as gold (AuNP), silver (AgNP), or platinum (PtNP), can be used for both diagnostic and therapeutic purposes.

Gold nanoparticles (AuNPs) are, usually, synthesized by chemical reduction and stabilization using a capping agent, electrochemical deposition on electrode surfaces (in the case of nanoparticles-based electrochemical sensors), or newly emerging biological-synthesis techniques [10,25]. The resulting gold nanoparticles can be easily functionalized, can be used for hyperthermia applications, and can easily be controlled regarding their size and surface properties [11]. Moreover, given their size-dependent visible-light-absorption behavior, their trajectories into cells can be tracked [11]. Although AuNPs are among the least toxic metallic nanoparticles, some studies have shown that high concentrations of AuNPs can, potentially, be genotoxic [26]. Further studies are needed to assess the biocompatibility and long-term effects of AuNPs. Until then, AuNPs and other metallic nanoparticles remain valuable candidates for diagnostic techniques, such as the electrochemical detection of different biomarkers. In the context of HCC, gold nanoparticles [17], gold–platinum [16], or platinum–palladium nanostructures [18] have been used in conjunction with carbon-based nanomaterials, for the detection of biomarkers such as AFP or glypican-3. Gold nanostructures can, also, be employed for the imaging and photothermal ablation of tumors [27].

2.4. Magnetic Nanoparticles

Superparamagnetic iron oxide nanoparticles (SPIONs) can be, successfully, employed for both diagnostic [28] and therapeutic applications [29]. These nano-carriers consist of a magnetic core and a coating. Their main advantage consists of their ability to be guided using an external magnetic field, thus being promising candidates for targeted drug delivery. Moreover, magnetic nanoparticles can be used as contrast agents in magnetic resonance imaging diagnosis, due to their superparamagnetic properties [30]. Magnetic nanoparticles can be obtained using a variety of different physical or chemical synthesis strategies and can be functionalized using inorganic or organic ligands to increase their stability [11]. Their versatility and high stability make them desirable nanomaterials for drug delivery and diagnosis [31].

2.5. Polymeric Micelles

Polymeric micelles are, also, promising candidates for targeted drug delivery applications. These nano-carriers consist of self-assembled amphiphilic block-copolymers that form a hydrophobic core and a hydrophilic corona [32]. Polymeric micelles can be obtained from a wide variety of polymers, such as poly (lactic-co-glycolic acid), poly (ε-caprolactone), and poly (lactic acid), and are formed when the concentration of the polymer reaches the critical micelle concentration [32]. Polymeric micelles have been employed in the treatment of HCC for the encapsulation of hydrophobic drugs, such as ursolic acid [33] or niclosamide [34], in order to increase their bioavailability.

Table 1. Synthesis methods, advantages, and disadvantages of different types of nanoparticles.

Nanoparticle Type	Synthesis Methods	Advantages	Disadvantages
Carbon-based NPs	Mechanical exfoliation (GRP) Chemical exfoliation (GRP) Chemical vapor deposition (GRP, CNT) Laser ablation (CNT, FUL) [8]	High surface area High electrical conductivity [15] Chemical stability	Bio-corona formation Toxicity Environmental toxicity [14]
Liposomes	Thin-film method Proliposome method Injection method Emulsification method [9]	Hydrophillic and hydrophobic drug encapsulation Biocompatibility Low imunogenicity	Low encapsulation efficiency [11] Short shelf-life Accelerated blood clearance
AuNPs	Chemical reduction Electrochemical reduction Seed-mediated growth Digestive ripening Biologic synthesis [10]	Easily functionable Hyperthermia applications Visible light absorption High surface area High electrical conductivity [15]	Potential genotoxicity High costs of raw materials
SPIONs	Chemical co-precipitation Thermal decomposition Gas-phase deposition Pulsed laser ablation Electron beam lithography Biologic methods [12]	External guidance Hyperthermia applications High stability	Potential toxicity Laborious synthesis
Polymeric micelles	Phase-inversion method Rehydration method Polymerization-induced self-assembly Microfluid method [13]	Hydrophillic and hydrophobic drug encapsulation Biocompatibility Prolonged circulation time	Low drug loading Low stability [32]

3. Nanoparticles Used for Diagnostics

In order to diagnose HCC, nanomedicine can be used in the following directions: imaging diagnosis using computed tomography (CT), magnetic resonance imaging (MRI), real-time up-conversion luminescence imaging, or a combination of these imaging techniques, as well as pre- or intra-operatory diagnosis and serological or histopathological diagnosis, by detecting biomarkers in patients' biological fluids or biopsy samples.

3.1. *Imaging Diagnosis*

Ultrasound imaging is a valuable tool for the surveillance of patients who are at risk of developing HCC, while CT and MRI are common diagnostic methods for HCC [1]. Although diagnosis is, relatively, specific, due to the existence of the "radiological hallmark of HCC" [1], certain limitations exist in regards to currently employed diagnostic methods. The contrast agents that are, typically, used for imaging diagnosis have short half-lives and do not present specificity for the tumor tissue [35].

Imaging diagnosis can be improved by developing tracers or contrast agents with superior qualities to those currently used by nano-candidacy. Several studies have shown significant progress in the field; however, the properties of these nano-tracers must, additionally, meet certain conditions: have increased stability, so that scanning devices (CT and MRI) can be set in certain parameters, and have comparable image characteristics. Nanoparticles must accumulate in tissues at certain time intervals after injection and should be rapidly eliminated at set time intervals [7]. Due to the versatility of nanomaterials, the majority of imaging approaches for HCC combine their diagnostic properties with therapeutic ones, and, thus, these approaches will be presented in the Section 5, while a few simple imaging diagnostic methods will be presented, herein.

An MRI-detection method for HCC was developed, based on the presence of high concentrations of AFP and glypican-3 in the cytoplasm and on the surface of tumor cells, respectively, compared to healthy cells. Ultra-small SPIONs (5 nm diameter) were functionalized with antibodies for the two biomarkers, by amide bond formation, and the modified SPIONs (SAG) were used for MRI imaging of HCC cells. In parallel, SPIONs functionalized only with AFP (SA) or glypican-3 antibodies (SG) were, also, obtained and tested. Prussian blue staining assay and in vitro MRI tests were carried out to investigate the accumulation of these carriers in tumor cells. The highest Prussian blue staining, as well as the highest reduction in T2 values in MRI, was obtained for SAG compared to SA and SG, indicating that double functionalization can enhance the detection of HCC compared to single functionalization, this being an advantage of the presented method [28].

An aptamer-conjugated fifth-generation dendrimer, for the combined visual near-nfrared fluorescence and MRI imaging of HCC in rats, was proposed. The aptamer was selected for endoglin, a biomarker present on the surface of vascular endothelial cells in tumors, thus increasing the specificity of the nanoprobe towards tumor tissue. The modified dendrimers were labeled with fluorophores, for near-infrared fluorescence detection, and with paramagnetic Gd^{3+}-DTPA chelators, which helped achieve T1-weighted relaxation and an MRI signal. The obtained nanoprobes displayed high fluorescence in tumor cells, compared to aptamer-free dendrimers. In vivo MRI imaging showed a clear delimitation of the tumor margins, indicating that this approach could be used for guiding surgical procedures for the resection of small hepatic tumors [35].

Another dual-imaging approach was reported by Lee et al. In their study, Nd^{3+}-doped up-conversion nanoparticles (UCNP) were employed for the MRI and real-time up-conversion luminescence imaging of HCC, in a simple rat model. Nd^{3+} was used, due to its lower potential for overheating upon irradiation, compared to the commonly used Yb^{3+}. The core-shell particles were obtained via thermal decomposition (the core), followed by shell-coating via a seed-mediated process. The surface of the nanoprobe was functionalized with anti-CD44 antibody, to increase selectivity for the tumor cells. Selective accumulation of the nanoprobes in tumor tissue was demonstrated by MRI scans as well as by the bright UCL signal observed in tumor tissue, compared to the surrounding healthy tissue [36].

Dual MRI and CT for HCC models in rats have, also, been reported. A polyethyleneimine substrate was modified with a gadolinium chelator and lactobionic acid, then used as a scaffold for the deposition of AuNPs. The obtained AuNPs presented selectivity for HCC tissue, due to the lactobionic acid, which acts as a ligand for the asialoglycoprotein receptor present on the surface of HCC cells. The AuNPs presented good biocompatibility, accumulation in the tumor site, and high r_1 relaxitivity and specificity towards HCC tissue [37].

3.2. Serologic and Histopathologic Diagnosis

There is abundant literature on serological biomarkers for the early detection of HCC [38–41], but relatively few of these biomarkers are used for nano-mediated detection. This may lead to the development of systems that will significantly improve the diagnosis of HCC [7]. The most commonly employed HCC biomarker, and the only one with phase 5 validation data, is AFP, but numerous other biomarkers (glypican-3, osteopontin, Golgi-protein 73, and PIVKA) have been, recently, reported [38].

The levels of serological tumor markers used in the detection and monitoring of HCC are usually low, and the detection of these biomarkers in biological samples requires the development of sensitive methods. Nanomaterials, such as metallic nanoparticles or carbon-based nanomaterials, can be used to increase the surface area of sensors, improve their sensing capabilities, and, thus, increase overall sensitivity. Moreover, they can also be functionalized with biological (antibodies and enzymes) or biomimetic (aptamers and molecularly imprinted polymers) recognition systems, which increase the selectivity of the detection method towards a selected target. Many nanomaterial-based detection systems have been reported for the detection of cancer biomarkers, such as proteins, circulating tumor cells, or miRNA [42–44], and some of those related to HCC detection are, briefly, presented herein.

An AFP sensing technique uses an electrochemical method and an AFP aptamer for detection [16]. Aptamers are short, single-stranded nucleic acids that can bind to specific targets [45]. Gold–platinum nanoparticles were deposited on the surface of an electrode, to increase its electrical conductivity, followed by the deposition of reduced graphene oxide, chitosan, and a redox mediator. Lastly, the aptamer was added to ensure selectivity towards AFP. The method showed some interferences with other proteins, but its sensitivity was sufficient for the detection of low concentrations of AFP [16].

Metallic nanoparticles were, also, employed for the electrochemical detection of glypican 3. Li's group reported two similar methods, using gold nanoparticles [17] and platinum–palladium structures [18], respectively. These nanomaterials were grafted onto electrode surfaces, together with reduced graphene oxide, hemin [17], or ferrocene [18], as redox mediators, and the glypican-3-specific aptamer, which was immobilized via cross-linking. The detection was done by measuring the decrease in the differential-pulse voltammetry oxidation peak of the redox mediator, when glypican-3 was added to the aptasensor. The platinum–palladium nanostructure approach presented higher stability and a better linear correlation coefficient, but its sensitivity was lower, compared to that of the AuNP approach.

A biosensor based on a nanopore structure was used for the multiplexed detection of three cancer biomarkers (AFP, carcinoembryonic antigen, and human epidermal growth factor receptor-2), by charge modulation ion current rectification. The system contained three different antibodies, to ensure selectivity towards each marker. This sensor exhibited excellent sensitivity, with limits of detection in the femtomolar level for undiluted serum samples [46].

3.3. *Intraoperatory Diagnosis*

A new trend that has, recently, emerged is intraoperative diagnosis. This fact refers to the highlighting of the remaining tumor tissue, usually on the hepatic resection trance, which is, often, difficult to visualize without special techniques.

Andreou et al. developed Surface-Enhanced Raman Scattering (SERS) silica-coated nanoparticles, as a contrast agent for the intraoperative detection of HCC in mice. Gold NPs were obtained, using a chemical reduction method, and were, then, coated with silica in the presence of a Raman reporter, using a modified Stöber method. The silica layer was obtained using a non-toxic protocol, ensuring the protection of NPs against aggregation. The detection of HCC, using these NPs, relied on their higher accumulation in healthy liver tissue compared to tumors, thus highlighting the tumor margins. The approach was compared to MRI intraoperatory imaging, and the SERS-based method was able to discern smaller tumors. Moreover, compared to fluorescence methods, the SERS-based approach has higher sensitivity and photostability. One of the limitations of this method is the lack of clinically available Raman imaging devices [47].

Fluorescence-based intraoperatory imaging was, also, reported. Tsunda et al. developed lactosomes, which encapsulated indocyanine green (ICG), a photosensitizing agent for near-infrared fluorescence imaging. Lactosomes are a type of core-shell polymeric micelle, obtained by the polymerization of lactic acid and sarcosine. These NPs were, intravenously, administered to nude mice and demonstrated prolonged retention in liver tumors, compared to ICG alone. This represents an advantage, in the case of long surgical procedures, where a single administration would be sufficient for the whole procedure. Another advantage is the ability of ICG-loaded lactosomes to perform photodynamic therapy of the tumor, upon laser irradiation [48].

Another fluorescence imaging-based nanoconditioned system is that of Dai et al., which is aimed at detecting AFP on the membrane of HCC cells. The system is a complex comprised of $CuInS_2$–ZnS quantum dots (QD), gold nanoflowers, and anti-AFP antibody. The QDs were synthesized using a hydrothermal method and were sonicated together with gold nanoflowers to yield Au/$CuInS_2$–ZnS, which was later incubated with anti-AFP antibody. The system has the advantage of being biocompatible and has low toxicity. The association between gold and QDs emits higher fluorescence compared to QDs alone, making the complex suitable for in vivo detection. The complex could successfully identify HepG2 HCC cells [49].

4. Nanoparticles Used for Therapy

Depending on HCC stage, different treatment strategies can be employed, with various degrees of efficacy and safety for the patient [4]. For very early and early stages of the disease, ablation, resection, or liver transplantation are the methods of choice, in the intermediate stage, chemoembolization is used, and in the advanced stage HCC benefits only from systemic therapy. Nanotechnology can be incorporated into many of these strategies (ablation, chemoembolization, and, especially, systemic drug delivery) in order to overcome their current limitations and provide better alternatives for the patient. Moreover, newly emerging treatment strategies, such as the administration of siRNA or short microRNAs (miRs), which lead to the inhibition of post-transcriptional expression of mRNA targets, have been recently reported [50,51]. Nanotechnology could play a crucial role in the use of these treatments, due to its ability to reduce some limitations of RNA, such as rapid degradation in intra- and extra-cellular environments and low stability.

For nanoconditioned carriers to be used for therapeutic purposes, they must meet certain requirements, which include biocompatibility, rapid degradation and elimination, reduced or, even, absent side effects, reduced immunogenicity, and sufficient stability until the delivery of active drugs.

4.1. Nanoparticle Targeting and Drug Release

The accumulation of nano-conditioned systems in the target tissue can be performed passively or actively [52]. Passive accumulation is achieved because of the effect of permeability and increased retention, due to the modified structure of the vascularization of intratumoral neoformation [52]. PEGylated nanoparticles have been, frequently, used for the passive targeting of tumors, due to their high permeability and retention effects [53]. Recent studies have demonstrated correlations between certain properties (hydration state and PEG surface density) of PEGylated dendrimers and their passive-tumor-targeting capacity, thus offering insight into new ways of optimizing synthesis of passive targeting carriers [53]. Passive targeting of tumor cells, using coacervate droplets formed from poly (2-methoxyethyl acrylate) (PMEA), has also been reported in the literature [54]. It was found that the addition of a PMEA solution to water leads to the formation of dense coacervate droplets, which selectively accumulate in certain cancerous cells lines but do not have affinity towards healthy cells. This could provide a facile alternative to the use of ligands for active accumulation.

Active accumulation is induced by the presence on the surface of loaded nanoparticles of ligand molecules, which lead to specific attachment in that tissue. The ligand molecules that can be employed for active targeting are represented by biological (antibodies and enzymes) or pseudo-biological (aptamers and molecularly-imprinted polymers) elements, which can specifically recognize targets that are over-expressed on the surface of cancer cells. Such molecules can be represented by: P-selectin [55], serotonin 1 B and 2 B receptors, gamma-aminobutyric acid A receptors (theta subunit), epidermal growth factor (EGRF), fibroblast growth factor receptor 3 (FGRF3), somatostatin receptor, insulin receptor (present in two isoforms), prostaglandin E2, adeno-self receptors (A2b), insulin-like growth factor, asialoglycoprotein receptor 1 (ASGPR1) for galactosamine [56], folate receptor [56], transferrin receptor [56], AFP, glypcian-3 [56], and retinoic acid receptor-alpha. Specific delivery to the target using this method can reduce the systemic toxicity of active molecules as well as bypass chemotherapeutic resistance systems [57].

Active release mechanisms at the tumor site rely on the use of external or internal factors that trigger drug release, such as ultrasound, pH, temperature, and ionizing radiation.

Through these mechanisms, the release of active drugs into the tumor tissue is increased and lowered into healthy tissues.

4.1.1. Systemic Therapy

The structure of the nanoparticles allows their loading with active substances for targeted drug delivery. This feature depends on the ability to make a connection between the nanoparticles and the drug, the porous inner structure, or the outer structure. In some cases, intermediate molecules are attached to the outer surface of the nanoparticles, creating a link between the basic and chemotherapeutic nanoparticles. The use of nanoparticles for the delivery of systemic therapy in HCC can enhance the therapeutic effects of drugs, reduce their systemic toxicity, and overcome some of their physio-chemical limitations. In Table 2, several nanoparticle-based approaches for targeted drug delivery in the treatment of HCC are presented.

The gold standard systemic oncological treatment for hepatocellular carcinoma is sorafenib (SOR), an antiangiogenic agent that binds to receptors on the surface of tumor cells: PDGFR-β, VEGFR 2, and VEGFR 3 [58]. Sorafenib has certain limitations, such as very poor water solubility and low bio-availability [59], which could be overcome by its encapsulation in different nanoparticle-based delivery systems. Different strategies for the encapsulation of SOR in nano-carriers have been reported.

A passive method for SOR delivery was developed using monoolein-based liquid crystalline nanoparticles (LCN). SOR-loaded LCNs were obtained by self-assembly and were, then, coated with polymeric layers of poly-l-lysine (PLL) and polyethylene glycol-b-polyaspartic acid (PEG-b-PAsp), using a layer-by-layer process. The polymeric layers increased the biocompatibility of the particles and contributed to the sustained release of SOR. The release of SOR was higher at pH 5.5, compared to physiological pH, indicating that the method could successfully be used to release SOR in the acidic tumor microenvironment. The SOR-loaded LCNs displayed higher toxicity towards HCC cells, compared to SOR alone [60].

Another passive method relies on the use of SOR-gold nanoconjugates for HCC treatment. Due to concerns related to the safety profiles of metallic nanoparticles, extensive tests were carried out to assess the biocompatibility of the resulted nanoconjugates. No organic solvents were used in the synthesis process, to reduce the potential toxicity of the complex. Biochemical and hematological parameters were determined in rats, after intraperitoneal administration of SOR-Au nanoconjugates, and no signs of toxicity were observed. Moreover, a blinded histological examination of liver, kidney, heart, and brain tissue revealed no pathologic changes after SOR-Au nanoconjuagate administration. The conjugates were tested on tumor cell cultures with induced SOR resistance, and the conjugates demonstrated higher inhibition of tumor cell growth, compared to SOR alone [61].

Active targeting of SOR was achieved, using folate-decorated bovine serum albumin (BSA) nanoparticles loaded with SOR (FA-SOR-BSANP). The nanoparticles were obtained by mixing BSA with an ethanolic solution of SOR, and folic acid was covalently bound on the surface of BSA nanoparticles using NHS/EDC coupling. Cytotoxicity studies demonstrated that FA-SOR-BSANP exhibited higher inhibition of tumor cell growth compared to SOR alone or non-folate-decorated SOR-BSANP. In vivo studies in rats showed that FA-SOR-BSANP had better tumor-targeting properties compared to SOR-BSANP, demonstrating the utility of folic acid as an active targeting ligand [62].

SPIONs have, also, been employed for the active delivery of SOR to tumor cells, using magnetic fields. SPIONs were obtained by the co-precipitation method and functionalized with polyvinyl alcohol [29] or zinc/aluminum-layered double hydroxide [63]. In both approaches, the functionalized nanoparticles demonstrated higher toxicity towards tumor cells and lower toxicity towards healthy hepatocytes, compared to free SOR.

Standard chemotherapeutics used as single agents, such as doxorubicin [64,65] or gemcitabine [66], have produced modest results in the treatment of HCC. Moreover, doxorubicin (DOX), a natural anthracycline, presents severe acute and long-term side effects, including multiple organ toxicity, in a dose-dependent manner [67]. In this regard, scientists have explored many paths to reduce the toxicity related to DOX therapy, without altering its efficacy. One of the many proposed solutions is the encapsulation of DOX in different nanoparticles. A further advantage of this strategy is overcoming chemotherapy resistance, thus rendering doxorubicin more efficient in HCC treatment. In a phase II study [68,69], in which doxorubicin was encapsulated in mixantrone-loaded polybutylcyanoacrylate nanoparticles, patient survival was significantly increased, compared to that of the unencapsulated drug. In another phase II study, in which DOX was encapsulated into poly isohexylcyanoacrylate polymer-based NPs (used as chemoembolizing therapy), significant survival was, also, observed compared to treatment with the native drug [70]. Multiple liposomal formulations of DOX were, also, reported in the literature [21–23,55]. ASGPR active targeting was, mostly, employed in these studies and the liposomes were functionalized with a variety of agents to increase circulation time and biocompatibility. Cytotoxicity of the liposomal formulation towards tumor cells was increased, while cytotoxicity towards healthy cells was reduced, compared to free DOX, in all approaches.

Table 2. Comparison of different targeted delivery strategies for the treatment of HCC.

Drug	Nanoparticle Type	Diameter (nm)	Targeting	Release	Test Method	Ref.
DOX	Lac-DOPE-L-DOX	96 ± 39	Active (ASGPR ligands)	Cellular uptake of drug	Cell cultures Mice xenografts	[21]
DOX	Lf-PEG-L-DOX	100	Active (ASGPR ligands)	Cellular uptake of drug	Cell cultures Mice xenografts	[22]
DOX	PAG-L-DOX	184.8 ± 1.7	Active (ASGPR ligands)	pH dependent	Cell cultures Mice xenografts	[23]
DOX	Fuc-L-BML-DOX	92.1 ± 12.5	Active (P-selectin ligands)	MW and pH triggered	Cell cultures Mice xenografts	[55]
SOR	LbL-LCN-SOR	165	Passive	pH triggered	Cell cultures	[60]
SOR	FA-SOR-BSANP	158.00	Active (FR ligands)	Cellular uptake of drug	Cell cultures Mice xenografts	[62]
SOR	Lac-SOR/CCM-NPs	115.5 ± 3.6	Active (ASGPR ligands)	pH triggered	Cell cultures Mice xenografts	[71]
SOR	AuNPs-SOR	10	Passive	Cellular uptake of drug	Cell cultures Mice xenografts Ex-vivo studies	[61]
SOR	SPION-PVA-SOR	15	Active (magnetic field)	Magnetic field	Cell cultures Mice xenografts	[29]
SOR	SPION-PEG-ZLDH-SOR	16	Active (magnetic field)	pH dependent	Cell cultures	[63]
GMB	AgNP-GMB	75.1 ± 7	Passive		Cell cultures In vivo toxicity study	[72]
DTX	TPSSNP-DTX	103.6 ± 9.2	Passive	pH and redox triggered	Cell cultures Mice xenografts	[73]
TRP	Gal-Chi-TP-NP	227.4 ± 3.7	Active (ASGPR ligands)	Cellular uptake of drug	Cell cultures Mice xenografts In vivo toxicity studies	[74]

DOX—doxorubicin; Lac—lactobionic acid; DOPE—dioleoylphosphatidylethanolamine; L—liposome; ASGPR—asialoglycoprotein receptors; Lf—lactoferin; PEG—polyethyleneglycol; L—liposome; PAG—palmitoylated arabinogalactan; Fuc—fucoidan; BML—1-butyl-3-methylimidazolium-L-lactate; MW—microwave; SOR—sorafenib; LbL-LCN—layer by layer polymer assembled liquid crystalline nanoparticles; FA—folic acid; BSANP—bovine serum albumin nanoparticles; FR—folate receptor; CCM—curcumin; PVA—polyvinyl alcohol; SPION—superparamagnetic iron oxide nanoparticles; ZLDH—zinc/aluminum layered double hydroxide; GMB—gemcitabine; AgNP—silver nanoparticles; DTX—docetaxel; TPSSNP—D-α-tocopheryl polyethylene glycol 1000-poly (β-amino ester) block copolymer containing disulfide linkages; TRP—triptolide; Gal—galactose; Chi—chitosan.

The combination of chemotherapeutics with molecules sensitive to various energy fields (ultrasound, magnetic, microwave, etc.) can lead to the development of molecular platforms that release targeted chemotherapy and, in addition, achieve additional therapeutic effects by in situ thermal ablation. The relevance of such therapies is important, especially in the case of tumors with resistance to conventional therapy, in which systemic toxicity is increased.

4.1.2. Nanoparticle-Mediated Nucleic Acid Delivery

Molecular-targeted therapy, based on genes uploaded to NPs, is a relatively new approach. siRNA is a small strain of RNA that can inhibit the expression of pro-oncogenic HCC genes, thus representing a promising treatment alternative. However, siRNAs have a short half-life, due to nuclease degradation in the blood stream [75], and present low bioavailability. Moreover, virus carriers, which are usually used for the transport of siRNAs, can cause immunogenicity and mutagenesis [76].

Biocompatible NPs aim to protect siRNA from degradation, increase bioavailability, and reduce immunogenic effects. Furthermore, by decorating them with molecules that bind, preferentially, to certain targets on the surface of tumor cells, the specificity of antitumor activity may be increased.

Selenium nanoparticles were obtained, using a reduction method, and were employed for the delivery of siRNA in the treatment of HCC. Hyaluronic acid (HA) was used to modify the surface of selenium NPs to increase biocompatibility and, selectively, target HCC cells. Then, positively charged polyethylenimine (PEI) was added onto the particles and anionic siRNA was loaded via electrostatic interaction. The HA-modified SeNPs demonstrated low toxicity, superior tumor accumulation, and apoptosis induction, compared to non-HA modified NPs [76]. A similar approach was developed by the same group, using folic acid (FA) as a targeting ligand for HCC. Similarly, the FA-modified SeNPs exhibited higher cytotoxicity in HepG2 cells and higher anti-tumor efficacy in vivo, compared to non-modified NPs [77].

Supermaramagnetic iron oxide NPs were obtained by co-precipitation and were, also, functionalized with PEI and siRNA and galactose was added as an ASGPR-specific ligand (Figure 3A). The modified SPIONS accumulated at the tumor site in orthothopic mice. Moreover, encapsulation into a nanoparticle system extended the half-life of siRNA [78].

Another system that targeted the ASGPR was developed by Huang et al. Their approach consisted of the development of galactose-modified lipid/calcium/phosphate nanoparticles (LCP NPs), incorporating anti-VEGF siRNA (Figure 3B). The calcium phosphate core of the nanoparticles dissolved into the acidic pH of the tumor microenvironment, thus delivering siRNA. The nano-system demonstrated enhanced gene silencing in vitro and in vivo [75].

Combined delivery of SOR and anti-VEGF siRNA, using pH responsive liposomes, was, also, reported. The liposomes were obtained via thin-film hydration and loaded with the two therapeutic agents. Higher cell internalization was demonstrated for tumor cells in media of pH 6.5, compared to physiological pH, indicating the utility of the prepared liposomes for targeted delivery in cancer tissue. In vitro down-regulation of VEGF was higher for the co-loaded liposomes, compared to SOR alone, siRNA alone, or single-loaded liposomes, indicating promising results [24].

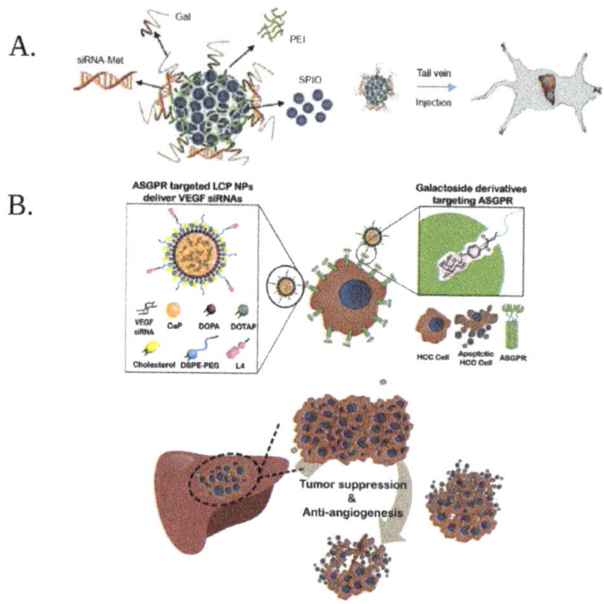

Figure 3. (**A**) Schematic representation of the development strategy for superparamagnetic iron oxide nanoparticles (SPIONs) modified with polyethylenimine (PEI), galactose (Gal), and siRNA as well as their administration in rats, for the treatment of HCC; (adapted from reference [78] (Open Access)); (**B**) Schematic representation of the development strategy of galactose-modified lipid/calcium/phosphate nanoparticles (LCP NPs) incorporating anti VEGF siRNA; DOPA—1,2-dioleoyl-sn-glycero-3-phosphate; DOTAP—1,2-dioleoyl-3-trimethylammonium-propane; DSPE-PEG—1,2-distearoyl-sn-glycero-3-phosphoethanolamine-N-[amino (polyethylene glycol)-2000]; L4—phenyl β-D-galactoside (adapted with permission from reference [75]). (Copyright 2022 American Chemical Society. Created with BioRender.com).

4.2. Thermo-Ablation Systems

Thermal ablation systems destroy tumor cells by the thermal effect generated by them after irradiation, with various types of waves, such as those generated by ultrasound and magnetic fields. In situ thermal ablation processes using nanostructures are derived from current radiofrequency ablation therapies or microwave ablation.

Glypican 3-targeting IR780 dye-loaded mesoporous silica NPs were proposed by Ma et al. [79] and were used to produce local hyperthermia upon laser irradiation in vitro and in vivo. Glypican-3 targeting was achieved using a complex protocol. First, chimeric antigen receptor T (CAR-T) cells, targeting glypican-3, were produced by transfection. CAR-T cells are cells that can specifically recognize different tumor-surface antigens, in this case glypican-3. Then, the cells were suspended in lysis buffer and CAR-T membranes were obtained through extrusion. The next step consisted of loading commercially available mesoporous silica NPs with IR780 dye and, then, coating the complex with the obtained membranes by sonication followed by extrusion. Membrane coating ensured selectivity towards glypican-3 and ensured the "stealth" of the NPs in vivo. The nano-complex showed great phototermal effect, due to the presence of IR780 [79].

Wang et al. proposed an NIR (near infrared)-powered synthetic high-density lipoprotein (sHDLs) platform, with which they performed photothermal therapy, with tumor destruction. The synthetic HDLs were obtained through self-assembly, using an ApoA-1 mimetic peptide and a mixture of lipids, and were, then, loaded with photothermal agent DiR and mertansine or vadimezan [80].

Chen et al. developed IR820-PEG-melanin NPs (MNPs) that demonstrated superior penetration of tumor tissues with properties useful in both MRI imaging and in situ tumor ablation. An important advantage of melanin is its biocompatibility, being a pigment that can be naturally found in the human body. The nano-carrier was obtained, by self-assembly of pegylated MNPs and the near-infrared dye IR820, and demonstrated selective tumor ablation in orthotopic mice models [81].

4.3. Trans-Catheter Arterial Chemoembolization with NPs

The use of NPs in TACE can lead to the improvement of this technique, by avoiding bead aggregation, extratumoral embolization, inefficiency of small particles, reduced degradation of embolizing agents, and dependence of embolization on vessel diameter, in tumor microcirculation [1].

A recombinant silk-elastin-like protein polymer (SELP) was developed by Poursaid et al., conditioned in the form of a hydrogel that can be injected through a microcatheter, which leads to obstruction of blood flow. In the future, these types of hydrogels can be combined with chemotherapeutics, to increase the antitumor effect of TACE [82].

Other authors have used paclitaxel nanocrystals, to be injected as embolizing agents by the TACE technique. The nanocrystalline structures were obtained by one-pot synthesis through paclitaxel self-assembly and crosslinking with glutaraldehyde. The resulting structures were water soluble, unlike paclitaxel, which displays poor water solubility. Another advantage of the nanocrystals was their intrinsic fluorescence, which could potentially allow real-time monitoring during TACE. Compared to paclitaxel alone, the nanocrystalline assemblies displayed prolonged release of paclitaxel. The structures, successfully, reduced the dimensions of tumor spheroids in vitro. The results of this study are promising, however, further in vivo testing needs to be done, in order to evaluate its safety and efficacy [83].

5. Nanoparticles Used for Theragnostic

Theragnostic approaches represent a modern strategy, which combines diagnosis and treatment into a single platform. The development of theragnostic strategies requires interdisciplinary collaborations between different specialists, in order to ensure faster implementation of treatment and better outcomes for the patient [45]. Numerous theragnostic strategies for HCC have been developed.

An example of this mechanism, which includes both diagnosis and treatment, is the system developed by Patra et al., which can be used to track targeted delivery using MRI imaging. The system was composed of SPIONs, which act as an MRI tracer and can, as well, be guided to the tumor site using a magnetic field. The SPIONs were modified with a block polymer for doxorubicin encapsulation and with folic acid for HCC targeting. The release of DOX was achieved, using a pH-dependent mechanism [84].

Another SPION-based approach was developed by Cai et al. In this study, sorafenib was encapsulated into polymer-modified SPIONs and active targeting was achieved using an anti-glypican 3 antibody. The release of sorafenib at the tumor site was done using a double pH and redox-dependent mechanism. The system was also used for MRI imaging of tumors in vivo [85].

Ferrous-platinum nanoparticles were grafted onto montmorillonite (MMT) porous clay to develop a tool for HCC MRI imaging. The magnetic modified-MMT was also loaded with mitoxantrone to endow the complex with therapeutic properties (Figure 4A). Another treatment strategy that was incorporated into this approach was the induction of magnetic hyperthermia upon magnetic field application. The theragnostic MMT complexes were tested in vivo in rats and displayed high tumor inhibition, compared to controls, as well as good tumor-imaging capacity [86].

Another example of theragnostic use of nanoparticles is the one proposed by Wang et al. that combined mesoporous silicon (loaded with doxorubicin) and Au-based Janus particles that show therapeutic effects and qualities of CT-scan contrast tracers [87].

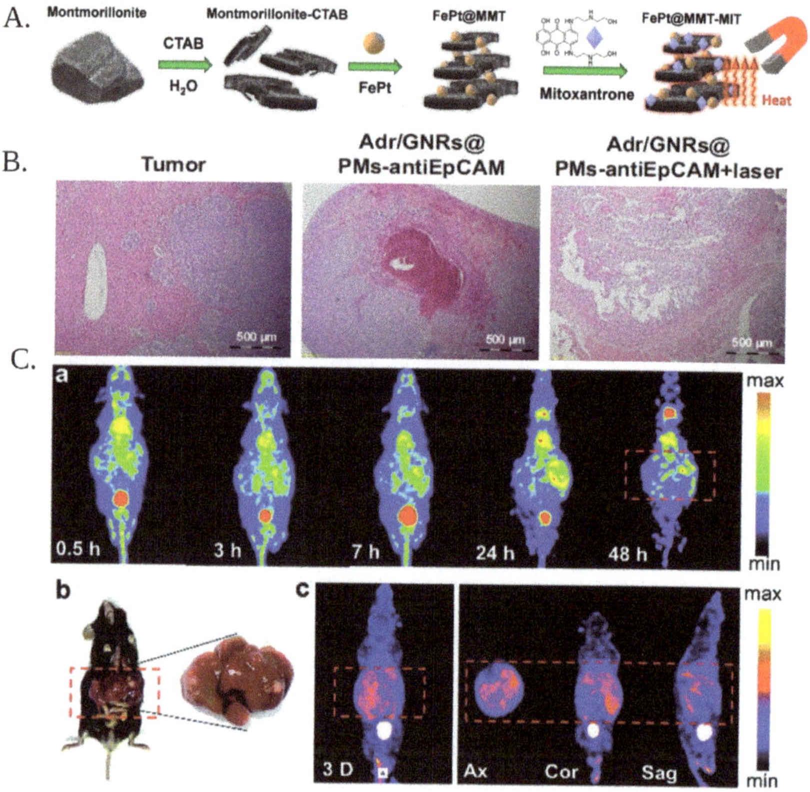

Figure 4. (**A**) Schematic representation of the synthesis of montmorillonite (MMT)-based nanoparticles for theragnostic applications. The MMT reacted with cetyltrimethylammonium bromide (CTAB), to form a layered structure in which iron-platinum nanoparticles and mitoxantrone were incorporated. The platform was used for magnetically induced hyperthermia (adapted from Reference [86] (Open access)). (**B**) Histology studies illustrating the anti-tumor of adryamicin-loaded, gold-nanorod polymeric nanomicelles modified with EpCAM antibody (Adr/GNR@PMs-antiEpCAM), and the same complex under laser irradiation (adapted from Reference [88] (Open Access)). (**C**) SPECT images of the radioactive iodide labeled Pd nanosheets at different time intervals in vivo (**a**) image of liver tissue and (**b**) SPECT images obtained using a conventional tracer used as control method (**c**) (adapted from Reference [89] (Open Access). Created with BioRender.com).

A gold-nanorod micelle complex was loaded with Adriamycin and employed as a theragnostic nano-carrier for HCC. The complex was conjugated with anti-Epithelial cell adhesion molecule (EpCAM) antibodies, to offer selectivity towards HCC stem cells. The nano-carriers were used for the selective destruction of HCC stem cells, by using the effects of Adriamycin-induced cytotoxicity and gold-nanorod-mediated photothermal effect. Upon laser irradiation, higher antitumor efficacy was highlighted, compared to the antitumor action of the complex alone, indicating the synergistic effects of Adriamycin and photothermal effect (Figure 4B). Moreover, the system was able to identify tumors by photoacoustic imaging in an ex vivo model [88].

Palladium nanosheets were, also, employed for photoacoustic imaging, combined with radiotherapy and photothermal effects. Palladium nanosheets were functionalized with iodide radioisotopes, via a simple synthesis method. The resulting modified Pd nanosheets displayed synergistic antitumor activity, and high-quality single photon emission computed tomography (SPECT) images were obtained using this approach in three different in vivo HCC models (Figure 4C) [89].

Carbon-based nanomaterials have, also, been proposed for theragnostic purposes. An antibody specific for glypican-3 was conjugated with reduced graphene oxide and, then, combined with nanobubbles to form a system that was used as an ultrasonic contrast agent, as well as for in vitro photothermal action [90].

Future trends of nanomaterials in drug delivery, such as magnetic nanoparticles (MNPs), have, recently, contributed to important developments in several fields, with a focus on medical applications and oncology. MNPs are widely applied in cancer diagnosis, cancer screening, targeted drug delivery, and cancer treatment and diagnosis [91]. Owing to their magnetism, in particular, MNPs (especially superparamagnetic iron oxide nanoparticles (SPIONs)) have been, mostly, used as contrast agents in cancer screening for magnetic resonance imaging (MRI), computed tomography in magneto-acoustic tomography, and near-infrared imaging. Hybrid nanomaterials have received tremendous interest, owing to the combination of the unique properties of organic and inorganic components in a single material. In this class, magnetic polymer nanocomposites are of particular interest because of their excellent magnetic properties, stability, and good biocompatibility of polymeric films. Several therapeutics based on nanoparticulate polymers and iron oxide drug carriers are, currently, under development. These include polymer–drug conjugates, single-crystal NPs, magnetic nanoclusters, micelles, dendritic polymer carriers, sophisticated stimuli-responsive systems, nanorobots, and diagnostic devices. In addition, a growing number of polymer conjugates and other organic or inorganic drug delivery systems are entering clinical development and evaluation [92].

Carbon-based nanomaterials have been exploited, extensively, in the delivery of therapeutics, due to their easy functionalization of their surface, rendering them capable of crossing biological membranes. Toxicity represents a main concern for their in situ use but, lately, the functionalization with an appropriate targeting ligand aids in the reduction in the cytotoxicity of loaded drugs toward healthy cells and augments therapeutic efficiency [93]. Carbon nanotubes as well as graphene have good mechanical strength, aspect ratio, conductivity, and chemical stability, possessing tunable physical properties, such as biocompatibility and high surface area. However, the lack of solubility in aqueous media, non-homogeneity in size (diameter and length), and the presence of metallic impurities are important drawbacks that, still, limit their use in targeted drug delivery. Moreover, the colloidal instability, the limited synthetic control of graphene or other carbon-based nanomaterials, such as fullerenes and carbon nano-anions, as well as the poor chemical stability in the biological environment and susceptibility to the oxidative environment, are other challenges that have to be overcame in order to expand their area of applications.

Metallic nanoparticles have, as main characteristics, a large surface-area-to-volume ratio as compared to the bulk equivalents, large surface energies, existence as a transition between molecular and metallic states providing specific electronic structure, plasmon excitation, quantum confinement, short range ordering, a large number of low-coordination sites such as corners and edges, a large number of "dangling bonds", and, consequently, specific and chemical properties as well as the ability to store excess electrons. The Au nanoparticles have, for instance, excellent properties for bioconjugation: they are biocompatible and able to fix a great variety of biomolecules [94].

Nanosomes are promising nanoparticles for drug delivery. Phospholipid nanosomes (small, uniform liposomes) have been developed to overcome some of the limitations associated with conventional liposome manufacturing [95]. Nanosomes can be actively targeted to specific cells by ligand labeling, such as folic acid, transcobalamin, iron transport proteins, hormones, and growth factors, to facilitate their active cellular import. Tumor cells can be passively targeted, to use the effect of retention and permeability.

It appears that drug delivery systems based on nanomaterials hold great potential to overcome some of the barriers to efficient targeting of cells and molecules in inflammation and cancer [96]. In addition, another huge advantage is the possibility to overcome problems of drug resistance in target cells and to facilitate the movement of drugs across barriers, such as those in the brain. However, there are some challenges, such as the precise characterization of molecular targets, the cytotoxicity of some nanomaterials, and the possibility of ensuring that these molecules are expressed only in the targeted organs, to prevent effects on healthy tissues.

6. Conclusions

HCC represents the fourth-most common cause of cancer-related death worldwide, with various risk factors. The diagnosis of HCC is, mainly, based on imaging and blood tumor markers detection. HCC is, usually, diagnosed at the clinical advanced stage, so current diagnostic and therapeutic methods do not meet the clinical needs, due to late detection and rapid tumor progressing, therefore, new technologies, drugs, and delivery methods are required in clinical practice.

Significant advances have been made in the field of nanomedicine for the diagnosis and therapy of HCC. The most commonly used nanomaterials for targeted drug delivery were liposomes, but also iron-oxide compounds or organic microspheres, the first being approved by the FDA [7].

Improving the imaging diagnosis of nanotechnology can significantly increase the detection of intrahepatic tumors and decrease the size limit at which these tumors can be identified. In addition, a number of nanostructures have useful properties in the diagnosis of CT, MRI, or ultrasound, and features that make them very useful in the treatment of these tumors, thus leading to the concept of theragnostic (diagnosis and therapy).

The use of nano-conditioned ICG-type intravital tracers can improve the intraoperative evaluation of the extent of liver resection for such tumors, helping surgeons to evaluate hepatic resection performance.

The treatments accessible using nanostructures vary, from the transport of cytostatics to the target tissue, the thermal destruction of tumor cells by activation in certain energy fields, and the introduction of genetic material into tumor cells.

The right diameters, the high specific surface area, and specific physicochemical properties made the nanomaterials important building blocks in imaging diagnosis, serological diagnosis, pathological detection, systematic therapy, thermal ablation, chemoembolization, immunotherapy, and many other aspects for HCC [7].

A number of limitations restrict researchers in using these nanostructures in clinical practice, but, progressively, these molecules will be modified, and their qualities improved; therefore, in the future, an important technological advance is foreseen, and nanomedicine will have a say in anticancer therapies.

Funding: This research and The APC were funded by a grant of the Romanian Ministry of Research and Innovation, CCCDI—UEFISCDI, project number PN-III-P2-2.1-PED-2019-4375/397/2020 (OnTarget), within PNCDI III.

Institutional Review Board Statement: Not applicable.

Informed Consent Statement: Not applicable.

Data Availability Statement: Data available upon request from the corresponding author.

Acknowledgments: This work was supported by a grant of the Ministry of Research, Innovation and Digitization, CNCS/CCCDI—UEFISCDI, project number 397PED—OnTarget, within PNCDI III.

Conflicts of Interest: The authors declare no conflict of interest.

References

1. Llovet, J.M.; Kelley, R.K.; Villanueva, A.; Singal, A.G.; Pikarsky, E.; Roayaie, S.; Lencioni, R.; Koike, K.; Zucman-Rossi, J.; Finn, R.S. Hepatocellular carcinoma. *Nat. Rev. Dis. Prim.* **2021**, *7*, 6. [CrossRef] [PubMed]
2. Kim, E.; Viatour, P. Hepatocellular carcinoma: Old friends and new tricks. *Exp. Mol. Med.* **2020**, *52*, 1898–1907. [CrossRef] [PubMed]

3. Ferrante, N.; Pillai, A.; Signal, A. Update on the Diagnosis and Treatment of Hepatocellular Carcinoma. *Gastroenterol. Hepatol.* **2020**, *18*, 506.
4. Galle, P.R.; Forner, A.; Llovet, J.M.; Mazzaferro, V.; Piscaglia, F.; Raoul, J.L.; Schirmacher, P.; Vilgrain, V. EASL Clinical Practice Guidelines: Management of hepatocellular carcinoma. *J. Hepatol.* **2018**, *69*, 182–236. [CrossRef]
5. Deng, M.; Li, S.H.; Guo, R.P. Recent Advances in Local Thermal Ablation Therapy for Hepatocellular Carcinoma. *Am. Surg.* **2021**, *88*, 31348211054532. [CrossRef] [PubMed]
6. Keshavarz, P.; Raman, S.S. Comparison of combined transarterial chemoembolization and ablations in patients with hepatocellular carcinoma: A systematic review and meta-analysis. *Abdom. Radiol.* **2022**, *47*, 1009–1023. [CrossRef]
7. Wu, H.; Wang, M.-D.; Liang, L.; Xing, H.; Zhang, C.-W.W.; Shen, F.; Huang, D.-S.S.; Yang, T. Nanotechnology for Hepatocellular Carcinoma: From Surveillance, Diagnosis to Management. *Small* **2021**, *17*, 1–17. [CrossRef]
8. Porto, L.S.; Silva, D.N.; de Oliveira, A.E.F.; Pereira, A.C.; Borges, K.B. Carbon nanomaterials: Synthesis and applications to development of electrochemical sensors in determination of drugs and compounds of clinical interest. *Rev. Anal. Chem.* **2020**, *38*, 1–16. [CrossRef]
9. Šturm, L.; Ulrih, N.P. Basic methods for preparation of liposomes and studying their interactions with different compounds, with the emphasis on polyphenols. *Int. J. Mol. Sci.* **2021**, *22*, 6547. [CrossRef]
10. Amina, S.J.; Guo, B. A review on the synthesis and functionalization of gold nanoparticles as a drug delivery vehicle. *Int. J. Nanomed.* **2020**, *15*, 9823–9857. [CrossRef]
11. Gavas, S.; Quazi, S.; Karpiński, T.M. Nanoparticles for Cancer Therapy: Current Progress and Challenges. *Nanoscale Res. Lett.* **2021**, *16*, 173. [CrossRef] [PubMed]
12. Samrot, A.V.; Sahithya, C.S.; Selvarani, A.J.; Purayil, S.K.; Ponnaiah, P. A review on synthesis, characterization and potential biological applications of superparamagnetic iron oxide nanoparticles. *Curr. Res. Green Sustain. Chem.* **2021**, *4*, 100042. [CrossRef]
13. Bai, K.; Wang, A. Polymeric Micelles: Morphology, Synthesis, and Pharmaceutical Application. *E3S Web Conf.* **2021**, *290*, 01029. [CrossRef]
14. Debnath, S.K.; Srivastava, R. Drug Delivery with Carbon-Based Nanomaterials as Versatile Nanocarriers: Progress and Prospects. *Front. Nanotechnol.* **2021**, *3*, 15. [CrossRef]
15. Gupta, S.; Murthy, C.N.; Prabha, C.R. Recent advances in carbon nanotube based electrochemical biosensors. *Int. J. Biol. Macromol.* **2018**, *108*, 687–703. [CrossRef]
16. Li, W.; Chen, M.; Liang, J.; Lu, C.; Zhang, M.; Hu, F.; Zhou, Z.; Li, G. Electrochemical aptasensor for analyzing alpha-fetoprotein using RGO-CS-Fc nanocomposites integrated with gold-platinum nanoparticles. *Anal. Methods* **2020**, *12*, 4956–4966. [CrossRef]
17. Li, G.; Feng, H.; Shi, X.; Chen, M.; Liang, J.; Zhou, Z. Highly sensitive electrochemical aptasensor for Glypican-3 based on reduced graphene oxide-hemin nanocomposites modified on screen-printed electrode surface. *Bioelectrochemistry* **2021**, *138*, 107696. [CrossRef]
18. Shi, X.; Chen, M.; Feng, H.; Zhou, Z.; Wu, R.; Li, W.; Liang, J.; Chen, J.; Li, G. Glypican-3 electrochemical aptasensor based on reduced graphene oxide-chitosan-ferrocene deposition of platinum–palladium bimetallic nanoparticles. *J. Appl. Electrochem.* **2021**, *51*, 781–794. [CrossRef]
19. Saraf, S.; Jain, A.; Tiwari, A.; Verma, A.; Panda, P.K.; Jain, S.K. Advances in liposomal drug delivery to cancer: An overview. *J. Drug Deliv. Sci. Technol.* **2020**, *56*, 101549. [CrossRef]
20. Miere (Groza), F.; Vicas, S.I.; Tima, A.V.; Ganea, M.; Zdrinca, M.; Cavalu, S.; Fritea, L.; Vicas, L.; Muresan, M.; Pallag, A.; et al. Preparation and characterization of two different liposomal formulations with bioactive natural extract for multiple applications. *Processes* **2021**, *9*, 432. [CrossRef]
21. Zhou, X.; Zhang, M.; Yung, B.; Li, H.; Zhou, C.; James Lee, L.; Lee, R.J. Lactosylated liposomes for targeted delivery of doxorubicin to hepatocellular carcinoma. *Int. J. Nanomed.* **2012**, *7*, 5465. [CrossRef]
22. Wei, M.; Guo, X.; Zou, Q.; Li, Q.; Tang, C.; Chen, B.; Xu, Y.; Wu, C. Lactoferrin-modified PEGylated liposomes loaded with doxorubicin for targeting delivery to hepatocellular carcinoma. *Int. J. Nanomed.* **2015**, *10*, 5123. [CrossRef]
23. Shah, S.M.; Goel, P.N.; Jain, A.S.; Pathak, P.O.; Padhye, S.G.; Govindarajan, S.; Ghosh, S.S.; Chaudhari, P.R.; Gude, R.P.; Gopal, V.; et al. Liposomes for targeting hepatocellular carcinoma: Use of conjugated arabinogalactan as targeting ligand. *Int. J. Pharm.* **2014**, *477*, 128–139. [CrossRef] [PubMed]
24. Yao, Y.; Wang, T.; Liu, Y.; Zhang, N. Co-delivery of sorafenib and VEGF-siRNA via pH-sensitive liposomes for the synergistic treatment of hepatocellular carcinoma. *Artif. Cells Nanomed. Biotechnol.* **2019**, *47*, 1374–1383. [CrossRef]
25. Bhattacharya, T.; Das, D.; Borges E Soares, G.A.; Chakrabarti, P.; Ai, Z.; Chopra, H.; Hasan, M.A.; Cavalu, S. Novel Green Approaches for the Preparation of Gold Nanoparticles and Their Promising Potential in Oncology. *Processes* **2022**, *10*, 426. [CrossRef]
26. Neha, D.; Momin, M.; Khan, T.; Gharat, S.; Ningthoujam, R.S.; Omri, A. Metallic nanoparticles as drug delivery system for the treatment of cancer. *Expert Opin. Drug Deliv.* **2021**, *18*, 1261–1290. [CrossRef]
27. Gobin, A.M.; Lee, M.H.; Halas, N.J.; James, W.D.; Rebekah, A.; Drezek, J.L.W. Near-Infrared Resonant Nanoshells for Combined Optical Imaging and Photothermal Cancer Therapy. *Nano Lett.* **2007**, *7*, 1929–1934. [CrossRef]
28. Ma, X.-H.; Wang, S.; Liu, S.-Y.; Chen, K.; Wu, Z.-Y.; Li, D.-F.; Mi, Y.-T.; Hu, L.-B.; Chen, Z.-W.; Zhao, X.-M. Development and in vitro study of a bi-specific magnetic resonance imaging molecular probe for hepatocellular carcinoma. *World J. Gastroenterol.* **2019**, *25*, 3030–3043. [CrossRef]

29. Tom, G.; Philip, S.; Isaac, R.; Praseetha, P.K.; Jiji, S.G.; Asha, V.V. Preparation of an efficient and safe polymeric-magnetic nanoparticle delivery system for sorafenib in hepatocellular carcinoma. *Life Sci.* **2018**, *206*, 10–21. [CrossRef]
30. Cristea, C.; Graur, F.; Gălătuş, R.; Vaida, C.; Pîslă, D.; Săndulescu, R. Nanobiomaterials for Cancer Diagnosis and Therapy. In *Nanobiomaterials*; Apple Academic Press: Palm Bay, FL, USA, 2018; pp. 329–375. ISBN 9781315204918.
31. Cernat, A.; Florea, A.; Rus, I.; Truta, F.; Dragan, A.-M.; Cristea, C.; Tertis, M. Applications of magnetic hybrid nanomaterials in Biomedicine. In *Biopolymeric Nanomaterials*; Elsevier: Amsterdam, The Netherlands, 2021; pp. 639–675. [CrossRef]
32. Majumder, N.; Das, N.G.; Das, S.K. Polymeric micelles for anticancer drug delivery. *Ther. Deliv.* **2020**, *11*, 613–635. [CrossRef]
33. Zhou, M.; Yi, Y.; Liu, L.; Lin, Y.; Li, J.; Ruan, J.; Zhong, Z. Polymeric micelles loading with ursolic acid enhancing anti-tumor effect on hepatocellular carcinoma. *J. Cancer* **2019**, *10*, 5820–5831. [CrossRef] [PubMed]
34. Zeyada, M.S.; Abdel-Rahman, N.; El-Karef, A.; Yahia, S.; El-Sherbiny, I.M.; Eissa, L.A. Niclosamide-loaded polymeric micelles ameliorate hepatocellular carcinoma in vivo through targeting Wnt and Notch pathways. *Life Sci.* **2020**, *261*, 118458. [CrossRef] [PubMed]
35. Yan, H.; Gao, X.; Zhang, Y.; Chang, W.; Li, J.; Li, X.; Du, Q.; Li, C. Imaging Tiny Hepatic Tumor Xenografts via Endoglin-Targeted Paramagnetic/Optical Nanoprobe. *ACS Appl. Mater. Interfaces* **2018**, *10*, 17047–17057. [CrossRef] [PubMed]
36. Lee, J.; Gordon, A.C.; Kim, H.; Park, W.; Cho, S.; Lee, B.; Larson, A.C.; Rozhkova, E.A.; Kim, D.-H. Targeted Multimodal Nano-Reporters for Pre-Procedural MRI and Intra-Operative Image-Guidance. *Biomaterials* **2016**, *109*, 69–77. [CrossRef] [PubMed]
37. Li, D.; Yang, J.; Wen, S.; Shen, M.; Zheng, L.; Zhang, G.; Shi, X. Targeted CT/MR dual mode imaging of human hepatocellular carcinoma using lactobionic acid-modified polyethyleneimine-entrapped gold nanoparticles. *J. Mater. Chem. B* **2017**, *5*, 2395–2401. [CrossRef] [PubMed]
38. Parikh, N.D.; Mehta, A.S.; Singal, A.G.; Block, T.; Marrero, J.A.; Lok, A.S. Biomarkers for the early detection of hepatocellular carcinoma. *Cancer Epidemiol. Biomark. Prev.* **2020**, *29*, 2495–2503. [CrossRef]
39. Pan, Y.; Chen, H.; Yu, J. Biomarkers in hepatocellular carcinoma: Current status and future perspectives. *Biomedicines* **2020**, *8*, 576. [CrossRef]
40. Zacharakis, G.; Aleid, A.; Aldossari, K.K. New and old biomarkers of hepatocellular carcinoma. *Hepatoma Res.* **2018**, *4*, 65. [CrossRef]
41. Abreu, P.; Ferreira, R.; Mineli, V.; Ribeiro, M.A.; Ferreira, F.G.; de Mello Vianna, R.M.; Tomasich, F.D.S.; Szutan, L.A. Alternative Biomarkers to Predict Tumor Biology in Hepatocellular Carcinoma. *Anticancer Res.* **2020**, *40*, 6573–6584. [CrossRef]
42. Dai, Y.; Han, B.; Dong, L.; Zhao, J.; Cao, Y. Recent advances in nanomaterial-enhanced biosensing methods for hepatocellular carcinoma diagnosis. *TrAC Trends Anal. Chem.* **2020**, *130*, 115965. [CrossRef]
43. Kaya, S.I.; Ozcelikay, G.; Mollarasouli, F.; Bakirhan, N.K.; Ozkan, S.A. Recent achievements and challenges on nanomaterial based electrochemical biosensors for the detection of colon and lung cancer biomarkers. *Sens. Actuators B Chem.* **2022**, *351*, 130856. [CrossRef]
44. Sun, Z.F.; Chang, Y.; Xia, N. Recent development of nanomaterials-based cytosensors for the detection of circulating tumor cells. *Biosensors* **2021**, *11*, 281. [CrossRef] [PubMed]
45. Ștefan, G.; Hosu, O.; De Wael, K.; Lobo-Castañón, M.J.; Cristea, C. Aptamers in biomedicine: Selection strategies and recent advances. *Electrochim. Acta* **2021**, *376*, 137994. [CrossRef]
46. Duan, L.; Yobas, L. Label-Free Multiplexed Electrical Detection of Cancer Markers on a Microchip Featuring an Integrated Fluidic Diode Nanopore Array. *ACS Nano* **2018**, *12*, 7892–7900. [CrossRef] [PubMed]
47. Andreou, C.; Neuschmelting, V.; Tschaharganeh, D.-F.; Huang, C.-H.; Oseledchyk, A.; Iacono, P.; Karabeber, H.; Colen, R.R.; Mannelli, L.; Lowe, S.W.; et al. Imaging of Liver Tumors Using Surface-Enhanced Raman Scattering Nanoparticles. *ACS Nano* **2016**, *10*, 5015–5026. [CrossRef] [PubMed]
48. Tsuda, T.; Kaibori, M.; Hishikawa, H.; Nakatake, R.; Okumura, T.; Ozeki, E.; Hara, I.; Morimoto, Y.; Yoshii, K.; Kon, M. Near-infrared fluorescence imaging and photodynamic therapy with indocyanine green lactosome has antineoplastic effects for hepatocellular carcinoma. *PLoS ONE* **2017**, *12*, e0183527. [CrossRef]
49. Dai, Y.W.; Zhu, L.X.; Zhang, Y.; Wang, S.H.; Chen, K.; Jiang, T.T.; Xu, X.L.; Geng, X.P. Au@SiO2@CuInS2–ZnS/Anti-AFP fluorescent probe improves HCC cell labeling. *Hepatobiliary Pancreat. Dis. Int.* **2019**, *18*, 266–272. [CrossRef]
50. Khan, A.A.; Alanazi, A.M.; Jabeen, M.; Chauhan, A.; Ansari, M.A. Therapeutic potential of functionalized siRNA nanoparticles on regression of liver cancer in experimental mice. *Sci. Rep.* **2019**, *9*, 303–310. [CrossRef]
51. Hajiasgharzadeh, K.; Hossein Somi, M.; Shanehbandi, D.; Mokhtarzadeh, A.; Baradaran, B. Small interfering RNA-mediated gene suppression as a therapeutic intervention in hepatocellular carcinoma. *J. Cell. Physiol.* **2018**, *234*, 3263–3276. [CrossRef]
52. Cheng, Z.; Li, M.; Chen, Y. Nanomaterials for cancer therapy: Current progress and perspectives. *J. Hematol. Oncol.* **2021**, *14*, 85. [CrossRef]
53. Tsujimoto, A.; Uehara, H.; Yoshida, H.; Nishio, M.; Furuta, K.; Inui, T.; Matsumoto, A.; Morita, S.; Tanaka, M.; Kojima, C. Different hydration states and passive tumor targeting ability of polyethylene glycol-modified dendrimers with high and low PEG density. *Mater. Sci. Eng. C* **2021**, *126*, 112159. [CrossRef] [PubMed]
54. Nishida, K.; Nishimura, S.N.; Tanaka, M. Selective Accumulation to Tumor Cells with Coacervate Droplets Formed from a Water-Insoluble Acrylate Polymer. *Biomacromolecules* **2021**, *23*, 1569–1580. [CrossRef] [PubMed]

55. Xu, J.; Cheng, X.; Tan, L.; Fu, C.; Ahmed, M.; Tian, J.; Dou, J.; Zhou, Q.; Ren, X.; Wu, Q.; et al. Microwave Responsive Nanoplatform via P-Selectin Mediated Drug Delivery for Treatment of Hepatocellular Carcinoma with Distant Metastasis. *Nano Lett.* **2019**, *19*, 2914–2927. [CrossRef]
56. Bakrania, A.; Zheng, G.; Bhat, M. Nanomedicine in hepatocellular carcinoma: A new frontier in targeted cancer treatment. *Pharmaceutics* **2022**, *14*, 41. [CrossRef] [PubMed]
57. Mohamed, N.K.; Hamad, M.A.; Hafez, M.Z.E.; Wooley, K.L.; Elsabahy, M. Nanomedicine in management of hepatocellular carcinoma: Challenges and opportunities. *Int. J. Cancer* **2017**, *140*, 1475–1484. [CrossRef]
58. Al-Rajabi, R.; Patel, S.; Ketchum, N.S.; Jaime, N.A.; Lu, T.-W.; Pollock, B.H.; Mahalingam, D. Comparative dosing and efficacy of sorafenib in hepatocellular cancer patients with varying liver dysfunction. *J. Gastrointest. Oncol.* **2015**, *6*, 259. [CrossRef]
59. Kong, F.H.; Ye, Q.F.; Miao, X.Y.; Liu, X.; Huang, S.Q.; Xiong, L.; Wen, Y.; Zhang, Z.J. Current status of sorafenib nanoparticle delivery systems in the treatment of hepatocellular carcinoma. *Theranostics* **2021**, *11*, 5464–5490. [CrossRef]
60. Thapa, R.K.; Choi, J.Y.; Poudel, B.K.; Hiep, T.T.; Pathak, S.; Gupta, B.; Choi, H.G.; Yong, C.S.; Kim, J.O. Multilayer-coated liquid crystalline nanoparticles for effective sorafenib delivery to hepatocellular carcinoma. *ACS Appl. Mater. Interfaces* **2015**, *7*, 20360–20368. [CrossRef]
61. Vishwakarma, S.K.; Sharmila, P.; Bardia, A.; Chandrakala, L.; Raju, N.; Sravani, G.; Sastry, B.V.S.; Habeeb, M.A.; Khan, A.A.; Dhayal, M. Use of Biocompatible Sorafenib-gold Nanoconjugates for Reversal of Drug Resistance in Human Hepatoblatoma Cells. *Sci. Rep.* **2017**, *7*, 8539. [CrossRef]
62. Wang, H.; Sun, S.; Zhang, Y.; Wang, J.; Zhang, S.; Yao, X.; Chen, L.; Gao, Z.; Xie, B. Improved drug targeting to liver tumor by sorafenib-loaded folate-decorated bovine serum albumin nanoparticles. *Drug Deliv.* **2019**, *26*, 89–97. [CrossRef]
63. Ebadi, M.; Buskaran, K.; Bullo, S.; Hussein, M.Z.; Fakurazi, S.; Pastorin, G. Synthesis and cytotoxicity study of magnetite nanoparticles coated with polyethylene glycol and sorafenib–zinc/aluminium layered double hydroxide. *Polymers* **2020**, *12*, 2716. [CrossRef] [PubMed]
64. Buschauer, S.; Koch, A.; Wiggermann, P.; Müller, M.; Hellerbrand, C. Hepatocellular carcinoma cells surviving doxorubicin treatment exhibit increased migratory potential and resistance to doxorubicin re-treatment in vitro. *Oncol. Lett.* **2018**, *15*, 4635–4640. [CrossRef] [PubMed]
65. Sanduzzi-Zamparelli, M.; Forner, A. Systemic doxorubicin and hepatocellular carcinoma: The end of an era never risen up. *Lancet Gastroenterol. Hepatol.* **2019**, *4*, 418–420. [CrossRef]
66. Fuchs, C.S.; Clark, J.W.; Ryan, D.P.; Kulke, M.H.; Kim, H.; Earle, C.C.; Vincitore, M.; Mayer, R.J.; Stuart, K.E. A phase II trial of gemcitabine in patients with advanced hepatocellular carcinoma. *Cancer* **2002**, *94*, 3186–3191. [CrossRef]
67. Rus, I.; Tertiș, M.; Barbălată, C.; Porfire, A.; Tomuță, I.; Săndulescu, R.; Cristea, C. An Electrochemical Strategy for the Simultaneous Detection of Doxorubicin and Simvastatin for Their Potential Use in the Treatment of Cancer. *Biosensors* **2021**, *11*, 15. [CrossRef]
68. Efficacy and Safety Doxorubicin Transdrug Study in Patients Suffering from Advanced Hepatocellular Carcinoma—Full Text View—ClinicalTrials.gov. Available online: https://clinicaltrials.gov/ct2/show/NCT01655693 (accessed on 2 March 2022).
69. Onxeo annonce les résultats principaux de l'étude de phase III de Livatag®, ReLive, dans le carcinome hépatocellulaire avancé—Onxeo. Available online: https://www.onxeo.com/onxeo-announces-top-line-results-relive-phase-iii-study-livatag-advanced-hepatocellular-carcinoma/ (accessed on 2 March 2022).
70. Merle, P.; Camus, P.; Abergel, A.; Pageaux, G.P.; Masliah, C.; Bronowicki, J.P.; Zarski, J.P.; Pelletier, G.; Bouattour, M.; Farloux, L.; et al. Safety and efficacy of intra-arterial hepatic chemotherapy with doxorubicin-loaded nanoparticles in hepatocellular carcinoma. *ESMO Open* **2017**, *2*, e000238. [CrossRef]
71. Bian, Y.; Guo, D. Targeted therapy for hepatocellular carcinoma: Co-delivery of sorafenib and curcumin using lactosylated ph-responsive nanoparticles. *Drug Des. Devel. Ther.* **2020**, *14*, 647–659. [CrossRef]
72. El-Sheikh, S.M.A.; Khairy, M.H.; Osama, E.; Metwally, M.M.M.; Galal, A.A.A. Nanotechnology improves the therapeutic efficacy of gemcitabine against a human hepatocellular carcinoma cell line and minimizes its in vivo side effects. *Naunyn. Schmiedebergs. Arch. Pharmacol.* **2021**, *394*, 631–643. [CrossRef]
73. Chen, F.; Zhang, J.; Wang, L.; Wang, Y.; Chen, M. Tumor pHe-triggered charge-reversal and redox-responsive nanoparticles for docetaxel delivery in hepatocellular carcinoma treatment. *Nanoscale* **2015**, *7*, 15763–15779. [CrossRef]
74. Zhang, Y.Q.; Shen, Y.; Liao, M.M.; Mao, X.; Mi, G.J.; You, C.; Guo, Q.Y.; Li, W.J.; Wang, X.Y.; Lin, N.; et al. *Galactosylated Chitosan Triptolide Nanoparticles for Overcoming Hepatocellular Carcinoma: Enhanced Therapeutic Efficacy, Low Toxicity, and Validated Network Regulatory Mechanisms*; Elsevier: Amsterdam, The Netherlands, 2019; Volume 15, ISBN 8610640144112.
75. Huang, K.W.; Lai, Y.T.; Chern, G.J.; Huang, S.F.; Tsai, C.L.; Sung, Y.C.; Chiang, C.C.; Hwang, P.B.; Ho, T.L.; Huang, R.L.; et al. Galactose Derivative-Modified Nanoparticles for Efficient siRNA Delivery to Hepatocellular Carcinoma. *Biomacromolecules* **2018**, *19*, 2330–2339. [CrossRef]
76. Xia, Y.; Guo, M.; Xu, T.T.; Li, Y.H.; Wang, C.B.; Lin, Z.F.; Zhao, M.Q.; Zhu, B. SiRNA-loaded selenium nanoparticle modified with hyaluronic acid for enhanced hepatocellular carcinoma therapy. *Int. J. Nanomed.* **2018**, *13*, 1539–1552. [CrossRef] [PubMed]
77. Xia, Y.; Zhao, M.; Chen, Y.; Hua, L.; Xu, T.; Wang, C.; Li, Y.; Zhu, B. Folate-targeted selenium nanoparticles deliver therapeutic siRNA to improve hepatocellular carcinoma therapy. *RSC Adv.* **2018**, *8*, 25932–25940. [CrossRef] [PubMed]
78. Yang, Z.; Duan, J.; Wang, J.; Liu, Q.; Shang, R.; Yang, X.; Lu, P.; Xia, C.; Wang, L.; Dou, K. Superparamagnetic iron oxide nanoparticles modified with polyethylenimine and galactose for siRNA targeted delivery in hepatocellular carcinoma therapy. *Int. J. Nanomed.* **2018**, *13*, 1851–1865. [CrossRef]

79. Ma, W.; Zhu, D.; Li, J.; Chen, X.; Xie, W.; Jiang, X.; Wu, L.; Wang, G.; Xiao, Y.; Liu, Z.; et al. Coating biomimetic nanoparticles with chimeric antigen receptor T cell-membrane provides high specificity for hepatocellular carcinoma photothermal therapy treatment. *Theranostics* **2020**, *10*, 1281–1295. [CrossRef]
80. Wang, J.; Meng, J.; Ran, W.; Lee, R.J.; Teng, L.; Zhang, P.; Li, Y. Hepatocellular Carcinoma Growth Retardation and PD-1 Blockade Therapy Potentiation with Synthetic High-density Lipoprotein. *Nano Lett.* **2019**, *19*, 5266–5276. [CrossRef] [PubMed]
81. Chen, K.; Li, Q.; Zhao, X.; Zhang, J.; Ma, H.; Sun, X.; Yu, Q.; Zhang, Y.; Fang, C.; Nie, L. Biocompatible melanin based theranostic agent for: In vivo detection and ablation of orthotopic micro-hepatocellular carcinoma. *Biomater. Sci.* **2020**, *8*, 4322–4333. [CrossRef]
82. Poursaid, A.; Price, R.; Tiede, A.; Olson, E.; Huo, E.; McGill, L.; Ghandehari, H.; Cappello, J. In situ gelling silk-elastinlike protein polymer for transarterial chemoembolization. *Biomaterials* **2015**, *57*, 142–152. [CrossRef] [PubMed]
83. Dev, A.; Sood, A.; Choudhury, S.R.; Karmakar, S. Paclitaxel nanocrystalline assemblies as a potential transcatheter arterial chemoembolization (TACE) candidate for unresectable hepatocellular carcinoma. *Mater. Sci. Eng. C. Mater. Biol. Appl.* **2020**, *107*, 110315. [CrossRef] [PubMed]
84. Patra, H.K.; Khaliq, N.U.; Romu, T.; Wiechec, E.; Borga, M.; Turner, A.P.F.; Tiwari, A. MRI-Visual Order–Disorder Micellar Nanostructures for Smart Cancer Theranostics. *Adv. Healthc. Mater.* **2014**, *3*, 526–535. [CrossRef]
85. Cai, M.; Li, B.; Lin, L.; Huang, J.; An, Y.; Huang, W.; Zhou, Z.; Wang, Y.; Shuai, X.; Zhu, K. A reduction and pH dual-sensitive nanodrug for targeted theranostics in hepatocellular carcinoma. *Biomater. Sci.* **2020**, *8*, 3485–3499. [CrossRef]
86. Chan, M.H.; Lu, C.N.; Chung, Y.L.; Chang, Y.C.; Li, C.H.; Chen, C.L.; Wei, D.H.; Hsiao, M. Magnetically guided theranostics: Montmorillonite-based iron/platinum nanoparticles for enhancing in situ MRI contrast and hepatocellular carcinoma treatment. *J. Nanobiotechnol.* **2021**, *19*, 1–16. [CrossRef] [PubMed]
87. Wang, Z.; Shao, D.; Chang, Z.; Lu, M.; Wang, Y.; Yue, J.; Yang, D.; Li, M.; Xu, Q.; Dong, W.F. Janus Gold Nanoplatform for Synergetic Chemoradiotherapy and Computed Tomography Imaging of Hepatocellular Carcinoma. *ACS Nano* **2017**, *11*, 12732–12741. [CrossRef] [PubMed]
88. Locatelli, E.; Li, Y.; Monaco, I.; Guo, W.; Maturi, M.; Menichetti, L.; Armanetti, P.; Martin, R.C.; Franchini, M.C. A novel theranostic gold nanorods- and adriamycin-loaded micelle for EpCA M targeting, laser ablation, and photoacoustic imaging of cancer stem cells in hepatocellular carcinoma. *Int. J. Nanomed.* **2019**, *14*, 1877–1892. [CrossRef] [PubMed]
89. Chen, M.; Guo, Z.; Chen, Q.; Wei, J.; Li, J.; Shi, C.; Xu, D.; Zhou, D.; Zhang, X.; Zheng, N. Pd nanosheets with their surface coordinated by radioactive iodide as a high-performance theranostic nanoagent for orthotopic hepatocellular carcinoma imaging and cancer therapy. *Chem. Sci.* **2018**, *9*, 4268–4274. [CrossRef]
90. Liu, Z.; Zhang, J.; Tian, Y.; Zhang, L.; Han, X.; Wang, Q.; Cheng, W. Targeted delivery of reduced graphene oxide nanosheets using multifunctional ultrasound nanobubbles for visualization and enhanced photothermal therapy. *Int. J. Nanomed.* **2018**, *13*, 7859–7872. [CrossRef]
91. Cristea, C.; Graur, F.; Galatus, R.; Vaida, C.; Pisla, D.; Sandulescu, R. *Nanobiomaterials, Applications in Drug Delivery*; Anil, K., Sharma Raj, K., Keservani, R.K.K., Eds.; Apple Academic Press: Palm Bay, FL, USA, 2018; pp. 329–375. ISBN 978-1-77188-591-1.
92. Hosu, O.; Florea, A.; Cristea, C.; Sandulescu, R. Functionalized Advanced Hybrid Materials for Biosensing Applications. In *Advanced Biosensors for Health Care Applications*; Elsevier: Amsterdam, The Netherlands, 2019; pp. 171–207. [CrossRef]
93. Mahor, A.; Singh, P.P.; Bharadwaj, P.; Sharma, N.; Yadav, S.; Rosenholm, J.M.; Bansal, K.K. Carbon-Based Nanomaterials for Delivery of Biologicals and Therapeutics: A Cutting-Edge Technology. *C* **2021**, *7*, 19. [CrossRef]
94. Ravalli, A.; Pilon Dos Santos, G.; Ferroni, M.; Faglia, G.; Yamanaka, H.; Marrazza, G. New label free CA125 detection based on gold nanostructured screen-printed electrode. *Sens. Actuators B Chem.* **2013**, *179*, 194–200. [CrossRef]
95. Castor, T. Phospholipid Nanosomes. *Curr. Drug Deliv.* **2005**, *2*, 329–340. [CrossRef]
96. Suri, S.S.; Fenniri, H.; Singh, B. Nanotechnology-based drug delivery systems. *J. Occup. Med. Toxicol.* **2007**, *2*, 1–6. [CrossRef]

materials

Article

Advantages of Cubosomal Formulation for Gatifloxacin Delivery in the Treatment of Bacterial Keratitis: In Vitro and In Vivo Approach Using Clinical Isolate of Methicillin-Resistant *Staphylococcus aureus*

Mohamed Nasr [1,2,*], Sameh Saber [3], Alaa Y. Bazeed [1], Heba A. Ramadan [4], Asmaa Ebada [4], Adela Laura Ciorba [5,*], Simona Cavalu [5] and Heba I. Elagamy [1]

1 Department of Pharmaceutics, Faculty of Pharmacy, Delta University for Science and Technology, Gamasa 35712, Egypt; alaayosf@gmail.com (A.Y.B.); hebaelagamy1985@yahoo.com (H.I.E.)
2 Department of Pharmaceutics and Industrial Pharmacy, Faculty of Pharmacy, Helwan University, Cairo 11790, Egypt
3 Department of Pharmacology, Faculty of Pharmacy, Delta University for Science and Technology, Gamasa 35712, Egypt; sampharm81@gmail.com
4 Department of Microbiology and Immunology, Faculty of Pharmacy, Delta University for Science and Technology, Gamasa 11152, Egypt; Hebaaa.aadel@gmail.com (H.A.R.); asmaaebada89@yahoo.com (A.E.)
5 Faculty of Medicine and Pharmacy, University of Oradea, 410087 Oradea, Romania; simona.cavalu@gmail.com
* Correspondence: m2nasr@yahoo.com (M.N.); adela.ciorba@yahoo.com (A.L.C.)

Abstract: The objective of this study was to enhance the corneal permeation of gatifloxacin (GTX) using cubosomal nanoparticle as a delivery system. Cubosomal nanoparticle loaded with GTX was prepared and subjected for in vitro and in vivo investigations. The prepared GTX-loaded cubosomal particles exhibited nanoparticle size of 197.46 ± 9.40 nm and entrapment efficiency of 52.8% ± 2.93. The results of ex vivo corneal permeation of GTX-loaded cubosomal dispersion show approximately 1.3-fold increase compared to GTX aqueous dispersion. The incorporation of GTX into cubosomal particles resulted in a fourfold reduction in the minimum inhibitory concentration (MIC) value for the GTX cubosomal particles relative to GTX aqueous dispersion. Furthermore, the enhanced corneal penetration of GTX-loaded cubosomal dispersion compared was evident by a significant decrease in the area % of corneal opacity in MRSA infected rats. Moreover, these results were confirmed by photomicrographs of histological structures of corneal tissues from rats treated with GTX-cubosomal dispersion which did not present any change compared to that of the normal rat corneas. In conclusion, treatment of ocular bacterial infections and reduction in the probability of development of new resistant strains of MRSA could be accomplished with GTX-loaded cubosomal nanoparticles.

Keywords: gatifloxacin; cubosomal dispersion; MIC; corneal permeation; methicillin resistant *Staphylococcus aureus*; keratitis

Citation: Nasr, M.; Saber, S.; Bazeed, A.Y.; Ramadan, H.A.; Ebada, A.; Ciorba, A.L.; Cavalu, S.; Elagamy, H.I. Advantages of Cubosomal Formulation for Gatifloxacin Delivery in the Treatment of Bacterial Keratitis: In Vitro and In Vivo Approach Using Clinical Isolate of Methicillin-Resistant *Staphylococcus aureus*. *Materials* **2022**, *15*, 3374. https://doi.org/10.3390/ma15093374

Academic Editor: Lucio Montanaro

Received: 23 March 2022
Accepted: 6 May 2022
Published: 8 May 2022

1. Introduction

The ocular drug delivery field is considered one of the most attractive, inspiring, and competitive fields owing to the many obstacles found in drug application due to the complex physiological and chemical nature of the eye [1,2]. Recently, many formulation strategies have been implemented to evolve new drug delivery systems which can overcome the inadequacies of the conventional ocular delivery systems including poor formula retention time, pulsed dosing of drug, frequent drug administration which leads to mechanical injury, corneal pigmentation, as well as sensitivity reactions of the conjunctiva and large drainage factor which leads to low ocular bioavailability [3–5].

The previously mentioned factors demand the development of a novel ocular drug delivery system that would extend the pre-ocular retention, and hence enhance the ocular penetration and absorption of drugs through ocular barriers providing new tools for the therapy of anterior and posterior eye segment diseases [6]. Novel ocular delivery systems based on nanoparticles have recently become alternative tools for traditional ocular delivery systems such as polymeric nanoparticles [7–9], polymeric micelles [10], solid lipid nanoparticles [11], nanostructured lipid carrier [12] and nano-emulsion [13].

Gatifloxacin (GTX) is a fourth generation fluoroquinolone antibacterial used for treatment of ocular infections through inhibiting bacterial DNA replication, transcription and repair [14,15]. Compared to other quinolones, GTX is more effective with broader spectrum especially in the case of staphylococcus and streptococcus infections [16]. However, the use of GTX in ocular infections is limited by its sub-therapeutic concentrations in the corneal aqueous humour and iris-clarity body due to the reduced GTX permeability. The lower GTX permeability might be attributed to efflux pumps such as P-glycoprotein (P-gp/MR1/ABCB1) and multi-drug resistance-associated protein 2 (MRP2/ABCC2/cMOAT) [17]. The multidrug resistance associated with these efflux pumps may limit the drug concentration at the target site and, subsequently, reduce drug uptake by the bacteria. Moreover, GTX suffers from pH-dependent solubility with maximum solubility ranging from 40 to 60 mg/mL at pH 2–5 [18]. These limitations necessitate the development of a novel drug delivery system rather than the traditional ophthalmic eye drops solution.

Glyceryl-monooleate (GMO) is used to form novel self-assembled thermodynamically stable cubic nanoparticles known as cubosomes which are stabilized using poloxamer 407 (P407). Cubosomes, as dispersed colloidal particles of cubic phase liquid crystals, have drawn attention to their use in controlled-release drug delivery systems. Cubosomes are advantageous due to their simple formulation and high loading capacity for different drugs with different nature. Moreover, extended drug release, biocompatibility with biological tissues, bio-adhesion, exceptional penetration ability and biodegradability give this drug delivery system superiority in ocular delivery. These features were previously reported upon employing cubosomes for many drugs such as dexamethasone and fluconazole as an ophthalmic drug delivery system [19,20].

Hence, the objective of the current study is to investigate the potential of cubosomal nanoparticles dispersion as a promising drug carrier for ocular administration of GTX for enhancing in vitro and in vivo performance of the drug. GTX-loaded cubosomal dispersions were in vitro evaluated for particle size, shape, drug entrapment, and ex vivo trans-corneal permeation characteristics using isolated rabbit cornea. The extent of treatment of induced bacterial keratitis in rats was used to evaluate the in vivo performance of the drug-loaded cubosomes compared to free drug dispersion. The cubosomal formulation is expected to improve the permeability of GTX and decrease the incidence of its resistance.

2. Materials and Methods

2.1. Materials

Pure GTX was gifted by Delta Grand Pharma for pharmaceutical industries (10th of Ramadan City, Egypt). Glyceryl monooleate (GMO) was supplied by Gattefosse, Lyon, France under the trade name Paceol®. Poloxamer 407 (P407) was acquired from Sigma-Aldrich (Milwaukee, WI, USA). Acetonitrile (HPLC grade) was purchased from VWR International Ltd., Poole, England. Potassium dihydrogen phosphate, sodium hydroxide, disodium hydrogen phosphate (pharmaceutical grade), phosphoric acid, and methanol (analytical grade) was bought from El Nasr, Cairo, Egypt.

2.2. GTX-Loaded Cubosomes Formulation

An emulsification technique was used to prepare the drug-loaded cubosomal dispersion reference to Nasr et al. [21]. At the beginning, GMO was melted at 60 °C by heating in a water bath. Then, 2 mL of deionized water was added drop wise with continuous stirring until homogenous gel formation. The obtained gel was left for 24 h at room temperature.

After which, a solution of GTX and 10% w/w of P407 (relative to GMO) in phosphate buffer (pH 5) was added to the gel using a high-speed vortex for 3 min. Probe sonication (Sonics vibra cell, Sonics & Materials INC, Newtown, CT, USA) was used in the obtained dispersion for 5 min (5 s on and 3 s off). The final weight of the dispersion was adjusted to 50 g using deionized water. The final GTX concentration in the obtained dispersion was 3 mg/g.

2.3. Particle Size Analysis, Polydispersity Index and Zeta Potential

A Zetasizer Nano series (Nano ZS, Malvern, UK) was implemented to measure the particle size distribution, polydispersity index (PDI) and zeta potential of cubosomal nanoparticles at 25 ± 0.5 °C in triplicate after being properly diluted with deionized water.

2.4. Transmission Electron Microscopy (TEM)

The morphology of GTX-cubosomal nanoparticles was observed by a transmission electron microscope (JEOL, Tokyo, Japan), model JEM-2100 with super twin lens. A drop of the cubosomal dispersion was put on carbon-coated copper grid and 1% sodium phosphotungstate solution as a stain. The stained sample was left to dry for 15 min at room temperature before monitoring.

2.5. Entrapment Efficiency (EE %)

Amicon Ultracentrifuge tubes (3000 MWCO, Millipore, St. Louis, MO, USA) were utilized to measure the entrapment efficiency by ultrafiltration centrifugation technique [22] as follows: 9 mL of deionized water was used to dilute 1 mL of GTX-loaded cubosomes, subsequently, a sample of 3 mL of the diluted dispersion was centrifuged at 6000 rpm for 10 min. Deionized water was used to properly dilute the acquired filtrate and measured spectrophotometrically at 290 nm (Jenway, Stafford, UK) [23] representing free GTX (Q Free). To determine the total amount of GTX existing in 1 mL cubosomal dispersion (Q Total), the formulation was appropriately diluted with methanol and measured spectrophotometrically as mentioned above using methanol as blank. EE % was calculated according to the following equation:

$$EE\ \% = [(Q\ Total - Q\ Free)\ /\ Q\ Total] \times 100 \quad (1)$$

2.6. Differential Scanning Calorimetry (DSC)

A thermal analysis study was conducted on the cubosomal dispersion, each of its components separately and with pure GTX. The DSC thermograms were achieved using DSC-60, Shimadzu, Kyoto, Japan. The aluminum pan was filled with 5 mg of each sample, heated with a 10 °C/min heating rate under a purgative nitrogen. A comparable empty pan was used as the control.

2.7. Ex Vivo Corneal Permeation Study

The Delta university ethics committee authorized the protocol of study (Approval Number: FPDU11621/3). Rabbits' corneas were utilized in this experiment after being freshly separated and submerged for 30 min in fabricated lacrimal fluid before the experiment. The corneas were mounted onto one end of a plastic syringe imitating a dialyzing tube with permeation area of nearly 1.54 cm^2 using a rubber band to ensure that they were watertight. The surface epithelium layer of each cornea was placed inward of the syringe, while the deepest endothelium layer was directed outwards facing the permeation medium. An equivalent amount (3 mg) of pure and tested formula was placed on the donor compartment. The acceptor compartment (50 mL of phosphate buffer of a pH 7.4 and a temperature of 37 ± 0.5 °C) was stirred at 50 rpm. The corneal surface of the donor compartment was adjusted just to touch the surface of permeation medium. At different time intervals (30 min, 1, 2, 3, 4, 5, 6, 8, 10, 11 and 12 h), 1 mL of the permeation media was removed, and an equivalent volume was used to replenish it. Withdrawn samples were analyzed using HPLC after being passed through a 0.45 μm syringe filter. The cumulative amount of GTX permeated per unit area was calculated and plotted versus time. The slope of the plot was calculated and taken as flux ($mg/cm^2/h$).

2.8. HPLC Assay of GTX

The HPLC technique was implemented using Shimadzu CTO-20A (Kyoto, Japan) with UV detector (VWD 1260) and ODS Hypersil column, with an average particle size of 5 µm, 150 mm tall and 4.6 mm internal diameter (Thermo Fisher Scientific, Waltham, MA, USA). Corresponding to a previously reported method [24] with minor changes modifications, acetonitrile and potassium phosphate buffer (pH 5) were used as a mobile phase with ratio (65:35 v/v), respectively. An isocratic elution was used with flow rate 1.5 mL/min and injection volume of 20 µL. The UV detector was set at 290 nm and 254 nm. The method was validated with reasonable linearity in the range of 1–8 µg/mL with 0.995 correlation coefficient. In addition, all samples were assayed for interday and intraday accuracy and precision.

2.9. Determination of Minimum Inhibitory Concentration (MIC) of GTX

The in vitro antibacterial activity of GTX-loaded cubosomal formulation against clinical isolate of methicillin-resistant Staphylococcus aureus (MRSA) was investigated compared to GTX aqueous dispersion using MIC test. Broth microdilution technique was employed for determination of MIC as described by Andrews [25] and according to CLSI M07-A9 guidelines [26]. The test was performed in 96-well rounded bottom microtiter plate filled with 100 µL of Muller Hinton broth medium. Two-fold serial dilutions of both treatments were prepared using sterile distilled water. The tested concentrations of GTX were ranged from 0.03 to 16 µg/mL. Aliquots (100 µL) of each concentration were loaded in 96-well flat-bottom microtiter plates. The bacterial inoculums were adjusted to the concentration of 5 × 105 CFU/mL. Positive and negative GTX free controls were formulated in the presence and absence of bacteria, respectively. All plates were incubated at 37 °C for 18 h. The lowest concentration that inhibits the bacterial growth is considered as the minimum inhibitory concentration.

2.10. In Vivo Studies

The in vivo study was conducted to assess the influence of cubosomal formulation on the antibacterial activity of GTX against methicillin-resistant strain (MRSA). Staphylococcus aureus is the main infective organism causing endophthalmitis and in some cases vision loss due to keratitis [27]. Clinical isolate of MRSA, isolated from a human corneal ulcer (Outpatient clinics, Mansoura University Ophthalmic Hospital, Egypt), was utilized as a pathogen to establish infection. Briefly, one colony of bacterial isolate was picked up from a fresh culture plate and injected into 5 mL of Luria Broth media, which was then incubated at 37 °C for 24 h with constant shaking at 150 rpm. The growth suspension was diluted until turbidity was matched with 0.5 McFarland standards to obtain approximately the organism number of 1 × 108 colony forming units (CFU) per mL.

Twenty-four adult male Wistar rats (270 ± 10 g) were enrolled in the study. The study and its procedures were accepted by the Institutional Animal Care and Use Committee (IACUC) at the Faculty of Pharmacy Delta University (Approval Number: FPDU11621/3) and were performed in agreement with relevant guidelines and regulations.

For all rats to be immunocompromised, they were treated with intraperitoneal injection of methyl prednisolone at a dose of 50 mg/kg in saline (300 µL/rat) for 3 consecutive days to facilitate the development of bacterial infection. Four groups (6 rats each) were divided as follows: group 1 (uninfected untreated rats); group 2 (infected rats with no treatment); group 3 (infected rats treated with GTX aqueous solution); and group 4 (infected rats treated with GTX-loaded cubosomal dispersion). A 27-gauge needle was used to scare the cornea of the right eye of each rat. Tear film break up was accomplished by pipetting 10 µL of 0.6% acetylcysteine onto the cornea and after that, normal saline solution was used for washing. After which, a 10 µL MRSA suspension (5 × 108 CFU) was applied and spread evenly. The left eyes of rats were scarred by the same pattern but were not subjected to infection to serve as a control.

The treatment began three days after infection. An ophthalmologist, who was blinded to the treatment protocol, confirmed the infection. A clear visual cloudy cornea was noticed at the 3rd day after the infection. Treatments were resumed for another 2 days until the vanishing of the cornea's cloudy appearance in one rat from treatment groups. On the 5th day after infection, the photographs of the affected eyes were processed using ImageJ 1.52i software (NIH, Bethesda, MD, USA). Images were captured at a 35 cm distance and processed using the lower threshold level at 95 and the upper threshold level at 145.

2.11. Histological Examination

At the end of the in vivo study, cervical dislocation was applied for euthanization of animals. From each group, some rats' eyes served as an autopsy sample where it was fixed in 10% formalin solution for 24 h, then rinsed with distilled water. Afterwards, dehydration was accomplished by using different concentrations of alcohol. Xylene was used for washing specimens followed by embedding it in paraffin at 56° in an oven for 24 h. Tissue paraffin beeswax was cut into blocks 4 µm in thickness using Sledge microtome. The final tissue sections were and stained by hematoxylin and eosin stain after removal of paraffin and examined by the light electric microscope [28].

2.12. Statistical Analysis

Differences between groups were analyzed by one-way analysis of variance followed by Tukey as a Post-Hoc Test using GraphPad Prism software, 8.0.2 (GraphPad Software Inc., La Jolla, CA, USA). Data are presented as the mean ± standard deviation (SD) ($p \leq 0.05$).

3. Results and Discussion

3.1. Particle Size, Polydispersity Index and Zeta Potential

The formed cubosomal dispersion displayed a narrow monomodal with mean particle size of 197.46 ± 9.40 nm. The PDI value of the cubosomal dispersion was 0.14 ± 0.05 indicating acceptable homogeneity. The zeta potential of the cubosomal nanoparticles was −21.90 ± 2.03 mV indicating high stability and lower tendency for particles aggregation.

3.2. Morphology of GTX-Cubosomal Nanoparticles

The TEM photomicrograph of GTX-cubosomal nanoparticles (Figure 1), revealed that the GTX-loaded cubosomal particles are nearly spherical polyangular non-aggregated particles with homogenous and a narrow size distribution. However, diameters of TEM photographed particles are smaller relative to the measured particle size by the dynamic light-scattering particle size analyzer.

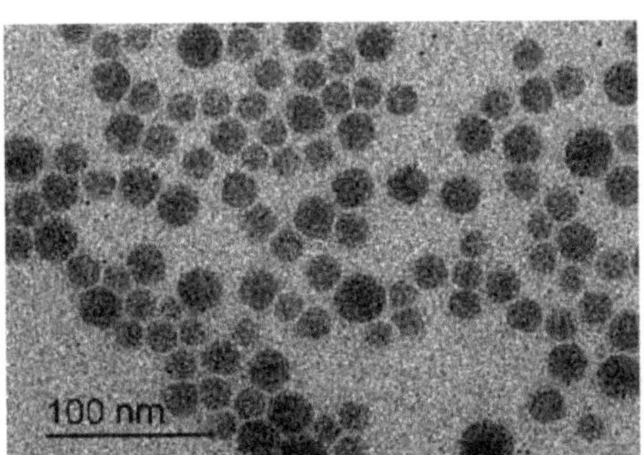

Figure 1. TEM photomicrograph of GTX loaded cubosomes.

3.3. Entrapment Efficiency

The EE % of GTX in cubosomal nanoparticle was 52.8% ± 2.93. The relatively lower % of drug entrapment might be attributed to the hydrophilic nature of GTX (log $p = -0.23$) and the higher % of the ionized form of GTX in phosphate buffer pH 5.5 that may lead to lower partitioning of the drug into the cubosomal lipid bilayer and favor the leakage of the drug from the water channels of cubosomal particles.

3.4. Differential Scanning Calorimetry

Figure 2 represents the thermograms of pure GTX, GTX-loaded cubosomal dispersion, blank cubosomal dispersion, P407 and GMO. The pure GTX powder presented a thermogram with a broad endothermic peak at 100–150 °C which may be due to dehydration and some structure rearrangement of GTX and a sharp endothermic peak of 191 °C indicating its crystalline nature [29]. In addition, the sharp characteristic peak disappeared in the thermogram of cubosomal dispersion indicating that GTX was not in a crystalline state but rather present in an amorphous state.

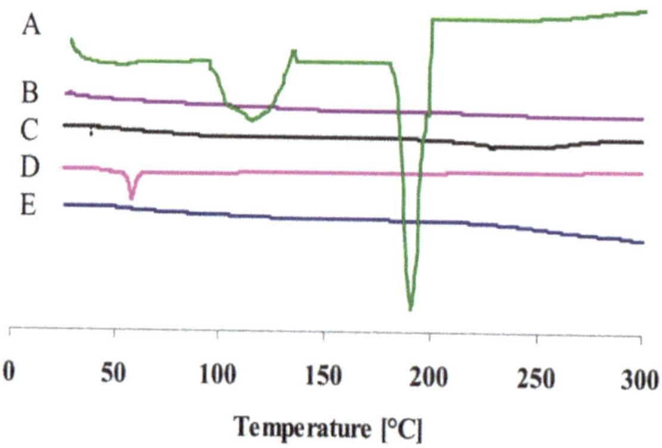

Figure 2. DSC thermograms of (A) pure GTX, (B) GTX-loaded cubosomal dispersion, (C) blank cubosomal dispersion, (D) P407 and (E) GMO.

3.5. Ex Vivo Corneal Permeation

Figure 3 illustrates the result of the rabbit corneal permeation of GTX from cubosomal dispersion compared to aqueous GTX dispersion. The cubosomal dispersion showed approximately a 1.3-fold increase in GTX corneal permeation. The slope of the graph (flux) of the cubosomal dispersion was significantly ($p < 0.05$) higher in the case of cubosomal dispersion (93.84 ± 7.89 $\mu gcm^{-2}\ h^{-1}$) in comparison with GTX aqueous dispersion (73.01 ± 3.45 $\mu gcm^{-2}\ h^{-1}$). The enhanced permeability could be attributed to the bioadhesive ability of cubosomes to attach to the large epithelium surface with lipophilic nature which will improve the delivery of the drug [30]. In addition, the cubosomal structure has similar features as the natural biological membrane which might facilitate drug permeability and consequently drug absorption. Besides, the components of cubosomal nanoparticles (GMO and P407) may play an important role in alteration of the activity of P-glycoprotein efflux pumps by both membrane permeability changes through a reduction in membrane micro-viscosity and depletion of ATP levels [31]. These results agree with previously reported studies that utilized cubosomes as delivery systems for enhanced ocular delivery of fluconazole, flurbiprofen and sertaconazole [20,32,33].

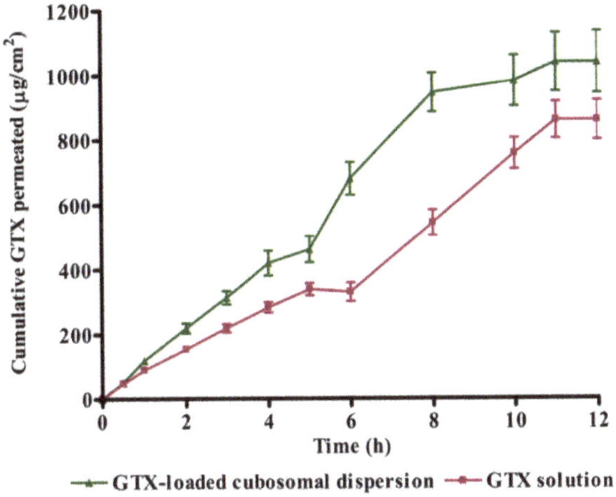

Figure 3. Comparison between rabbit corneal permeation of GTX delivered from cubosomal dispersion and aqueous GTX dispersion.

3.6. *Influence of GTX-Loaded Cubosomal Nanoparticles on MIC*

MICs of GTX-loaded cubosomal dispersion and free GTX aqueous dispersion were determined to evaluate the influence of cubosomal formulation on values of MIC against MRSA. The values of MIC were 4 and 16 μg/mL for GTX-loaded cubosomal dispersion and GTX aqueous dispersion, respectively. The result reveals that the incorporation of GTX into cubosomal particles resulted in a significant reduction in MIC value. The fourfold reduction in MIC value of GTX cubosomal formulation is expected to enhance the antibacterial activity of GTX and reduce the chance for development of new resistant bacterial strains. The obtained results could be attributed to the composition of cubosomal particles with the presence of GMO as a penetration enhancer in addition to its bioadhesiveness that facilitate a possible adhesion and/or fusion to bacterial cell membrane and consequently improve the drug permeability into bacterial cells [34].

3.7. *In Vivo Studies*

3.7.1. Efficacy of GTX-Loaded Cubosomal Nanoparticles in the Treatment of MRSA-Induced Ocular Keratitis

Figure 4 shows the ImageJ-processed photographs of eyes from different groups. The areas of focal lesions were indicated in the processed pictures by red color. In addition, the area % of corneal opacity was calculated and considered a measure of the intensity of corneal MRSA infection (Figure 5). Statistical analysis of the obtained results revealed that GTX-loaded cubosomal dispersion significantly reduced the area % of corneal opacity compared to GTX solution. This result may be explained based on the enhanced corneal tissue penetration of GTX-loaded cubosomal nanoparticles, as previously indicated in the in vivo permeation study.

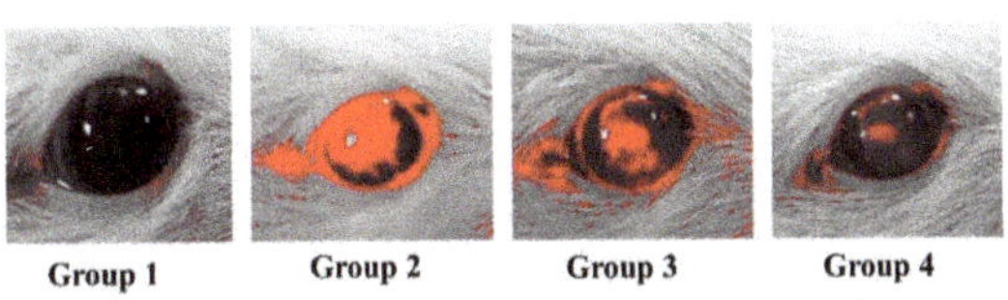

Figure 4. ImageJ-processed photographs of eyes from different groups: group 1 (uninfected untreated rats); group 2 (infected rats without treatment); group 3 (rats treated with GTX aqueous dispersion); and group 4 (rats treated with GTX-loaded cubosomal dispersion). The red color in the processed pictures designates the areas of focal lesions.

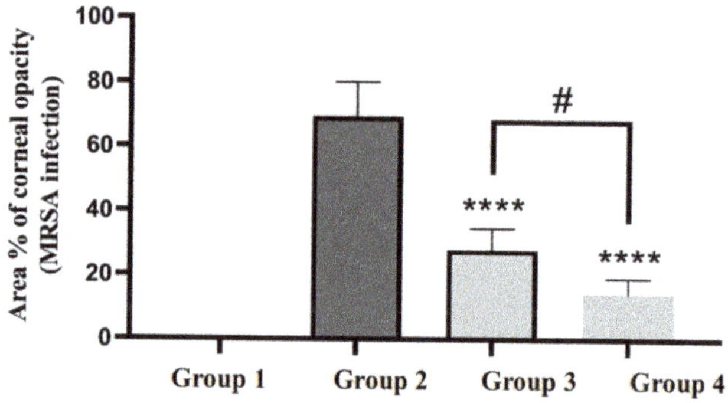

Figure 5. The area % of corneal opacity in different groups of rats. ****, group 4 and 3 compared to group 2 ($p < 0.0001$); #, group 4 compared to group 3 ($p < 0.05$).

3.7.2. Histological Evaluation

Figure 6 displays photomicrographs of corneal tissue sections stained with H&E from different groups. The photograph of group 1 (uninfected untreated) showed normal histological structure of rat corneal epithelium (filled arrowhead), stromal connective tissue layer with no inflammatory-cell infiltration (arrow), and endothelial layer (open arrowhead). However, corneal tissue sections from infected untreated rats (group 2) showed a dense focal stratified epithelial layer (filled arrowhead), a stromal layer with apparent edema, fibrotic changes, massive inflammatory-cell infiltration (arrows), and a dense focal endothelial layer (open arrowhead). The corneal tissue sections from rats of group 3 and 4 treated with GTX dispersion or GTX-loaded dispersion, respectively, showed a lower extent of focal stratification of the epithelial layer (filled arrowhead), few or absent lymphocyte infiltration (arrows), lower degree or absence of edema in the stromal layer and normal endothelium (open arrowhead). The histological presentation of corneal tissues from rats treated with GTX-cubosomal dispersion was almost more or less similar to that of the normal rat corneas. These findings were previously illustrated upon calculation of area % of corneal opacity.

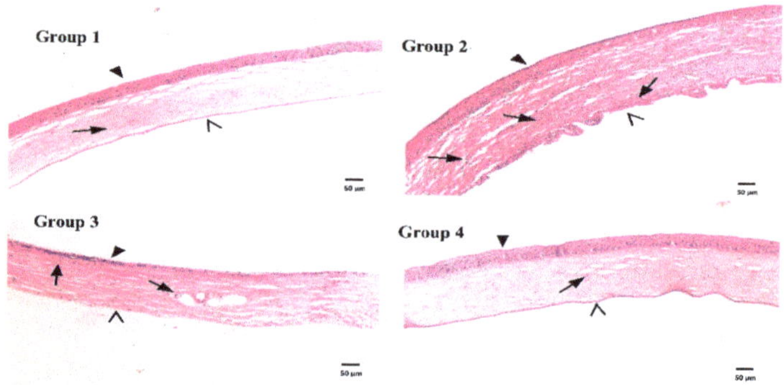

Figure 6. Histological structures of rat corneas stained with H&E of normal rat corneas (group 1), untreated rat corneas infected with MRSA (group 2), rat corneas infected with MRSA and treated with GTX aqueous dispersion (group 3) and rat corneas infected with MRSA and treated with GTX-loaded cubosomal dispersion (group 4).

4. Conclusions

Incorporating GTX in cubosomal nanoparticles comprised of GMO and P407 as a stabilizer improved both in vitro and in vivo performance compared to GTX aqueous dispersion. The size of prepared formula was in the nano-range with a relatively high GTX entrapment. In vivo and in vitro experiments demonstrated that cubosomal formulation achieved higher corneal permeation and a fourfold reduction in the MIC of GTX against clinically isolated MRSA strain. The results reveal that cubosomal nanoparticle may be a promising delivery system for GTX for the treatment of ocular bacterial infections and reduce the probability of development of new resistant strains of MRSA.

Author Contributions: Conceptualization, M.N.; methodology, H.I.E. and A.Y.B.; investigation, S.S.; S.C., H.A.R., A.E. and A.L.C.; writing—original draft preparation, M.N. and S.S.; writing—review and editing, M.N., S.S., H.I.E., A.Y.B., H.A.R., A.E., S.C. and A.L.C.; supervision, M.N. and S.S. All authors have read and agreed to the published version of the manuscript.

Funding: This research received no external funding.

Institutional Review Board Statement: Ethical approval for animal studies was obtained from the Institutional Animal Care and Use Committee (IACUC) at the Faculty of Pharmacy Delta University (Approval Number: FPDU11621/3) and were performed in agreement with relevant guidelines and regulations.

Data Availability Statement: Data will be available upon request.

Conflicts of Interest: The authors declare no conflict of interest.

References

1. Dey, S.; Mitra, A.K. Transporters and receptors in ocular drug delivery: Opportunities and challenges. *Expert Opin. Drug Deliv.* **2005**, *2*, 201–204. [CrossRef] [PubMed]
2. Civiale, C.; Licciardi, M.; Cavallaro, G.; Giammona, G.; Mazzone, M. Polyhydroxyethylaspartamide-based micelles for ocular drug delivery. *Int. J. Pharm.* **2009**, *378*, 177–186. [CrossRef] [PubMed]
3. Ramsay, E.; del Amo, E.M.; Toropainen, E.; Tengvall-Unadike, U.; Ranta, V.-P.; Urtti, A.; Ruponen, M. Corneal and conjunctival drug permeability: Systematic comparison and pharmacokinetic impact in the eye. *Eur. J. Pharm. Sci.* **2018**, *119*, 83–89. [CrossRef] [PubMed]
4. Nguyen, D.D.; Lai, J.-Y. Synthesis, bioactive properties, and biomedical applications of intrinsically therapeutic nanoparticles for disease treatment. *Chem. Eng. J.* **2022**, *435*, 134970. [CrossRef]
5. Luo, L.-J.; Nguyen, D.D.; Lai, J.-Y. Harnessing the tunable cavity of nanoceria for enhancing Y-27632-mediated alleviation of ocular hypertension. *Theranostics* **2021**, *11*, 5447. [CrossRef]
6. Pepic, I.; Lovric, J.; Filipovic-Grcic, J. Polymeric micelles in ocular drug delivery: Rationale, strategies and challenges. *Chem. Biochem. Eng. Q.* **2012**, *26*, 365–377.
7. Nagarwal, R.C.; Kant, S.; Singh, P.N.; Maiti, P.; Pandit, J.K. Polymeric nanoparticulate system: A potential approach for ocular drug delivery. *J. Control. Release* **2009**, *136*, 2–13. [CrossRef]
8. Aksungur, P.; Demirbilek, M.; Denkbaş, E.B.; Vandervoort, J.; Ludwig, A.; Ünlü, N. Development and characterization of Cyclosporine A loaded nanoparticles for ocular drug delivery: Cellular toxicity, uptake, and kinetic studies. *J. Control. Release* **2011**, *151*, 286–294. [CrossRef]
9. Katara, R.; Majumdar, D.K. Eudragit RL 100-based nanoparticulate system of aceclofenac for ocular delivery. *Colloids Surf. B Biointerfaces* **2013**, *103*, 455–462. [CrossRef]
10. Gupta, A.K.; Madan, S.; Majumdar, D.; Maitra, A. Ketorolac entrapped in polymeric micelles: Preparation, characterisation and ocular anti-inflammatory studies. *Int. J. Pharm.* **2000**, *209*, 1–14. [CrossRef]
11. Cavalli, R.; Gasco, M.R.; Chetoni, P.; Burgalassi, S.; Saettone, M. Solid lipid nanoparticles (SLN) as ocular delivery system for tobramycin. *Int. J. Pharm.* **2002**, *238*, 241–245. [CrossRef]
12. Bose, S.; Michniak-Kohn, B. Preparation and characterization of lipid based nanosystems for topical delivery of quercetin. *Eur. J. Pharm. Sci.* **2013**, *48*, 442–452. [CrossRef] [PubMed]
13. Gallarate, M.; Chirio, D.; Bussano, R.; Peira, E.; Battaglia, L.; Baratta, F.; Trotta, M. Development of O/W nanoemulsions for ophthalmic administration of timolol. *Int. J. Pharm.* **2013**, *440*, 126–134. [CrossRef] [PubMed]
14. Gorle, A.P.; Gattani, S.G. Design and Evaluation of Polymeric Ocular Drug Delivery System. *Chem. Pharm. Bull.* **2009**, *57*, 914–919. [CrossRef] [PubMed]
15. Perry, C.M.; Barman Balfour, J.A.; Lamb, H.M. GTX. *Drugs* **1999**, *58*, 683–696. [CrossRef]
16. Perry, C.M.; Ormrod, D.; Hust, M. GTX, a review of its use in the management of bacterial infections. *Drugs* **2002**, *62*, 169–207. [CrossRef]
17. Kwatra, D.; Vadlapatla, R.K.; Vadlapudi, A.D.; Pal, D.; Mitra, A.K. Interaction of gatifloxacin with efflux transporters: A possible mechanism for drug resistance. *Int. J. Pharm.* **2010**, *395*, 114–121. [CrossRef]

18. Maaz, A.; Abdelwahed, W.; Tekko, I.; Trefi, S. Influence of nanoprecipitation method parameters on nanoparticles loaded with gatifloxacin for ocular drug delivery. *IJASR* **2015**, *3*, 101–112.
19. Gan, L.; Han, S.; Shen, J.; Zhu, J.; Zhu, C.; Zhang, X.; Gan, Y. Self-assembled liquid crystalline nanoparticles as a novel ophthalmic delivery system for dexamethasone: Improving preocular retention and ocular bioavailability. *Int. J. Pharm.* **2010**, *396*, 179–187. [CrossRef]
20. Nasr, M.; Teiama, M.; Ismail, A.; Ebada, A.; Saber, S. In vitro and in vivo evaluation of cubosomal nanoparticles as an ocular delivery system for fluconazole in treatment of keratomycosis. *Drug Deliv. Transl. Res.* **2020**, *10*, 1841–1852. [CrossRef]
21. Nasr, M.; Ghorab, M.K.; Abdelazem, A. In vitro and in vivo evaluation of cubosomes containing 5-fluorouracil for liver targeting. *Acta Pharm. Sin. B* **2014**, *5*, 79–88. [CrossRef] [PubMed]
22. Saber, S.; Nasr, M.; Kaddah, M.M.; Mostafa-Hedeab, G.; Cavalu, S.; Mourad, A.A.; Gaafar, A.G.A.; Zaghlool, S.S.; Saleh, S.; Hafez, M.M.; et al. Nifuroxazide-loaded cubosomes exhibit an advancement in pulmonary delivery and attenuate bleomycin-induced lung fibrosis by regulating the STAT3 and NF-κB signaling: A new challenge for unmet therapeutic needs. *Biomed. Pharmacother.* **2022**, *148*, 112731. [CrossRef] [PubMed]
23. Khurana, G.; Arora, G.; Pawar, P. Ocular insert for sustained delivery of GTX sesquihydrate: Preparation and evaluations. *Int. J. Pharm. Investig.* **2012**, *2*, 70–77. [CrossRef] [PubMed]
24. Kalam, M.A.; Sultana, Y.; Ali, A.; Aqil, M.; Mishra, A.K.; Chuttani, K.; Aljuffali, I.A.; Alshamsan, A. Part II: Enhancement of transcorneal delivery of gatifloxacin by solid lipid nanoparticles in comparison to commercial aqueous eye drops. *J. Biomed. Mater. Res. Part A* **2012**, *101*, 1828–1836. [CrossRef]
25. Andrews, J.M. Determination of minimum inhibitory concentrations. *J. Antimicrob. Chemother.* **2001**, *48*, 5–16. [CrossRef]
26. Clinical and Laboratory Standardrs Institute. CLSI Document M07-A9. In *Methods for Dilution Antimicrobial Susceptibility Tests for Bacteria That Grow Aerobically: Approved Standard-Ninth Edition*; CLSI: Wayne, PA, USA, 2012.
27. Ong, S.J.; Huang, Y.-C.; Tan, H.-Y.; Ma, D.H.K.; Lin, H.-C.; Yeh, L.-K.; Chen, P.Y.F.; Chen, H.-C.; Chuang, C.-C.; Chang, C.-J.; et al. Staphylococcus aureus Keratitis: A Review of Hospital Cases. *PLoS ONE* **2013**, *8*, e80119. [CrossRef]
28. Banchroft, J.; Stevens, A.; Turner, D. *Theory and Practice of Histological Techniques*, 4th ed.; Churchill Livingstone: New York, NY, USA, 1996.
29. Arthanari, S.; Mani, G.; Jang, J.H.; Choi, J.O.; Cho, Y.H.; Lee, J.H.; Cha, S.E.; Oh, H.S.; Kwon, D.H.; Jang, H.T. Preparation and characterization of gatifloxacin-loaded alginate/poly (vinyl alcohol) electrospun nanofibers. *Artif. Cells Nanomed. Biotechnol.* **2014**, *44*, 847–852. [CrossRef]
30. Sharma, R.; Kaur, G.; Kapoor, D.K. Fluconazole loaded cubosomal vesicles for topical delivery. *Int. J. Drug Dev. Res.* **2015**, *7*, 32–41.
31. Cornaire, G.; Woodley, J.; Hermann, P.; Cloarec, A.; Arellano, C.; Houin, G. Impact of excipients on the absorption of P-glycoprotein substrates in vitro and in vivo. *Int. J. Pharm.* **2004**, *278*, 119–131. [CrossRef]
32. Han, S.; Shen, J.-Q.; Gan, Y.; Geng, H.-M.; Zhang, X.-X.; Zhu, C.-L.; Gan, L. Novel vehicle based on cubosomes for ophthalmic delivery of flurbiprofen with low irritancy and high bioavailability. *Acta Pharmacol. Sin.* **2010**, *31*, 990–998. [CrossRef]
33. Younes, N.F.; Abdel-Halim, S.A.; Elassasy, A.I. Corneal targeted Sertaconazole nitrate loaded cubosomes: Preparation, statistical optimization, in vitro characterization, ex vivo permeation and in vivo studies. *Int. J. Pharm.* **2018**, *553*, 386–397. [CrossRef]
34. Abdelaziz, A.A.; Elbanna, T.E.; Sonbol, F.I.; Gamaleldin, N.M.; El Maghraby, G.M. Optimization of niosomes for enhanced antibacterial activity and reduced bacterial resistance: In vitro and in vivo evaluation. *Expert Opin. Drug Deliv.* **2015**, *12*, 163–180. [CrossRef] [PubMed]

materials

MDPI

Review

Exploring the Journey of Zinc Oxide Nanoparticles (ZnO-NPs) toward Biomedical Applications

Fahadul Islam [1,†], Sheikh Shohag [2,†], Md. Jalal Uddin [2], Md. Rezaul Islam [1], Mohamed H. Nafady [3], Aklima Akter [1], Saikat Mitra [4], Arpita Roy [5], Talha Bin Emran [1,6,*] and Simona Cavalu [7,*]

1 Department of Pharmacy, Faculty of Allied Health Sciences, Daffodil International University, Dhaka 1207, Bangladesh; fahadulislamdiu@gmail.com (F.I.); md.rezaulislam100ds@gmail.com (M.R.I.); aklima.ph@diu.edu.bd (A.A.)

2 Department of Biochemistry and Molecular Biology, Faculty of Life Science, Bangabandhu Sheikh Mujibur Rahman Science and Technology University, Gopalganj 8100, Bangladesh; sheikhshohag.bmb@gmail.com (S.S.); jalalrana86@gmail.com (M.J.U.)

3 Faculty of Applied Health Science Technology, Misr University for Science and Technology, Giza 12568, Egypt; mohamed.nafady@must.edu.eg

4 Department of Pharmacy, Faculty of Pharmacy, University of Dhaka, Dhaka 1000, Bangladesh; saikatmitradu@gmail.com

5 Department of Biotechnology, School of Engineering & Technology, Sharda University, Greater Noida 201310, India; arbt2014@gmail.com

6 Department of Pharmacy, BGC Trust University Bangladesh, Chittagong 4381, Bangladesh

7 Faculty of Medicine and Pharmacy, University of Oradea, 400087 Oradea, Romania

* Correspondence: talhabmb@bgctub.ac.bd (T.B.E.); simona.cavalu@gmail.com (S.C.); Tel.: +880-1819-942214 (T.B.E.)

† These authors contributed equally to this work.

Citation: Islam, F.; Shohag, S.; Uddin, M.J.; Islam, M.R.; Nafady, M.H.; Akter, A.; Mitra, S.; Roy, A.; Emran, T.B.; Cavalu, S. Exploring the Journey of Zinc Oxide Nanoparticles (ZnO-NPs) toward Biomedical Applications. *Materials* **2022**, *15*, 2160. https://doi.org/10.3390/ma15062160

Academic Editor: Gang Wei

Received: 2 February 2022
Accepted: 3 March 2022
Published: 15 March 2022

Abstract: The field of nanotechnology is concerned with the creation and application of materials having a nanoscale spatial dimensioning. Having a considerable surface area to volume ratio, nanoparticles have particularly unique properties. Several chemical and physical strategies have been used to prepare zinc oxide nanoparticles (ZnO-NPs). Still, biological methods using green or natural routes in various underlying substances (e.g., plant extracts, enzymes, and microorganisms) can be more environmentally friendly and cost-effective than chemical and/or physical methods in the long run. ZnO-NPs are now being studied as antibacterial agents in nanoscale and microscale formulations. The purpose of this study is to analyze the prevalent traditional method of generating ZnO-NPs, as well as its harmful side effects, and how it might be addressed utilizing an eco-friendly green approach. The study's primary focus is on the potential biomedical applications of green synthesized ZnO-NPs. Biocompatibility and biomedical qualities have been improved in green-synthesized ZnO-NPs over their traditionally produced counterparts, making them excellent antibacterial and cancer-fighting drugs. Additionally, these ZnO-NPs are beneficial when combined with the healing processes of wounds and biosensing components to trace small portions of biomarkers linked with various disorders. It has also been discovered that ZnO-NPs can distribute and sense drugs. Green-synthesized ZnO-NPs are compared to traditionally synthesized ones in this review, which shows that they have outstanding potential as a potent biological agent, as well as related hazardous properties.

Keywords: ZnO-NPs; traditional synthesis; green synthesis; biomedical applications; toxicity

1. Introduction

Nanotechnology is a rapidly developing discipline of science and technology concerned with producing and developing nanomaterials with particle sizes ranging from 1 to 100 nanometers [1]. Recently, the scientific research community worldwide expressed interest in synthesizing metal and metal oxide nanoparticles (NPs) [2]. The ZnO-NPs are of huge importance due to their wide variety of applications in photocatalysis, antimicrobial defense, and water purification. ZnO-NPs display properties that are distinct from those of typical NPs [3]. Additionally, these NPs are employed in the cosmetics industry to produced sunblock creams, which guard the human body against ultraviolet radiations [4]. Due to ZnO-NPs' characteristics, such as their biocompatibility and non-toxicity, they are particularly well-suited for specialized biomedical applications [5–7]. Metal oxide NPs are important components in a wide range of consumer goods, including electronic equipment and cosmetics. ZnO-NPs are versatile materials with distinct chemical, optoelectronic, and wettability properties. They are easily made and widely used in a variety of industries, including wastewater treatment [8].

ZnO-NPs are manufactured using nanotechnology and are extensively used in various nanotechnology disciplines involving gas sensors [9], biosensors [10,11], cosmetics [12], ceramics [13], optical devices [14], display window materials for solar cells [15], and drug delivery [16,17]. Solar cells may directly transform light energy into electricity with their photovoltaic impact on ZnO-NPs [18].

ZnO-NPs absorb and scatter light very efficiently, making them excellent materials for optoelectronics applications that operate in the ultraviolet and visible spectrum areas. ZnO-NPs offer excellent photoluminescence properties, making them suitable for emission display systems, such as televisions [14]. In terms of photocatalytic degradation, ZnO-NPs seem to be the most promising choices [19]. The detection of gas leakage and the checking of gaseous contaminants in the environment may both benefit from semiconductor nano ZnO gas sensors [9]. ZnO-NPs are used to protect fabrics and wood from UV damage [20]. ZnO-NPs are made in a way that does not harm the environment, and they can control harmful microbes. Moreover, ZnO-NPs may be utilized as a treatment activator and a cross-linking agent in rubber treating, and can promote the vulcanization procedure in rubbers used to produce industrial and medical gloves, balloons, tires, and other rubber goods [21]. These substances have excellent antimicrobial and UV absorption properties and are commonly utilized in sunscreens, lotions, and ointments because of their versatility [12]. Antimicrobial ZnO-NPs are used in food and in can linings to keep fish, pork, peas, and maize safe from spoilage. ZnO-NPs have been proposed for next-generation biological applications, such as the delivery of medication, use as antimicrobial agents, and use as bioimaging probes [22].

The two ways that can be used to synthesize NPs are the top-down approach and the bottom-up approach (Figure 1). Electro-explosion, etching, sputtering, and mechanical milling are examples of top-down approaches, whereas bottom-up approaches comprise three basic methods for producing NPs: physical, chemical, and biological processes [23,24]. It is possible to produce pure, high-quality nanoparticles using conventional methods. Still, the process is expensive and sometimes results in the development of hazardous byproducts that may have detrimental consequences when employed for medical purposes. Furthermore, additional capping and stabilizing chemicals are required for these procedures [25]. This problem exists when NPs are produced using the green pathway, a bottom-up strategy that results in an oxidation/reduction reaction [26].

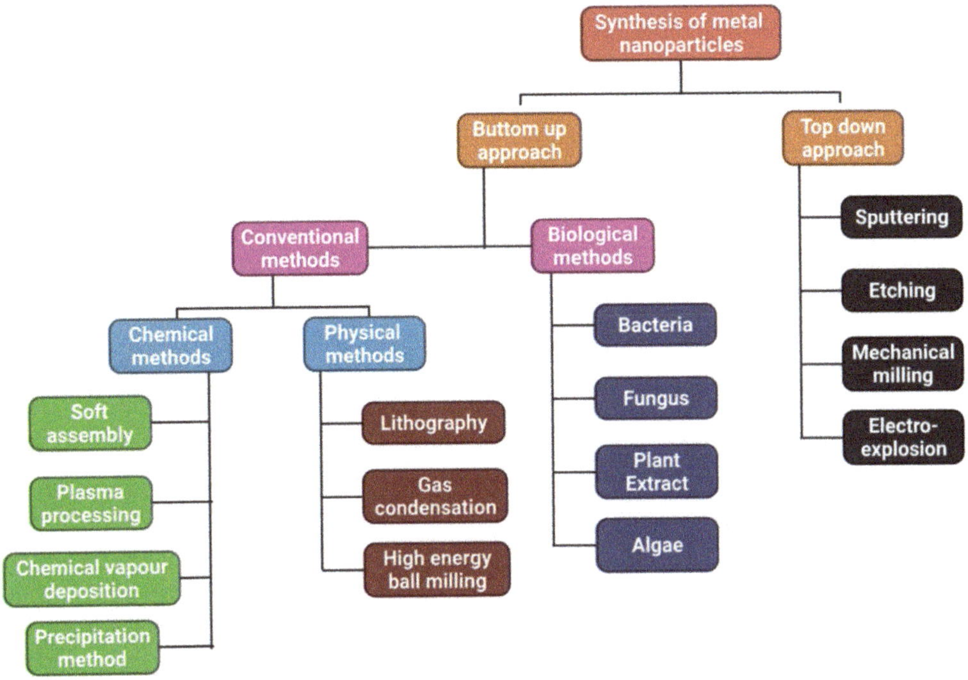

Figure 1. Methods to synthesize NPs from the bottom up and the top down.

Green synthesis can be accomplished using plants, bacteria, fungi, and algae. They enable the significant manufacturing of pure ZnO-NPs [27]. During green synthesis, a mixture of different parts of medicinal plants is used to produce NPs. The phytochemicals play a role as a biocatalyst, capping agent, and organic stabilizer for NPs [28]. The process does not require high temperatures, pressures, expensive tools, or toxic chemicals [29]. The green synthesis of NPs is more cost-efficient, toxic-free, and environmentally beneficial than the expensive and hazardous procedures used before [30,31].

This review highlights the prevalent traditional method of generating ZnO-NPs, as well as its harmful side effects, and discusses how it might utilize an eco-friendly green approach. The study's primary focus is on the potential biomedical applications of green-synthesized ZnO-NPs.

2. Methodology

To identify the most relevant articles (available in the most well-known medical/biology/chemical databases, such as Scopus, PubMed, and Web of Science) for this review as precisely as possible, "zinc oxide nanoparticles," "traditional techniques," and "biomedical applications" were used as primary keywords, and "plant extracts" and "green synthesis" were used as secondary keywords. An algorithm shown by the flow chart displayed in Figure 2 (according to the recommendations of Page et al. [32,33]) was used, which inserted all of the steps/selections requisite for identifying the necessary material in the literature.

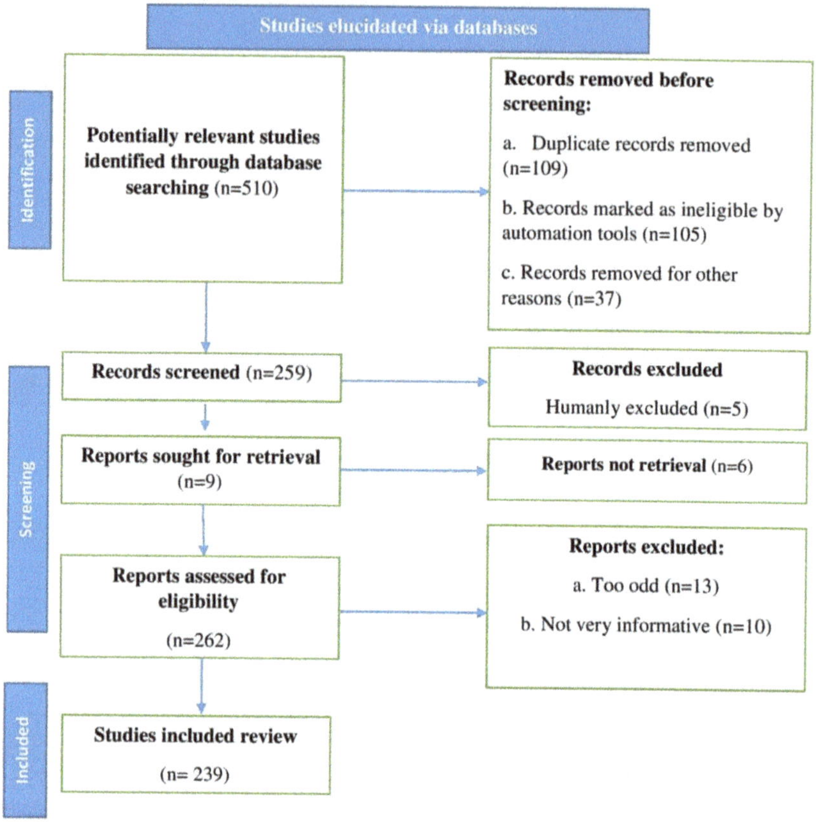

Figure 2. Stages involved in selecting published data for inclusion in the current study are depicted in a flow chart; n = number of literature reports.

3. Traditional Synthesis of ZnO Nanomaterials

Traditional methods for producing metallic NPs, such as ZnO-NPs, include mechanochemical and chemical processes. Sol-gel, hydrothermal, microemulsion procedures, and co-precipitation are all considered classic chemical synthesis approaches. Mechanochemical synthesis includes high-energy ball milling and laser ablation techniques [34–38]. The benefits and drawbacks of typical ZnO-NPs synthesis methods and particular innovative and noteworthy examples are briefly explored in the following sections of this paper.

3.1. Sol-Gel Technique

The transformation of a sol (e.g., a solution comprising inorganic metallic salts) progressively into a solid "gel" phase over a succession of hydrolysis and polymerization reactions is most commonly utilized to synthesize metal oxide NPs. Subsequently, the gel is treated to vaporize the solvents and heated to produce the final product [39–41]. Figure 3 depicts the sol-gel process in a simplified form. Using the sol-gel technique, it is possible to produce ZnO-NPs in a fine powder-like structure with a controlled chemical composition [42,43]. This process also has inherent drawbacks, including shrinkage, breaking while drying, and an inability to manage porosity [41]. Since the protocol is easy to follow and the critical material is generated quickly, it is frequently discussed in the relevant literature. Despite its shortcomings, this technique remains one of the most regularly employed. Zinc acetate dehydrates ($Zn(CH_3COO)_2.2H_2O$), and ethanol were utilized as solvents to create rod-shaped ZnO-NPs in the range of 81.28–84.98 nanometers [23].

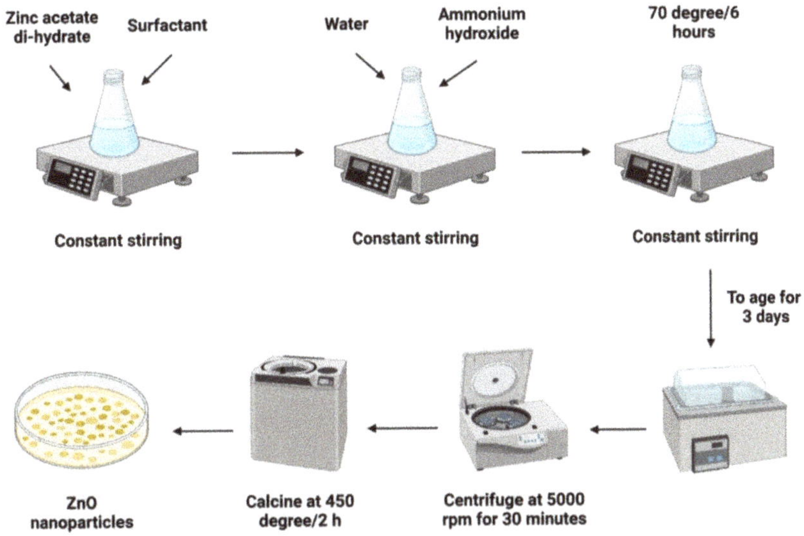

Figure 3. Diagrammatic representation of the stages required for the synthesis of metallic NPs (for example ZnO-NPs) employing the sol-gel process.

ZnO-NPs that averaged 28 nm in diameter with a spherical structure was produced by Jurablu et al. [44] using the sol-gel process. Zinc sulfate heptahydrate ($ZnSO_4.7H_2O$) and diethylene glycol ($C_4H_{10}O_3$) surfactants were utilized in this method. Additionally, ZnO-NPs with a mean particle size between 12 and 30 nm were produced using a mixture of ammonia and methanol, as well as $Zn(CH_3COO)_2.2H_2O$ as the precursor in a sol-gel process, which resulted in spherical ZnO-NPs in the range of 50–60 nm [43,45].

3.2. *Hydrothermal Technique*

An autoclave is a closed reaction vessel with high pressure and high temperature, where hydrothermal activities are carried out. Under high temperature and high pressure, the insoluble or challenging-to-dissolve materials are dissolved [46]. Various solvents can be utilized in these reactions, such as water or organic solvents such as ethanol or polyols, known as hydrothermal or solvothermal techniques, respectively [47,48]. An example of a hydrothermal procedure is shown in Figure 4. In addition to high product purity and crystallinity, hydrothermal methods control the final nanostructure size, shape, and crystal phase with little pollution due to the closed system environment [37,48]. Since the procedure is deemed environmentally beneficial, it is included in the green methods for producing ZnO-NPs. This method has some negatives; for example, it requires an extremely expensive autoclave and it has limitations for studies because the reactor cannot be kept open. There are also potential safety hazards throughout the autoclave procedure, which can be a concern [35,49,50]. Hydrothermal/solvothermal techniques, like the sol-gel strategy, are simple to start up. Examples include a study by Bharti and Bharati [51], which used a hydrothermal method to manufacture a length scale of 15.8–25 nm ZnO-NPs and various morphologies. ZnO-NPs with cylinder-shaped pores ranging from 9 to 12 nm in width were also manufactured by Reddy et al. [52], with the help of zinc nitrate hexahydrate ($Zn(NO_3)_2.6H_2O$) and sodium hydroxide (NaOH). Similarly, by utilizing an autoclave unit, Wirunmongkol et al. [52] produced ZnO-NPs in which NaOH and $Zn(NO_3)_2.6H_2O$ were used as the initial precursors. Shaped like tiny prisms and flowers, the NPs were between 30 and 80 nm wide and 0.5–0.1 μm long, depending on the type of material used to make them.

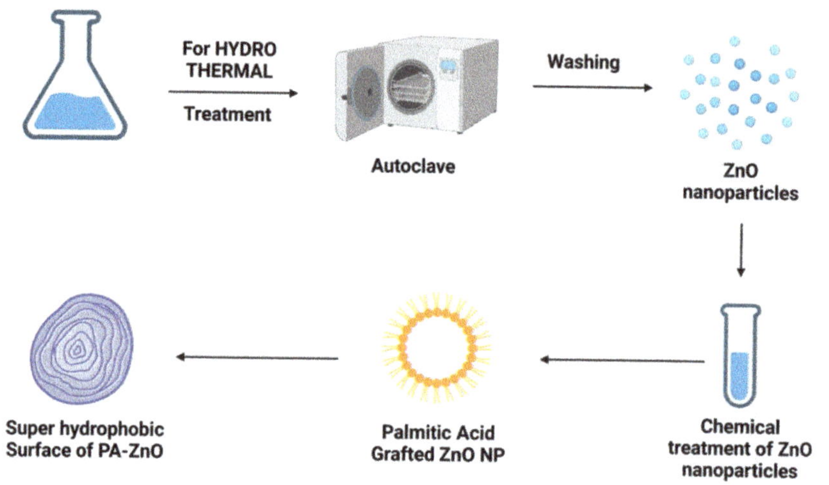

Figure 4. Diagrammatic representation of the stages required for metallic ZnO-NPs synthesis employing the hydrothermal technique.

3.3. Co-Precipitation Technique

The co-precipitation technique creates metallic NPs by simultaneous nucleation followed by growing and then agglomerating tiny nuclei. The co-precipitation process is depicted in Figure 5. This process has several advantages, including ease of use, minimal need for high temperatures, and ease of overall energy management [35]. As a side note, this approach has one significant drawback: it produces NPs with large quantities of water molecules attached to them [53]. Additionally, batch-to-batch repeatability issues, a wide range of particle sizes, and severe agglomeration are negatives [35,54,55]. However, remarkable instances include zinc acetate solution in methanol, in which spherical ZnO-NPs were synthesized by co-precipitation ranging between 2 and 10 nm in particle size. In contrast, a co-precipitation method using zinc acetate dihydrate, hydrochloric acid, and ammonia as reactants was used to manufacture ZnO-NPs. The ZnO-NPs was discovered to have a pseudo-spherical form with an average particle size between 11 and 20 nm [56,57]. A similar co-precipitation approach was used by Adam et al. [58] to produce ZnO-NPs with an average diameter of 140 nm.

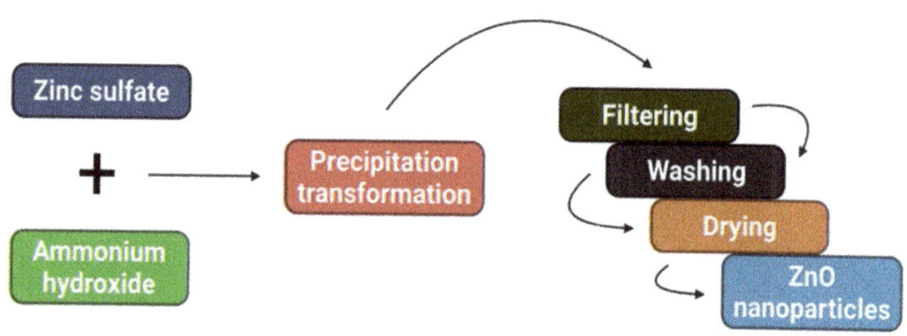

Figure 5. Diagrammatic representation of the stages required for ZnO-NPs synthesis employing the co-precipitation method.

3.4. Microemulsion Technique

Water droplets colliding with each other in a microemulsion environment resulted in a precipitation reaction, which led to the formation of NPs with surfactant-stabilized nucleation. The microemulsion process is depicted in Figure 6. The rewards of this approach include its simplicity, thermodynamic stability, and low accumulation. Microemulsion techniques have several drawbacks, including the impact of temperature and pH on the stability of the microemulsion and the persistent demand for highly concentrated surfactants and/or cosurfactants that may irritate [35]. ZnO-NPs have been manufactured in microchannel reactor systems with an average diameter of 16 nanometers by Wang et al. [59]. Following a drying period of 2 h at 130 °C, the ZnO-NPs were then calcined at 550 °C for 3 h. ZnO-NPs were also produced by Li et al. [60] via a simple procedure of microemulsion, with diverse morphologies including columnar and spherical.

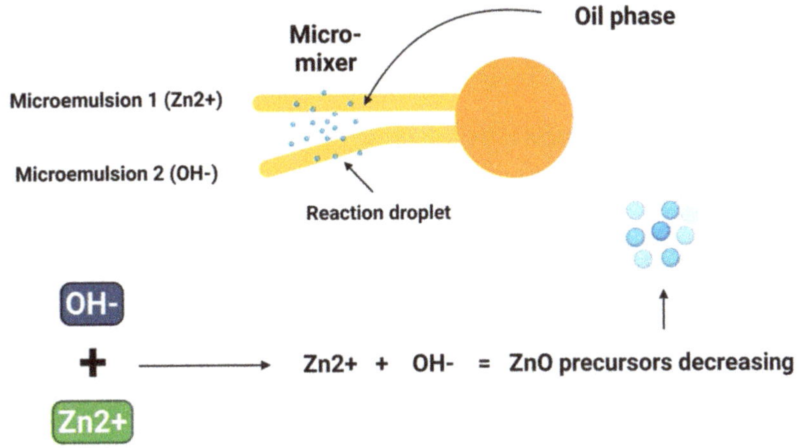

Figure 6. Diagrammatic representation of the stages required for metallic ZnO-NPs synthesis employing the microemulsion method.

3.5. Laser Ablation Technique

A typical laser ablation technique can remove metallic ions from metal surfaces by employing a laser beam and a small liquid quantity of methanol, ethanol, and purified water. The surface is immersed in the liquid. A schematic representation of the laser ablation approach is shown in Figure 7. Simplicity, and a quite safe procedure from an environmental standpoint, are two of the approach's advantages, resulting in a process that is both efficient and simple to carry out [61]. Pyrolysis byproducts (the result of laser ablation when organic substances are present) have yet to be fully clarified and need to be addressed [62]. The works of Al-Dahash et al. [63] are exciting: with laser ablation in NaOH aqueous solution, they could produce ZnO-NPs from 80.76 to 102.54 nm with a spherical structure. In addition, Farahani et al. [64] used a zinc target in a solution of methanol and distilled water to generate ZnO-NPs with a roughly spherical morphology ranging from 1 to 30 nm by laser ablation. In the same way, Mintcheva et al. [61] indicated that they made ZnO-NPs that were rod-shaped, 30 nm in diameter, and 40–110 nm in size.

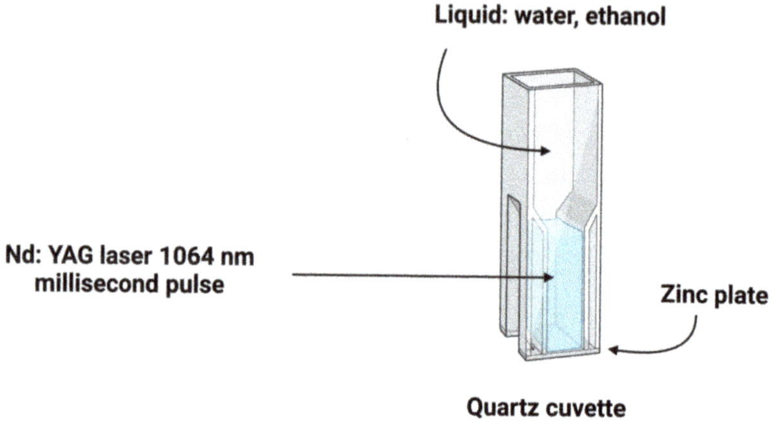

Figure 7. Diagrammatic representation of the stages required for metallic NPs synthesis (for example ZnO-NPs) employing laser ablation.

3.6. High-Energy Ball Milling Techniques

The high-energy ball milling technique is a manufacturing process that produces fine metal NPs in an elevated shaker mill [65]. This technology is depicted in Figure 8. Its key advantage is the ability to generate vast quantities of material simultaneously. Its downsides include contaminants from milling balls and/or from the environment and irregularly shaped NPs that result from this process [66–68]. ZnO-NPs may still be synthesized using commercially available ZnO powder with a mean of 0.8 m particle size, as demonstrated by Prommalikit et al. [69], who used high-energy ball milling to manufacture ZnO-NPs. Particles with a final size of 200–400 nm were obtained through milling. In a similar vein, Mohammadi et al. [70] synthesized rod-shaped ZnO-NPs in the 20–90 nm range using a high-energy ball milling technique. Salah et al. [70] employed the same high-intensity ball milling procedure to make ZnO-NPs from ZnO microcrystalline powder. The samples were ground into a ball mill for 2, 10, 20, and 50 h. The size of the particles changed over time, according to the results. The smaller the particle size is, the longer the ball milling process lasts. Spherical ZnO-NPs with approximately 30 nm particle sizes were found in the milled sample.

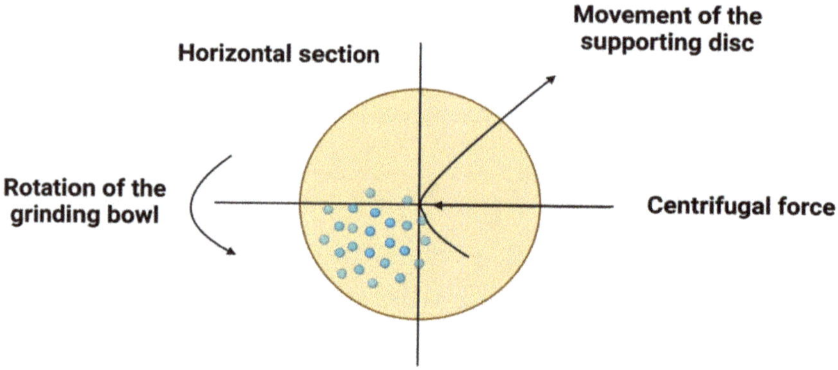

Figure 8. Diagrammatic representation of the stages required for metallic NPs synthesis (for example ZnO-NPs) employing high-energy ball milling procedures.

4. Green Synthesis of ZnO-NPs

4.1. Green Synthesis of ZnO-NPs Using Plant Extract

Because of the unique phytochemicals that they generate, plant components, for instance, the root, stem, leaf, seed, and fruit, have been employed to fabricate ZnO-NPs. The use of organic isolates of plant parts is a highly eco-friendly, cost-efficient method that does not need intermediary base groups. It takes a fraction of the time, requires no expensive equipment or precursors, and produces a highly natural and magnitude-enriched product devoid of contaminants [71]. Plants are considered a popular source of NP synthesis because they allow for the significant production of NPs with various shapes and sizes [72].

Phytochemicals, such as polysaccharides, vitamins, alkaloids, polyphenolic compounds, amino acids, and terpenoids released by plants, decrease metal oxides or metal ions to around 0 valence metal NPs [71,72]. The plant portion's manufacturing of ZnO-NPs extracted from flowers or leaves is mainly processed via being bathed in running tap water and sterilized double distilled water. The plant portion is then allowed to dry at room temperature before being weighed and crushed with a mortar and pestle. The necessary amount of Milli-Q H_2O is added to the plant component and boiled under vigorous agitation using a magnetic stirrer [71–75]. The plant's extractions are made by filtration through Whatman filter paper (sample). To ensure efficient mixing, the mixture is heated to the necessary temperature for the necessary time to integrate the extract into 0.5 mm of hydrous zinc sulfate or zinc nitrate, or ZnO or solution [74,75]. At this point, some experiments were done with extract concentration, temperature, duration, and pH to see what works best. An incubation period causes the mixture to turn yellow as visual proof of the newly produced NPs [74,75].

Next, the mixture is centrifuged and dried in a hot oven to obtain the crystal NPS from the synthesized NPs, and confirmed by UV-Vis spectrometry [76]. To further characterize the synthesized NPs, various techniques, such as Field Emission Scanning Electron Microscopy (FE-SEM, JEOL IT800 series, New York, NY, USA), X-ray Photoelectron Microscopy (XPS, Phadke Instruments Private Limited, Maharashtra, India), Energy Dispersion Analysis of X-ray (EDAX, Nunes Instruments, Tamil Nadu, India),Scanning Electron Microscopy (SEM, Analytical Technologies Limited, Gujarat, India), X-ray diffractometer (XRD, Expert Vision Labs Pvt. Ltd., Maharashtra, India), UV-Visible Diffuse Reflectance Spectroscopy (UV-DRS, Nunes Instruments, Tamil Nadu, India), Fourier Transform Infrared Spectroscopy (FTIR, Alliance Enterprise, Mumbai, India), Transmission Electron Microscopy (TEM, Expert Vision Labs Pvt. Ltd., Maharashtra, India), and Atomic Force Microscopy (AFM, V Instek Analytical, Gujarat, India), are propagated [75–77]. Microwave irradiation (MI, V Instek Analytical, Gujarat, India) takes less time than conventional heating (CH), according to an experiment by Jafarirad et al. [78], and this is due to the higher level of heating provided by MI and a consequently faster response rate. *Anisochilus carnosus* [79], *Plectranthus amboinicus* [80], and *Vitex negundo* [81], members of the Lamiaceae family, have been widely investigated; the size of produced NPs reduces as the content of a plant extract increases [79–81].

Additionally, results comparing the size ranges recorded using other techniques, such as FE-SEM, TEM, and XRD, revealed similar range values [80,81]. SEM and EDAX yielded results that differed slightly from those of XRD. According to the Debye-Scherrer equation, NPs synthesized from the leaves and flowers of *Vitex negundo* had the same diameter of 38.17 nm, validated by XRD analysis [81]. For the fabrication of ZnO-NPs, the leaves of the *Azadirachta indica* of the Meliaceae family were the ones most typically employed [82,83]. XRD and TEM examination verified that the NPs in all trials were in the same size range, with spherical and hexagonal disc-shaped NPs and Nano buds carboxylic acid, alkane, amine alcohol, carbonate moieties, and amide were involved in the synthesis of NPs, as evidenced by FTIR investigations. *Aloe vera* leaf extract and leaf peel belong to the Liliaceae family [84,85]. The size of synthesized NPs differed (NPs synthesized from peel were more extensive, as validated by SEM and TEM studies), but the forms were similar (hexagonal and spherical). *Agathosma betulina, Pongamia pinnata, Plectranthus amboinicus, Nephelium lappaceum,* and *Calatropis gigantea* were extracted for synthesized NPs, which form aggregates [86]. Plants employed to synthesize ZnO-NPs up to these points are included in Table 1.

Table 1. ZnO-NPs synthesized using a plant-mediated process.

Common Name	Plant (Family)	Extraction Part	Functional Group	Shape	Size (nm)	References
Coptis Rhizome	*Coptidis rhizoma* (Ranunculaceae)	Dried Rhizome	Primary and secondary amine, aromatic, aliphatic amine, alcohol, carboxylic acid, alkyl halide, and alkynes.	Spherical, rod-shaped	2.9–25.2 (TEM)	[87]
Neem	*Azadirachta indica* (Meliaceae)	Fresh leaves	Amine, alcohol, ketone, carboxylic acid	Spherical	18 (XRD)	[88]
Indian beech	*Pongamia pinnata* (Legumes)	Fresh leaves	O-H stretching, C=O spreading carboxylic acid or their ester, C-O-H bending mode.	Spherical, hexagonal, nanorod	26 (XRD), agglomeration of 100 (DLS, SEM, TEM)	[89]
Red Rubin basil	*Ocimum basilicum* (Lamiaceae)	Leaf extract	-	Hexagonal (wurtzite)	50 (TEM, EDS), 14.28 (XRD)	[90]
Bhuiamla, stone breaker	*Phyllanthus niruri* (Phyllanthaceae)	Leaf extract	O-H, C-H, C-O stretching, aromatic aldehyde.	Hexagonal wurtzite, quasi-spherical	25.61 (FE-SEM & XRD)	[91]
Buchu	*Agathosma betulina* (Rutaceae)	Dry leaves	O-H of hydroxyl group, Zn-O stretching band	Quasi-spherical agglomerates	15.8 (TEM), 12–26 (HRTEM)	[92]
Red clover	*Trifolium pratense* (Legumes)	Flower	Hydroxyl, -C-O, -C-O-C, C=C stretching mode.	Spherical	60–70 (XRD)	[93]
Kapurli	*Anisochilus carnosus* (Lamiaceae)	Leaf extract	O-H of water, alcohol, phenol C-H of alkane, O-H of carboxylic acid, C=O of the nitro group.	Hexagonal wurtzite, quasi-spherical	56.14 (30 mL of extract), 49.55 (40 mL), 38.59 (50 mL) [XRD], 20–40 (FE-SEM), 30–40 (TEM)	[79]
Water hyacinth	*E. crassipes* (Pontederiaceae)	Leaf extract	-	Spherical without aggregation	32–36 (SEM & TEM), 32 (XRD)	[94]
Dog rose	*Rosa canina* (Rosaceae)	Fruit extract	C-O and C=O of esters, hydroxyl, C-H stretching.	Spherical	[13.3 (CH), 11.3 (MI)] (XRD), [25–204 (CH), 21–243 (MI)] (DLS),	[7]
Black nightshade	*Solanum nigrum* (Solanaceae)	Leaf extract	O-H, aldehydic C-H, amide III bands of protein, carboxyl side group, C-N of amine, the carbonyl group	Wurtzite hexagonal, quasi-spherical	20–30(XRD and FE-SEM),29.79(TEM)	[95]
Aloe vera	*Aloe vera* (Liliaceae)	Freeze-dried leaf peel	-	Spherical, hexagonal	25–65 (SEM & TEM)	[84]
Neem	*Azadirachta indica* (Meliaceae)	Leaf	Amide II was stretching band, C-N stretching band of aliphatic, aromatic amide, an aliphatic amine, alcohol, phenol, secondary amine, C-H of alkane and aromatics, C=C-H of alkynes, C=O, C-C of an alkane.	Spherical	9.6–25.5 (TEM)	[82]
Drumstick tree	*Moringa oleifera* (Moringaceae)	Leaf	O-H, C-H of alkane, C=O of alcohol, carboxylic acid	Spherical and granular nano-sized shape with a group of aggregates	24 (XRD), 16–20 (FE-SEM)	[96]
Coconut	*Cocus nucifera* (Arecaceae)	Coconut water	O-H of alcohol and a carboxylic acid, C=O of ketones, C-N of aromatic and aliphatic amines,	Spherical and predominantly hexagonal without any agglomeration	20–80 (TEM), 21.2 (XRD)	[97]
Cotton	*Gossypium* (Malvaceae)	Cellulosic fiber	O-H, [C=O, C-O, C-O-C] (due to Zn precursor)	Wurtzite, spherical, nanorod	13 (XRD)	[98]

Table 1. *Cont.*

Common Name	Plant (Family)	Extraction Part	Functional Group	Shape	Size (nm)	References
Santa maria feverfew, carrot grass, congress weed	*Parthenium hysterophorus* (Asteraceae)	Leaf extract	N-H bending & N-H stretching mode, a phosphorus compound, secondary sulfonamide, monosubstituted alkyne, amine salt, vinyl cis-tri substituted	Spherical, hexagonal	22–35 (50% plant extract), 75–90 (25% plant extract) (XRD, TEM)	[99]
Neem	*Azadirachta indica* (Meliaceae)	Fresh leaves	O-H between H_2O and CO_2, carbonate moieties	Hexagonal disk, nanobuds	10–30 (TEM), 9–40 (XRD)	[83]
Mexican mint	*Plectranthusamboinicus* (Lamiaceae)	Leaf extract	Zn-O, C-O of C-O-SO_3, phosphorus compound	Rod-shaped nanoparticles with agglomerates	50–180 (SEM)	[100]
Crown flower	*Calatropis gigantea* (Apocynaceae)	Fresh leaves	-	Spherical-shaped forming agglomerates	30–35 (SEM)	[101]
Nochi	*Vitex negundo* (Lamiaceae)	Flowers	-	Hexagonal	38.17 (XRD), 10–130 (DLS)	[30]
Sandalwood	*S. album* (Santalaceae)	Leaves	N-H stretching of amide II, carboxylate group, carbonyl stretching, O-H of alcohol	Nano rods	100 (DLS & SEM), 70–140 (TEM)	[102]
Nochi	*Vitex negundo* (Lamiaceae)	Leaf	OH, C-H, C=C stretching band.	Spherical	75–80 (SEM & EDX), 38.17 (XRD)	[103]
Rambutan	*Nephelium lappaceum* (Sapindaceae)	Fruit peels	O-H stretching, H-O-H bending	Needle-shaped forming agglomerate	50.95 (XRD)	[104]
Aloe Vera	*Aloe Vera* (Liliaceae)	Leaf extract	O-H of phenol, amines, O-H of alcohol, and C-H of alkanes, the amide of protein and enzymes.	Spherical, oval, hexagonal	8–20 (XRD)	[85]
African tulip tree	*Sphathodea campanulata* (Bignoniaceae)	Leaf extract	O-H stretching of polyphenols, nitrile group, C-H, C=O group	Spherical	30–50 (TEM)	[105]

4.2. ZnO-NPs Green Synthesis Using Bacteria

There are various drawbacks to employing bacteria to synthesize NPs, including the time and effort necessary to screen microbes, the need for constant observing of culture broth and the entire process, the NPs' shape and size, and the expense of the media used to grow bacteria. Using an eco-friendly technique, the photocatalytic activity and degradation of nanoflowers ZnO were demonstrated by *B. licheniformis*. The photocatalytic action for these nanoflowers was shown to be improved when compared to existing photocatalytic materials. It has been speculated that the more considerable oxygen vacancy in the produced NPs provides this property. It is possible to employ photocatalysis as a bioremediation method because it generates active species by absorbing light. Synthesized nanoflowers based on *B. licheniformis* were 40 nm wide by 400 nm long [106].

Rhodococcus can persist in unfavorable conditions and metabolize hydrophobic substances, which enables it to contribute to biodegradation [107]. *Rhodococcus pyridinivorans* and zinc sulfate were used to manufacture spherical NPs with a 100–130 nm size range, which XRD and FE-SEM assessment confirmed. In addition, FTIR examination indicated the existence of mononuclear benzene band, secondary sulfornamide, lactone, amine salt, monosubstituted alkyne, enol of 1-3-di ketone, hydroxy aryl ketone, amide I bending band, alkane, amide II stretching band, and phosphorus compound [108]. NPs of ZnO were created using *Aeromonas hydrophilla* as a substrate for ZnO synthesis. AFM and XRD analyses showed that the NPs produced had a size range of 42–64 nm and diverse forms including oval and spherical [109]. Because it is difficult for rhamnolipid to make micelle aggregate on carboxymethyl cellulose, this helps keep ZnO-NPs from breaking apart into micelle groups, making them more stable [110]. Because of its lengthy carbon chain, it works as a better capping agent [111]. The TEM, XRD, and DLS analyses revealed the synthesis of spherical NPs with a nano size range of 27–81 nm [111]. The properties of ZnO-NPs produced utilizing bacterial strains are shown in Table 2.

Table 2. Synthesis of ZnO-NP using bacterial strain.

Family	Bacterial Strain	Functional Group	Shape	Size (nm)	References
Bacillaceae	*Lactobacillus sporogens*	-	Hexagonal unit cell	5–15 (TEM), 11 (XRD)	[112]
Pseudomonadaceae	*Pseudomonas aeruginosa*	O-H stretching vibration, -CH of aliphatic stretching vibration, ester carbonyl group.	Spherical	35–80 (TEM), 27 (XRD), 81 (DLS)	[113]
Pseudomonadaceae	*Aeromonas hydrophila*	Phosphorus compound, vinyl cis-trisubstituted, monosubstituted alkyne	Spherical, oval	57.72 (AFM), 42–64 (XRD)	[114]
Bacillaceae	*B.licheniformis*	0-H, N-H,-C-O (carbonyl stretching in the amide I and amide II linkage of protein), C-N stretching bond.	Nanoflowers	200 with nanopetals 40 in width and 400 in length (TEM)	[108]
Nocardiaceae	*Rhodococcus pyridinivorans*	Phosphorus compound, secondary sulfornamide, monosubstituted alkyne, β-lactone, amine salt, amide II stretching band, enol of 1-3-di ketone, a hydroxy aryl ketone, amide I bending band, alkane, mononuclear benzene band.	Hexagonal phase, roughly spherical	100–120 (FE-SEM), 120–130 (XRD)	[110]
Enterobacteriacea	*Serratia ureilytica* (HM475278)	-	Spherical- to nanoflower-shaped	170–250 (30 min), 300–600 (60 min), 185–365 (90 min) [SEM]	[115]

4.3. ZnO-NPs Green Synthesis Using Microalgae and Macroalgae

Unicellular algae (chlorella) and multicellular algae (chlorophyll) are examples of photosynthetic organisms (for instance, brown algae). Basic plant structures, such as leaves and roots, are absent from algae. Marine algae are classified according to the pigments they contain, such as Rhodophyta, Phaeophyta, and chlorophytes, which have red, brown, and green pigments, respectively. For the formation of Au and Ag NPs, algae have been extensively exploited. However, their utilization for ZnO-NPs synthesis has been limited and documented in relatively few works [92]. The potential of microalgae to break down hazardous metals and transform them into less harmful forms has drawn significant attention [116]. *S. muticum* and *S. myriocystum*, both Sargassaceae plants, were employed to synthesize ZnO-NPs. Sulfated polysaccharides were present in the NPs investigated by XRD and FE SEM, revealing similar NP sizes and hexagonal wurtzite structure. For *S. myriocystum*, DLS and AFM measurements demonstrated varied size ranges, with carbonyl and hydroxol stretching in NPs that vary substantially in form [99]. The micro- and macro-algae listed in Table 3 were used to synthesize ZnO-NPs.

Table 3. Synthesis of ZnO-NPs using algae.

Algal Strain	Family	Size (nm)	Shape	Functional Group	Reference
Chlamydomonas reinhardtii	Chlamydomonaceae	55–80 (HR-SEM), 21 (XRD)	Nanorod, nanoflower, porous nanosheet	C=O stretching, N-H bending band of amide I and amide II, C=O stretch of zinc acetate, C-O-C of polysaccharide	[117]
S. myriocystum	Sargassaceae	46.6 (DLS), 20–36 (AFM)	Spherical, radial, triangle, hexagonal, rod	O-H and C=O stretching band, carboxylic acid	[118]
Sargassum muticum	Sargassaceae	30–57 (FE-SEM), 42 (XRD)	Hexagonal wurtzite	Asymmetric stretching band of the sulfate group, an asymmetric C-O band associated with C-O-SO_3 & -OH group, sulfated polysaccharides	[119]

4.4. ZnO-NPs Green Synthesis Using Fungus Theorem

Extensive production, easy downstream processing, and commercial feasibility make extracellular NPs from fungi beneficial [120]. Because of their higher tolerance and their ability to bioaccumulate metals, fungi are preferred over bacteria [121]. Mycelia of *Aspergillus fumigatus* were used to produce ZnO-NPs. According to the DLS study, NPs ranged from 1.2 to 6.8 in area size, with a 3.8 average size. AFM established the average height of NPs to be 8.56 nm for 90 days, with a significant particle size of more than 100 nm. After 90 days, they developed an agglomeration with an average particle size of 100 nm, indicating that the produced NPs were stable for 90 days [122]. SEM confirmed a size range between 54.8–82.6 nm for NPs produced from *Aspergillus terreus* that belong to the Trichocomaceae family. XRD investigation results revealed a 29 nm average size, which was determined using the Debye-Sherrer equation. FTIR analyses indicated the formation of primary alcohol, aromatic nitro compounds, and amine in the produced NPs [123]. SEM, TEM, and XRD analysis verified that NPs generated with *Candida albicans* had a comparable size range of 15–25 nm [124]. In most cases, ZnO-NPs developed from *Aspergillus* species were spherical. Table 4 lists the fungi most typically employed for ZnO-NPs production

Table 4. Synthesis of ZnO NPs using fungi.

Family	Fungal Strain	Functional Group	Shape	Size (nm)	Reference
Trichocomaceae	*Aspergillus* strain	-	Spherical forming aggregates	50–120 (SEM)	[125]
Trichocomaceae	*Aspergillus terreus*	C-N bond of primary amine, C-O of a primary alcohol, primary and secondary alcohol, N=O aromatic nitro compound, alkyl C=C, amide, open-chain amino group	Spherical	54.8–82.6 (SEM), 29 (XRD)	[126]
	Candida albicans	-	Quasi-spherical, hexagonal phase (wurtzite structure)	25 (XRD), 15–25 (SEM), 20 (TEM)	[124]
Trichocomaceae	*Aspergillus fumigatus* TFR-8	-	Oblate spherical and hexagonal forming aggregates	1.2–6.8 (DLS), 100 (agglomerate)	[106]

4.5. ZnO-NPs Green Synthesis Using Other Green Sources

NPs can be synthesized using biocompatible chemicals and alternative green sources. NP nucleation and synthesis reactions can be carried out within a short time and cost-efficiently. They result in the production of NPs with a well-dispersed nature that may be precisely regulated in shape and size [127]. Antibacterial capabilities improved in 99.9% of NPs produced using a wet chemical method when layered on cotton fabric [128]. Table 5 summarizes a few more green resources used to synthesize ZnO-NPs.

Table 5. Synthesis of ZnO-NPs by proteins.

Others	Size (nm)	Shape	Functional Group	References
Egg albumin	16 (XRD), 10–20 (TEM), 8–22 (AFM)	Spherical, Hexagonal wurtzite	Hydroxyl group	[129]
L-alanine	50–110 (TEM, SEM)	-	Hydroxyl group, C-O vibration of Schiff- base.	[130]
Soluble starch	50 (SEM)	-	-	[131]

5. Biomedical Applications of Green-Synthesized ZnO-NPs

There has been a sharp rise in attention to NP research in the past decade, particularly in regard to biological applications [132]. Since nanotechnology has been integrated into medical research, a more excellent grasp of molecular biology has been achieved. As a result, innovative treatment strategies may be possible for illnesses that were previously impossible to address due to size limits [133]. For biomedical applications, the formulation of biofunctional NPs has attracted various research groups that are continually addressing this subject [134]. Biomedical applications of ZnO-NPs are now under investigation using a wide range of materials and chemical synthesis processes, as we have discussed in this study. As an ecological element and part of nature's intrinsic materials, zinc has a vital role in human, animal, and plant metabolism. Zinc is required for all living species, which must be exposed to environmentally appropriate amounts of zinc in the biosphere. ZnO is extensively utilized in cosmetic, pharmaceutical, and medicinal applications, and as a nutritional supplement. Even though ZnO dust and fumes are typically considered harmless, breathing them should be avoided. Regulations have been put in place to limit the risk of exposure [135]. Figure 9 depicts the green production and uses of ZnO-NPs.

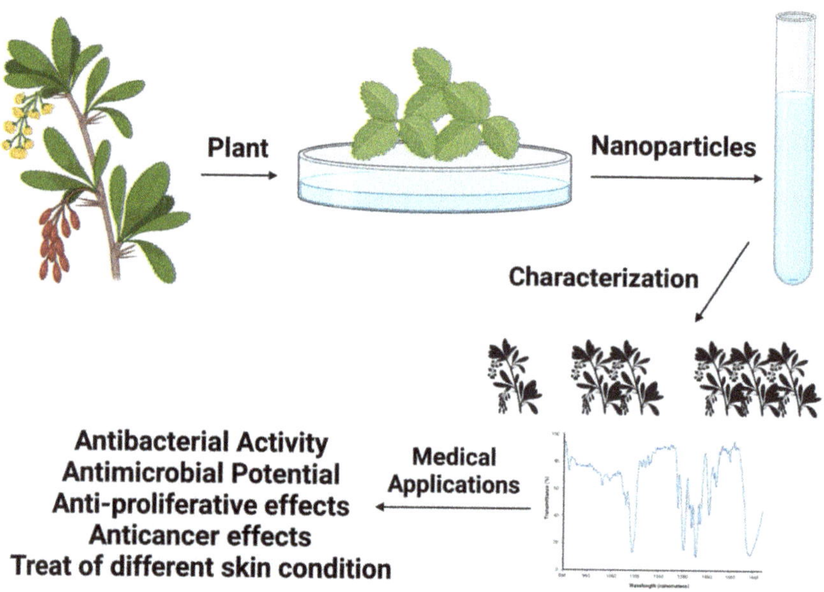

Figure 9. An illustration of the green synthesis and use of ZnO-NPs.

5.1. *ZnO-NPs Antibacterial Activity*

Organic and inorganic materials are the most common divisions in pharmaceutical medicinal agents. Organic medicinal drug substances have been found to be less stable at high temperatures and high pressures, when compared with inorganic medicinal drug substances [136]. ZnO-NPs are powerful pharmacological agents for therapeutic applications. ZnO-NPs seem to have a significant therapeutic drug activity when compared with microparticles. It is noteworthy that the specific mechanisms of medicinal drug action have not been wholly established [137]. Both gram-positive and gram-negative bacteria are germicidal to ZnO-NPs [138], and the ZnO-NPs also include medical therapeutic actions against high temperature and high pressure-resistant spores. Research shows that their extent and concentration influence ZnO-NPs' medicinal properties, but not their crystalline structure or particle type. Therefore, the more NPs there are, the more potent the medical medicine [139].

Synthesized ZnO-NPs, which have natural antibacterial effects and are photocatalytic in the ultraviolet (UV-B) light range, can create potent hydroxyl (-OH) free radicals to kill dangerous pathogens and germs at wound sites [140]. This observation led to the development of a 3D printed customized wound-healing template made of ZnO-NPs that were uniformly scattered within an alginate template, which can be easily created and contour-printed to the exact size and depth of a wound. 3D printing consist of the adding of material layer by layer, allowing for the fabrication of unique shapes and customizability, which are crucial in biomedical areas such as tissue engineering and pharmaceuticals [141].

ZnO-NPs' medical medication action mechanism is still a mystery. Hydrogen peroxide emission may be the essential factor in the action of therapeutic drugs. It is also possible that the mechanism is due to the binding of particles on the bacterial surface, owing to static tensions [142]. According to the results, the antibacterial activity of ZnO-NPs seems to be stronger than that of tiny particles. Particle dosage, treatment duration, and the NP production process influence NPs' efficacy. Furthermore, the surface area and the size of particle variation, which are noteworthy in green-synthesized ZnO-NPs, are responsible for enhanced antibacterial activity. Future medical difficulties might benefit from green-synthesized ZnO-NPs applications in food safety and agriculture that have not yet been confirmed [143]. Table 6 provides applications of green-synthesized ZnO-NPs for antibacterial purposes.

Table 6. Green-synthesized ZnO-NPs applications for antibacterial purposes.

Platform	Raw Material	Size	System	Targeted Bacteria	Reference
Bacteria-mediated	*Bacillus megaterium*	45–95 nm	ZnO-NPs	• *H. pylori*	[144]
	Bacillus licheniformis	10–100 nm	ZnO-NPs	• *P. aeruginosa* • *Proteus vulgaris* • *Bacillus subtilis* • *Bacillus pumilus*	[145]
Plant-mediated	*Cassia fistula*	5–15 nm	ZnO-NPs	• *Klebsiella aerogenes* • *E. coli* • *Plasmodium desmolyticum*	[146]
	Trifolium pretense	60–70 nm	ZnO-NPs	• *P. aeruginosa* • *E. coli* • *S. aureus*	[93]
	Boerhavia diffusa	140 nm	ZnO-NPs	• *MRSA*	[147]
	Artocarpus gomezianus	39, 35, 31 nm prepared with 5, 10 and 15 mL of 10% extract	ZnO-NPs	• *S. aureus*	[148]
	Sechium edule	30–70 nm	ZnO-NPs	• *Bacillus subtilis* • *Klebsiella pneumonia*	[149]
	Azadirachta indica	9.6–25.5 nm	ZnO-NPs	• *Streptococcus pyogenes* • *E. coli* • *S. aureus*	[82]
	Azadirachta indica	9–40 nm	ZnO-NPs	• *Klebsiella aerogenes* • *S. aureus*	[83]
	Acalypha indica	20 nm	ZnO-NPs	• *E. coli* • *S. aureus*	[150]

Table 6. *Cont.*

Platform	Raw Material	Size	System	Targeted Bacteria	Reference
	Tabernaemontana divaricata	20–50 nm	ZnO-NPs	• *E. coli* • *S. aureus* • *Salmonella paratyphi*	[151]
	Laurus nobilis	47.27 nm	ZnO-NPs	• *P. aeruginosa* • *S. aureus*	[152]
	Ruta graveolens	28 nm	ZnO-NPs	• *Klebsiella aerogenes* • *P. aeruginosa* • *E. coli* • *S. aureus*	[31]
	Aristolochia indica	22.5 nm	ZnONPs	• Multi-drug resistant organisms (MDROs) isolated from pus samples of DFU patients	[153]
	Allium sativum	14 and 27 nm	ZnO-NPs	• *S. aureus* • *Bacillus subtilis* • *L. monocytogenes* • *E. coli* • *Salmonella typhimurium* • *P. aeruginosa*	[153]
	Bauhinia tomentosa	22–94 nm	ZnO-NPs	• *E. coli* • *P. aeruginosa*	[154]
	Ulva lactuca	10–50 nm	ZnO-NPs	• *Bacillus licheniformis* • *Bacillus pumilis* • *E. coli* • *Proteus vulgaris*	[155]
	Amaranthus spinosus	243 nm undoped/197 nm 1%-Fe-ZnO-NPs	Undoped and Fe-doped ZnO-NPs	• *E. coli* • *Bacillus safensis*	[156]
	Hibiscus rosa-sinensis	15–170 nm	Fe-doped ZnO-NPs	• *E. coli*	[157]
	G. sylvestre	138 nm, 52 nm, 59 nm, and 63 nm for undoped, La-, Ce-, and Nd-doped	Lanthanum-, cerium-, and neodymium-doped ZnO-NPs	• *S. aureus* • *Streptococcus pneumonia*	[158]

5.2. ZnO-NPs Antimicrobial Potential

ZnO is explored as a potential drug carrier in micro-and nanoscale formulations. Even though the medicament-specific mechanisms are not fully understood, it has been proposed that the ROS produced on the particle's surface, membrane dysfunction, zinc ion release, and the NPs' acquisition area unit are the common causes of cell swelling. Management of ZnO-NPs at elevated temperatures significantly affects their therapeutic activity, whereas treatment at lower temperatures reduces activity. The mechanisms underlying ZnO-NPs' medicament activity are unknown. While it is hypothesized that oxide generation contributes to such activity, it is indicated that the binding between particles and microorganism surface, due to electrical forces, could be a mechanism for ZnO-NPs' medicament behavior. This could be accomplished using oxygen electrode analysis and chemiluminescence. Metal NPs are highly ionic and can be generated with exceptional crystal and high surface, and morphologies with varying edge/corner and reactive surface sites. The ZnO-NPs area unit is subject to current research concerning therapeutic procedures with ablation regimens. Despite having a more significant thermal effect on neoplasm ablation, NPs will provide an antineoplastic medical specialty with a synergetic anticancer impact at the time of heat presence. They may even be imaged to achieve precise medical assistance. Numerous experiments revealed that understanding the molecular mechanism underlying tumor-mediated NP ablation will aid in the development of NPs with appropriate composition and characteristics to induce the ablation property [159–161].

5.3. Proliferating Cells Selective Killers

ZnO cancerous cells are killed by ZnO-NPs, whereas healthy cells are unaffected [162,163]. Before ZnO-NPs can be used medically, a slew of issues must be addressed, including a lack of biocompatible dispersion procedures and a more profound knowledge of the mechanism underlying their selective cytotoxicity [142]. To date, there have only been a few investigations on the ZnO-NPs cytotoxicity mammal cells, and experts are divided about the importance of the results that have been published. A study found that ZnO-NPs have no influence on T cells' viability in both gram-negative and gram-positive microorganism concentrations [164]. According to various publications, these NPs are harmless with respect to the culture of human dermal fibroblasts; still, they are harmful to metastatic tumor cells [165] and the cells of vascular endothelial [166], triggering programmed cell death in neural stem cells. It has been stated that the NPs' size can affect cell viability. Jones et al. [34] discovered that ZnO-NPs with a diameter of eight nm were more hazardous than were larger zinc oxide particles (50–70 nm) in *Staphylococcus aureus*. Hanley et al. [167] recently established a reverse relationship between class cells' toxicity and NP size, such as reactive oxygen species (ROS) production. In contrast, Deng et al. [168] showed ZnO-NPs' toxic influence on nervous stem cells in a dose-dependent manner, regardless of particle size.

5.4. ZnO-NPs Anticancer Effects

ZnO cancer nanotechnology has vast implications for molecular identification, molecular imaging, and tailored medical treatment, according to the nursing knowledge domain area of analysis in engineering, science, and medications. To put it simply, nanometer-sized particles, such as semiconductor quantum dots and iron chemical complex nanocrystals, exhibit optical, magnetic, or structural features that are not found in molecules or bulk materials. As soon as these NPs are attached to antigen-targeting ligands, such as antibodies or peptides, they can target neoplasm antigens as biomarkers as well as neoplasm vessels with significant similarity and specificity. Because of their large surface areas and functional groups, many diagnostic and therapeutic substances can be conjugated to NPs in the 5–100 nm diameter range. A junction rectifier to bio-affinity NPs for molecular and cell imaging can provide customized medical treatment using NP medication. Researchers have recently developed and incorporated nano-devices to detect and screen cancer in early stages. Biomarkers for cancer diagnosis and treatment based on individualized molecular profiles and tailored genetic and super molecular biomarkers are now possible because of these breakthroughs in personalized medicine [169].

Several types of research have indicated that ZnO-NPs positively influence cancer cell growth. It was found that the cell response to ZnO-NPs was dynamic. Hence, the final composition was affected by multiple challenging or intersecting signals in the microenvironment, as revealed by Premanathan et al. [142]. ZnO-NPs were more hazardous to HL60 cancer cells than to normal PBMCs with a therapeutic index, according to the findings (i.e., hepatotoxic dose) [142]. The inability to distinguish between traditional and changed tissues in malignant neoplasm medicine may be of essential clinical interest and the biggest hurdle in treatment [170]. Even though various commonly prescribed drugs can slow down the rate at which cells divide, many of these treatments have a low therapeutic index [171,172]. Table 7 summarizes the anticancer uses of ZnO-NPs synthesized by the green synthesis technique, whereas Figure 10 illustrates the molecular mechanisms underpinning green ZnO-NPs' anticancer action.

Table 7. Anticancer applications of ZnO-NPs generated in the green synthesis process.

Platform	Raw Material	Size	System	Targeted Cell Line	Reference
Fungi-mediated	*Pichia kudriavzevii* yeast	10–61 nm	ZnO-NPs	MCF-7, breast	[173]
	Penicillium chrysogenum fungus	29–37 nm	ZnO-NPs	MCF-7, breast HCT-116, colon	[174]
	Aspergillus niger fungus	80–130 nm	ZnO-NPs	HepG2, liver	[175]
	Aspergillus niger fungus	11.8–17.6 nm	ZnO-NPs	A549, lung	[176]
	Aspergillus terreus fungus	28–63 nm	L-asparginase—ZnO-NPs	MCF-7, breast	[177]
Algae and plant-mediated	*Sargassum muticum* algae extract	30–57 nm	ZnO-NPs	HepG2, liver	[178]
	Sargassum muticum algae extract	50–100 nm	ZnO-NPs	WEHI-3, murine leukemia	[179]
	Sargassum muticum algae extract	3–8 nm	ZnO-NPs	PANC-1, pancreas CaOV-3, ovarian COLO205, colon HL-60, leukemia	[180]
	Gracilaria edulis algae extract	4.04 ± 1.81 nm; length 1.39 ± 0.6 nm; width	ZnO-NPs	SiHa, cervical	[181]
	Rehmanniae radix plant extract	10–12 m	ZnONPs	MG-63 bone	[182]
	Myristica fragans plant extract	100–200 nm	ZnONPs	HeoG2, liver	[183]
	Albizia lebbeck stem bark	66.25 nm	ZnONPs	MCF-7, breast MDAMB231, breast	[184]
	Mangifera infica leaves	45–60 nm	ZnO-NPs	A549, lung	[185]
	Pongamia pinnata seeds	30.4–40.8 nm	ZnO-NPs	MCF-7, breast	[186]
	Eclipta prostrata leaves	20–1.3 nm	ZnO-NPs	HepG2, liver	[187]
	Borassus flabellifer fruit extract	110 nm	ZnO-NPs loaded with DOX	MDAMB231, breast	[188]
	Ziziphus nummalaria leaves	17.33 m	ZnO-NPs	HeLa, cervical	[189]
	Laurus nobilis leaves	47.27 nm	ZnO-NPs	A549, lung	[152]
	Nephelium lappaceum peel	-	ZnO-NPs	HepG2, liver	[190]
	Tecoma castanifolia flower	70–75 nm	ZnO-NPs	A549, lung	[191]
	Gymnema sylvestre, plant extract	38 nm 33/27/23 nm	ZnO-NPs La/Nd/Ce—ZnO-NPs	A498, kidney	[158]
	Costus pictus, leaves	20–80 nm	ZnO-NPs	DLA, Daltons lymphoma ascites	[5]
Protein mediated	Collagen protein	20–50 nm	ZnO-NPs	HepG2, liver	[192]
	Milk casein protein	9.3–13.7 nm	ZnO-NPs loaded with curcumin	MCF-7, breast HeLa, cervical MDAMB231, breast MG-63, bone	[193]
	Tocopherol lipid	100 nm	Chitosan coated ZnO-NPs	HeLa, cervical	[194]

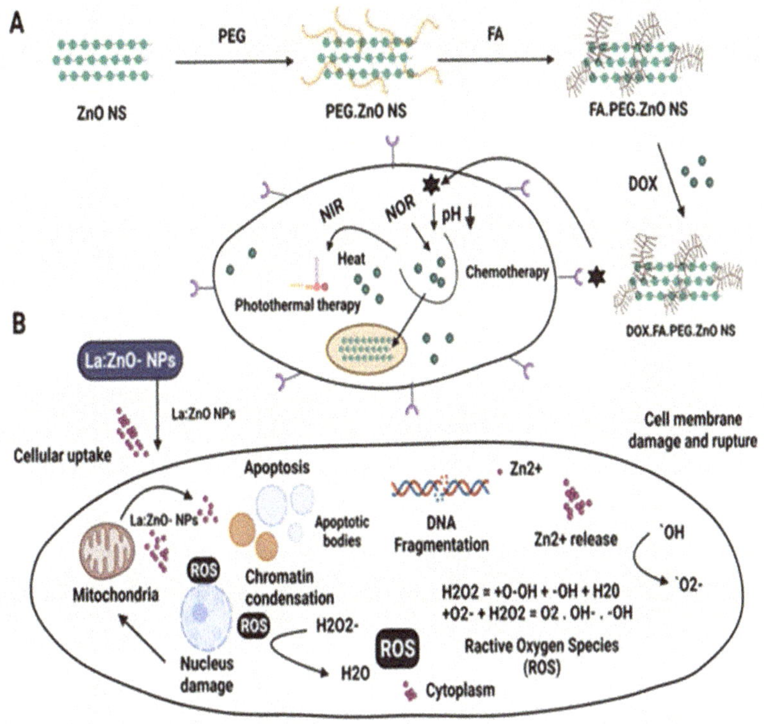

Figure 10. The mechanisms underlying the anticancer activity of green ZnO-NPs. (**A**) Cytotoxic action of La-doped ZnO-NPs causes cell death [195]. (**B**) The development of DOX-FA-ZnO NS is a unique breast cancer treatment drug delivery system [195]. ZnO nanostructures (ZnO-NS), doxorubicin (DOX), folic acid (FA), near-infrared (NIR), polyethylene glycol (PEG), and lanthanum (La) are all components of ZnO-NS.

5.5. Treatment of Different Skin Conditions

ZnO is frequently used to treat skin diseases, including diaper rashes, and in shampoos, anti-dandruff treatments, hemimorphite creams, and antibacterial ointments. Additionally, it is a component of tape that athletes use as a bandage to prevent soft-tissue injuries during workouts [196]. It is possible to use ZnO-NPs in the form of an ointment, cream, or lotion to guard against UV-induced skin damage and the resulting sunburn. Only this UVA/UVB reflector, which is entirely photo-stable, has been authorized for use as a sunscreen [197]. As a sunscreen component, ZnO inhibits all UV-A (320–400 nm) and UV-B wavelengths' ultraviolet radiation. Additionally, ZnO-NPs are considered to be common diverse conventional physical sun blocks, protecting pigments and area units that need to be free from irritations, allergens, and acne-causing properties [198].

5.6. Drug Delivery

Among several nanotechnology implementations, drug delivery via ZnO-NPs has developed into a highly effective method for treating various disorders such as cancer [199,200]. Nanomaterials are one of the essential mechanisms in the delivery of drugs. ZnO-NPs have been used for drug delivery for multiple diseases [199,200]. ZnO quantum dots were employed in a study by Yuan et al. to administer doxorubicin to HeLa cells [201]. ZnO-NPs were stabilized by encasing them in chitosan. According to the results of their study, this drug delivery method could be utilized to target cancer cells with doxorubicin [201]. It is also important to note that one of the primary uses of NPs is the transport of genetic material to distinct cells, particularly tumor cells [200]. This technology for gene delivery has several benefits. For instance, the appearance of a plasmid-encoded gene on NPs' surfaces could assure reliable and effective gene delivery to the receiving tissues [199,200].

Consequently, NPs can be an effective instrument for directing genes to various cells, including tumor cells. Nie et al. [202] reported that they had created ZnO tetrapod-like nanostructures that might be employed as innovative gene-delivery vectors. They revealed that ZnO-NSs, such as a silica-coated amino-modified tetra pod, could bind effectively to DNA through electrostatic interactions, potentially increasing the efficacy of melanoma cell transfection [202]. In another investigation, Zhang et al. [203] showed that polycation-capped ZnO quantum dots might transmit DNA into COS-7 cells. Additionally, the usage of this method allows for the instantaneous visualization of gene delivery [203]. Several investigations have employed metal oxide NPs for gene silencing and gene delivery. However, it is vital that further knowledge be obtained [199,200]. ZnO-NPs-based drug delivery methods are shown in Table 8.

Table 8. ZnO-NPs-based drug delivery methods.

Materials	Cell Line	Drug	References
ZnO (Tetrapod)	CHO-K1, HeLa, Vero, VK2/E6	-	[204]
ZnO@PMAA-co-PDMAEMA-NPs	COS-7	DNA	[205]
ZnO/Carboxymethyl Cellulose (CMC)	L929, MA104	Curcumin	[206]
Curcumin/O-CMCS/n-ZnO nanocomposites	MA 104	Curcumin	[207]
Mesoporous ZnO	-	DOX	[208]
ZnO@PNIPAM-NPs	-	DOX	[209]
ZnO-NPs	T47D	PPDME	[210]
ZnO-NPs	HeLa	DOX	[211]
ZnO/PEG-NPs	Gram-positive microorganisms	DOX	[212]
ZnO/Au-NPs	Hela	Camptothecin	[213]
ZnO-QDs	HepG2	-	[214]
Chitosan/ZnO-NPs	-	DOX	[201]
ZnO	cancerous T, activated human T	-	[215]
ZnO@Polymer-NPs	U251	DOX	[216]

5.7. Bioimaging

ZnO is a common semiconductor material that can completely replace the typical Cd-related species found in biological and optical environments [199,217]. At this point, a variety of ZnO-NPs types have been identified. The bioimaging potential of ZnO-NPs is intriguing to researchers [199,217]. A wide range of biological and medicinal uses are possible for this feature. For instance, luminous ZnO-NPs may have excellent photophysical qualities [199,217]. The surfaces of these NPs have been demonstrated to be easily manipulated. For ZnO-NPs, it has been discovered that their quantum yield (QY) may be increased to about 30% following careful tweaking [199,200,217]. According to the common consensus, ZnO is a safe material. ZnO has been used in sunblock goods and in diet packing as a food preservative. This means that many biological and medicinal applications could use the luminous features of ZnO-NPs [199,200,217]. The bioimaging uses of ZnO-NPs are shown in Table 9.

Table 9. Bioimaging Uses of ZnO-NPs.

Model	Type of Material	Size (nm)	Reference
Skin tissue/cellular architecture	ZnO-NPs	15–30	[218]
KB cells	ZnO Nanocrystals	<100	[219]
S. oneidensis	CdSe(S)/ZnO-QDs	2–4	[220]
Human skin and rat liver cells	ZnO-NPs	26–30	[221]
Plants tissues cell implosion	ZnO-NPs	2–200	[222]
Blood cells of zebrafish; roots and shoots of *Arabidopsis* plants	ZnO-NPs	10–300	[223]
Hela cells	ZnO@silica-NPs	2–5	[224]
Skin	ZnO-NPs	21	[225]
B16F10 cells	ZnO/Au@PEG-NPs	45–98	[226]

6. Toxicity Associated with ZnO-NPs

ZnO is a nanomaterial that is widely employed in a variety of applications [227]. Using a well-known photocatalyst, the degradation of environmental pollutants has garnered considerable attention from researchers [19]. Zinc salts have been utilized as an active ingredient in lubricants for a long time [228] and used by the pharmaceutical industry to make emollients [229]. In wound care, anti-infection therapeutic goods, and disinfectants, ZnO-NPs containing medicines are extensively employed. ZnO-NPs have many applications in cosmetics, hair and skincare formulations, protective sunblocks, food additives, and vitamins, among others [230,231]. ZnO is used as an antibacterial compound commonly used in lotions, ointments, body washes, and surface coatings to prevent the growth of microorganisms [146,232]. As nutritional supplements, ZnO-NPs have also been utilized by humans and livestock to stimulate the body's reaction to inflammation and to enhance the immune system [233]. The expanding use of ZnO-NPs in consumer goods and pharmaceuticals has prompted researchers to look into the potentially hazardous consequences of ZnO-NPs for human health [230]. The advantages must be carefully balanced against the potential disadvantages of other NPs.

According to the available research, the inhalation of ZnO-NPs has the most harmful effects on human lungs [230,234]. The size and surface area of ZnO-NPs have been linked to the severity of inflammatory illness caused by their exposure [235]. Previous research has shown that ZnO-NPs elicit a more severe inflammatory response than liquid zinc ions [236]. In various investigations, ZnO-NPs' cytotoxic characteristics have been tested on human red and white blood cells. A cytotoxic effect has been seen at concentrations more than 50 ppm, likely due to increased oxidative stress [237,238]. At more significant concentrations than predicted in the environment, ZnO-NPs can produce acute impacts on fish [239]. Therefore, a thorough evaluation of ZnO-NPs' characteristics, routes of administration, target cells, and related physiological processes is required to better understand the therapeutic advantages and to minimize unwanted harmful consequences and negative clinical diagnostic potential. Long-term effects must still be investigated for the better and safer use of these NPs.

7. Conclusions and Future Perspectives

Because of its environmentally friendly nature, the green synthesis of ZnO-NPs is favored. The use of diverse plant components, bacteria, fungi, and algae to synthesize ZnO-NPs is an efficient, simple, and environmentally friendly approach. Plant extracts contain a variety of biomolecules that act as reducing, capping, and stabilizing agents, including amino acids, proteins, and a variety of additional primary and secondary metabolites that serve as reducing, capping, and stabilizing agents during the synthesis process.

The synthesis of these critical nanomaterials has some risks for the environment and for civilization. As a result, the biological qualities of these materials are directly affected. The use of biomolecules and living organisms as nanomaterials' capping agents in green nanotechnology is a powerful option as a potential solution to minimize the development of toxic products and undesirable reactions with various biologic membranes. NPs biogenesis with minimal impact on the environment has been the focus of research for the past decade. These NPs can be precisely sized and shaped using green synthesis methods. Medical practitioners are increasingly using antimicrobial NPs bandages. Medicine delivery and clinical diagnostics have produced a growing demand for these technologies. A rising number of people are interested in environmentally friendly nanomaterials such as ZnO-NPs, which can be produced with minimal danger and expense. Green synthesis technologies appear to be increasing in popularity in recent years. ZnO-NPs generated from plants may be an essential research topic in the biomedical sectors. The green synthesis of ZnO-NPs using plants and microbes has been highlighted in this review, as it is a rapid, simple, environmentally friendly, and relatively low-cost process. Biosynthesized ZnO-NPs for biomedical applications, especially against pathogenic germs, have also been addressed, to overcome the limitations of conventional chemical and physical methods. The biological source affects the size of ZnO-NPs and, consequently, their biological activities. However, additional study is needed to standardize synthesis procedures, as a critical limitation of green chemistry is the variability of the end products. Further in vitro and in vivo experiments are expected to elucidate the mechanism of action involved at the cellular level, with applications in various biomedical fields.

Author Contributions: Conceptualization, F.I. and S.S.; investigation and resources, F.I., M.J.U. and S.C.; writing—original draft preparation, F.I., S.S., M.J.U., M.R.I. and M.H.N.; writing—review, updating, and editing, A.A., S.M., A.R. and T.B.E.; visualization and supervision, A.A., S.C. and T.B.E.; formal analysis, S.M., A.R. and A.A.; resources, F.I. and M.J.U.; project administration, A.A. and T.B.E.; funding acquisition, S.C. All authors have read and agreed to the published version of the manuscript.

Funding: This research received no external funding.

Institutional Review Board Statement: Not applicable.

Informed Consent Statement: Not applicable.

Data Availability Statement: Available data are presented in the manuscript.

Conflicts of Interest: The authors declare no conflict of interest.

References

1. Fouad, H.; Yang, G.; El-Sayed, A.A.; Mao, G.; Khalafallah, D.; Saad, M.; Ga'al, H.; Ibrahim, E.; Mo, J. Green synthesis of AgNP-ligand complexes and their toxicological effects on Nilaparvata lugens. *J. Nanobiotechnol.* **2021**, *19*, 318. [CrossRef] [PubMed]
2. Amendola, V.; Amans, D.; Ishikawa, Y.; Koshizaki, N.; Scirè, S.; Compagnini, G.; Reichenberger, S.; Barcikowski, S. Room-Temperature Laser Synthesis in Liquid of Oxide, Metal-Oxide Core-Shells, and Doped Oxide Nanoparticles. *Chemistry* **2020**, *26*, 9206–9242. [CrossRef] [PubMed]
3. Faisal, S.; Jan, H.; Shah, S.A.; Shah, S.; Khan, A.; Akbar, M.T.; Rizwan, M.; Jan, F.; Wajidullah; Akhtar, N.; et al. Green Synthesis of Zinc Oxide (ZnO) Nanoparticles Using Aqueous Fruit Extracts of Myristica fragrans: Their Characterizations and Biological and Environmental Applications. *ACS Omega* **2021**, *6*, 9709–9722. [CrossRef] [PubMed]
4. Długosz, O.; Szostak, K.; Staroń, A.; Pulit-Prociak, J.; Banach, M. Methods for Reducing the Toxicity of Metal and Metal Oxide NPs as Biomedicine. *Materials* **2020**, *13*, 279. [CrossRef]
5. Suresh, J.; Pradheesh, G.; Alexramani, V.; Sundrarajan, M.; Hong, S.I. Green synthesis and characterization of zinc oxide nanoparticle using insulin plant (Costus pictus D. Don) and investigation of its antimicrobial as well as anticancer activities. *Adv. Nat. Sci. Nanosci. Nanotechnol.* **2018**, *9*, 015008. [CrossRef]
6. Fadeeva, I.V.; Goldberg, M.A.; Preobrazhensky, I.I.; Mamin, G.V.; Davidova, G.A.; Agafonova, N.V.; Fosca, M.; Russo, F.; Barinov, S.M.; Cavalu, S.; et al. Improved cytocompatibility and antibacterial properties of zinc-substituted brushite bone cement based on β-tricalcium phosphate. *J. Mater. Sci. Mater. Med.* **2021**, *32*, 99. [CrossRef] [PubMed]
7. Jafarirad, S.; Mehrabi, M.; Divband, B.; Kosari-Nasab, M. Biofabrication of zinc oxide nanoparticles using fruit extract of Rosa canina and their toxic potential against bacteria: A mechanistic approach. *Mater. Sci. Eng. C* **2016**, *59*, 296–302. [CrossRef]
8. Huang, Z.; Pan, C.; Huang, P.; Si, P.; Wu, W.; Xu, C.; Zhou, J.; Li, X. Effects of ZnO nanoparticles on the microstructure, mechanical properties and wettability of polypyrrole–polydopamine nanocomposites coated on W substrate. *Mater. Today Commun.* **2021**, *28*, 102620. [CrossRef]

9. Waclawik, E.R.; Chang, J.; Ponzoni, A.; Concina, I.; Zappa, D.; Comini, E.; Motta, N.; Faglia, G.; Sberveglieri, G. Functionalised zinc oxide nanowire gas sensors: Enhanced NO(2) gas sensor response by chemical modification of nanowire surfaces. *Beilstein J. Nanotechnol.* **2012**, *3*, 368–377. [CrossRef]
10. Xu, J.; Pan, Q.; Shun, Y.; Tian, Z. Grain size control and gas sensing properties of ZnO gas sensor. *Sens. Actuators B Chem.* **2000**, *66*, 277–279. [CrossRef]
11. Fritea, L.; Banica, F.; Costea, T.O.; Moldovan, L.; Dobjanschi, L.; Muresan, M.; Cavalu, S. Metal Nanoparticles and Carbon-Based Nanomaterials for Improved Performances of Electrochemical (Bio)Sensors with Biomedical Applications. *Materials* **2021**, *14*, 6319. [CrossRef] [PubMed]
12. Cross, S.E.; Innes, B.; Roberts, M.S.; Tsuzuki, T.; Robertson, T.A.; McCormick, P. Human skin penetration of sunscreen nanoparticles: In-vitro assessment of a novel micronized zinc oxide formulation. *Skin Pharmacol. Physiol.* **2007**, *20*, 148–154. [CrossRef] [PubMed]
13. Grigorjeva, L.; Millers, D.; Grabis, J.; Monty, C.; Kalinko, A.; Smits, K.; Pankratov, V.; Łojkowski, W. Luminescence Properties of ZnO Nanocrystals and Ceramics. *IEEE Trans. Nucl. Sci.* **2008**, *55*, 1551–1555. [CrossRef]
14. Ko, Y.H.; Lee, S.H.; Yu, J.S. Zinc Oxide Nanostructures for Optoelectronic and Energy Devices. Available online: https://spie.org/news/5270-zinc-oxide-nanostructures-for-optoelectronic-and-energy-devices (accessed on 23 December 2021).
15. Gordillo, G. New materials used as optical window in thin film solar cells. *Surf. Rev. Lett.* **2012**, *9*, 1675–1680. [CrossRef]
16. Thirumal, S.; Senthilkumar, S.R.; Sivakumar, T. Green tea (Camellia sinensis) mediated synthesis of zinc oxide (ZnO) nanoparticles and studies on their antimicrobial activities. *Artic. Int. J. Pharm. Pharm. Sci.* **2014**, *6*, 461–465.
17. Cavalu, S.; Antoniac, I.V.; Mohan, A.; Bodog, F.; Doicin, C.; Mates, I.; Ulmeanu, M.; Murzac, R.; Semenescu, A. Nanoparticles and Nanostructured Surface Fabrication for Innovative Cranial and Maxillofacial Surgery. *Materials* **2020**, *13*, 5391. [CrossRef]
18. Beek, W.J.E.; Wienk, M.M.; Janssen, R.A.J. Efficient hybrid solar cells from zinc oxide nanoparticles and a conjugated polymer. *Adv. Mater.* **2004**, *16*, 1009–1013. [CrossRef]
19. Rajamanickam, D.; Shanthi, M. Photocatalytic degradation of an organic pollutant by zinc oxide—Solar process. *Arab. J. Chem.* **2016**, *9*, S1858–S1868. [CrossRef]
20. Farouk, A.; Textor, T.; Schollmeyer, E.; Tarbuk, A.; Grancacic, A.M. Sol-gel-derived inorganic-organic hybrid polymers filled with ZnO nanoparticles as an ultraviolet protection finish for textiles. *Autex Res. J.* **2010**, *10*, 58–63.
21. Przybyszewska, M.; Zaborski, M. Effect of ionic liquids and surfactants on zinc oxide nanoparticle activity in crosslinking of acrylonitrile butadiene elastomer. *J. Appl. Polym. Sci.* **2010**, *116*, 155–164. [CrossRef]
22. Padmavathy, N.; Vijayaraghavan, R. Enhanced bioactivity of ZnO nanoparticles-an antimicrobial study. *Sci. Technol. Adv. Mater.* **2008**, *9*, 035004. [CrossRef] [PubMed]
23. Hasnidawani, J.N.; Azlina, H.N.; Norita, H.; Bonnia, N.N.; Ratim, S.; Ali, E.S. Synthesis of ZnO Nanostructures Using Sol-Gel Method. *Procedia Chem.* **2016**, *19*, 211–216. [CrossRef]
24. Prakash, V.; Niyogi, U.K.; Diwan, R.K. Characterization of synthesized copper oxide nanopowders and their use in nanofluids for enhancement of thermal conductivity. *Indian J. Pure Appl. Phys.* **2015**, *53*, 753–758.
25. Suntako, R. Effect of zinc oxide nanoparticles synthesized by a precipitation method on mechanical and morphological properties of the CR foam. *Bull. Mater. Sci.* **2015**, *38*, 1033–1038. [CrossRef]
26. Vijayakumar, S.; Mahadevan, S.; Arulmozhi, P.; Sriram, S.; Praseetha, P.K. Green synthesis of zinc oxide nanoparticles using Atalantia monophylla leaf extracts: Characterization and antimicrobial analysis. *Mater. Sci. Semicond. Process.* **2018**, *82*, 39–45. [CrossRef]
27. Sawai, J.; Yoshikawa, T. Quantitative evaluation of antifungal activity of metallic oxide powders (MgO, CaO and ZnO) by an indirect conductimetric assay. *J. Appl. Microbiol.* **2004**, *96*, 803–809. [CrossRef] [PubMed]
28. Tirla, A.; Vesa, C.M.; Cavalu, S. Severe Cardiac and Metabolic Pathology Induced by Steroid Abuse in a Young Individual. *Diagnostics* **2021**, *11*, 1313. [CrossRef]
29. Parthiban, C.; Sundaramurthy, N. Biosynthesis, Characterization of ZnO Nanoparticles by Using Pyrus Pyrifolia Leaf Extract and Their Photocatalytic Activity. *Int. J. Innov. Res. Sci. Eng. Technol.* **2015**, *4*, 9710–9718.
30. Sundrarajan, M.; Ambika, S.; Bharathi, K. Plant-extract mediated synthesis of ZnO nanoparticles using Pongamia pinnata and their activity against pathogenic bacteria. *Adv. Powder Technol.* **2015**, *26*, 1294–1299. [CrossRef]
31. Lingaraju, K.; Raja Naika, H.; Manjunath, K.; Basavaraj, R.B.; Nagabhushana, H.; Nagaraju, G.; Suresh, D. Biogenic synthesis of zinc oxide nanoparticles using Ruta graveolens (L.) and their antibacterial and antioxidant activities. *Appl. Nanosci.* **2016**, *6*, 703–710. [CrossRef]
32. Page, M.J.; McKenzie, J.E.; Bossuyt, P.M.; Boutron, I.; Hoffmann, T.C.; Mulrow, C.D.; Shamseer, L.; Tetzlaff, J.M.; Moher, D. Updating guidance for reporting systematic reviews: Development of the PRISMA 2020 statement. *J. Clin. Epidemiol.* **2021**, *134*, 103–112. [CrossRef] [PubMed]
33. Page, M.J.; McKenzie, J.E.; Bossuyt, P.M.; Boutron, I.; Hoffmann, T.C.; Mulrow, C.D.; Shamseer, L.; Tetzlaff, J.M.; Akl, E.A.; Brennan, S.E.; et al. The PRISMA 2020 statement: An updated guideline for reporting systematic reviews. *BMJ* **2021**, *372*, n160. [CrossRef] [PubMed]
34. Cavalu, S.; Kamel, E.; Laslo, V.; Fritea, L.; Costea, T.; Antoniac, I.V.; Vasile, E.; Antoniac, A.; Semenescu, A.; Mohan, A.; et al. Eco-Friendly, Facile and Rapid Way for Synthesis of Selenium Nanoparticles Production, Structural and Morphological Characterisation. *Rev. Chim.* **2018**, *68*, 2963–2966. [CrossRef]

35. Rane, A.V.; Kanny, K.; Abitha, V.K.; Thomas, S. Methods for Synthesis of Nanoparticles and Fabrication of Nanocomposites. *Synth. Inorg. Nanomater.* **2018**, 121–139. [CrossRef]
36. Shah, M.; Fawcett, D.; Sharma, S.; Tripathy, S.K.; Poinern, G.E.J. Green Synthesis of Metallic Nanoparticles via Biological Entities. *Materials* **2015**, *8*, 7278–7308. [CrossRef] [PubMed]
37. Kolodziejczak-Radzimska, A.; Jesionowski, T. Zinc Oxide-From Synthesis to Application: A Review. *Materials* **2014**, *7*, 2833–2881. [CrossRef]
38. Nadaroğlu, H.; Alayli Güngör, A.; İnce, S. Synthesis of Nanoparticles by Green Synthesis Method. *Int. J. Innov. Res. Rev.* **2017**, *1*, 6–9.
39. Somiya, S. *Handbook of Advanced Ceramics: Materials, Applications, Processing, and Properties*, 2nd ed.; Academic Press: Cambridge, MA, USA; Elsevier: Amsterdam, The Netherlands, 2013; pp. 1–1229. [CrossRef]
40. Pooyan, S.S. Sol-gel process and its application in Nanotechnology. *J. Polym. Eng. Technol.* **2005**, *13*, 38–41.
41. Simon, V.; Cavalu, S.; Simon, S.; Mocuta, H.; Vanea, E.; Prinz, M.; Neumann, M. Surface functionalisation of sol-gel derived aluminosilicates in simulated body fluids. *Solid State Ionics* **2009**, *180*, 764–769. [CrossRef]
42. Carter, B.C.; Norton, M.G. *Ceramic Materials: Science and Engineering*; Springer Science & Business Media: Berlin/Heidelberg, Germany, 2007; 716p.
43. Al Abdullah, K.; Awad, S.; Zaraket, J.; Salame, C. Synthesis of ZnO Nanopowders By Using Sol-Gel and Studying Their Structural and Electrical Properties at Different Temperature. *Energy Procedia* **2017**, *119*, 565–570. [CrossRef]
44. Jurablu, S.; Farahmandjou, M.; Firoozabadi, T.P. Sol-Gel Synthesis of Zinc Oxide (ZnO) Nanoparticles: Study of Structural and Optical Properties. *J. Sci. Islam. Repub. Iran* **2015**, *26*, 281–285.
45. Alwan, R.M.; Kadhim, Q.A.; Sahan, K.M.; Ali, R.A.; Mahdi, R.J.; Kassim, N.A.; Jassim, A.N. Synthesis of Zinc Oxide Nanoparticles via Sol—Gel Route and Their Characterization. *Nanosci. Nanotechnol.* **2015**, *5*, 1–6. [CrossRef]
46. Yang, G.; Park, S.J. Conventional and Microwave Hydrothermal Synthesis and Application of Functional Materials: A Review. *Materials* **2019**, *12*, 1177. [CrossRef] [PubMed]
47. Komarneni, S. Nanophase Materials by Hydrothermal, Microwave-Hydrothermal and Microwave-Solvothermal Methods. Available online: https://www.jstor.org/stable/24109979 (accessed on 24 December 2021).
48. Cavalu, S.; Antoniac, I.V.; Fritea, L.; Mates, I.M.; Milea, C.; Laslo, V.; Vicas, S.; Mohan, A. Surface Modifications of the Titanium Mesh for Cranioplasty Using Selenium Nanoparticles Coating. *J. Adhes. Sci. Technol.* **2018**, *32*, 2509–2522. [CrossRef]
49. Parhi, P.; Kramer, J.; Manivannan, V. Microwave initiated hydrothermal synthesis of nano-sized complex fluorides, KMF3 (K = Zn, Mn, Co, and Fe). *J. Mater. Sci.* **2008**, *43*, 5540–5545. [CrossRef]
50. Sonawane, G.H.; Patil, S.P.; Sonawane, S.H. Nanocomposites and Its Applications. *Appl. Nanomater.* **2018**, *2018*, 1–22. [CrossRef]
51. Bharti, D.B.; Bharati, A.V. Synthesis of ZnO nanoparticles using a hydrothermal method and a study its optical activity. *Luminescence* **2017**, *32*, 317–320. [CrossRef]
52. Wirunmongkol, T.; O-Charoen, N.; Pavasupree, S. Simple Hydrothermal Preparation of Zinc Oxide Powders Using Thai Autoclave Unit. *Energy Procedia* **2013**, *34*, 801–807. [CrossRef]
53. Pudovkin, M.S.; Zelenikhin, P.V.; Shtyreva, V.; Morozov, O.A.; Koryakovtseva, D.A.; Pavlov, V.V.; Osin, Y.N.; Evtugyn, V.G.; Akhmadeev, A.A.; Nizamutdinov, A.S.; et al. Coprecipitation Method of Synthesis, Characterization, and Cytotoxicity of Pr3+:LaF3 (CPr = 3, 7, 12, 20, 30%) Nanoparticles. *J. Nanotechnol.* **2018**, *2018*, 8516498. [CrossRef]
54. Mostafavi, E.; Babaei, A.; Ataie, A. Synthesis of Nano-Structured La0.6Sr0.4Co0.2Fe0.8O3 Perovskite by Co-Precipitation Method. *J. Ultrafine Grained Nanostructured Mater.* **2015**, *48*, 45–52. [CrossRef]
55. Wang, B.; Wei, Q.; Qu, S. Synthesis and Characterization of Uniform and Crystalline Magnetite Nanoparticles via Oxidation-precipitation and Modified co-precipitation Methods. *Int. J. Electrochem. Sci.* **2013**, *8*, 3786–3793.
56. Costenaro, D.; Carniato, F.; Gatti, G.; Marchese, L.; Bisio, C. Preparation of luminescent ZnO nanoparticles modified with aminopropyltriethoxy silane for optoelectronic applications. *New J. Chem.* **2013**, *37*, 2103–2109. [CrossRef]
57. Purwaningsih, S.Y.; Pratapa, S.; Triwikantoro; Darminto. Synthesis of nano-sized ZnO particles by co-precipitation method with variation of heating time. *AIP Conf. Proc.* **2016**, *1710*, 030040. [CrossRef]
58. Adam, R.E.; Pozina, G.; Willander, M.; Nur, O. Synthesis of ZnO nanoparticles by co-precipitation method for solar driven photodegradation of Congo red dye at different pH. *Photonics Nanostruct.-Fundam. Appl.* **2018**, *32*, 11–18. [CrossRef]
59. Wang, Y.; Zhang, X.; Wang, A.; Li, X.; Wang, G.; Zhao, L. Synthesis of ZnO nanoparticles from microemulsions in a flow type microreactor. *Chem. Eng. J.* **2014**, *235*, 191–197. [CrossRef]
60. Li, X.; He, G.; Xiao, G.; Liu, H.; Wang, M. Synthesis and morphology control of ZnO nanostructures in microemulsions. *J. Colloid Interface Sci.* **2009**, *333*, 465–473. [CrossRef] [PubMed]
61. Mintcheva, N.; Aljulaih, A.A.; Wunderlich, W.; Kulinich, S.A.; Iwamori, S. Laser-Ablated ZnO Nanoparticles and Their Photocatalytic Activity toward Organic Pollutants. *Materials* **2018**, *11*, 1127. [CrossRef]
62. Amendola, V.; Meneghetti, M. Laser ablation synthesis in solution and size manipulation of noble metal nanoparticles. *Phys. Chem. Chem. Phys.* **2009**, *11*, 3805–3821. [CrossRef]
63. Al-Dahash, G.; Mubder Khilkala, W.; Abd Alwahid, S.N. Preparation and Characterization of ZnO Nanoparticles by Laser Ablation in NaOH Aqueous Solution. *Iran. J. Chem. Chem. Eng.* **2018**, *37*, 11–16. [CrossRef]
64. Farahani, S.V.; Mahmoodi, A.; Goranneviss, M. The effect of laser environment on the characteristics of ZnO nanoparticles by laser ablation. *Int. Nano Lett.* **2015**, *6*, 45–49. [CrossRef]

65. Hodaei, A.; Ataie, A.; Mostafavi, E. Intermediate milling energy optimization to enhance the characteristics of barium hexaferrite magnetic nanoparticles. *J. Alloys Compd.* **2015**, *640*, 162–168. [CrossRef]
66. Piras, C.C.; Fernández-Prieto, S.; De Borggraeve, W.M. Ball milling: A green technology for the preparation and functionalisation of nanocellulose derivatives. *Nanoscale Adv.* **2019**, *1*, 937–947. [CrossRef]
67. Yang, L. Nanotechnology-enhanced metals and alloys for orthopedic implants. *Nanotechnol.-Enhanced Orthop. Mater.* **2015**, *2015*, 27–47. [CrossRef]
68. Edelstein, A.S. Nanomaterials. In *Encyclopedia of Materials: Science and Technology*, 2nd ed.; Naval Research Laboratory: Washington, DC, USA, 2001; pp. 5916–5927. [CrossRef]
69. Dargatz, B.; Gonzalez-Julian, J.; Guillon, O.; Rinaldi, A.; Prommalikit, C.; Mekprasart, W.; Pecharapa, W. Effect of Milling Speed and Time on Ultrafine ZnO Powder by High Energy Ball Milling Technique. *J. Phys. Conf. Ser.* **2019**, *1259*, 012023. [CrossRef]
70. Mohammadi, N.; Mirhosseini, M.; Shirzad, M.; Dehghan Hamdan, A.; Yazdani, N. Synthesizing Zno Nanoparticles by High-Energy Milling and Investigating Their Antimicrobial Effect. *SSU_Journals* **2015**, *23*, 2070–2082.
71. Heinlaan, M.; Ivask, A.; Blinova, I.; Dubourguier, H.C.; Kahru, A. Toxicity of nanosized and bulk ZnO, CuO and TiO2 to bacteria Vibrio fischeri and crustaceans Daphnia magna and Thamnocephalus platyurus. *Chemosphere* **2008**, *71*, 1308–1316. [CrossRef]
72. Qu, J.; Yuan, X.; Wang, X.; Shao, P. Zinc accumulation and synthesis of ZnO nanoparticles using *Physalis alkekengi* L. *Environ. Pollut.* **2011**, *159*, 1783–1788. [CrossRef]
73. Qu, J.; Luo, C.; Hou, J. Synthesis of ZnO nanoparticles from Zn-hyperaccumulator (Sedum alfredii Hance) plants. *Micro Nano Lett.* **2011**, *6*, 174–176. [CrossRef]
74. Ochieng, P.E.; Iwuoha, E.; Michira, I.; Masikini, M.; Ondieki, J.; Githira, P.; Kamau, G.N. Green Route Synthesis and Characterization of ZnO Nanoparticles using spathodea campanulata. *Int. J. Biochem.* **2015**, *23*, 53–61.
75. Rajeshkumar, S.; Malarkodi, C.; Vanaja, M.; Annadurai, G. Anticancer and enhanced antimicrobial activity of biosynthesizd silver nanoparticles against clinical pathogens. *J. Mol. Struct.* **2016**, *1116*, 165–173. [CrossRef]
76. Yasmin, A.; Ramesh, K.; Rajeshkumar, S. Optimization and stabilization of gold nanoparticles by using herbal plant extract with microwave heating. *Nano Converg.* **2014**, *1*, 1–7. [CrossRef] [PubMed]
77. Arfat, Y.A.; Benjakul, S.; Prodpran, T.; Sumpavapol, P.; Songtipya, P. Properties and antimicrobial activity of fish protein isolate/fish skin gelatin film containing basil leaf essential oil and zinc oxide nanoparticles. *Food Hydrocoll.* **2014**, *41*, 265–273. [CrossRef]
78. Vickers, N.J. Animal Communication: When I'm Calling You, Will You Answer Too? *Curr. Biol.* **2017**, *27*, R713–R715. [CrossRef] [PubMed]
79. Anbuvannan, M.; Ramesh, M.; Viruthagiri, G.; Shanmugam, N.; Kannadasan, N. Anisochilus carnosus leaf extract mediated synthesis of zinc oxide nanoparticles for antibacterial and photocatalytic activities. *Mater. Sci. Semicond. Process.* **2015**, *39*, 621–628. [CrossRef]
80. Fu, L.; Fu, Z. Plectranthus amboinicus leaf extract-assisted biosynthesis of ZnO nanoparticles and their photocatalytic activity. *Ceram. Int.* **2015**, *41*, 2492–2496. [CrossRef]
81. Rancan, F.; Wiehe, A.; Nöbel, M.; Senge, M.O.; Al Omari, S.; Böhm, F.; John, M.; Röder, B. Influence of substitutions on asymmetric dihydroxychlorins with regard to intracellular uptake, subcellular localization and photosensitization of Jurkat cells. *J. Photochem. Photobiol. B Biol.* **2005**, *78*, 17–28. [CrossRef]
82. Bhuyan, T.; Mishra, K.; Khanuja, M.; Prasad, R.; Varma, A. Biosynthesis of zinc oxide nanoparticles from Azadirachta indica for antibacterial and photocatalytic applications. *Mater. Sci. Semicond. Process.* **2015**, *32*, 55–61. [CrossRef]
83. Madan, H.R.; Sharma, S.C.; Udayabhanu; Suresh, D.; Vidya, Y.S.; Nagabhushana, H.; Rajanaik, H.; Anantharaju, K.S.; Prashantha, S.C.; Sadananda Maiya, P. Facile green fabrication of nanostructure ZnO plates, bullets, flower, prismatic tip, closed pine cone: Their antibacterial, antioxidant, photoluminescent and photocatalytic properties. *Spectrochim. Acta Part A Mol. Biomol. Spectrosc.* **2016**, *152*, 404–416. [CrossRef]
84. Qian, Y.; Yao, J.; Russel, M.; Chen, K.; Wang, X. Characterization of green synthesized nano-formulation (ZnO–A. vera) and their antibacterial activity against pathogens. *Environ. Toxicol. Pharmacol.* **2015**, *39*, 736–746. [CrossRef]
85. Ali, K.; Dwivedi, S.; Azam, A.; Saquib, Q.; Al-Said, M.S.; Alkhedhairy, A.A.; Musarrat, J. Aloe vera extract functionalized zinc oxide nanoparticles as nanoantibiotics against multi-drug resistant clinical bacterial isolates. *J. Colloid Interface Sci.* **2016**, *472*, 145–156. [CrossRef]
86. Agarwal, H.; Venkat Kumar, S.; Rajeshkumar, S. A review on green synthesis of zinc oxide nanoparticles—An eco-friendly approach. *Resour. Technol.* **2017**, *3*, 406–413. [CrossRef]
87. Nagajyothi, P.C.; Sreekanth, T.V.M.; Tettey, C.O.; Jun, Y.I.; Mook, S.H. Characterization, antibacterial, antioxidant, and cytotoxic activities of ZnO nanoparticles using Coptidis Rhizoma. *Bioorg. Med. Chem. Lett.* **2014**, *24*, 4298–4303. [CrossRef] [PubMed]
88. Singh, A.; Neelam; Kaushik, M. Physicochemical investigations of zinc oxide nanoparticles synthesized from Azadirachta Indica (Neem) leaf extract and their interaction with Calf-Thymus DNA. *Results Phys.* **2019**, *13*, 102168. [CrossRef]
89. Malaikozhundan, B.; Vinodhini, J. Nanopesticidal effects of Pongamia pinnata leaf extract coated zinc oxide nanoparticle against the Pulse beetle, Callosobruchus maculatus. *Mater. Today Commun.* **2018**, *14*, 106–115. [CrossRef]
90. Abbes, N.; Bekri, I.; Cheng, M.; Sejri, N.; Cheikhrouhou, M.; Xu, J. Green Synthesis and Characterization of Zinc Oxide Nanoparticles Using Mulberry Fruit and Their Antioxidant Activity. *Mater. Sci.* **2021**, *2021*, 1–7. [CrossRef]

91. Wary, R.R.; Baglari, S.; Brahma, D.; Gautam, U.K.; Kalita, P.; Baruah, M.B. Synthesis, characterization, and photocatalytic activity of ZnO nanoparticles using water extract of waste coconut husk. *Environ. Sci. Pollut. Res. Int.* **2022**. [CrossRef]
92. Thema, F.T.; Manikandan, E.; Dhlamini, M.S.; Maaza, M. Green synthesis of ZnO nanoparticles via Agathosma betulina natural extract. *Mater. Lett.* **2015**, *161*, 124–127. [CrossRef]
93. Dobrucka, R.; Długaszewska, J. Biosynthesis and antibacterial activity of ZnO nanoparticles using Trifolium pratense flower extract. *Saudi J. Biol. Sci.* **2016**, *23*, 517–523. [CrossRef]
94. Akintelu, S.A.; Folorunso, A.S. A Review on Green Synthesis of Zinc Oxide Nanoparticles Using Plant Extracts and Its Biomedical Applications. *BioNanoScience* **2020**, *10*, 848–863. [CrossRef]
95. Ramesh, M.; Anbuvannan, M.; Viruthagiri, G. Green synthesis of ZnO nanoparticles using Solanum nigrum leaf extract and their antibacterial activity. *Spectrochim. Acta Part A Mol. Biomol. Spectrosc.* **2015**, *136*, 864–870. [CrossRef]
96. Abel, S.; Tesfaye, J.L.; Nagaprasad, N.; Shanmugam, R.; Dwarampudi, L.P.; Krishnaraj, R. Synthesis and Characterization of Zinc Oxide Nanoparticles Using Moringa Leaf Extract. *J. Nanomater.* **2021**, *2021*, 4525770. [CrossRef]
97. Albukhaty, S.; Al-Karagoly, H.; Dragh, M.A. Synthesis of zinc oxide nanoparticles and evaluated its activity against bacterial isolates. *J. Biotech Res.* **2020**, *11*, 47–53.
98. Noorian, S.A.; Hemmatinejad, N.; Navarro, J.A.R. Ligand modified cellulose fabrics as support of zinc oxide nanoparticles for UV protection and antimicrobial activities. *Int. J. Biol. Macromol.* **2020**, *154*, 1215–1226. [CrossRef] [PubMed]
99. Rajiv, P.; Rajeshwari, S.; Venckatesh, R. Bio-Fabrication of zinc oxide nanoparticles using leaf extract of Parthenium hysterophorus L. and its size-dependent antifungal activity against plant fungal pathogens. *Spectrochim. Acta Part A Mol. Biomol. Spectrosc.* **2013**, *112*, 384–387. [CrossRef]
100. Zheng, Y.; Huang, Y.; Shi, H.; Fu, L. Green biosynthesis of ZnO nanoparticles by plectranthus amboinicus leaf extract and their application for electrochemical determination of norfloxacin. *Inorg. Nano-Metal Chem.* **2019**, *49*, 277–282. [CrossRef]
101. Rajashekara, S.; Shrivastava, A.; Sumhitha, S.; Kumari, S. Biomedical Applications of Biogenic Zinc Oxide Nanoparticles Manufactured from Leaf Extracts of Calotropis gigantea (L.) Dryand. *BioNanoScience* **2020**, *10*, 654–671. [CrossRef]
102. Kavithaa, K.; Paulpandi, M.; Ponraj, T.; Murugan, K.; Sumathi, S. Induction of intrinsic apoptotic pathway in human breast cancer (MCF-7) cells through facile biosynthesized zinc oxide nanorods. *Karbala Int. J. Mod. Sci.* **2016**, *2*, 46–55. [CrossRef]
103. Vaishnavi Devi, B.; Rajasekar, A.; Rajeshkumar, S. Antiinflammatory Activity of Zinc Oxide Nanoparticles Synthesised Using Grape Seed Extract: An In Vitro Study. Available online: https://www.ikppress.org/index.php/PCBMB/article/view/5364 (accessed on 24 February 2022).
104. Normah, N.; Juleanti, N.; Palapa, N.R.; Taher, T.; Siregar, P.M.S.B.N.; Wijaya, A.; Lesbani, A. Hydrothermal carbonization of rambutan peel (Nephelium lappaceum L.) as a Green and low-cost adsorbent for Fe(II) removal from aqueous solutions. *Chem. Ecol.* **2022**, *2022*, 1–17. [CrossRef]
105. Antonescu, A.-I.; Miere, F.; Fritea, L.; Ganea, M.; Zdrinca, M.; Dobjanschi, L.; Antonescu, A.; Vicas, S.I.; Bodog, F.; Sindhu, R.K.; et al. Perspectives on the Combined Effects of *Ocimum basilicum* and *Trifolium pratense* Extracts in Terms of Phytochemical Profile and Pharmacological Effects. *Plants* **2021**, *10*, 1390. [CrossRef]
106. Raliya, R.; Tarafdar, J.C. ZnO Nanoparticle Biosynthesis and Its Effect on Phosphorous-Mobilizing Enzyme Secretion and Gum Contents in Clusterbean (Cyamopsis tetragonoloba L.). *Agric. Res.* **2013**, *2*, 48–57. [CrossRef]
107. Reddy, K.O.; Maheswari, C.U.; Shukla, M.; Rajulu, A.V. Chemical composition and structural characterization of Napier grass fibers. *Mater. Lett.* **2012**, *67*, 35–38. [CrossRef]
108. Tripathi, R.M.; Bhadwal, A.S.; Gupta, R.K.; Singh, P.; Shrivastav, A.; Shrivastav, B.R. ZnO nanoflowers: Novel biogenic synthesis and enhanced photocatalytic activity. *J. Photochem. Photobiol. B Biol.* **2014**, *141*, 288–295. [CrossRef] [PubMed]
109. Mehta, S.K.; Kumar, S.; Chaudhary, S.; Bhasin, K.K. Effect of cationic surfactant head groups on synthesis, growth and agglomeration behavior of ZnS nanoparticles. *Nanoscale Res. Lett.* **2009**, *4*, 1197–1208. [CrossRef] [PubMed]
110. Kundu, D.; Hazra, C.; Chatterjee, A.; Chaudhari, A.; Mishra, S. Extracellular biosynthesis of zinc oxide nanoparticles using Rhodococcus pyridinivorans NT2: Multifunctional textile finishing, biosafety evaluation and in vitro drug delivery in colon carcinoma. *J. Photochem. Photobiol. B Biol.* **2014**, *140*, 194–204. [CrossRef] [PubMed]
111. Xiao, X.; Wu, Z.C.; Chou, K.C. A Multi-Label Classifier for Predicting the Subcellular Localization of Gram-Negative Bacterial Proteins with Both Single and Multiple Sites. *PLoS ONE* **2011**, *6*, e20592. [CrossRef]
112. Mittal, H.; Morajkar, P.P.; Al Alili, A.; Alhassan, S.M. In-Situ Synthesis of ZnO Nanoparticles using Gum Arabic Based Hydrogels as a Self-template for Effective Malachite Green Dye Adsorption. *J. Polym. Environ.* **2020**, *28*, 1637–1653. [CrossRef]
113. Khan, F.; Kang, M.G.; Jo, D.M.; Chandika, P.; Jung, W.K.; Kang, H.W.; Kim, Y.M. Phloroglucinol-Gold and -Zinc Oxide Nanoparticles: Antibiofilm and Antivirulence Activities towards Pseudomonas aeruginosa PAO1. *Mar. Drugs* **2021**, *19*, 601. [CrossRef]
114. Gudkov, S.V.; Burmistrov, D.E.; Serov, D.A.; Rebezov, M.B.; Semenova, A.A.; Lisitsyn, A.B. A Mini Review of Antibacterial Properties of ZnO Nanoparticles. *Front. Phys.* **2021**, *9*, 641481. [CrossRef]
115. Verbič, A.; Brenčič, K.; Primc, G.; Gorjanc, M. Importance of Protocol Design for Suitable Green In Situ Synthesis of ZnO on Cotton Using Aqueous Extract of Japanese Knotweed Leaves as Reducing Agent. *Forests* **2022**, *13*, 143. [CrossRef]
116. Bird, S.M.; El-Zubir, O.; Rawlings, A.E.; Leggett, G.J.; Staniland, S.S. A novel design strategy for nanoparticles on nanopatterns: Interferometric lithographic patterning of Mms6 biotemplated magnetic nanoparticles. *J. Mater. Chem. C* **2016**, *4*, 3948–3955. [CrossRef]

117. Rao, M.D.; Gautam, P. Synthesis and characterization of ZnO nanoflowers using Chlamydomonas reinhardtii: A green approach. *Environ. Prog. Sustain. Energy* **2016**, *35*, 1020–1026. [CrossRef]
118. Deekala, V.; Babu, B.K.; Rudraraju, R. Pharmacological studies of zinc oxide nanoparticles. *Indian J. Biochem. Biophys.* **2019**, *56*, 500–505.
119. Sanaeimehr, Z.; Javadi, I.; Namvar, F. Antiangiogenic and antiapoptotic effects of green-synthesized zinc oxide nanoparticles using Sargassum muticum algae extraction. *Cancer Nanotechnol.* **2018**, *9*, 3. [CrossRef] [PubMed]
120. Azizi, S.; Ahmad, M.B.; Namvar, F.; Mohamad, R. Green biosynthesis and characterization of zinc oxide nanoparticles using brown marine macroalga Sargassum muticum aqueous extract. *Mater. Lett.* **2014**, *116*, 275–277. [CrossRef]
121. Pati, R.; Mehta, R.K.; Mohanty, S.; Padhi, A.; Sengupta, M.; Vaseeharan, B.; Goswami, C.; Sonawane, A. Topical application of zinc oxide nanoparticles reduces bacterial skin infection in mice and exhibits antibacterial activity by inducing oxidative stress response and cell membrane disintegration in macrophages. *Nanomed. Nanotechnol. Biol. Med.* **2014**, *10*, 1195–1208. [CrossRef]
122. Jaidev, L.R.; Narasimha, G. Fungal mediated biosynthesis of silver nanoparticles, characterization and antimicrobial activity. *Colloids Surfaces B Biointerfaces* **2010**, *81*, 430–433. [CrossRef]
123. Hoffmann, M.R.; Martin, S.T.; Choi, W.; Bahnemann, D.W. Environmental Applications of Semiconductor Photocatalysis. *Chem. Rev.* **2002**, *95*, 69–96. [CrossRef]
124. Shamsuzzaman; Mashrai, A.; Khanam, H.; Aljawfi, R.N. Biological synthesis of ZnO nanoparticles using C. albicans and studying their catalytic performance in the synthesis of steroidal pyrazolines. *Arab. J. Chem.* **2017**, *10*, S1530–S1536. [CrossRef]
125. Najibi Ilkhechi, N.; Mozammel, M.; Yari Khosroushahi, A. Antifungal effects of ZnO, TiO 2 and ZnO-TiO 2 nanostructures on Aspergillus flavus. *Pestic. Biochem. Physiol.* **2021**, *176*, 104869. [CrossRef]
126. Chandrasekaran, R.; Gnanasekar, S.; Seetharaman, P.; Keppanan, R.; Arockiaswamy, W.; Sivaperumal, S. Formulation of Carica papaya latex-functionalized silver nanoparticles for its improved antibacterial and anticancer applications. *J. Mol. Liq.* **2016**, *219*, 232–238. [CrossRef]
127. Nagarajan, S.; Arumugam Kuppusamy, K. Extracellular synthesis of zinc oxide nanoparticle using seaweeds of gulf of Mannar, India. *J. Nanobiotechnol.* **2013**, *11*, 39. [CrossRef] [PubMed]
128. Asokan, A.; Ramachandran, T.; Ramaswamy, R.; Koushik, C.V.; Muthusamy, M. Preparation and Characterization of Zinc Oxide Nanoparticles and a Study of the Anti-microbial Property of Cotton Fabric Treated with the Particles. *J. Text. Apparel. Technol. Manag.* **2010**, *6*, 1–7.
129. Ambika, S.; Sundrarajan, M. Green biosynthesis of ZnO nanoparticles using Vitex negundo L. extract: Spectroscopic investigation of interaction between ZnO nanoparticles and human serum albumin. *J. Photochem. Photobiol. B Biol.* **2015**, *149*, 143–148. [CrossRef] [PubMed]
130. Mukhtar, S.S.; Hassan, A.S.; Morsy, N.M.; Hafez, T.S.; Hassaneen, H.M.; Saleh, F.M. Overview on Synthesis, Reactions, Applications, and Biological Activities of Schiff Bases. *Egypt. J. Chem.* **2021**, *64*, 6541–6554. [CrossRef]
131. Shehabeldine, A.M.; Hashem, A.H.; Wassel, A.R.; Hasanin, M. Antimicrobial and Antiviral Activities of Durable Cotton Fabrics Treated with Nanocomposite Based on Zinc Oxide Nanoparticles, Acyclovir, Nanochitosan, and Clove Oil. *Appl. Biochem. Biotechnol.* **2021**, *194*, 783–800. [CrossRef]
132. Burgess, R. Medical applications of nanoparticles and nanomaterials. *Stud. Health Technol. Inform.* **2009**, *149*, 257–283. [CrossRef]
133. Salata, O.V. Applications of nanoparticles in biology and medicine. *J. Nanobiotechnol.* **2004**, *2*, 3. [CrossRef]
134. Zhang, L.; Gu, F.X.; Chan, J.M.; Wang, A.Z.; Langer, R.S.; Farokhzad, O.C. Nanoparticles in medicine: Therapeutic applications and developments. *Clin. Pharmacol. Ther.* **2008**, *83*, 761–769. [CrossRef]
135. Newman, M.D.; Stotland, M.; Ellis, J.I. The safety of nanosized particles in titanium dioxide- and zinc oxide-based sunscreens. *J. Am. Acad. Dermatol.* **2009**, *61*, 685–692. [CrossRef]
136. Zhang, L.; Jiang, Y.; Ding, Y.; Daskalakis, N.; Jeuken, L.; Povey, M.; O'Neill, A.J.; York, D.W. Mechanistic investigation into antibacterial behaviour of suspensions of ZnO nanoparticles against E. coli. *J. Nanoparticle Res.* **2010**, *12*, 1625–1636. [CrossRef]
137. Huang, Z.; Zheng, X.; Yan, D.; Yin, G.; Liao, X.; Kang, Y.; Yao, Y.; Huang, D.; Hao, B. Toxicological effect of ZnO nanoparticles based on bacteria. *Langmuir* **2008**, *24*, 4140–4144. [CrossRef] [PubMed]
138. Limbach, L.K.; Wick, P.; Manser, P.; Grass, R.N.; Bruinink, A.; Stark, W.J. Exposure of engineered nanoparticles to human lung epithelial cells: Influence of chemical composition and catalytic activity on oxidative stress. *Environ. Sci. Technol.* **2007**, *41*, 4158–4163. [CrossRef] [PubMed]
139. Husseiny, M.I.; El-Aziz, M.A.; Badr, Y.; Mahmoud, M.A. Biosynthesis of gold nanoparticles using Pseudomonas aeruginosa. *Spectrochim. Acta-Part A Mol. Biomol. Spectrosc.* **2007**, *67*, 1003–1006. [CrossRef] [PubMed]
140. Cleetus, C.M.; Primo, F.A.; Fregoso, G.; Raveendran, N.L.; Noveron, J.C.; Spencer, C.T.; Ramana, C.V.; Joddar, B. Alginate Hydrogels with Embedded ZnO Nanoparticles for Wound Healing Therapy. *Int. J. Nanomed.* **2020**, *15*, 5097–5111. [CrossRef] [PubMed]
141. Kumar, R.; Singh, R.; Singh, M.; Kumar, P. ZnO nanoparticle-grafted PLA thermoplastic composites for 3D printing applications: Tuning of thermal, mechanical, morphological and shape memory effect. *J. Thermoplast. Compos. Mater.* **2020**. [CrossRef]
142. Premanathan, M.; Karthikeyan, K.; Jeyasubramanian, K.; Manivannan, G. Selective toxicity of ZnO nanoparticles toward Gram-positive bacteria and cancer cells by apoptosis through lipid peroxidation. *Nanomed. Nanotechnol. Biol. Med.* **2011**, *7*, 184–192. [CrossRef]

143. Sangeetha, G.; Rajeshwari, S.; Venckatesh, R. Green synthesis of zinc oxide nanoparticles by aloe barbadensis miller leaf extract: Structure and optical properties. *Mater. Res. Bull.* **2011**, *46*, 2560–2566. [CrossRef]
144. Saravanan, M.; Gopinath, V.; Chaurasia, M.K.; Syed, A.; Ameen, F.; Purushothaman, N. Green synthesis of anisotropic zinc oxide nanoparticles with antibacterial and cytofriendly properties. *Microb. Pathog.* **2018**, *115*, 57–63. [CrossRef]
145. Abinaya, M.; Vaseeharan, B.; Divya, M.; Sharmili, A.; Govindarajan, M.; Alharbi, N.S.; Kadaikunnan, S.; Khaled, J.M.; Benelli, G. Bacterial exopolysaccharide (EPS)-coated ZnO nanoparticles showed high antibiofilm activity and larvicidal toxicity against malaria and Zika virus vectors. *J. Trace Elem. Med. Biol.* **2018**, *45*, 93–103. [CrossRef]
146. Suresh, D.; Nethravathi, P.C.; Udayabhanu; Rajanaika, H.; Nagabhushana, H.; Sharma, S.C. Green synthesis of multifunctional zinc oxide (ZnO) nanoparticles using Cassia fistula plant extract and their photodegradative, antioxidant and antibacterial activities. *Mater. Sci. Semicond. Process.* **2015**, *31*, 446–454. [CrossRef]
147. Kalappurakkal, R. Molecular Taxonomic Identification, Biosynthesis and in vitro Antibacterial Activity of ZNO Nanoparticles Using Boerhavia diffusa Against MRSA In vitro screening and characterization of mycoendophytes residing in Withania somnifera for potent anticancer agents View project Bioprospecting of hydrolytic bacteria from mangroves View project. *Artic. Int. J. Toxicol. Pharmacol. Res.* **2016**, *8*, 40–44.
148. Anitha, R.; Ramesh, K.V.; Ravishankar, T.N.; Sudheer Kumar, K.H.; Ramakrishnappa, T. Cytotoxicity, antibacterial and antifungal activities of ZnO nanoparticles prepared by the Artocarpus gomezianus fruit mediated facile green combustion method. *J. Sci. Adv. Mater. Devices* **2018**, *3*, 440–451. [CrossRef]
149. Elavarasan, N.; Kokila, K.; Inbasekar, G.; Sujatha, V. Evaluation of photocatalytic activity, antibacterial and cytotoxic effects of green synthesized ZnO nanoparticles by Sechium edule leaf extract. *Res. Chem. Intermed.* **2016**, *43*, 3361–3376. [CrossRef]
150. Karthik, S.; Siva, P.; Balu, K.S.; Suriyaprabha, R.; Rajendran, V.; Maaza, M. Acalypha indica–mediated green synthesis of ZnO nanostructures under differential thermal treatment: Effect on textile coating, hydrophobicity, UV resistance, and antibacterial activity. *Adv. Powder Technol.* **2017**, *28*, 3184–3194. [CrossRef]
151. Raja, A.; Ashokkumar, S.; Pavithra Marthandam, R.; Jayachandiran, J.; Khatiwada, C.P.; Kaviyarasu, K.; Ganapathi Raman, R.; Swaminathan, M. Eco-friendly preparation of zinc oxide nanoparticles using Tabernaemontana divaricata and its photocatalytic and antimicrobial activity. *J. Photochem. Photobiol. B Biol.* **2018**, *181*, 53–58. [CrossRef] [PubMed]
152. Vijayakumar, S.; Vaseeharan, B.; Malaikozhundan, B.; Shobiya, M. Laurus nobilis leaf extract mediated green synthesis of ZnO nanoparticles: Characterization and biomedical applications. *Biomed. Pharmacother.* **2016**, *84*, 1213–1222. [CrossRef]
153. Steffy, K.; Shanthi, G.; Maroky, A.S.; Selvakumar, S. Enhanced antibacterial effects of green synthesized ZnO NPs using Aristolochia indica against Multi-drug resistant bacterial pathogens from Diabetic Foot Ulcer. *J. Infect. Public Health* **2018**, *11*, 463–471. [CrossRef]
154. Sharmila, G.; Muthukumaran, C.; Sandiya, K.; Santhiya, S.; Sakthi Pradeep, R.; Manoj Kumar, N.; Suriyanarayanan, N.; Thirumarimurugan, M. Biosynthesis, characterization, and antibacterial activity of zinc oxide nanoparticles derived from Bauhinia tomentosa leaf extract. *J. Nanostructure Chem.* **2018**, *8*, 293–299. [CrossRef]
155. Ishwarya, R.; Vaseeharan, B.; Kalyani, S.; Banumathi, B.; Govindarajan, M.; Alharbi, N.S.; Kadaikunnan, S.; Al-anbr, M.N.; Khaled, J.M.; Benelli, G. Facile green synthesis of zinc oxide nanoparticles using Ulva lactuca seaweed extract and evaluation of their photocatalytic, antibiofilm and insecticidal activity. *J. Photochem. Photobiol. B Biol.* **2018**, *178*, 249–258. [CrossRef]
156. Aiswarya Devi, S.; Harshiny, M.; Udaykumar, S.; Gopinath, P.; Matheswaran, M. Strategy of metal iron doping and green-mediated ZnO nanoparticles: Dissolubility, antibacterial and cytotoxic traits. *Toxicol. Res.* **2017**, *6*, 854–865. [CrossRef]
157. Chai, H.Y.; Lam, S.M.; Sin, J.C. Green synthesis of magnetic Fe-doped ZnO nanoparticles via Hibiscus rosa-sinensis leaf extracts for boosted photocatalytic, antibacterial and antifungal activities. *Mater. Lett.* **2019**, *242*, 103–106. [CrossRef]
158. Chopra, H.; Bibi, S.; Islam, F.; Ahmad, S.U.; Olawale, O.A.; Alhumaydhi, F.A.; Marzouki, R.; Baig, A.A.; Emran, T.B. Emerging Trends in the Delivery of Resveratrol by Nanostructures: Applications of Nanotechnology in Life Sciences. *J. Nanomater.* **2022**, *2022*, 3083728. [CrossRef]
159. Wang, X.; Chen, H.; Zheng, Y.; Ma, M.; Chen, Y.; Zhang, K.; Zeng, D.; Shi, J. Au-nanoparticle coated mesoporous silica nanocapsule-based multifunctional platform for ultrasound mediated imaging, cytoclasis and tumor ablation. *Biomaterials* **2013**, *34*, 2057–2068. [CrossRef] [PubMed]
160. Ramamurthy, C.H.; Sampath, K.S.; Arunkumar, P.; Kumar, M.S.; Sujatha, V.; Premkumar, K.; Thirunavukkarasu, C. Green synthesis and characterization of selenium nanoparticles and its augmented cytotoxicity with doxorubicin on cancer cells. *Bioprocess Biosyst. Eng.* **2013**, *36*, 1131–1139. [CrossRef]
161. Zhang, H.; Shan, Y.; Dong, L. A comparison of TiO2and ZnO nanoparticles as photosensitizers in photodynamic therapy for cancer. *J. Biomed. Nanotechnol.* **2014**, *10*, 1450–1457. [CrossRef]
162. Taccola, L.; Raffa, V.; Riggio, C.; Vittorio, O.; Iorio, M.C.; Vanacore, R.; Pietrabissa, A.; Cuschieri, A. Zinc oxide nanoparticles as selective killers of proliferating cells. *Int. J. Nanomed.* **2011**, *6*, 1129–1140. [CrossRef]
163. Akhtar, M.J.; Ahamed, M.; Kumar, S.; Majeed Khan, M.A.; Ahmad, J.; Alrokayan, S.A. Zinc oxide nanoparticles selectively induce apoptosis in human cancer cells through reactive oxygen species. *Int. J. Nanomed.* **2012**, *7*, 845–857. [CrossRef]
164. Reddy, K.M.; Feris, K.; Bell, J.; Wingett, D.G.; Hanley, C.; Punnoose, A. Selective toxicity of zinc oxide nanoparticles to prokaryotic and eukaryotic systems. *Appl. Phys. Lett.* **2007**, *90*, 213902. [CrossRef]

165. Ivask, A.; Titma, T.; Visnapuu, M.; Vija, H.; Kakinen, A.; Sihtmae, M.; Pokhrel, S.; Madler, L.; Heinlaan, M.; Kisand, V.; et al. Toxicity of 11 Metal Oxide Nanoparticles to Three Mammalian Cell Types in vitro. *Curr. Top. Med. Chem.* **2015**, *15*, 1914–1929. [CrossRef]
166. Gojova, A.; Guo, B.; Kota, R.S.; Rutledge, J.C.; Kennedy, I.M.; Barakat, A.I. Induction of inflammation in vascular endothelial cells by metal oxide nanoparticles: Effect of particle composition. *Environ. Health Perspect.* **2007**, *115*, 403–409. [CrossRef]
167. Hanley, C.; Thurber, A.; Hanna, C.; Punnoose, A.; Zhang, J.; Wingett, D.G. The influences of cell Type and ZnO nanoparticle size on immune cell cytotoxicity and cytokine induction. *Nanoscale Res. Lett.* **2009**, *4*, 1409–1420. [CrossRef] [PubMed]
168. Deng, X.; Luan, Q.; Chen, W.; Wang, Y.; Wu, M.; Zhang, H.; Jiao, Z. Nanosized zinc oxide particles induce neural stem cell apoptosis. *Nanotechnology* **2009**, *20*, 115101. [CrossRef] [PubMed]
169. Nie, S.; Xing, Y.; Kim, G.J.; Simons, J.W. Nanotechnology applications in cancer. *Annu. Rev. Biomed. Eng.* **2007**, *9*, 257–288. [CrossRef] [PubMed]
170. Hellman, S. Improving the Therapeutic Index in Breast Cancer Treatment: The Richard and Hinda Rosenthal Foundation Award Lecture. Available online: https://pubmed.ncbi.nlm.nih.gov/7438067/ (accessed on 27 December 2021).
171. Huang, P.S.; Oliff, A. Drug-targeting strategies in cancer therapy. *Curr. Opin. Genet. Dev.* **2001**, *11*, 104–110. [CrossRef]
172. Bosanquet, A.G.; Bell, P.B. Ex vivo therapeutic index by drug sensitivity assay using fresh human normal and tumor cells. *J. Exp. Ther. Oncol.* **2004**, *4*, 145–154. [PubMed]
173. Moghaddam, A.B.; Moniri, M.; Azizi, S.; Rahim, R.A.; Ariff, A.B.; Navaderi, M.; Mohamad, R. Eco-Friendly Formulated Zinc Oxide Nanoparticles: Induction of Cell Cycle Arrest and Apoptosis in the MCF-7 Cancer Cell Line. *Genes* **2017**, *8*, 281. [CrossRef]
174. Hussain, A.; Oves, M.; Alajmi, M.F.; Hussain, I.; Amir, S.; Ahmed, J.; Rehman, M.T.; El-Seedi, H.R.; Ali, I. Biogenesis of ZnO nanoparticles using Pandanus odorifer leaf extract: Anticancer and antimicrobial activities. *RSC Adv.* **2019**, *9*, 15357–15369. [CrossRef]
175. Gao, Y.; Anand, M.A.V.; Ramachandran, V.; Karthikkumar, V.; Shalini, V.; Vijayalakshmi, S.; Ernest, D. Biofabrication of Zinc Oxide Nanoparticles from Aspergillus niger, Their Antioxidant, Antimicrobial and Anticancer Activity. *J. Clust. Sci.* **2019**, *30*, 937–946. [CrossRef]
176. Majeed, S.; Danish, M.; Norazmi, F.S.B. Fungal Derived Zinc Oxide Nanoparticles and Their Antibacterial and Anticancer Activities Against Human Alveoli Lung Cancer A-549 Cell Line. *Adv. Sci. Eng. Med.* **2018**, *10*, 551–556. [CrossRef]
177. Baskar, G.; Chandhuru, J.; Sheraz Fahad, K.; Praveen, A.S.; Chamundeeswari, M.; Muthukumar, T. Anticancer activity of fungal l-asparaginase conjugated with zinc oxide nanoparticles. *J. Mater. Sci. Mater. Med.* **2015**, *26*, 43. [CrossRef]
178. Chopra, H.; Bibi, S.; Mishra, A.K.; Tirth, V.; Yerramsetty, S.V.; Murali, S.V.; Ahmad, S.U.; Mohanta, Y.K.; Attia, M.S.; Algahtani, A.; et al. Nanomaterials: A Promising Therapeutic Approach for Cardiovascular Diseases. *J. Nanomater.* **2022**, *2022*, 4155729. [CrossRef]
179. Safavi, E.; Homayouni-Tabrizi, M.; Karimi, E.; Mohammad, G.R.K.S. Biosynthesis of zinc oxide nanoparticles using anjbar (root of Persicaria bistorta) extract and their cytotoxic effects on human breast cancer cell line (MCF-7). *IET Nanobiotechnol.* **2019**, *13*, 736–741. [CrossRef] [PubMed]
180. Namvar, F.; Azizi, S.; Rahman, H.S.; Mohamad, R.; Rasedee, A.; Soltani, M.; Rahim, R.A. Green synthesis, characterization, and anticancer activity of hyaluronan/zinc oxide nanocomposites. *Onco. Targets. Ther.* **2016**, *9*, 4549–4559. [CrossRef] [PubMed]
181. Asik, R.M.; Gowdhami, B.; Jaabir, M.S.M.; Archunan, G.; Suganthy, N. Anticancer potential of zinc oxide nanoparticles against cervical carcinoma cells synthesized via biogenic route using aqueous extract of Gracilaria edulis. *Mater. Sci. Eng. C* **2019**, *103*, 109840. [CrossRef]
182. Cheng, J.; Wang, X.; Qiu, L.; Li, Y.; Marraiki, N.; Elgorban, A.M.; Xue, L. Green synthesized zinc oxide nanoparticles regulates the apoptotic expression in bone cancer cells MG-63 cells. *J. Photochem. Photobiol. B Biol.* **2020**, *202*, 111644. [CrossRef] [PubMed]
183. Ashokan, A.P.; Paulpandi, M.; Dinesh, D.; Murugan, K.; Vadivalagan, C.; Benelli, G. Toxicity on Dengue Mosquito Vectors Through Myristica fragrans-Synthesized Zinc Oxide Nanorods, and Their Cytotoxic Effects on Liver Cancer Cells (HepG2). *J. Clust. Sci.* **2017**, *28*, 205–226. [CrossRef]
184. Umar, H.; Kavaz, D.; Rizaner, N. Biosynthesis of zinc oxide nanoparticles using *Albizia lebbeck* stem bark, and evaluation of its antimicrobial, antioxidant, and cytotoxic activities on human breast cancer cell lines. *Int. J. Nanomed.* **2018**, *14*, 87–100. [CrossRef]
185. Rajeshkumar, S.; Kumar, S.V.; Ramaiah, A.; Agarwal, H.; Lakshmi, T.; Roopan, S.M. Biosynthesis of zinc oxide nanoparticles usingMangifera indica leaves and evaluation of their antioxidant and cytotoxic properties in lung cancer (A549) cells. *Enzyme Microb. Technol.* **2018**, *117*, 91–95. [CrossRef]
186. Malaikozhundan, B.; Vaseeharan, B.; Vijayakumar, S.; Pandiselvi, K.; Kalanjiam, M.A.R.; Murugan, K.; Benelli, G. Biological therapeutics of Pongamia pinnata coated zinc oxide nanoparticles against clinically important pathogenic bacteria, fungi and MCF-7 breast cancer cells. *Microb. Pathog.* **2017**, *104*, 268–277. [CrossRef]
187. Chung, I.M.; Rahuman, A.A.; Marimuthu, S.; Kirthi, A.V.; Anbarasan, K.; Rajakumar, G. An Investigation of the Cytotoxicity and Caspase-Mediated Apoptotic Effect of Green Synthesized Zinc Oxide Nanoparticles Using Eclipta prostrata on Human Liver Carcinoma Cells. *Nanomaterials* **2015**, *5*, 1317–1330. [CrossRef]
188. Vimala, K.; Shanthi, K.; Sundarraj, S.; Kannan, S. Synergistic effect of chemo-photothermal for breast cancer therapy using folic acid (FA) modified zinc oxide nanosheet. *J. Colloid Interface Sci.* **2017**, *488*, 92–108. [CrossRef] [PubMed]
189. Padalia, H.; Chanda, S. Characterization, antifungal and cytotoxic evaluation of green synthesized zinc oxide nanoparticles using Ziziphus nummularia leaf extract. *Artif. Cells Nanomed. Biotechnol.* **2017**, *45*, 1751–1761. [CrossRef] [PubMed]

190. Yuvakkumar, R.; Suresh, J.; Saravanakumar, B.; Joseph Nathanael, A.; Hong, S.I.; Rajendran, V. Rambutan peels promoted biomimetic synthesis of bioinspired zinc oxide nanochains for biomedical applications. *Spectrochim. Acta Part A Mol. Biomol. Spectrosc.* **2015**, *137*, 250–258. [CrossRef] [PubMed]
191. Sharmila, G.; Thirumarimurugan, M.; Muthukumaran, C. Green synthesis of ZnO nanoparticles using Tecoma castanifolia leaf extract: Characterization and evaluation of its antioxidant, bactericidal and anticancer activities. *Microchem. J.* **2019**, *145*, 578–587. [CrossRef]
192. Vijayakumar, S.; Vaseeharan, B. Antibiofilm, anti cancer and ecotoxicity properties of collagen based ZnO nanoparticles. *Adv. Powder Technol.* **2018**, *29*, 2331–2345. [CrossRef]
193. Somu, P.; Paul, S. A biomolecule-assisted one-pot synthesis of zinc oxide nanoparticles and its bioconjugate with curcumin for potential multifaceted therapeutic applications. *New J. Chem.* **2019**, *43*, 11934–11948. [CrossRef]
194. Wu, H.; Zhang, J. Chitosan-based zinc oxide nanoparticle for enhanced anticancer effect in cervical cancer: A physicochemical and biological perspective. *Saudi Pharm. J.* **2018**, *26*, 205–210. [CrossRef]
195. Karthikeyan, M.; Jafar Ahamed, A.; Karthikeyan, C.; Vijaya Kumar, P. Enhancement of antibacterial and anticancer properties of pure and REM doped ZnO nanoparticles synthesized using Gymnema sylvestre leaves extract. *SN Appl. Sci.* **2019**, *1*, 355. [CrossRef]
196. Hughes, G.; McLean, N.R. Zinc oxide tape: A useful dressing for the recalcitrant finger-tip and soft-tissue injury. *Arch. Emerg. Med.* **1988**, *5*, 223–227. [CrossRef]
197. Mitchnick, M.A.; Fairhurst, D.; Pinnell, S.R. Microfine zinc oxide (Z-cote) as a photostable UVA/UVB sunblock agent. *J. Am. Acad. Dermatol.* **1999**, *40*, 85–90. [CrossRef]
198. Agren, M.S. Percutaneous Absorption of Zinc from Zinc Oxide Applied Topically to Intact Skin in Man. *Dermatology* **1990**, *180*, 36–39. [CrossRef]
199. Asharani, P.V.; Lian Wu, Y.; Gong, Z.; Valiyaveettil, S. Toxicity of silver nanoparticles in zebrafish models. *Nanotechnology* **2008**, *19*, 255102. [CrossRef] [PubMed]
200. Taylor, E.; Webster, T.J. Reducing infections through nanotechnology and nanoparticles. *Int. J. Nanomed.* **2011**, *6*, 1463. [CrossRef]
201. Yuan, Q.; Hein, S.; Misra, R.D.K. New generation of chitosan-encapsulated ZnO quantum dots loaded with drug: Synthesis, characterization and in vitro drug delivery response. *Acta Biomater.* **2010**, *6*, 2732–2739. [CrossRef] [PubMed]
202. Nie, L.; Gao, L.; Feng, P.; Zhang, J.; Fu, X.; Liu, Y.; Yan, X.; Wang, T. Three-dimensional functionalized tetrapod-like ZnO nanostructures for plasmid DNA delivery. *Small* **2006**, *2*, 621–625. [CrossRef] [PubMed]
203. Zhang, P.; Liu, W. ZnO QD@PMAA-co-PDMAEMA nonviral vector for plasmid DNA delivery and bioimaging. *Biomaterials* **2010**, *31*, 3087–3094. [CrossRef]
204. Antoine, T.E.; Mishra, Y.K.; Trigilio, J.; Tiwari, V.; Adelung, R.; Shukla, D. Prophylactic, therapeutic and neutralizing effects of zinc oxide tetrapod structures against herpes simplex virus type-2 infection. *Antivir. Res.* **2012**, *96*, 363–375. [CrossRef]
205. Kalpana, V.N.; Devi Rajeswari, V. A Review on Green Synthesis, Biomedical Applications, and Toxicity Studies of ZnO NPs. *Bioinorg. Chem. Appl.* **2018**, *2018*, 3569758. [CrossRef]
206. Irfan, M.; Munir, H.; Ismail, H. Moringa oleifera gum based silver and zinc oxide nanoparticles: Green synthesis, characterization and their antibacterial potential against MRSA. *Biomater. Res.* **2021**, *25*, 17. [CrossRef]
207. Upadhyaya, L.; Singh, J.; Agarwal, V.; Pandey, A.C.; Verma, S.P.; Das, P.; Tewari, R.P. Efficient water soluble nanostructured ZnO grafted O-carboxymethyl chitosan/curcumin-nanocomposite for cancer therapy. *Process Biochem.* **2015**, *50*, 678–688. [CrossRef]
208. Barick, K.C.; Nigam, S.; Bahadur, D. Nanoscale assembly of mesoporous ZnO: A potential drug carrier. *J. Mater. Chem.* **2010**, *20*, 6446–6452. [CrossRef]
209. Tan, L.; Liu, J.; Zhou, W.; Wei, J.; Peng, Z. A novel thermal and pH responsive drug delivery system based on ZnO@PNIPAM hybrid nanoparticles. *Mater. Sci. Eng. C. Mater. Biol. Appl.* **2014**, *45*, 524–529. [CrossRef] [PubMed]
210. Kishwar, S.; Asif, M.H.; Nur, O.; Willander, M.; Larsson, P.O. Intracellular ZnO Nanorods Conjugated with Protoporphyrin for Local Mediated Photochemistry and Efficient Treatment of Single Cancer Cell. *Nanoscale Res. Lett.* **2010**, *5*, 1669–1674. [CrossRef] [PubMed]
211. Muhammad, F.; Guo, M.; Guo, Y.; Qi, W.; Qu, F.; Sun, F.; Zhao, H.; Zhu, G. Acid degradable ZnO quantum dots as a platform for targeted delivery of an anticancer drug. *J. Mater. Chem.* **2011**, *21*, 13406–13412. [CrossRef]
212. Hong, E.J.; Sivakumar, P.; Ravichandran, V.; Choi, D.G.; Kim, Y.S.; Shim, M.S. Pro-Oxidant Drug-Loaded Au/ZnO Hybrid Nanoparticles for Cancer-Specific Chemo-Photodynamic Combination Therapy. *ACS Biomater. Sci. Eng.* **2019**, *5*, 5209–5217. [CrossRef]
213. Chen, T.; Zhao, T.; Wei, D.; Wei, Y.; Li, Y.; Zhang, H. Core-shell nanocarriers with ZnO quantum dots-conjugated Au nanoparticle for tumor-targeted drug delivery. *Carbohydr. Polym.* **2013**, *92*, 1124–1132. [CrossRef]
214. Ahmad, J.; Wahab, R.; Siddiqui, M.A.; Musarrat, J.; Al-Khedhairy, A.A. Zinc oxide quantum dots: A potential candidate to detain liver cancer cells. *Bioprocess Biosyst. Eng.* **2015**, *38*, 155–163. [CrossRef]
215. Stepankova, H.; Swiatkowski, M.; Kruszynski, R.; Svec, P.; Michalkova, H.; Smolikova, V.; Ridoskova, A.; Splichal, Z.; Michalek, P.; Richtera, L.; et al. The Anti-Proliferative Activity of Coordination Compound-Based ZnO Nanoparticles as a Promising Agent Against Triple Negative Breast Cancer Cells. *Int. J. Nanomed.* **2021**, *16*, 4431–4449. [CrossRef]

216. Liu, M.; Sun, X.; Liao, Z.; Li, Y.; Qi, X.; Qian, Y.; Fenniri, H.; Zhao, P.; Shen, J. Zinc oxide end-capped Fe 3 O 4@mSiO 2 core-shell nanocarriers as targeted and responsive drug delivery system for chemo-/ions synergistic therapeutics. *Drug Deliv.* **2019**, *26*, 732–743. [CrossRef]
217. Zhang, Z.Y.; Xiong, H.M. Photoluminescent ZnO nanoparticles and their biological applications. *Materials* **2015**, *8*, 3101–3127. [CrossRef]
218. Mohammed, Y.H.; Barkauskas, D.S.; Holmes, A.; Grice, J.; Roberts, M.S. Noninvasive in vivo human multiphoton microscopy: A key method in proving nanoparticulate zinc oxide sunscreen safety. *J. Biomed. Opt.* **2020**, *25*, 1–19. [CrossRef] [PubMed]
219. Jaatinen, E.A.; Fernando, J.F.S.; Shortell, M.P.; Walden, S.L. Accurate determination of nonlinear refraction in ZnO and Au composite nanostructures. *Opt. Mater. Express* **2020**, *10*, 653–661. [CrossRef]
220. Prasanna, A.P.S.; Venkataprasanna, K.S.; Pannerselvam, B.; Asokan, V.; Jeniffer, R.S.; Venkatasubbu, G.D. Multifunctional ZnO/SiO_2 Core/Shell Nanoparticles for Bioimaging and Drug Delivery Application. *J. Fluoresc.* **2020**, *30*, 1075–1083. [CrossRef] [PubMed]
221. Couto, N.; Newton, J.R.A.; Russo, C.; Karunakaran, E.; Achour, B.; Al-Majdoub, Z.M.; Sidaway, J.; Rostami-Hodjegan, A.; Clench, M.R.; Barber, J. Label-Free Quantitative Proteomics and Substrate-Based Mass Spectrometry Imaging of Xenobiotic Metabolizing Enzymes in Ex Vivo Human Skin and a Human Living Skin Equivalent Model. *Drug Metab. Dispos.* **2021**, *49*, 39–52. [CrossRef] [PubMed]
222. Kilin, V.; Campargue, G.; Fureraj, I.; Sakong, S.; Sabri, T.; Riporto, F.; Vieren, A.; Mugnier, Y.; Mas, C.; Staedler, D.; et al. Wavelength-Selective Nonlinear Imaging and Photo-Induced Cell Damage by Dielectric Harmonic Nanoparticles. *ACS Nano* **2020**, *14*, 4087–4095. [CrossRef] [PubMed]
223. Chattopadhyay, S.; Kumawat, A.; Misra, K.P.; Halder, N.; Bandyopadhyay, A.; Antony, A.; Rao, A.; Poornesh, P.; Jedryka, J.; Ozga, K.; et al. Micro-strain administered SHG intensity enhancement by heavy Ce doping in co-precipitated ZnO nanoparticles. *Mater. Sci. Eng. B Solid-State Mater. Adv. Technol.* **2021**, *266*, 115041. [CrossRef]
224. Zhang, H.J.; Xiong, H.M.; Ren, Q.G.; Xia, Y.Y.; Kong, J.L. ZnO@silica core-shell nanoparticles with remarkable luminescence and stability in cell imaging. *J. Mater. Chem.* **2012**, *22*, 13159–13165. [CrossRef]
225. Prabhu, S.; Binu Parvathy, A.; Ashok Kumar, R.; Sandhya, V. Potential Effect of Zinc Oxide Nanoparticle as a Natural Antibiotic Drug against Bacterial Species. Available online: https://www.ikprress.org/index.php/PCBMB/article/view/5442 (accessed on 25 February 2022).
226. Jiang, J.; Pi, J.; Cai, J. The Advancing of Zinc Oxide Nanoparticles for Biomedical Applications. *Bioinorg. Chem. Appl.* **2018**, *2018*, 1062562. [CrossRef]
227. Santhoshkumar, J.; Kumar, S.V.; Rajeshkumar, S. Synthesis of zinc oxide nanoparticles using plant leaf extract against urinary tract infection pathogen. *Resour. Technol.* **2017**, *3*, 459–465. [CrossRef]
228. Arfat, Y.A.; Ahmed, J.; Al Hazza, A.; Jacob, H.; Joseph, A. Comparative effects of untreated and 3-methacryloxypropyltrimethoxysilane treated ZnO nanoparticle reinforcement on properties of polylactide-based nanocomposite films. *Int. J. Biol. Macromol.* **2017**, *101*, 1041–1050. [CrossRef]
229. Mirzaei, H.; Darroudi, M. Zinc oxide nanoparticles: Biological synthesis and biomedical applications. *Ceram. Int.* **2017**, *43*, 907–914. [CrossRef]
230. Osmond, M.J.; McCall, M.J. Zinc oxide nanoparticles in modern sunscreens: An analysis of potential exposure and hazard. *Nanotoxicology* **2010**, *4*, 15–41. [CrossRef] [PubMed]
231. Markowska-Szczupak, A.; Endo-Kimura, M.; Paszkiewicz, O.; Kowalska, E. Are Titania Photocatalysts and Titanium Implants Safe? Review on the Toxicity of Titanium Compounds. *Nanomaterials* **2020**, *10*, 2065. [CrossRef] [PubMed]
232. Siddiqi, K.S.; ur Rahman, A.; Tajuddin; Husen, A. Properties of Zinc Oxide Nanoparticles and Their Activity against Microbes. *Nanoscale Res. Lett.* **2018**, *13*, 141. [CrossRef] [PubMed]
233. Singh, T.; Shukla, S.; Kumar, P.; Wahla, V.; Bajpai, V.K. Application of nanotechnology in food science: Perception and overview. *Front. Microbiol.* **2017**, *8*, 1501. [CrossRef] [PubMed]
234. Freire, K.; Ordóñez Ramos, F.; Soria, D.B.; Pabón Gelves, E.; Di Virgilio, A.L. Cytotoxicity and DNA damage evaluation of TiO 2 and ZnO nanoparticles. Uptake in lung cells in culture. *Toxicol. Res.* **2021**, *10*, 192–202. [CrossRef] [PubMed]
235. Cavalu, S.; Fritea, L.; Brocks, M.; Barbaro, K.; Murvai, G.; Costea, T.O.; Antoniac, I.; Verona, C.; Romani, M.; Latini, A.; et al. Novel Hybrid Composites Based on PVA/SeTiO2 Nanoparticles and Natural Hydroxyapatite for Orthopedic Applications: Correlations between Structural, Morphological and Biocompatibility Properties. *Materials* **2020**, *13*, 2077. [CrossRef]
236. Alghsham, R.S.; Satpathy, S.R.; Bodduluri, S.R.; Hegde, B.; Jala, V.R.; Twal, W.; Burlison, J.A.; Sunkara, M.; Haribabu, B. Zinc Oxide Nanowires Exposure Induces a Distinct Inflammatory Response via CCL11-Mediated Eosinophil Recruitment. *Front. Immunol.* **2019**, *10*, 2604. [CrossRef]
237. Khan, M.; Naqvi, A.H.; Ahmad, M. Comparative study of the cytotoxic and genotoxic potentials of zinc oxide and titanium dioxide nanoparticles. *Toxicol. Rep.* **2015**, *2*, 765–774. [CrossRef]
238. Liu, J.; Kang, Y.; Yin, S.; Song, B.; Wei, L.; Chen, L.; Shao, L. Zinc oxide nanoparticles induce toxic responses in human neuroblastoma SHSY5Y cells in a size-dependent manner. *Int. J. Nanomed.* **2017**, *12*, 8085. [CrossRef]
239. Keerthana, S.; Kumar, A. Potential risks and benefits of zinc oxide nanoparticles: A systematic review. *Crit. Rev. Toxicol.* **2020**, *50*, 47–71. [CrossRef] [PubMed]

materials

MDPI

Review

Electrospun Nanofibers Revisited: An Update on the Emerging Applications in Nanomedicine

Nehal E. Elsadek [1], Abdalrazeq Nagah [2], Tarek M. Ibrahim [3], Hitesh Chopra [4], Ghada A. Ghonaim [2], Sherif E. Emam [3], Simona Cavalu [5,*] and Mohamed S. Attia [3,*]

1 Department of Pharmacokinetics and Biopharmaceutics, Institute of Biomedical Sciences, Tokushima University, 1-78-1 Sho-machi, Tokushima 770-8505, Japan; nehal.emam92@yahoo.com
2 Faculty of Pharmacy, Zagazig University, Zagazig 44519, Egypt; abdalrazeqnagah@gmail.com (A.N.); ghada99mahmoud@gmail.com (G.A.G.)
3 Department of Pharmaceutics, Faculty of Pharmacy, Zagazig University, Zagazig 44519, Egypt; tarekmetwally333@gmail.com (T.M.I.); sherif.emam86@yahoo.com (S.E.E.)
4 Chitkara College of Pharmacy, Chitkara University, Punjab 140401, India; chopraontheride@gmail.com
5 Faculty of Medicine and Pharmacy, University of Oradea, P-ta 1 Decembrie 10, 410087 Oradea, Romania
* Correspondence: simona.cavalu@gmail.com (S.C.); msalahatia@zu.edu.eg (M.S.A.)

Abstract: Electrospinning (ES) has become a straightforward and customizable drug delivery technique for fabricating drug-loaded nanofibers (NFs) using various biodegradable and non-biodegradable polymers. One of NF's pros is to provide a controlled drug release through managing the NF structure by changing the spinneret type and nature of the used polymer. Electrospun NFs are employed as implants in several applications including, cancer therapy, microbial infections, and regenerative medicine. These implants facilitate a unique local delivery of chemotherapy because of their high loading capability, wide surface area, and cost-effectiveness. Multi-drug combination, magnetic, thermal, and gene therapies are promising strategies for improving chemotherapeutic efficiency. In addition, implants are recognized as an effective antimicrobial drug delivery system overriding drawbacks of traditional antibiotic administration routes such as their bioavailability and dosage levels. Recently, a sophisticated strategy has emerged for wound healing by producing biomimetic nanofibrous materials with clinically relevant properties and desirable loading capability with regenerative agents. Electrospun NFs have proposed unique solutions, including pelvic organ prolapse treatment, viable alternatives to surgical operations, and dental tissue regeneration. Conventional ES setups include difficult-assembled mega-sized equipment producing bulky matrices with inadequate stability and storage. Lately, there has become an increasing need for portable ES devices using completely available off-shelf materials to yield highly-efficient NFs for dressing wounds and rapid hemostasis. This review covers recent updates on electrospun NFs in nanomedicine applications. ES of biopolymers and drugs is discussed regarding their current scope and future outlook.

Keywords: electrospinning; nanofibers; implants; wound healing; nanomedicine; targeted delivery; biopolymers

Citation: Elsadek, N.E.; Nagah, A.; Ibrahim, T.M.; Chopra, H.; Ghonaim, G.A.; Emam, S.E.; Cavalu, S.; Attia, M.S. Electrospun Nanofibers Revisited: An Update on the Emerging Applications in Nanomedicine. *Materials* **2022**, *15*, 1934. https://doi.org/10.3390/ma15051934

Academic Editors: Alessandro Pistone and Silvia Farè

Received: 10 January 2022
Accepted: 8 February 2022
Published: 4 March 2022

1. Introduction

Electrospinning (ES) is an old concept for fiber fabrication introduced in 1897 by Rayleigh and Kundu [1]. It aims at producing ultrafine non-woven nanofibers (NFs) by utilizing the action of high voltage with low current on the shape of the electroconductive liquid in a polymeric solution. The polymeric solution is pumped through a spinneret or syringe nozzle while voltage is applied between the spinneret and a collector; hence, it causes the surface of the liquid to form a cone (Taylor cone) with a rounded tip. This mechanism can be attributed to electrostatic repulsion on a droplet. When wiring electricity in the spinneret setup, charges start to build upon the droplet's surface, creating an electrostatic repulsion of similar magnitude to the surface tension and causing it to stretch and form the Taylor cone [2]. When the voltage exceeds a certain threshold, the electrostatic repulsion overcomes the surface tension causing the rounded tip to emit a jet of charged liquid floating in the space between the spinneret and the collector surface. Eventually, the solvent evaporates, leaving the desired product on the collector [1,3].

The ultrathin NFs have multiple applications in various fields [4], such as cosmetics [5], active food packaging [6], fabricated fabricating nanofibrous sensors [7], textiles [8], and in particular, their applications in drug delivery [9–11], wound healing [12,13], nanomedicine and tissue engineering [14,15].

The emerging role of ES in drug delivery systems offers a safe and easy encapsulation technique for the therapeutic cargos, maintains their structural integrity and biological activity, and accordingly, ensures a safe release at the target site [16]. At the same time, it provides a cost-effective procedure starting from fiber fabrication to the formulation of the final pharmaceutical product [17]. In addition, ES can be utilized in loading one drug alone or multiple drugs, depending on the characters of the nozzle setup. Loading one drug alone is simple and uses a one-nozzle setup which is not the case with numerous and sometimes incompatible drugs as it requires the use of other ES setups such as the coaxial nozzle setup that provides a core-shell NF, controlling the releasing kinetics of the loaded drugs, see Figure 1 [18,19]. Their dynamic ability to tailor their characters and match every drug requirement gives electrospun NFs an edge over other delivery systems. A tailored design for the NF's diameter, porosity, and thickness is achieved by adjusting solution parameters, environmental conditions, and ES setup parameters to the point that makes the system efficient enough for its intended release mode [20,21], which can be summed up into fast, sustained, and combined or customized release.

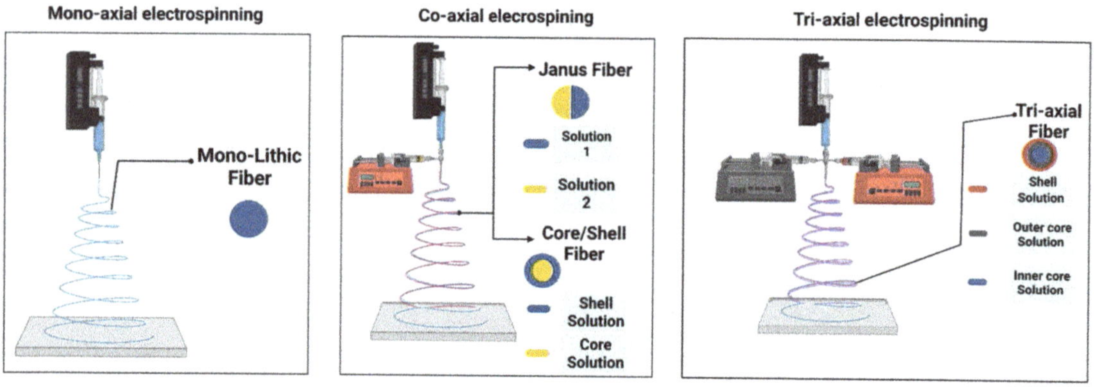

Figure 1. Techniques for nanofiber production.

Electrospun NFs have been the research subject to develop fibrous scaffolds that allow skin regeneration, primarily because of their compatibility with the skin's extracellular matrix (ECM) and biodegradability. Moreover, they can provide enhanced healing rates and be loaded with a bactericidal agent to prevent tissue infection and bioactive molecules and help the regeneration process [22]. Skin wounds are generally associated with increasingly high morbidity and mortality rates that can be regarded as a matter of life and death. Despite the availability of different treatment options and delivery systems, they utterly fail to recover normal tissue states and structures, exacerbating wound infection and tissue necrosis. Tissue-engineered scaffolds aim to regenerate diseased or damaged tissue by using fabricated scaffolds that mimic the complex human tissue's ECM. Fabricated scaffolds are designed to be porous and permit nutrient diffusion and cell migration while providing slow and steady degradation rates so that the natural ECM completely replaces it as the tissue recovers [14,15].

Recently, the emergence of single, handheld, and easy-to-use portable ES devices, has revolutionized the production of electrospun scaffolds. A portable device can produce personalized fiber matrices tailored for each case where the wounded skin tissue acts as the collector of the fiber matrices [23], in a process that reduces the usual pain associated with conventional wound dressings application or removal and enhances the patient's compliance to treatment [24].

In this regard, this review seeks to provide an overview of nanofiber-based drug delivery and its related applications in nanomedicine on the most recent literature research. Searching literature was through Reaxys database using the keywords "Implant", "Drug delivery", "electrospun", "nanofiber", and "wound healing" interchangeably. The recent literature (last five years) has been considered relevant, but discussions were also addressed to earlier reports in this field.

2. Electrospun NF Implants

Advances in biomedical sciences and the increasing complexity of diagnosis and therapeutic processes necessitate the development of a unique drug delivery system to improve the bioavailability of pharmaceuticals reduce their adverse effects. Nanotechnology makes these advancements possible, which allows for the creation and design of nanoparticle-based drug delivery systems. These systems can then be improved using nanoengineered implants, which ensure controlled and sustainable release, reduce side effects, and increase the drug's bioavailability at the target site.

ES technology has gained great attention for its potential biomedical applications since electrospun NFs are becoming more applied in treating various conditions like inflammations, neoplasia, infections, and many other diseases. Electrospun nanofibrous implants represent an ideal pharmaceutical option for the sustained release of medications in the body, which one of its roles is to improve the bioavailability of the used drug. Electrospun NFs offer a perfect therapeutic choice for various conditions as they provide a stable drug-release level in addition to the extended duration and controllable dosage [25].

2.1. Electrospun NF Implants for Cancer Therapy

Cancer is one of the leading causes of mortality worldwide, and cancer therapy still has a narrow margin of safety and efficacy since it affects healthy cells. Following diagnostic investigations, therapeutic techniques, and technology progress, the number of cancer survivors will increase while enhancing their quality of life by delivering more effective and efficient treatment regimens. Radiation therapy, chemotherapy, and surgical procedures are among the current cancer treatment options. Current therapy plans are insufficient and have numerous drawbacks [26,27].

Anticancer drugs are no exception since they have drawbacks such as high toxicity for normal cells, cancer therapy failure, or limited dissolvability, thus reducing their bioavailability and effectiveness [28,29]. According to the previous limitations, new therapeutic approaches and technologies become necessary to advance cancer treatment with minimum damage to normal healthy cells. Nanotechnology poses itself as a promising therapeutic approach since it provides novel treatment approaches. Electrospun NFs offer a unique drug delivery system in cancer therapy because they have high loading potential, large surface area, and are cost-effective and reliable [30].

Cancer patient survival rates are low due to some variables, including late-stage detection, the toxicity of anticancer treatment, and the poor response of many cancer types to therapy [31,32]. Targeted therapeutic approaches have been investigated in research and clinical practice to reduce the harmful side effects of conventional cancer therapies while maximizing the desired therapeutic benefits [33]. Nanotechnology provides two main therapeutic options. The first one involves encapsulating the drug with nanoparticles and targeting a specific site with this compound using a positive or a negative targeting system [34]. The second strategy entails implanting a drug-loaded system at the neoplasia site to locally deliver the required drug dosage to limit injury to normal tissues [34].

Electrospun NFs have a high surface-to-volume ratio, which improves the drug's solubility in an aqueous environment and enhances its efficacy [35]. In addition, electrospun NF implants, in particular, can be used for localized cancer treatment, which is one of the most effective and safe treatment strategies. ES has been proven to be one of the most competitive nanotechnologies since it allows for the mass production of NFs at reasonable costs. Moreover, electrospun NFs are very promising as drug carriers, particularly for local delivery of chemotherapy [31].

Extensive research has been done to study electrospun NF implants' efficacy and safety in cancer therapy. Zong et al. [36] examined the potential role of fibrous implants in treating cervix cancer. They employed cisplatin (CIS)-loaded poly(ethylene oxide)/poly(lactide) (PLA) composite NFs for vaginal implantation to deliver CIS to cervical cancer cells (U14) in mice. In comparison to intravenous injection, vaginal implantation significantly induced more CIS accumulation in the vaginal and cervix regions than in the kidney, liver, or blood (Figure 2).

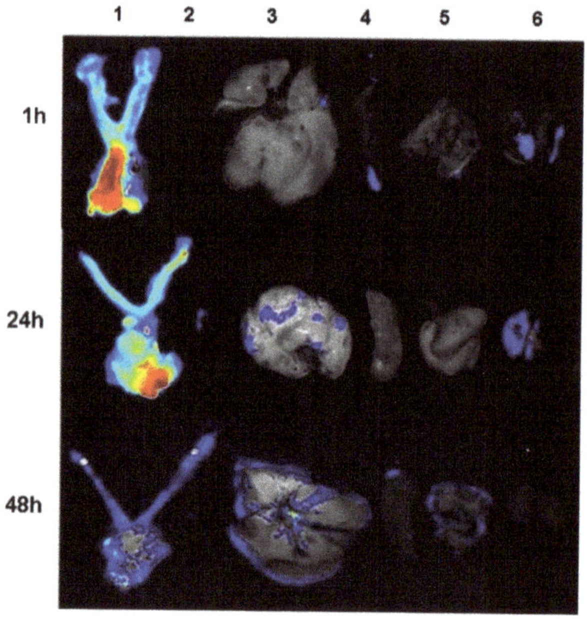

Figure 2. Ex vivo imaging of collected organs (1 tumor, 2 hearts, 3 livers, 4 spleens, 5 lungs and 6 kidneys) at 1, 24 and 48 h following the implantation of electrospun nanofibers into the mice's vaginas. Reprinted with permission from ref. [36].

Furthermore, Liu et al. [37] have studied the role of electrospun NFs in treating secondary hepatic carcinoma (SHCC). After explorative laparotomy in mice, they applied doxorubicin-loaded PLA (DOX-PLA) electrospun NFs as implants. Results showed an increased survival of the affected mice from 14 days to 38 days, in addition to a remarkable inhibition of the SHCC. These significant effects hold the promise of treating SHCC in vivo [37]. Similarly, another group of researchers has applied dacarbazine-loaded poly(vinyl alcohol) (PVA) electrospun NFs as brain implants to treat glioblastoma [38]. This system showed high encapsulation efficiency (>80%) and controlled dacarbazine release, thus significantly improving its anti-tumor effects, such as apoptosis-mediated cell death and DNA damage.

One promising approach that has been created through coaxial electrospinning is core-shell structured electrospun NFs, which allow more sustained drug release by inhibiting the initial burst release. Two different drugs can be encapsulated simultaneously using core-shell electrospun NFs: one in the core and the other in the shell, and released at varying rates depending on the drug nature and polymers utilized and the degree of degradability of the outer layer [39].

The core-shell NFs system introduced by Yang et al. [40] has increased nanoparticle stability, such as micelles. They developed an implantable drug delivery system for the localized delivery of anticancer drug through the coaxial ES of PVA/active targeting DOX-encapsulating micelles (composed through the self-assembly of folate-conjugated poly(ε-caprolactone) (PCL)/poly(ethylene glycol) (PEG) copolymer as the core and cross-linked gelatin as the shell (Figure 3). The degradation of NF matrix causes these active-targeting micelles to be released from the devices in a time-dependent fashion, quickly accumulating around the tumor tissue via interstitial transport and the enhanced permeation and retention (EPR) effect, lowering the dosage of the chemotherapeutic agent, frequency of administration, and adverse effects.

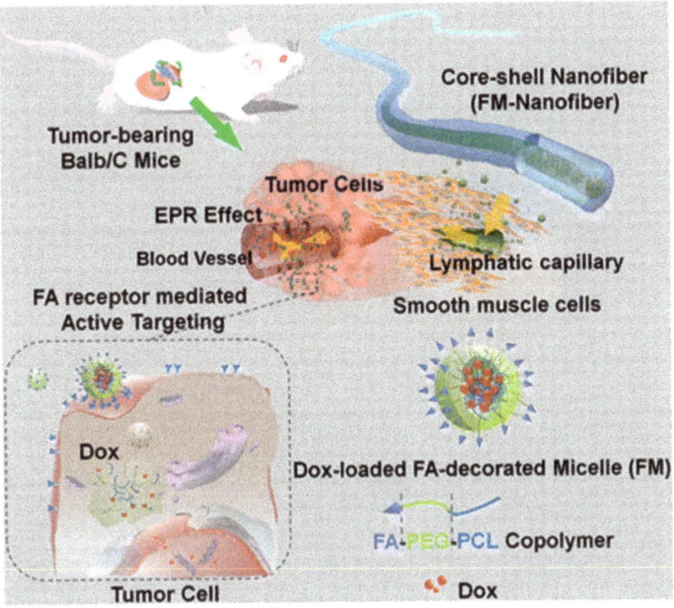

Figure 3. Coaxial electrospinning-prepared implantable Doxorubicin-loaded micelles in NFs for effective cancer therapy. Reprinted with permission from ref. [40].

Multilayering ES is another approach developed to gain more control of the therapeutic agent's discharge kinetics in the core-shell electrospun NFs, in which additional layers with different diffusion pathways and degradation behaviors are included to stop the first burst release and provide sustained release as well. Falde et al. used this approach to create an implantable multilayered 3D electrospun NF mat for the localized delivery of the chemotherapeutic agent SN-38. The system was constructed of a first barrier mesh, a second drug-loaded layer and a third barrier mesh. The in vitro release studies showed the system's success in reducing the burst release of the drug by ten times and the sustained release of SN-38 for more than 30 days. Furthermore, the developed system was cytotoxic to Lewis Lung Carcinoma (LLC) cells for more than 20 days [41].

The use of single chemotherapy is uncommon due to the toxicity of chemotherapeutic agents at high doses and the risk of developing resistance if single chemotherapy is used for a long time. As a result, multidrug-loaded electrospun NFs are one of the most promising options for overcoming the adverse effects of single therapy, lowering the applied dose and improving chemotherapy effectiveness [42]. Table 1 summarizes some of the reported implanted multidrug-loaded electrospun NFs for localized chemotherapeutic agent delivery into the tumor.

Table 1. Multidrug-loaded electrospun NF implants for cancer treatment.

Polymeric Fiber	ES Technique	Drugs	Type of Cancer	In Vivo	In Vitro	Ref.
BIC/(PLGA)	Blend ES	CAR, CIS and CPT-11	Brain cancer (C6 glioma)	+		[43]
	Blend ES	5-FU and oxaliplatin	Colorectal cancer (HCT8 and CT26 cell lines)	+	+	[44]
PGC-C18/PCL	Blend ES	CPT-11 and SN-38	Human colorectal (HT-29) cell line		+	[45]
PLGA/gelatin	Blend ES	DOX-encapsulated mesoporous zinc oxide microspheres /camptothecin	Liver cancer (HepG2) cell line		+	[46]
PEG/PLA	Blend ES	CBT A-4 and HCPT	Breast (4T1) tumor model	+	+	[47]
	Emulsion ES	Paclitaxel and DOX	Brain cancer (C6 glioma)		+	[48]
Dextran/PLGA	Emulsion ES	HCPT and tea polyphenol	Orthotopic liver (H22) carcinoma cell line			[49]
PCL/gelatin	Second carrier ES (core/shell silica nanoparticles)	DOX and Indomethacin	L929 fibroblast cells			[50]
PLGA	Sequential ES	CAR, CPT-11, CIS and CBT	Brain cancer (C6 glioma)	+		[51]
	Emulsion ES	Paclitaxel and Brefeldin A	Human liver (HepG2) cancer cell line		+	[52]
PLLA	Sequential ES	DCA and oxaliplatin	Cervical cancer (Hela cancer and U14 cancer cell lines)	+	+	[53]
	Second carrier ES (mesoporous silica nanoparticles)	DOX/Ibuprofen	Cervical cancer (Hela cell line)		+	[54]
	Blend ES	DOX and HCPT	Human cervical cancer (HeLa cells)		+	[55]
		DCA and diisopropylamine dichloroacetate	Colorectal cancer (C26 cells)	+	+	[56]
		Oxaliplatin and cyclophosphamide	Human hepatocellular cancer (HCC cells)	+	+	[57]
PCL	Blend ES	(−)-epigallocatechin-3-O-gallate and caffeic acid	Human gastric cancer MKN28 cells		+	[58]
		Cisplatin and CUR	Human cervical cancer (HeLa cells)	+	+	[59]
		CUR and aloe-vera or neem-extract	Lung carcinoma (A549) and breast cancer (MCF-7)		+	[60]
	Core-sheath ES	Ibuprofen and DOX	Human hepatocellular carcinoma cell line (HuH-7)		+	[61]
		5-FU and paclitaxel	TNBC cells human triple-negative breast cancer	+	+	[62]
PLLA/PCL	Microfluidic ES	DOX and angiogenesis inhibitor apatinib	Breast cancer (4T1 cells)	+	+	[63]
PVA	Second carrier ES (mPEG-PCL micelles)	DOX and CUR	Cervical cancer (Hela) cell line		+	[64]
	Blend ES	Dichloroacetate and Pt(IV) prodrug-backboned micelle	HeLa human cervical cancer cells	+	+	[65]

5-FU: 5-fluorouracil; CAR: Carmustine; CBT: combretastatin; CIS: Cisplatin; CPT-11: irinotecan; CUR: Curcumin; DCA: Sodium dichloroacetate; DOX: Doxorubicin; HCPT: Hydroxycamptothecin; PGC-C18: Poly(glycerol monostearate-co-ε-caprolactone); PLLA: Poly (l-lactic acid); PLGA: (D, L-lactic acid-co-glycolic acid); and SN-38: irinotecan metabolite.

Combination therapy is another recently adopted cancer therapy technique, in which two therapeutic strategies are integrated into one platform to enhance the antitumor effect while avoiding the negative side effects of monotherapy. Hyperthermia is one of the used approaches for targeted cancer therapy, with malignant cells being more thermosensitive than healthy ones. Using blend ES, Zhao and coworkers co-loaded DOX and MoS2, an effective photothermal agent, into chitosan/PVA electrospun NF mat to combine the photo hyperthermia and chemotherapy [66]. The heat generated from the near-infrared irradiation of 808 nm laser could sensitize the chemotherapeutic efficacy of DOX via controlling its discharge rate. The combined therapy efficiently boosted cytotoxicity against HT29 cells in vitro and in vivo, where surgical implantation of the nanofibrous mat into the tumor site could completely inhibit cancer recurrence. Similarly, another group combined the chemotherapeutic DOX with photothermal treatment: Cu9S5 through incorporating DOX-loaded Cu9S5@mSiO2 nanoparticles into PCL/gelatin nanofibrous mat. Under 980 nm laser irradiation, the combined chemo/photothermal therapy showed a more efficient tumour-suppressive effect in vivo than either single photothermal or chemo treatment [67].

Magnetic thermal therapy is another protocol that is used as a co-therapy with chemotherapy, in which iron oxide (Fe_3O_4) magnetic nanoparticles (MNPs) are applied as a heat source under an alternating medical field (AMF) [68]. In a similar way to photothermal gents, the magnetic thermal agents can synergize the antitumor effect of the chemotherapeutic agent that is co-loaded with them in the nanofibrous system. In a related study, Niiyama et al. loaded paclitaxel and MNPS into PCL polymer using the ES technique. The AMF irradiation could provide a sustained release of paclitaxel over six weeks. In vivo, the locally created implantable NF system showed a superior anticancer impact against the human lung (NCI-H23) cancer cell line than chemotherapy or thermal therapy alone while using an AMF of 166 kHz and 192 A [69].

Gene therapy is also one of the most widely used approaches for cancer treatment that is targeted, safe and successful. Electrospun NFs are one of the nanotechnology techniques used as a platform for cancer gene therapy delivery. They are also employed for the dual delivery of gene therapy and chemotherapy, resulting in a super additive antiproliferative effect [70]. In a related study, for the treatment of liver cancer, Che et al. fabricated nanoparticles based on the electrostatic interaction between the tumor suppressor miRNA-145, as anticancer gene therapy and cross-linked branched polyethyleneimine. The paclitaxel/PCL NFs were then coated with these nanoparticles. Upon applying the devised technology, both therapeutic compounds were released sustainably. Moreover, the concomitant delivery of paclitaxel and miRNA showed synergistic antiproliferative efficacy against hepatocellular carcinoma [71].

Nanotechnology is a highly active field as it has been evolving rapidly and becoming more and more reliable in clinical practice [72]. Electrospun NFs hold extraordinary promise to represent a breakthrough in cancer therapy as they show high loading capacity, large surface area, cost-effective production and direct delivery of various treatments. Moreover, those electrospun NFs represent a highly accurate diagnosis method such as ultra-sensitive sensing systems for cancer detection and migrating cancer cell targeting. However, electrospun NFs still need further research before introducing them to clinical practice and the market [31].

2.2. Electrospun NF Implants as Antimicrobial Agents

Antimicrobial agents work against bacteria and other microorganisms by preventing wall synthesis or limiting microbial development. Traditional antibiotic administration routes, such as topical and systemic administration, have some limitations in terms of bioavailability and dosage levels. Furthermore, poor antimicrobial agent distribution may reduce treatment efficacy in addition to its local and systemic side effects, such as inflammation and decreased normal gut microbiota. As a result, developing an effective antimicrobial drug delivery method is critical [73–75].

Electrospun NF implants are being studied for their use as novel delivery systems for antimicrobial agents using tetracycline hydrochloride with PLA, poly (ethylene-co-vinyl acetate) (PEVA) or a mixture of both [76].

Research showed that electrospun PCL/poly(trimethylene-carbonate) (PTMC) ultra-fine composite fiber mats perform as drug carriers that enclose the herbal antibacterial agent: Shikonin (SKN), a highly liposoluble naphthoquinone pigment isolated from the roots of lithospermum erythrorhizon. Drug release is then controlled by PCL/PTMC blend ratio in addition to drug-loading concentration. Those mats have free radical scavenging effects in addition to the antibacterial ones. Therefore, they also represent a compelling choice for treating dermal bacterial infections or wound healing [77].

Treating infections caused by Methicillin-resistant Staphylococcus aureus (MRSA) represents a serious challenge in the healthcare field as it requires accuracy in delivering the appropriate antibiotic at the correct dose to assure efficacy. Researchers developed a biodegradable localized delivery system for combining fusidic acid, sodium fusidic and rifampicin into PLGA polymer using the ES method for creating antimicrobial drug-loaded mats. Results showed that this delivery system was effective against Staphylococcus epidermidis and two MRSA strains [78]. These drug-loaded mats were proven to be an effective therapeutic option for implant-related infections that would otherwise necessitate orthopaedic surgery; thus, they would let patients avoid surgery wherever possible.

Guided bone regeneration and guided tissue regeneration (GBR/GTR) have become standard techniques in bone/tissue regeneration therapy to enhance bone/tissue regeneration through the application of GBR/GTR membranes that act as a barrier for the epithelial migration into the defective site and increase cell proliferation and attachment into the defect location [79]. In clinical applications, infections can be blamed for most GBR/GTR failure cases [80]. Accordingly, developing an anti-infective GBR-GTR membrane is a practical approach to solve this burden. By loading metronidazole into PCL/gelatin NF scaffolds, Xue and coworkers developed an efficient anti-infective GBR-GTR membrane with antibacterial activity delivered locally using the ES technique (Figure 4). The metronidazole release rate depended on the loaded content in the scaffold and the presence of gelatin in the membrane [81].

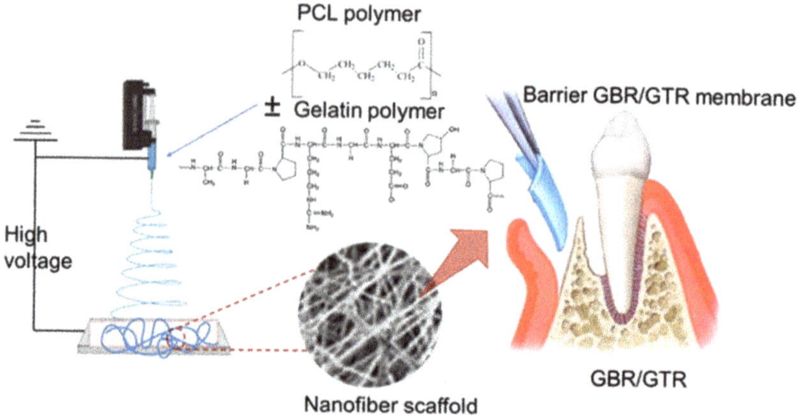

Figure 4. Development of a barrier membrane using the ES technique for the localized delivery of antibacterial agent in GBR/GTR applications.

Kataria et al. [82] had studied electrospun composite NFs transdermal patches for wound healing. Their research concluded that electrospun NF of sodium alginate (NaAlg)/PVA—ciprofloxacin-based transdermal patches were greatly beneficial for the local and fast delivery of medications to limit and control infections. These patches can successfully provide an efficient and fast drug delivery system to control infections. These patches show sustained drug release levels following the Korsmeyer-Peppas and Higuchi model.

2.3. *Electrospun NF Implants Used in Regenerative Medicine*

2.3.1. Wound Healing

Skin is the largest organ of the human body [83]. It represents the first barrier blocking pathogens from entering the body and limiting water loss [84,85]. Acute wounds (such as surgical or traumatic wounds, as well as burns and abrasions) and chronic wounds (which do not show a normal and correctly sequenced repair) are the two main types of skin wounds. These wounds are common in diabetic or decubitus ulcers [86]. According to National Center for Health Research and Statistics, trauma cases exceed 40 million in the United States each year, costing more than USD 670 billion. Chronic wounds affect approximately 6.5 million people in the United States, with treatment costs totaling more than USD 25 billion [87]. The situation is even worse in developing countries [88,89].

There are several classification systems for skin wounds, based on the size, depth and tissues involved, as well as the cause, such as burns, traumas or chronic illnesses. Wound dressings are essential for preventing infections and ensuring a healthy environment for wound healing. Sometimes, wounds may need dermal substitutes to trigger cells migration to the afflicted location [90,91]. Full-thickness wounds are more difficult to cure than superficial wounds because they require artificial skin substitutes or autografts, as well as a healthy environment [92]. The majority of wound healing materials cannot imitate the cutaneous extracellular matrix (ECM) [93]. On the other hand, ES technology has created a unique and sophisticated way for wound healing since it can produce biomimetic nanofibrous materials from various natural and synthetic polymers with clinically relevant properties [94]. Electrospun NFs made from type I and III collagen, which are the most common collagens in the skin, are now possible thanks to advances in nanotechnology and nanoengineering [95].

Electrospun NF implants are also an effective and efficient way to treat wounds as they feature ease of application and the great potential of loading them with different pharmaceutical agents such as antibiotics or regenerative agents [96].as shown in Table 2.

Table 2. Therapeutic agents loaded into electrospun NF implants to promote wound healing.

Polymeric Fiber	Therapeutic Agent	Purpose	Ref.
PLLA/PVA PLGA	Cefazoline	Antibacterial	[97] [98]
Chitosan/PVA	Lysozyme	Antimicrobial	[99]
PCL	Chloramphenicol	Antibacterial	[100]
PLGA	Quercetin	Cell proliferation and adhesion/antibacterial	[101]
PCL	Rifampicin	Antimicrobial	[102]
PCL/Gelatin	Metronidazole	Antibacterial	[81]
PCL	Tauroursodeoxycholic acid	Angiogenesis	[103]

Electrospun NF implants represent an excellent choice for regulating skin cell behavior through drug loading or using transmembrane receptors or intracellular signaling pathways [104]. Electrospun NFs encapsulated by type I collagen, laminin and integrin ligands prompted normal epidermal keratinocyte adhesion in humans, as well as a spreading morphology in 50% of proliferating cells, according to a study conducted by Rho and coworkers [105]. Furthermore, the effects of PCL-blended collagen nanofibrous membranes were investigated, and it was discovered that this membrane stimulates the adhesion and proliferation of human dermal fibroblasts [106].

Skin grafts are still one of the most popular and reliable methods for skin replacement or restoration because skin substitutes currently lack many key characteristics that allow them to mimic normal skin in terms of architecture and composition [107].

Ma et al. developed a novel method combining tissue engineering technologies with clinically viable techniques such as autologous skin grafting. This method involved creating a multifunctional NF skin graft in the shape of a sandwich (Figure 5) [108].

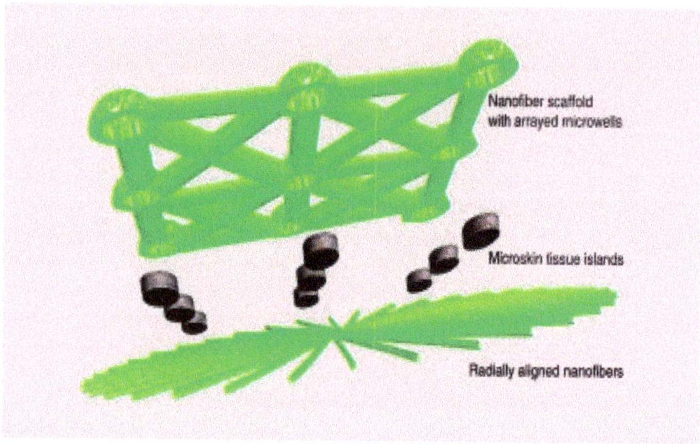

Figure 5. Sandwich-type NF skin grafts. Reprinted with permission from ref. [104].

This NF implant has several functions, including guiding cell migration to ensure proper and organized cell localization and accelerating the healing process in general by providing a higher expansion ratio and a sustained and stable release of antibiotics, which inhibits infections locally and limits antibiotic overuse toxicity.

This method was tested by implanting these sandwich-type scaffolds on skin excision in rats. Results showed a good acceptance of the scaffold in the wound site in all transplanted microskin tissue islands with a consistent distribution seven days after the surgery. The wound was completely closed 21 days after surgery, thanks to re-epithelialization via microskin grafts. This procedure will aid in healing severe burns and treating chronic injuries.

2.3.2. Dental Applications

One of the major applications of electrospun NF implants is in the dental field, as they provide an excellent option for replacing dental tissues. Figure 6 shows the standard approach for utilizing electrospun NFs implants for dental tissue regeneration [109].

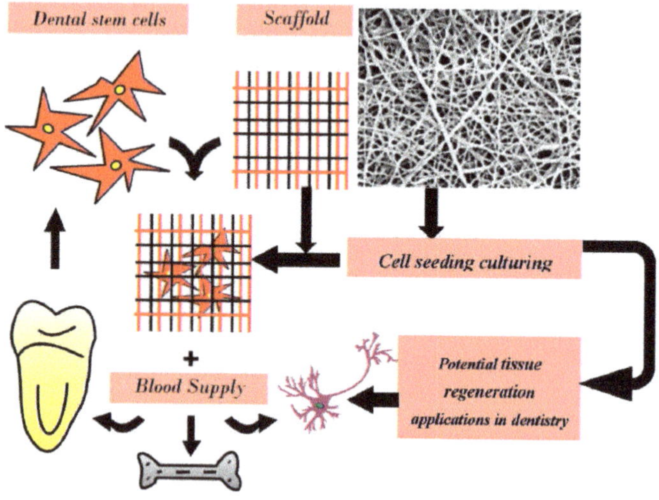

Figure 6. Schematic presentation for implementing electrospun NF implants in dental tissue regeneration. Republished under permission from ref. [109].

Numerous conditions such as dental caries and traumatic accidents may lead to losing dental tissues. Conventional treatment methods involve using different restorative materials such as ferric sulphate, hydroxide and mineral trioxide aggregates. However, these conventional methods may still result in internal resorption of teeth [110,111]. This is where electrospun NF implants make the difference as they were studied in order to assess their potential in dental regeneration. Kim e al. had created electrospun implants from PVA and hydroxyapatite (HA), which may have dentin regenerative features [112]. Additionally, electrospun NF meshes created with PCL have exhibited great potential for promoting odontogenic regeneration and differentiation. Evidence on that is the increased turnover of collagen I and other proteins when tested on human pulpal cells in vitro [113].

2.3.3. THE Role of Electrospun NF Implants for the Treatment of Pelvic Organ Prolapse

Pelvic Organ Prolapse (POP) is a deteriorating urogynecological pelvic floor chronic illness that affects the quality of life of 50% of parous women over the age of 50 [114]. Surgical repair with transvaginal meshes is one of the most common treatments for POP. However, such meshes have been linked to significant problems such as long-term chronic inflammation and poor tissue integration [115]. Nanotechnology has proposed a potential and unique solution to this problem. Mukherjee et al. created electrospun NF poly L-lactic acid-co-poly-caprolactone (PLCL) meshes loaded with endometrial mesenchymal stromal cells (MSCs) that alter foreign body reactions through several mechanisms and evaluated these meshes in mice in their study. According to the findings, the use of electrospun NF meshes loaded with MSCs in the treatment of POP in mice resulted in a significant increase in angiogenesis, cell adhesion and immunological response after 6 weeks post implant. Anti-inflammatory genes were also expressed more frequently, which helped to control inflammation and promote healing. Those findings point to a high likelihood of a successful implant as well as a viable alternative to surgical operations [115].

2.3.4. Electrospun NF Implants for Cardiac Tissue Engineering

Cardiovascular disorders are one of the leading causes of morbidity, reducing patients' quality of life and resulting in a huge number of untimely deaths around the world. Heart transplantation is the only drastic treatment for end-stage cardiac failure currently available [116]. This treatment has several disadvantages, including the restricted availability of donated organs and the problems associated with immunosuppressing the patient after the procedure to prevent transplant rejection.

Cardiac cells are non-renewable; thus, cardiac cell therapy is critical for addressing the majority of heart diseases [117]. It has a lot of potential in terms of treating myocardial infarctions, for example. One of the novel therapeutic methods is integrating cells into 3D biodegradable scaffolds, which appears to be viable for preserving cell viability and promoting cell embedding following transplantation [118]. This is a significant advancement in the field of cardiac cell therapy, as it is a more advanced treatment strategy than simply injecting isolated cells.

Cardiac tissue cells impose a complex challenge because of its unique characteristics and compound structure. However, there is a huge potential for designing new techniques and conducting further research to develop ultimate novel methods of cardiac cell therapy. Electrospun NF scaffolds, which blend natural and synthetic polymers, are a viable therapy technique for cardiac diseases. The major goal is to combine the benefits of both sources to create the best design that provides the required elasticity and electrical properties and typical healthy ECM protein adhesion cellular ligands [119].

2.3.5. Orthopaedic Applications

ES is a flexible technique that allows us to create NFs of a web-like design mimicking ECM, which significantly improves cell adhesion and integration [120,121]. This feature has been studied to assess its possible applications in the biomedical field. Hard tissues, such as bones and cartilages, are no exception since the polymers utilized in the electrospun

NF implants play an essential role in guiding them to their intended purpose. PLGA, PCL, PLLA and their copolymers have been widely employed because they are biocompatible and biodegradable [122–125]. Because of its biodegradability, low cost, flexibility and good biocompatibility with osteoblasts, the FDA has approved the use of PCL polymers in durable implants [126]. On the other hand, most other polymers have limits in the orthopedic sector because they are not strong enough, hydrophobic or bioactive [127,128]. To overcome their low stiffness, researchers have proposed mixing these polymers with bioactive chemicals such as HA, bioactive glass (BG) and tricalcium phosphate.

Magnesium alloys were reported as promising candidates for bone fixation and tissue regeneration [129]. Furthermore, electrospun NFs loaded with HA and simvastatin had a dual effect through controlling degradation and triggering bone regeneration. Overall, in a study conducted by Kim's group, magnesium alloy ES with a covering of PCL/HA/simvastatin NFs demonstrated promise as a drug delivery system and a hard-tissue engineering technology [130].

2.3.6. Electrospun NFs as Antibacterial Coatings to Prevent Implant-Related Infections

The development of an infection and formation of bacterial biofilm over an implant surface are considered the major impediments to the success of orthopedic and dental biomaterials implantation that may require the reoperation of patients. Accordingly, there is a clear need to develop a tunable implant coating with excellent bactericidal properties. However, there are only a few available techniques for coating the required contaminants, and most of them lack effectiveness [131]. ES technology has shown its ability to tackle the limitations of the other technologies due to its high antibiotic loading efficiency and sustained drug release properties [132] (Table 3).

Kranthi et al. used the ES technique to incorporate HA and rifampicin nanoparticles into PCL composite NFs and then coated them into titanium bone implants to tackle related infections [133]. Such coating was effective even against MRSA. HA ceramic was demonstrated to resemble the bone material and increase the osteogenic cell activity [134]. The presence of HA nanoparticles within the PCL composite NFs was shown to reinforce the mechanical properties of the composite scaffold.

Furthermore, most of the approaches used to prevent implant-related infections are designed to release only one antibiotic, increasing the risk of antibiotic resistance development [135,136]. As a result, many researchers used the ES technique to apply combinatorial antibacterial therapy to implant coatings. Jahanmard et al. developed PLGA/PCL bilayer coating structure with a combinatorial antibacterial activity using ES technology [137]. Vancomycin and rifampicin co-delivery in a bilayer structure could provide a sustained antibacterial effect over several weeks. Hence, it could cover both early and delayed onset of infection in comparison with the single coating structure.

Similarly, Ashbaugh et al. used the same two polymeric fibers, PCL/PGLA, for the tunable delivery of different antibody combinations [138]. Rifampicin was co-loaded into the composite coating with either vancomycin, linezolid or daptomycin. In the in vivo mouse model of *Staphylococcus aureus* prosthetic joint infections, all three antibody combinations could inhibit the infection and biofilm growth over the implant entirely. It is worth noting that by adding rifampicin within the PCL layer, this PCL/PGLA composite coating allows the tunable delivery of the antibiotic combination, with rifampicin diffusing faster than the other antibiotic. Accordingly, the development of the known rifampicin resistance [139,140] during therapy can be overcome by avoiding the existence of rifampicin as a single agent during the treatment period.

Table 3. Electrospun NF coatings for implant-related infections.

Polymeric Fiber	Antibacterial Agent	Implant	Antibacterial Action against	Ref.
PCL/HA (Polymer/Ceramic)	Rifampicin	Titanium (orthopedic)	*Staphylococcus aureus, Staphylococcus epidermidis, Pseudomonas aeruginosa* and *MRSA*	[133]
PLGA/PCL	Vancomycin and Rifampicin	Titanium (orthopedic)	*Staphylococcus aureus*	[137]
PLGA/PCL	Vancomycin/Rifampicin Linezolid/rifampicin Daptomycin/rifampicin	Titanium (orthopedic)	*MRSA*	[138]
PLGA	Vancomycin	Titanium (orthopedic)	*Staphylococcus aureus*	[132]
PLA/PCL/gelatin	Tetracycline	Titanium (dental)	*Prevotella intermedia, Porphyromonas gingivalis, Fusobacterium nucleatum* and *Porphyromonas gingivalis*	[141]
Chitosan/polyethylene oxide (PEO)	Vancomycin	Titanium (orthopedic)	*Staphylococcus aureus*	[142]
Keratin	Silver	Titanium (dental)	*Staphylococcus aureus*	[143]
PCL/PVA	Doxycycline	Titanium (orthopedic)	*Staphylococcus aureus*	[144]
Carboxymethylcellulose (CMC)/ PEO	Clindamycin	AISI 316LVM (stainless steel) and Ti90Al6V4 (alloy) (orgopedic)	Staphylococci, streptococci, pneumococci, and bacteroides species	[145]
PLGA/PEO	Gentamycin	Titanium (orthopedic)	*Staphylococcus aureus*	[146]

ES is a promising technology that has been investigated extensively for its various biomedical applications and advantages. Electrospun NF goods, particularly electrospun NF implants, have undergone extensive research to be introduced into clinical practice. A lot of research has shown promising results for finding novel treatments to various serious medical conditions and cosmetic purposes. These conditions involved cancer therapy, regenerative medicine, sustained release of medications, wound healing and other surgical uses. However, this field has not been fully assessed and studied. We still need to conduct further research to achieve the most possible out of this amazing technology and fully integrate it into clinical practice.

2.3.7. NF for Wound Healing Applications

Skin wounds are associated with increasingly high morbidity and mortality rates that can be regarded as a matter of life and death. While there are some available treatment options, they utterly fail to recover the damaged skin tissue's normal state and structure, exacerbating wound infection and/or tissue necrosis. Electrospun NFs have been the subject of research to develop wound dressings that can allow for skin regeneration primarily because they are compatible with the skin's ECM, improve skin healing rates, are degradable over time and can be quickly loaded with a bactericidal agent to prevent tissue infection and/or bioactive molecules and help with the regeneration process [22,147,148].

Water Resistance and Breathable Wound Dressings

NFs are currently one of the leading materials used for wound healing applications. The NFs can be molded as membranes supply excellent water resistance properties, and they also have a sufficient pore number on the surface, making them breathable, causing the moisture from the wound to escape [149–151]. However, a lower pore diameter guarantees water resistance [152–154].

Many researchers have tried to develop breathable and water-resistant wound dressings. Yue et al. developed a waterproof and breathable (W&B) membrane for wound healing [154]. Yue et al. electrospun the membrane with a perfusion speed of 0.05 mL/h to 10 mL/h and a voltage of 11 kV. The thymol-loaded ethanol-soluble polyurethane skin-like membranes were prepared. The membrane showed waterproofness to 17.6 cm of water and a comparatively better breathability rate. Apart from this, the membrane also showed antibacterial action against the wound.

Similarly, another group of researchers developed an asymmetric wettable gradient nanofibrous membrane, consisting of polyvinyl butyral (PVB)-polydimethylsiloxane (PDMS) as the top layer followed by a layer of PVB-PDMS/gelatin and gelatin as a hydrophilic layer [155]. The upper layer showed a better contact angle (CA) than the gelatin lower layer. The membrane showed water resistance to about 58.21 kPa with a water vapor transmission rate (WVTR) of 8.80 kg $m^{-2}d^{-1}$. Additionally, the membrane improved mesenchymal cells with the immobilization of stromal cell-derived factors to increase the wound healing rate.

The stretchable PDMS-embedded PVB fibrous membranes based on ethanol solvent system for W&B wound dressings were prepared by Guo et al. using the ES technique [156]. The presence of PDMS caused a reduction in pore size and enhanced surface hydrophobicity by four times. The membranes showed a WVTR of 8.98 kg $m^{-2}d^{-1}$. An increase in CA from 127.80° to 133.16° was observed after increasing PDMS content by 45%; however, the increase was not significant. This might be attributed to the presence of PDMS as a low surface energy agent on the fiber surface. However, due to surface roughness, the increase was not so prominent. In an experiment conducted with pure water, coffee, milk, methyl orange, methylene blue and rhodamine were dripped on the membrane surface. It was found that the sphere-like droplets were formed due to lyophobic properties. The membrane was tilted at an angle of $30 \pm 5°$ so that drops could roll, and the drops tolled off smoothly without wetting the membranes indicating the antifouling property. However, in the case of soybean oil and petroleum ether, poor oleophobicity was observed due to the oleophilic nature of PDMS. Researchers developed hydrophobic microporous NF membrane via fluorosilane-modified silica (F-SiO_2) nanoparticles, polyurethane and polyacrylonitrile solution [157]. The presence of flurosilane led to the hydrophobic behavior in the membrane. The formed solution was electrospun, and results showed that incorporating F-SiO_2 could promote the hydrophobic nature of the membrane; thus, the membrane exhibited excellent waterproof and mechanical properties with good WVTR results. In addition, they can be used in the medical field for drug delivery applications. Another group developed a nanofibrous membrane using polyacrylonitrile, polyurethane and titanium dioxide. Apart from W&B resistance, the formed films also showed UV resistance due to UV531 and fluorinated acrylic copolymer [158]. At the same time, the hydrophobic coating showed superhydrophobicity and wettability with a CA of 152.1°.

Electrospun NFs as a Drug Delivery Carrier for Wound Healing

Electrospun NFs are the first choice for practitioners owing to their mimicking nature with ECM and low or nil adverse effects. Researchers have used natural polymers for wound healing applications either alone or combined with synthetic polymeric counterparts to increase their mechanical and physical properties. Depending on the need and the wound getting treated, the NFs can be impregnated with drugs and stem cells as indicated in Table 4.

The NFs containing the active moiety have been shown by various studies and are the primarily used approach during the past decade. Mohammadi et al. prepared electrospun CUR-loaded PCL /gum tragacanth NFs to treat diabetic rats [159]. The scaffolds also possessed antimicrobial action against various Gram-positive and Gram-negative bacteria. After 15 days of application of NFs, remarkably faster wound healing was shown.

Morphologically, the wound showed proliferation of fibroblasts, deposition of collagen and formation of sweat glands and hair follicles, confirming the wound healing.

Similarly, another research group reported the formulation of CUR-based PCL NFs formed via the ES process [160]. The fibers were able to release the active ingredient for about three days in vitro. The human foreskin fibroblast cell lines used showed 70% enhancement in the bioavailability of CUR. Additionally, the fibers reduced the levels of interleukin-6 release from the in vitro model and showed an acceleration in wound closure rate in the streptozotocin-induced diabetic model. Similar membranes were prepared by ES technique using the gum tragacanth, PLGA and tetracycline as the active agent. These NFs showed smooth, beadles morphology and controlled release of tetracycline for over 75 days, with an initial burst of 19% for the first two days [161].

Golchin et al. prepared CUR incorporated CS/PVA/Carbopol/PCL nanofibrous composite to deliver buccal fat pad-derived mesenchymal cells and CUR [162]. The formed nanofibrous allowed cell seeding and enhanced their growth and proliferation. Mussel-inspired electrospun NFs doped with silver nanoparticles were formulated for wound healing applications by GhavamiNejad et al. using the catechol redox chemistry [163]. The silver nanoparticles themselves have antibacterial properties, and their activity was found to be size- and shape-dependent and the degree of aggregation is also engaged. Due to the absence of an external reducing agent, the arising tissue toxicity was minimized. The presence of 1% nanoparticles showed optimized antibacterial action against different strains of bacteria. The fibers offered a rapid release of nanoparticles in the first 24 h and could sustain release for about five days, which was the main requirement for effective antibacterial action (Figure 7).

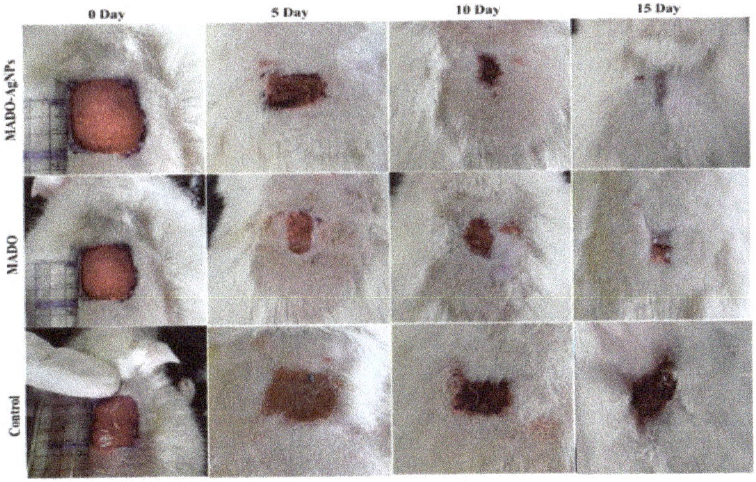

Figure 7. Wound appearance at 0, 5, 10 and 15 days after grafting with poly(dopamine methacrylamide-co-methyl methacrylate) (MADO)-AgNPs, MADO NF, and control [163].

Similarly, other researchers developed wound dressing based on metallic ions such as silver and magnesium [164]. The wound dressing consisted of gelatin, PCL NFs loaded with metallic ions. The in vitro results show that the dressings showed good cytocompatibility and had good antibacterial activity against different bacterial strains. The presence of magnesium promoted the tube formation of vascular endothelial cells, epidermal formation, vascularization and collagen deposition (Figure 8).

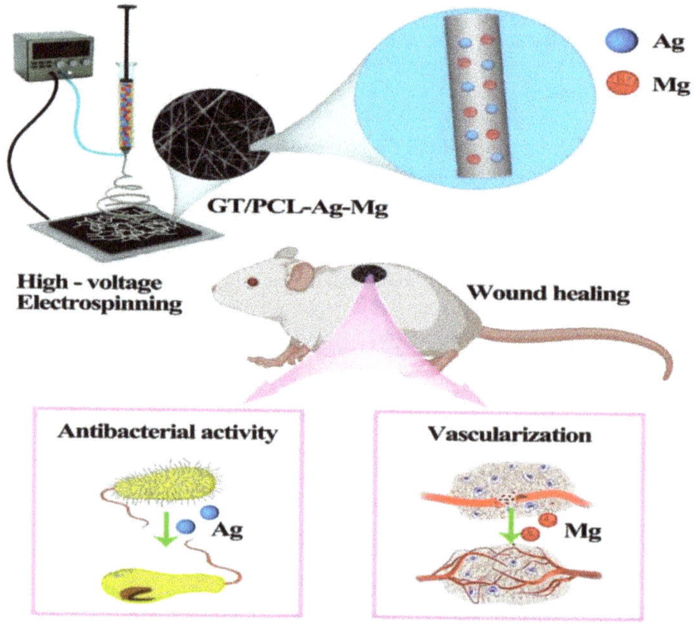

Figure 8. A gelatin (GT)/PCL NF membrane with Ag and Mg ions (GT/PCL-Ag-Mg) was fabricated, and its antibacterial and angiogenesis functions were demonstrated using in vitro and in vivo studies.

Many attempts were made to encapsulate and deliver the natural extracts via electrospun NFs. Vargas et al. followed this approach using the *Calendula officinalis* and hyperbranched polyglycerol (HPGL) since the active ingredients had wound healing and anti-inflammatory action. The electrospun fibers showed rapid release owing to their high swelling ability and high porosity [165]. By increasing the *Calendula officinalis* concentration, a higher release rate from the fibers was observed. The in vivo experiments conducted on rats showed their potential wound healing characteristics.

Table 4. A literature review of various antibacterial wound dressings prepared using the ES technique.

Polymeric Chain	Active Agents	Antibacterial Action against	Ref.
PLA/PCL	Tetracycline hydrochloride	*Staphylococcus aureus*, *Escherichia coli* and *Pseudomonas aeruginosa*	[166]
PLGA	Cefoxitin sodium Amoxicillin	*Staphylococcus aureus* *Staphylococcus aureus*	[167] [168]
PLA	Mupirocin	*Staphylococcus aureus*	[169]
coPLA/PEG	Ciprofloxacin hydrochloride, levofloxacin hemihydrate or moxifloxacin hydrochloride	*Staphylococcus aureus*	[170]
PLA, PLA/Collagen	Gentamicin	*Escherichia coli*, *Staphylococcus epidermidis*, *Pseudomonas aeruginosa*	[171]
PLLACL	Tetracycline hydrochloride	*Escherichia coli*	[172]
PMMA/Nylon 6	Ampicillin	*Listeria innocua*	[173]
PAA	Doxycycline hyclate	*Staphylococcus aureus*, *Streptococcus agalactiae*, *Pseudomonas aeruginosa*	[174]
Cyclodextrin complex	Triclosan	*Staphylococcus aureus*, *Escherichia coli*	[175]

Table 4. *Cont.*

Polymeric Chain	Active Agents	Antibacterial Action against	Ref.
PCL/PLA	N-halamine	*Escherichia coli, Staphylococcus epidermidis*	[176]
PLA		*Staphylococcus aureus, Escherichia coli*	[177]
PAN		*Staphylococcus aureus, Escherichia coli*	[178]
CAc	Quaternary ammonium salts	*Staphylococcus aureus, Escherichia coli*	[179]
	Chlorhexidine	*Escherichia coli, Staphylococcus epidermidis*	[180]
CAc/PEU	Polyhexamethylene biguanide	*Escherichia coli*	[181]
PEO/Chitosan	Potassium 5-nitro-8-quinolinolate	*Staphylococcus aureus, Escherichia coli, Candida albicans*	[182]
PAN	Silver NPs	*Staphylococcus aureus, Escherichia coli, Bacillus subtilis*	[183]
PLA/Chitosan		*Staphylococcus aureus, Escherichia coli*	[184]
PEO/Chitosan	Antimicrobial peptides (Plantaricin 423 and bacteriocin ST4SA)	*Escherichia coli*	[185]
PVA/Chitosan		*Escherichia coli*	[186]
PDLLA/PEO		*Enterococcus faecium*	[187]

Abbreviations. CAc: Cellulose acetate; coPLA: poly(l-lactide-co-d,l-lactide); PAA: poly(acrylic acid); PAN: polyacrylonitrile; PEU: polyester urethan; PCL: poly(ε-caprolactone); PDLLA: poly(d,l-lactide); PEG: polyethylene glycol; PEO: poly(ethylene oxide); PLA: poly(lactic acid); PLGA: poly(lactide-co-glycolide); PLLACL: poly(l-lactid-co-ε-caprolactone); PMMA: poly(methyl methacrylate); and PVA: polyvinyl alcohol.

3. Mechanical Properties of Electrospun NF Implants

The mechanical properties of electrospun nanofiber implants are critical in applications such as implants that require long-term endurance and structural integrity. To be useful in any application, fibers must have acceptable mechanical properties. Traditional tensile tests can be used to assess the electrospun nanofiber materials' tensile strength. However, in most of the published research, these tests are performed on electrospun mats, scaffolds and other materials rather than the fiber itself because measuring these parameters for fibers in the nano or submicron range is very difficult [1,2].

During the service of electrospun nanofibers as implants, nanofibers are continuously subjected to stresses from the surrounding environment, which can significantly result in their permanent deformation. The mechanical limitations of electrospun nanofibers can hinder their use in implants and other applications. Despite of the high compatibility of gelatin fibers, they have poor mechanical properties (tensile strength, Young's modulus, elongation, elasticity). Therefore, the applications and further development of gelatin-based nanofibers are compromised [3]. Similarly, PLA and PLGA-based scaffolds' low mechanical properties resulted in their permanent deformation after a prolonged duration of cyclic strain [4]. On the other hand, the flexible mechanical proprieties of PCL-based nanofiber membranes resulted in their extensive study in various applications [5].

As a result, many approaches have been proposed to enhance the mechanical properties of electrospun nanofibers. Polymer blending is one of the used approaches to provide mechanically competent electrospun nanofiber-based implants. The tensile modulus of PCL polymer is 2.05 $\pm$ 0.15 MPa, which is much higher than the stiffness range reported for human myocardium in many literatures (20–50 KPa) [6]. Accordingly, the electrospun PCL-based implant does not successfully imitate the host ECM and thus-does not support effective implant regeneration and proliferation into the myocardium. Castilho and colleagues created electrospun nanofibers made of miscible blend of PCL and poly(hydroxymethylglycolide-co-caprolactone) (pHMGCL) in various ratios to overcome the failure of PCL-based myocardial implant in cardiac tissue engineering [7]. The tensile modulus of a blend of pHMGCL and PCL (40:60) decreased from 2.05 $\pm$ 0.15 MPa to 1.18 $\pm$ 0.08 MPa, bringing it closer to the range of normal myocardium stiffness. Other approaches can be also used to provide mechanically competent electrospun nanofibers, such as the modification of the polymer structure and the modification of the processing method.

4. Portable ES Device Applications

Introducing portable ES devices to the field of nanomedicine has emerged, primarily to overcome possible drawbacks associated with conventional ES setups that usually consist of mega-sized equipment that is difficult to be assembled for instant applications in a one-time use manner. The mobility of this device and its deposited fibers have further expanded the promise of in situ electrospinning for applications in wound healing and other fields. Only the spinneret component of these devices is portable since the rest of the machine is too huge and heavy to be hand handled, as with most other portable gadgets. In addition, the resulting matrices are bulky with possible inadequate stability restricting their ease of transportation and storage, accounting for a higher risk of degradation. Portable ES devices, on the other hand, are easily assembled using completely available off-shelf materials while producing highly efficient fibers for point of need application [188], mainly implemented in wound dressings with skin regenerating qualities [189–191] and rapid hemostasis [192] either in minimally invasive surgeries [193] or outdoor settings [194].

Brako et al. [188] presented a simple, portable electrospinning device to overcome this constraint and the high cost of the equipment that was previously used to address it. This consists of a miniaturized high-precision microsyringe pump (Micrel mph+) as well as a minuscule high-voltage power supply (EMCO 4330+) that is capable of producing up to 33 kV at a power of 10 W (as shown in Figure 9). By completely combining the spinneret and the power supply unit into a portable device, the whole apparatus becomes lighter in weight and smaller, allowing it to be deployed almost everywhere where there is mains electricity. In this technology, CAc and CAc doped with silver nanoparticles have been electrospun into fibers that have been effectively applied directly to simulated wound sites. Such portable devices may point the way to the more widespread and practical use of functional nanofibers in advanced wound care that is more individualized.

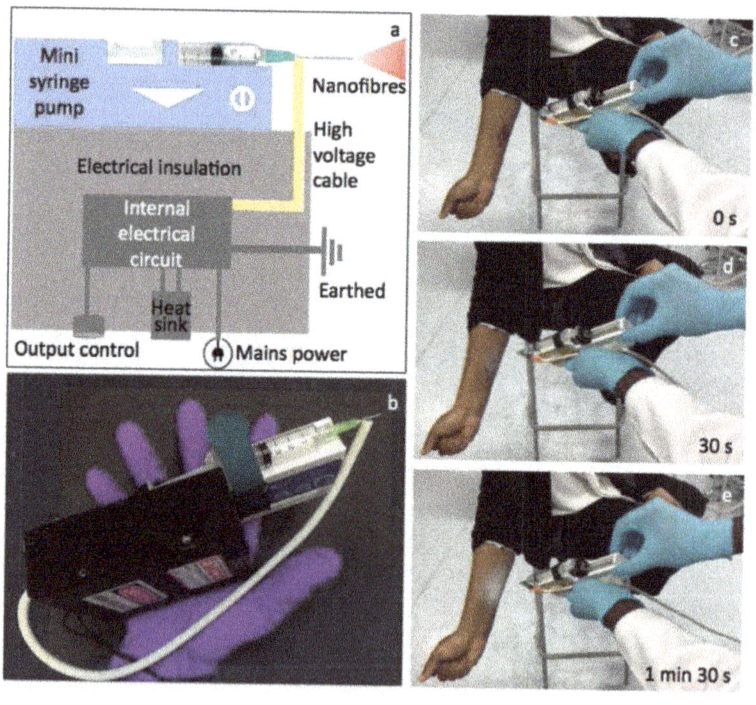

Figure 9. (**a**) Schematic diagram of the portable EHD device. (**b**) A photograph showing the assembly of the mini device held in hand. (**c–e**) Snapshots of real-time recording of the operation of the mini EHD device generating nanofibers onto a mock wound in situ [188] (Reproduced CC BY license).

Long and colleagues developed a simple portable battery-operated electrospinning apparatus (BOEA) that uses a finger-pressed syringe to generate electrospinning [195]. This device's high-voltage supply was provided by a combination of batteries and a high-voltage converter. It has been stated that the charge may be passed via the body (hand) by contacting the metal foil, hence preventing charge buildup from occurring. It was found that two AAA batteries (3 V) were sufficient for the practical use of the BOEA and that it was capable of continuous operation for more than 15 h with insignificant current and an effective working distance ranging from 2 to 10 cm with little current consumption. One benefit to note is using a battery and a high-voltage converter; thus, the device was lightweight (about 120 g) and had small overall dimensions, thus enhancing its portability. Several polymers, including polyvinyl pyrrolidone, polyvinylidene fluoride and PCL, were successfully electrospun into fibers with diameters in the hundreds of nanometers range using the BOEA, which was also tested for solution electrospinning performance.

4.1. Wound Dressings with Skin Regenerating Properties

Edirisinghe and colleagues suggested a portable electrospinning system, including a portable handheld spinneret, to get in situ electrospinning results. Essentially, the instrument consisted of a spray pistol equipped with either a single needle or two coaxial needles. This device was used to create PLGA particles and fibers and PLGA/Polymethysilsesquioxane core-shell fibers deposited onto a limited target region at various angles [196]. Following that, the portable electrospinning apparatus was used for in situ deposition of bioproducts throughout the wound healing process, and the results were promising. In this case, the portable device offered a handy way of depositing PLGA fibers onto a target location, such as an incised wound or the surface of a grazed wound. Furthermore, due to the electrostatic attraction forces, fibers may be sprayed onto a burn wound site in less than 300 s, resulting in a thin protective coating on the surface of the burn wound. A simple hand flexing might be used to remove the film [197].

Yue et al., 2021 introduced an enhanced portable ES device matching the shape of the gun and imports an attached battery (3 V), voltage transformer module (from 3 V to 5 V), high voltage converter (amplifying the 5 V up to 11 kV), stepper motor syringe pump, and a speed control module allowing for high voltage supply and controllable perfusion speed (from 0.05 mL/h to 10 mL/h) to prepare waterproof and breathable (W&D) fiber matrices. In addition, the current values were much smaller than the perception current, the minimum current value (of about 1 mA) that was perceivable by the human body senses, that they were ignored and ruled as safe for human application [154]. They used ethanol-soluble polyurethane (EPU), fluorinated polyurethane (FPU) and thymol (2-isopropyl-5-methylphenol) to prepare the polymer solution, as shown in Figure 10. EPU/FPU formed the constructive basis of the fiber-matrix. At the same time, thymol added an antibacterial property to the generated skin-like matrices that were tuned precisely for every wound structure with excellent mechanical properties, pertaining to its porosity, thickness and moisture permeability with a high water vapor transmittance (WVT) rate of about 13.1 Kg $m^{-2}d^{-1}$, robust hydrophobicity with a moderate hydrostatic pressure of 17.6 cm H_2O and excellent breathability of 3.56 kg $m^{-2}d^{-1}$ allowing for an overall enhanced patient experience [198].

Other possible candidates for the electrospun polymer solution used for preparing wound dressings include, but are not limited to, zein/thyme essential oil (TEO) [199], zein/clove essential oil (CEO) [200], zein and cinnamon oil [201] and PLA/gelatin [202]. The essential oil in every combination has antibacterial activity, and some even are active against Escherichia coli and Staphylococcus aureus, as in cinnamon oil with an inhibition zone of around 5 cm.

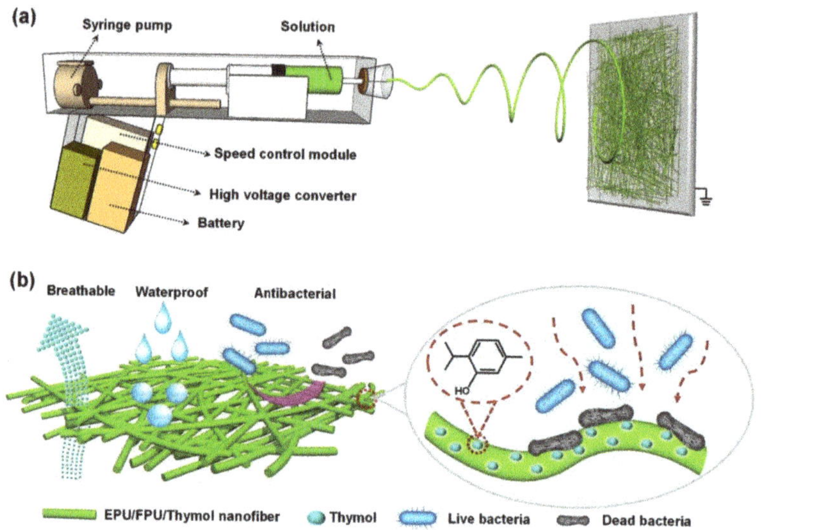

Figure 10. (**a**) Fabrication of EPU/FPU/Thymol nanofibrous membranes using the ES portable device. (**b**) Demonstrative scheme of the breathable, waterproof and antibacterial action of EPU/FPU/Thymol nanofibers. Republished under the permission of Yue et al. [154].

4.2. *Rapid Hemostasis*

The application of conventional ES fibers in internal organs such as the liver and meningeal closures requires their fixation and the fibers to be thick enough to mediate its adhesion to the organ lining; however, this limits its use in sensitive wounds such as in visceral bleeding. Among the proposed materials, PCL appears to be the most convenient for visceral application superior to chitosan, alginate and cyanoacrylates (CA) [203]. Zhou et al., 2020 proposed that it is was possible to use a portable ES device for in situ simultaneous application of PCL ES fibers in rapid intestinal hemostasis, as shown in Figure 11. The generated fiber matrices had strong adsorptive forces to wound exudates and were able to resist hydrostatic pressure; additionally, they had an overall good sealing effect [192].

Laparoscopic aided ES was introduced by Zhang et al., 2020 when they applied a handheld ES device with a long needle (40 cm long, 0.5 cm in diameter) inserted in a laparoscopic tube (0.6 cm in diameter) during a minimally invasive operation (MIO) on a pig liver. This technique showed rapid hemostasis (5 s) consistent with reduced inflammatory responses and accelerated recovery rates [193]. However, safety issues are still a concern, and clinical trials should evaluate the extent to which a portable ES device is applied within critical conditions, as in surgeries, instead of sutures and other conventional techniques.

The use of the BOEA for direct electrospinning of the medical adhesive, N-octyl-2-cyanoacrylate (NOCA), to create fibrous media suitable for soft tissue hemostasis, which is related to liver hemostasis, has been investigated more recently by the Luo et al. [204]. Controlling tissue adhesion after surgery is critical in this specific application, and accurate deposition of the hemostatic medical adhesive is often required to achieve this goal. It was necessary to attach a metal cone to the spinning nozzle to achieve precise deposition (as shown in Figure 12). It was discovered that by varying the metal cone's size, the electrospun fibers' deposition range could be modified to the desired value. Further, this modified BOEA was used to deposit NOCA fibers onto the resection site of the rat liver, resulting in rapid hemostasis within 10 s. When comparing the outcomes of this procedure to those of the airflow-assisted group, postoperative pathological findings revealed that the inflammatory response and tissue adhesion were decreased. A modified BOEA might be designed for emergency medical operations, community patient care or home care scenarios where the mobility and simplicity of operation would be especially helpful, as shown by this possibility. In this device, the voltage was fixed, limiting the extent to which the physical properties of dressings and the dimensions of dressings produced in situ could be modified. Additionally, the delivery of the spinning solution was dependent on the pressure applied by the fingers to the trigger, which could vary during use, resulting in inconsistent fiber deposition.

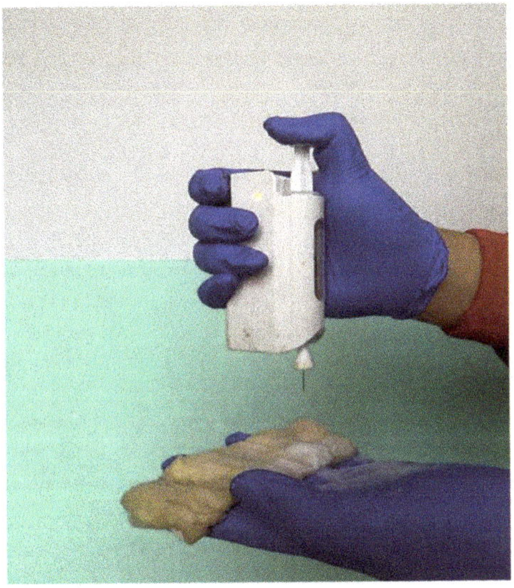

Figure 11. Implementing the portable ES device in an intestinal incision. With permission from ref. [192].

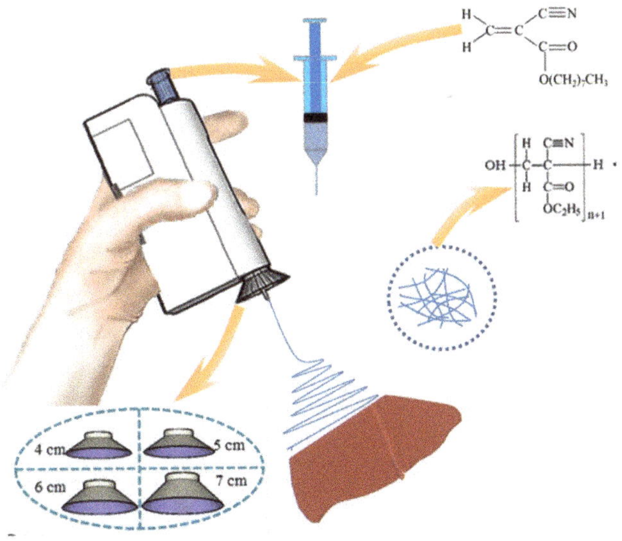

Figure 12. Schematic diagram of the electric field-modified e-spinning NOCA fibers for liver resection hemostasis.

Future Perspectives

The recently introduced portable devices have increasingly superseded the usage of homemade electrospinning equipment with portable spinnerets in recent years. These devices are comprised of a spinneret in the form of a pistol, and an airflow became commercially available. Therefore, the airflow constrains the electrospinning jets, resulting in the exact placement of electrospun fibers on the target region. Furthermore, the interest in using in situ fiber in mildly invasive surgeries over the conventional methods is increasing. However, this technology is in the early phase of its applications in clinical studies [192]. The problem with the approach is that the human body's conductive nature could precipitate short circuits and creepage. Therefore, further studies are necessary to establish cost-effective and less detrimental techniques required to meet the market demand. Besides that, in wound dressings level, rapid hemostasis, or even minimally invasive surgeries, fast, safe, and effective techniques are essential.

5. Conclusions

Advances in all biomedical domains and the increasing complexity of diagnosis and therapeutic processes necessitate the development of a unique drug delivery system to improve the bioavailability of pharmaceuticals and their actions, as well as their safety and adverse effects. Nanofiber biomaterials material fabricated by electrospinning possess extremely high specific surface area, high porosity, adjustable fiber morphology and surface function. Due to these properties, the electrospun fiber material can meet the requirements for advanced biomedicine and tissue engineering. Specifically, this paper aimed to explore the efficacy of electrospun nanofibers for application in wound treatment, antibacterial coating to prevent implant-related infections in orthopedic and dental applications, and implants for cancer therapy, by updating with information collected from the most recent literature. The ES of biopolymers and APIs is also discussed regarding their current scope, limits and future outlook. In the context of a nanotechnological approach for novel wound dressing fabrication, reducing scar formation and drug-loading with efficient therapeutic agents, this review may offer new perspectives and technical support for the development of advanced, functional, soft biomaterials from lab-scale to industrial-scale production.

Author Contributions: Conceptualization, N.E.E., A.N., T.M.I., H.C., G.A.G., S.E.E., S.C. and M.S.A.; data curation, M.S.A., S.C. and H.C.; writing—original draft preparation, N.E.E., A.N., T.M.I., H.C., G.A.G., S.E.E., S.C. and M.S.A.; writing—review and editing, H.C., M.S.A. and S.C.; funding acquisition, S.C. All authors have read and agreed to the published version of the manuscript.

Funding: No external funding was received.

Institutional Review Board Statement: Not applicable.

Informed Consent Statement: Not applicable.

Data Availability Statement: Not applicable.

Acknowledgments: The authors N.E.E., A.N., T.M.I., H.C., G.A.G., S.E., M.S.A. wish to extend their sincere thanks to Fakhr El-Din Soliman Ghazy (Dept. of Pharmaceutics, Faculty of Pharmacy, Zagazig University). In addition, the authors are thankful for biorender.com for creating the graphical abstract and Figure 1 with license number (SF23F59GTA, CG23IW3YWG).

Conflicts of Interest: The authors declare no conflict of interest.

References

1. Bhardwaj, N.; Kundu, S.C. Electrospinning: A Fascinating Fiber Fabrication Technique. *Biotechnol. Adv.* **2010**, *28*, 325–347. [CrossRef] [PubMed]
2. Taylor, G.I. Disintegration of Water Drops in an Electric Field. *Proc. R. Soc. Lond. Ser. A. Math. Phys. Sci.* **1964**, *280*, 383–397. [CrossRef]
3. Islam, M.S.; Ang, B.C.; Andriyana, A.; Afifi, A.M. A Review on Fabrication of Nanofibers via Electrospinning and Their Applications. *SN Appl. Sci.* **2019**, *1*, 1248. [CrossRef]

4. Luraghi, A.; Peri, F.; Moroni, L. Electrospinning for Drug Delivery Applications: A Review. *J. Control. Release* **2021**, *334*, 463–484. [CrossRef] [PubMed]
5. Miletić, A.; Pavlić, B.; Ristić, I.; Zeković, Z.; Pilić, B. Encapsulation of Fatty Oils into Electrospun Nanofibers for Cosmetic Products with Antioxidant Activity. *Appl. Sci.* **2019**, *9*, 2955. [CrossRef]
6. Zhang, C.; Li, Y.; Wang, P.; Zhang, H. Electrospinning of Nanofibers: Potentials and Perspectives for Active Food Packaging. *Compr. Rev. Food Sci. Food Saf.* **2020**, *19*, 479–502. [CrossRef]
7. Han, W.; Wang, Y.; Su, J.; Xin, X.; Guo, Y.; Long, Y.-Z.; Ramakrishna, S. Fabrication of Nanofibrous Sensors by Electrospinning. *Sci. China Technol. Sci.* **2019**, *62*, 886–894. [CrossRef]
8. Mirjalili, M.; Zohoori, S. Review for Application of Electrospinning and Electrospun Nanofibers Technology in Textile Industry. *J. Nanostruct. Chem.* **2016**, *6*, 207–213. [CrossRef]
9. Contreras-Cáceres, R.; Cabeza, L.; Perazzoli, G.; Díaz, A.; López-Romero, J.M.; Melguizo, C.; Prados, J. Electrospun Nanofibers: Recent Applications in Drug Delivery and Cancer Therapy. *Nanomaterials* **2019**, *9*, 656. [CrossRef]
10. Calori, I.R.; Braga, G.; de Jesus, P.d.C.C.; Bi, H.; Tedesco, A.C. Polymer Scaffolds as Drug Delivery Systems. *Eur. Polym. J.* **2020**, *129*, 109621. [CrossRef]
11. Bunggulawa, E.J.; Wang, W.; Yin, T.; Wang, N.; Durkan, C.; Wang, Y.; Wang, G. Recent Advancements in the Use of Exosomes as Drug Delivery Systems. *J. Nanobiotechnol.* **2018**, *16*, 81. [CrossRef] [PubMed]
12. Sandri, G.; Rossi, S.; Bonferoni, M.C.; Caramella, C.; Ferrari, F. Electrospinning Technologies in Wound Dressing Applications. In *Therapeutic Dressings and Wound Healing Applications*; John Wieley & Sons Ltd.: Chichester, UK, 2020; pp. 315–336, ISBN 978-1-119-43331-6.
13. Dong, Y.; Zheng, Y.; Zhang, K.; Yao, Y.; Wang, L.; Li, X.; Yu, J.; Ding, B. Electrospun Nanofibrous Materials for Wound Healing. *Adv. Fiber Mater.* **2020**, *2*, 212–227. [CrossRef]
14. Chahal, S.; Kumar, A.; Hussian, F.S.J. Development of Biomimetic Electrospun Polymeric Biomaterials for Bone Tissue Engineering. A Review. *J. Biomater. Sci. Polym. Ed.* **2019**, *30*, 1308–1355. [CrossRef]
15. Tan, G.Z.; Zhou, Y. Electrospinning of Biomimetic Fibrous Scaffolds for Tissue Engineering: A Review. *Int. J. Polym. Mater. Polym. Biomater.* **2020**, *69*, 947–960. [CrossRef]
16. Venugopal, J.; Prabhakaran, M.P.; Low, S.; Choon, A.T.; Zhang, Y.Z.; Deepika, G.; Ramakrishna, S. Nanotechnology for Nanomedicine and Delivery of Drugs. *Curr. Pharm. Des.* **2008**, *14*, 2184–2200. [CrossRef]
17. Hu, X.; Liu, S.; Zhou, G.; Huang, Y.; Xie, Z.; Jing, X. Electrospinning of Polymeric Nanofibers for Drug Delivery Applications. *J. Control. Release* **2014**, *185*, 12–21. [CrossRef]
18. Thakkar, S.; Misra, M. Electrospun Polymeric Nanofibers: New Horizons in Drug Delivery. *Eur. J. Pharm. Sci.* **2017**, *107*, 148–167. [CrossRef]
19. Su, Y.; Su, Q.; Liu, W.; Jin, G.; Mo, X.; Ramakrishn, S. Dual-Drug Encapsulation and Release from Core–Shell Nanofibers. *J. Biomater. Sci. Polym. Ed.* **2012**, *23*, 861–871. [CrossRef]
20. Rošic, R.; Kocbek, P.; Baumgartner, S.; Kristl, J. Electro-Spun Hydroxyethyl Cellulose Nanofibers: The Relationship between Structure and Process. *J. Drug Deliv. Sci. Technol.* **2011**, *21*, 229–236. [CrossRef]
21. Pelipenko, J.; Kocbek, P.; Kristl, J. Critical Attributes of Nanofibers: Preparation, Drug Loading, and Tissue Regeneration. *Int. J. Pharm.* **2015**, *484*, 57–74. [CrossRef]
22. Miguel, S.P.; Figueira, D.R.; Simões, D.; Ribeiro, M.P.; Coutinho, P.; Ferreira, P.; Correia, I.J. Electrospun Polymeric Nanofibres as Wound Dressings: A Review. *Colloids Surf. B Biointerfaces* **2018**, *169*, 60–71. [CrossRef] [PubMed]
23. Yan, X.; Yu, M.; Ramakrishna, S.; Russell, S.J.; Long, Y.-Z. Advances in Portable Electrospinning Devices for in Situ Delivery of Personalized Wound Care. *Nanoscale* **2019**, *11*, 19166–19178. [CrossRef] [PubMed]
24. Keirouz, A.; Chung, M.; Kwon, J.; Fortunato, G.; Radacsi, N. 2D and 3D Electrospinning Technologies for the Fabrication of Nanofibrous Scaffolds for Skin Tissue Engineering: A Review. *WIREs Nanomed. Nanobiotechnol.* **2020**, *12*, e1626. [CrossRef] [PubMed]
25. Padmakumar, S.; Paul-Prasanth, B.; Pavithran, K.; Vijaykumar, D.K.; Rajanbabu, A.; Sivanarayanan, T.B.; Kadakia, E.; Amiji, M.M.; Nair, S.V.; Menon, D. Long-Term Drug Delivery Using Implantable Electrospun Woven Polymeric Nanotextiles. *Nanomed. Nanotechnol. Biol. Med.* **2019**, *15*, 274–284. [CrossRef]
26. Ngweniform, P.; Abbineni, G.; Cao, B.; Mao, C. Self-Assembly of Drug-Loaded Liposomes on Genetically Engineered Target-Recognizing M13 Phage: A Novel Nanocarrier for Targeted Drug Delivery. *Small* **2009**, *5*, 1963–1969. [CrossRef]
27. Li, W.; Peng, J.; Tan, L.; Wu, J.; Shi, K.; Qu, Y.; Wei, X.; Qian, Z. Mild Photothermal Therapy/Photodynamic Therapy/Chemotherapy of Breast Cancer by Lyp-1 Modified Docetaxel/IR820 Co-Loaded Micelles. *Biomaterials* **2016**, *106*, 119–133. [CrossRef]
28. Jain, R.K.; Stylianopoulos, T. Delivering Nanomedicine to Solid Tumors. *Nat. Rev. Clin. Oncol.* **2010**, *7*, 653–664. [CrossRef]
29. Albanese, A.; Tang, P.S.; Chan, W.C. The Effect of Nanoparticle Size, Shape, and Surface Chemistry on Biological Systems. *Annu. Rev. Biomed. Eng.* **2012**, *14*, 1–16. [CrossRef]
30. Chen, P.; Wu, Q.S.; Ding, Y.P.; Chu, M.; Huang, Z.M.; Hu, W. A Controlled Release System of Titanocene Dichloride by Electrospun Fiber and Its Antitumor Activity In Vitro. *Eur. J. Pharm. Biopharm.* **2010**, *76*, 413–420. [CrossRef]
31. Chen, Z.; Chen, Z.; Zhang, A.; Hu, J.; Wang, X.; Yang, Z. Electrospun Nanofibers for Cancer Diagnosis and Therapy. *Biomater. Sci.* **2016**, *4*, 922–932. [CrossRef]

32. Iannazzo, D.; Espro, C.; Celesti, C.; Ferlazzo, A.; Neri, G. Smart Biosensors for Cancer Diagnosis Based on Graphene Quantum Dots. *Cancers* **2021**, *13*, 3194. [CrossRef] [PubMed]
33. Davis, M.E.; Chen, Z.; Shin, D.M. Nanoparticle Therapeutics: An Emerging Treatment Modality for Cancer. *Nat. Rev. Drug Discov.* **2008**, *7*, 771–782. [CrossRef]
34. Wolinsky, J.B.; Colson, Y.L.; Grinstaff, M.W. Local Drug Delivery Strategies for Cancer Treatment: Gels, Nanoparticles, Polymeric Films, Rods, and Wafers. *J. Control. Release* **2012**, *159*, 14–26. [CrossRef]
35. Torres-Martínez, E.J.; Bravo, J.M.C.; Medina, A.S.; González, G.L.P.; Gómez, L.J.V. A Summary of Electrospun Nanofibers as Drug Delivery System: Drugs Loaded and Biopolymers Used as Matrices. *Curr. Drug Deliv.* **2018**, *15*, 1360–1374. [CrossRef]
36. Zong, S.; Wang, X.; Yang, Y.; Wu, W.; Li, H.; Ma, Y.; Lin, W.; Sun, T.; Huang, Y.; Xie, Z.; et al. The Use of Cisplatin-Loaded Mucoadhesive Nanofibers for Local Chemotherapy of Cervical Cancers in Mice. *Eur. J. Pharm. Biopharm.* **2015**, *93*, 127–135. [CrossRef]
37. Liu, S.; Zhou, G.; Liu, D.; Xie, Z.; Huang, Y.; Wang, X.; Wu, W.; Jing, X. Inhibition of Orthotopic Secondary Hepatic Carcinoma in Mice by Doxorubicin-Loaded Electrospun Polylactide Nanofibers. *J. Mater. Chem. B* **2013**, *1*, 101–109. [CrossRef]
38. Steffens, L.; Morás, A.M.; Arantes, P.R.; Masterson, K.; Cao, Z.; Nugent, M.; Moura, D.J. Electrospun PVA-Dacarbazine Nanofibers as a Novel Nano Brain-Implant for Treatment of Glioblastoma: In Silico and In Vitro Characterization. *Eur. J. Pharm. Sci.* **2020**, *143*, 105183. [CrossRef] [PubMed]
39. Cui, W.; Zhou, Y.; Chang, J. Electrospun Nanofibrous Materials for Tissue Engineering and Drug Delivery. *Sci. Technol. Adv. Mater.* **2010**, *11*, 014108. [CrossRef]
40. Yang, G.; Wang, J.; Wang, Y.; Li, L.; Guo, X.; Zhou, S. An Implantable Active-Targeting Micelle-in-Nanofiber Device for Efficient and Safe Cancer Therapy. *ACS Nano* **2015**, *9*, 1161–1174. [CrossRef]
41. Falde, E.J.; Freedman, J.D.; Herrera, V.L.M.; Yohe, S.T.; Colson, Y.L.; Grinstaff, M.W. Layered Superhydrophobic Meshes for Controlled Drug Release. *J. Control. Release* **2015**, *214*, 23–29. [CrossRef] [PubMed]
42. Khodadadi, M.; Alijani, S.; Montazeri, M.; Esmaeilizadeh, N.; Sadeghi-Soureh, S.; Pilehvar-Soltanahmadi, Y. Recent Advances in Electrospun Nanofiber-Mediated Drug Delivery Strategies for Localized Cancer Chemotherapy. *J. Biomed. Mater. Res. Part A* **2020**, *108*, 1444–1458. [CrossRef] [PubMed]
43. Tseng, Y.-Y.; Huang, Y.-C.; Yang, T.-C.; Yang, S.-T.; Liu, S.-C.; Chang, T.-M.; Kau, Y.-C.; Liu, S.-J. Concurrent Chemotherapy of Malignant Glioma in Rats by Using Multidrug-Loaded Biodegradable Nanofibrous Membranes. *Sci. Rep.* **2016**, *6*, 30630. [CrossRef] [PubMed]
44. Zhang, J.; Wang, X.; Liu, T.; Liu, S.; Jing, X. Antitumor Activity of Electrospun Polylactide Nanofibers Loaded with 5-Fluorouracil and Oxaliplatin against Colorectal Cancer. *Drug Deliv.* **2016**, *23*, 794–800. [CrossRef]
45. Sridhar, R.; Ravanan, S.; Venugopal, J.R.; Sundarrajan, S.; Pliszka, D.; Sivasubramanian, S.; Gunasekaran, P.; Prabhakaran, M.; Madhaiyan, K.; Sahayaraj, A.; et al. Curcumin- and Natural Extract-Loaded Nanofibres for Potential Treatment of Lung and Breast Cancer: In Vitro Efficacy Evaluation. *J. Biomater. Sci. Polym. Ed.* **2014**, *25*, 985–998. [CrossRef]
46. Zhang, Z.; Liu, S.; Qi, Y.; Zhou, D.; Xie, Z.; Jing, X.; Chen, X.; Huang, Y. Time-Programmed DCA and Oxaliplatin Release by Multilayered Nanofiber Mats in Prevention of Local Cancer Recurrence Following Surgery. *J. Control. Release* **2016**, *235*, 125–133. [CrossRef] [PubMed]
47. Zhao, X.; Zhao, J.; Lin, Z.Y.; Pan, G.; Zhu, Y.; Cheng, Y.; Cui, W. Self-Coated Interfacial Layer at Organic/Inorganic Phase for Temporally Controlling Dual-Drug Delivery from Electrospun Fibers. *Colloids Surf. B Biointerfaces* **2015**, *130*, 1–9. [CrossRef] [PubMed]
48. Xu, X.; Chen, X.; Wang, Z.; Jing, X. Ultrafine PEG-PLA Fibers Loaded with Both Paclitaxel and Doxorubicin Hydrochloride and Their In Vitro Cytotoxicity. *Eur. J. Pharm. Biopharm.* **2009**, *72*, 18–25. [CrossRef]
49. Yohe, S.T.; Herrera, V.L.M.; Colson, Y.L.; Grinstaff, M.W. 3D Superhydrophobic Electrospun Meshes as Reinforcement Materials for Sustained Local Drug Delivery against Colorectal Cancer Cells. *J. Control. Release* **2012**, *162*, 92–101. [CrossRef]
50. Wei, J.; Hu, J.; Li, M.; Chen, Y.; Chen, Y. Multiple Drug-Loaded Electrospun PLGA/Gelatin Composite Nanofibers Encapsulated with Mesoporous ZnO Nanospheres for Potential Postsurgical Cancer Treatment. *RSC Adv.* **2014**, *4*, 28011–28019. [CrossRef]
51. Luo, X.; Zhang, H.; Chen, M.; Wei, J.; Zhang, Y.; Li, X. Antimetastasis and Antitumor Efficacy Promoted by Sequential Release of Vascular Disrupting and Chemotherapeutic Agents from Electrospun Fibers. *Int. J. Pharm.* **2014**, *475*, 438–449. [CrossRef]
52. Li, J.; Xu, W.; Li, D.; Liu, T.; Zhang, Y.S.; Ding, J.; Chen, X. Locally Deployable Nanofiber Patch for Sequential Drug Delivery in Treatment of Primary and Advanced Orthotopic Hepatomas. *ACS Nano* **2018**, *12*, 6685–6699. [CrossRef] [PubMed]
53. Liu, W.; Wei, J.; Wei, Y.; Chen, Y. Controlled Dual Drug Release and In Vitro Cytotoxicity of Electrospun Poly(Lactic-Co-Glycolic Acid) Nanofibers Encapsulated with Micelles. *J. Biomed. Nanotechnol.* **2015**, *11*, 428–435. [CrossRef] [PubMed]
54. Yang, G.; Wang, J.; Li, L.; Ding, S.; Zhou, S. Electrospun Micelles/Drug-Loaded Nanofibers for Time-Programmed Multi-Agent Release. *Macromol. Biosci.* **2014**, *14*, 965–976. [CrossRef] [PubMed]
55. Hou, Z.; Li, X.; Li, C.; Dai, Y.; Ma, P.; Zhang, X.; Kang, X.; Cheng, Z.; Lin, J. Electrospun Upconversion Composite Fibers as Dual Drugs Delivery System with Individual Release Properties. *Langmuir* **2013**, *29*, 9473–9482. [CrossRef]
56. Tseng, Y.Y.; Yang, T.C.; Wang, Y.C.; Lee, W.H.; Chang, T.M.; Kau, Y.C.; Liu, S.J. Targeted Concurrent and Sequential Delivery of Chemotherapeutic and Antiangiogenic Agents to the Brain by Using Drug-Loaded Nanofibrous Membranes. *Int. J. Nanomed.* **2017**, *12*, 1265–1276. [CrossRef]

57. Chen, M.; Feng, W.; Lin, S.; He, C.; Gao, Y.; Wang, H. Antitumor Efficacy of a PLGA Composite Nanofiber Embedded with Doxorubicin@ MSNs and Hydroxycamptothecin@ HANPs. *RSC Adv.* **2014**, *4*, 53344–53351. [CrossRef]
58. Liu, D.; Wang, F.; Yue, J.; Jing, X.; Huang, Y. Metabolism Targeting Therapy of Dichloroacetate-Loaded Electrospun Mats on Colorectal Cancer. *Drug Deliv.* **2015**, *22*, 136–143. [CrossRef]
59. Kim, Y.-J.; Park, M.R.; Kim, M.S.; Kwon, O.H. Polyphenol-Loaded Polycaprolactone Nanofibers for Effective Growth Inhibition of Human Cancer Cells. *Mater. Chem. Phys.* **2012**, *133*, 674–680. [CrossRef]
60. Liu, S.; Wang, X.; Zhang, Z.; Zhang, Y.; Zhou, G.; Huang, Y.; Xie, Z.; Jing, X. Use of Asymmetric Multilayer Polylactide Nanofiber Mats in Controlled Release of Drugs and Prevention of Liver Cancer Recurrence after Surgery in Mice. *Nanomed. Nanotechnol. Biol. Med.* **2015**, *11*, 1047–1056. [CrossRef]
61. Ma, Y.; Wang, X.; Zong, S.; Zhang, Z.; Xie, Z.; Huang, Y.; Yue, Y.; Liu, S.; Jing, X. Local, Combination Chemotherapy in Prevention of Cervical Cancer Recurrence after Surgery by Using Nanofibers Co-Loaded with Cisplatin and Curcumin. *RSC Adv.* **2015**, *5*, 106325–106332. [CrossRef]
62. Yuan, Z.; Zhao, X.; Zhao, J.; Pan, G.; Qiu, W.; Wang, X.; Zhu, Y.; Zheng, Q.; Cui, W. Synergistic Mediation of Tumor Signaling Pathways in Hepatocellular Carcinoma Therapy via Dual-Drug-Loaded PH-Responsive Electrospun Fibrous Scaffolds. *J. Mater. Chem. B* **2015**, *3*, 3436–3446. [CrossRef] [PubMed]
63. Iqbal, S.; Rashid, M.H.; Arbab, A.S.; Khan, M. Encapsulation of Anticancer Drugs (5-Fluorouracil and Paclitaxel) into Polycaprolactone (PCL) Nanofibers and In Vitro Testing for Sustained and Targeted Therapy. *J. Biomed. Nanotechnol.* **2017**, *13*, 355–366. [CrossRef] [PubMed]
64. Li, X.; He, Y.; Hou, J.; Yang, G.; Zhou, S. A Time-Programmed Release of Dual Drugs from an Implantable Trilayer Structured Fiber Device for Synergistic Treatment of Breast Cancer. *Small* **2020**, *16*, 1902262. [CrossRef] [PubMed]
65. Zhang, Z.; Wu, Y.; Kuang, G.; Liu, S.; Zhou, D.; Chen, X.; Jing, X.; Huang, Y. Pt (Iv) Prodrug-Backboned Micelle and DCA Loaded Nanofibers for Enhanced Local Cancer Treatment. *J. Mater. Chem. B* **2017**, *5*, 2115–2125. [CrossRef] [PubMed]
66. Zhao, J.; Zhu, Y.; Ye, C.; Chen, Y.; Wang, S.; Zou, D.; Li, Z. Photothermal Transforming Agent and Chemotherapeutic Co-Loaded Electrospun Nanofibers for Tumor Treatment. *Int. J. Nanomed.* **2019**, *14*, 3893–3909. [CrossRef]
67. Yao, Y.; Wei, W.; Sun, J.; Chen, L.; Deng, X.; Ma, L.; Hao, S. Proteomic Analysis of Exosomes Derived from Human Lymphoma Cells. *Eur. J. Med. Res.* **2015**, *20*, 8. [CrossRef]
68. Giustini, A.J.; Petryk, A.A.; Cassim, S.M.; Tate, J.A.; Baker, I.; Hoopes, P.J. Magnetic nanoparticle hyperthermia in cancer treatment. *Nano Life* **2010**, *1*, 17–32. [CrossRef]
69. Niiyama, E.; Uto, K.; Lee, C.M.; Sakura, K.; Ebara, M. Hyperthermia Nanofiber Platform Synergized by Sustained Release of Paclitaxel to Improve Antitumor Efficiency. *Adv. Healthc. Mater.* **2019**, *8*, 1900102. [CrossRef]
70. Roma-Rodrigues, C.; Rivas-García, L.; Baptista, P.V.; Fernandes, A.R. Gene Therapy in Cancer Treatment: Why Go Nano? *Pharmaceutics* **2020**, *12*, 233. [CrossRef]
71. Che, H.L.; Lee, H.J.; Uto, K.; Ebara, M.; Kim, W.J.; Aoyagi, T.; Park, I.K. Simultaneous Drug and Gene Delivery from the Biodegradable Poly(ε-Caprolactone) Nanofibers for the Treatment of Liver Cancer. *J. Nanosci. Nanotechnol.* **2015**, *15*, 7971–7975. [CrossRef]
72. Cavalu, S.; Roiu, G.; Pop, O.; Heredea, D.A.P.; Costea, T.O.; Costea, C.F. Nano-Scale Modifications of Amniotic Membrane Induced by UV and Antibiotic Treatment: Histological, AFM and FTIR Spectroscopy Evidence. *Materials* **2021**, *14*, 863. [CrossRef] [PubMed]
73. Zhang, L.; Pornpattananangku, D.; Hu, C.M.; Huang, C.M. Development of Nanoparticles for Antimicrobial Drug Delivery. *Curr. Med. Chem.* **2010**, *17*, 585–594. [CrossRef] [PubMed]
74. Rizvi, S.A.A.; Saleh, A.M. Applications of Nanoparticle Systems in Drug Delivery Technology. *Saudi Pharm. J.* **2018**, *26*, 64–70. [CrossRef] [PubMed]
75. Cavalu, S.; Banica, F.; Gruian, C.; Vanea, E.; Goller, G.; Simon, V. Microscopic and Spectroscopic Investigation of Bioactive Glasses for Antibiotic Controlled Release. *J. Mol. Struct.* **2013**, *1040*, 47–52. [CrossRef]
76. Kenawy, E.-R.; Bowlin, G.L.; Mansfield, K.; Layman, J.; Simpson, D.G.; Sanders, E.H.; Wnek, G.E. Release of Tetracycline Hydrochloride from Electrospun Poly(Ethylene-Co-Vinylacetate), Poly(Lactic Acid), and a Blend. *J. Control. Release* **2002**, *81*, 57–64. [CrossRef]
77. Han, J.; Chen, T.-X.; Branford-White, C.J.; Zhu, L.-M. Electrospun Shikonin-Loaded PCL/PTMC Composite Fiber Mats with Potential Biomedical Applications. *Int. J. Pharm.* **2009**, *382*, 215–221. [CrossRef]
78. Gilchrist, S.E.; Lange, D.; Letchford, K.; Bach, H.; Fazli, L.; Burt, H.M. Fusidic Acid and Rifampicin Co-Loaded PLGA Nanofibers for the Prevention of Orthopedic Implant Associated Infections. *J. Control. Release* **2013**, *170*, 64–73. [CrossRef]
79. Ratiu, C.; Brocks, M.; Costea, T.; Moldovan, L.; Cavalu, S. PRGF-Modified Collagen Membranes for Guided Bone Regeneration: Spectroscopic, Microscopic and Nano-Mechanical Investigations. *Appl. Sci.* **2019**, *9*, 1035. [CrossRef]
80. Cavalu, S.; Ratiu, C.; Ponta, O.; Simon, V.; Rugina, D.; Miclaus, V.; Akin, I.; Goller, G. Improving osseointegration of alumina/zirconia ceramic implants by fluoride surface treatment. *Dig. J. Nanomater. Biostructures* **2014**, *9*, 797–808.
81. Xue, J.; He, M.; Niu, Y.; Liu, H.; Crawford, A.; Coates, P.; Chen, D.; Shi, R.; Zhang, L. Preparation and in Vivo Efficient Anti-Infection Property of GTR/GBR Implant Made by Metronidazole Loaded Electrospun Polycaprolactone Nanofiber Membrane. *Int. J. Pharm.* **2014**, *475*, 566–577. [CrossRef]

82. Kataria, K.; Gupta, A.; Rath, G.; Mathur, R.B.; Dhakate, S. In Vivo Wound Healing Performance of Drug Loaded Electrospun Composite Nanofibers Transdermal Patch. *Int. J. Pharm.* **2014**, *469*, 102–110. [CrossRef] [PubMed]
83. Chen, S.; Zhang, M.X.; Shao, X.; Wang, X.; Zhang, L.; Xu, P.; Zhong, W.; Zhang, L.; Xing, M.M.Q.; Zhang, L. A Laminin Mimetic Peptide SIKVAV-Conjugated Chitosan Hydrogel Promoting Wound Healing by Enhancing Angiogenesis, Re-Epithelialization and Collagen Deposition. *J. Mater. Chem. B* **2015**, *3*, 6798–6804. [CrossRef]
84. Madison, K.C. Barrier Function of the Skin: "La Raison d'être" of the Epidermis. *J. Investig. Dermatol.* **2003**, *121*, 231–241. [CrossRef] [PubMed]
85. Proksch, E.; Brandner, J.M.; Jensen, J.M. The Skin: An Indispensable Barrier. *Exp. Dermatol.* **2008**, *17*, 1063–1072. [CrossRef] [PubMed]
86. Broughton, G., 2nd; Janis, J.E.; Attinger, C.E. The Basic Science of Wound Healing. *Plast. Reconstr. Surg.* **2006**, *117*, 12s–34s. [CrossRef] [PubMed]
87. Sen, C.K.; Gordillo, G.M.; Roy, S.; Kirsner, R.; Lambert, L.; Hunt, T.K.; Gottrup, F.; Gurtner, G.C.; Longaker, M.T. Human Skin Wounds: A Major and Snowballing Threat to Public Health and the Economy. *Wound Repair Regen.* **2009**, *17*, 763–771. [CrossRef]
88. Dagogo-Jack, S. Primary Prevention of Type-2 Diabetes in Developing Countries. *J. Natl. Med. Assoc.* **2006**, *98*, 415–419.
89. O'Reilly, G.; Cameron, P.; Joshipura, M. Global Trauma Registry Mapping: A Scoping Review. *Injury* **2012**, *43*, 1148–1153. [CrossRef]
90. Kim, T.G.; Park, T.G. Biomimicking Extracellular Matrix: Cell Adhesive RGD Peptide Modified Electrospun Poly(D,L-Lactic-Co-Glycolic Acid) Nanofiber Mesh. *Tissue Eng.* **2006**, *12*, 221–233. [CrossRef]
91. Keshel, S.H.; Biazar, E.; Rezaei Tavirani, M.; Rahmati Roodsari, M.; Ronaghi, A.; Ebrahimi, M.; Rad, H.; Sahebalzamani, A.; Rakhshan, A.; Afsordeh, K. The Healing Effect of Unrestricted Somatic Stem Cells Loaded in Collagen-Modified Nanofibrous PHBV Scaffold on Full-Thickness Skin Defects. *Artif. Cells Nanomed. Biotechnol.* **2014**, *42*, 210–216. [CrossRef]
92. Singer, A.J.; Clark, R.A.F. Cutaneous Wound Healing. *N. Engl. J. Med.* **1999**, *341*, 738–746. [CrossRef] [PubMed]
93. Meyer, U.; Meyer, T.; Handschel, J.; Wiesmann, H.-P. *Fundamentals of Tissue Engineering and Regenerative Medicine*; Springer: Berlin/Heidelberg, Germany, 2009; p. 1076, ISBN 978-3-540-77754-0.
94. Greiner, A.; Wendorff, J.H. Electrospinning: A Fascinating Method for the Preparation of Ultrathin Fibers. *Angew. Chem. Int. Ed.* **2007**, *46*, 5670–5703. [CrossRef] [PubMed]
95. Matthews, J.A.; Wnek, G.E.; Simpson, D.G.; Bowlin, G.L. Electrospinning of Collagen Nanofibers. *Biomacromolecules* **2002**, *3*, 232–238. [CrossRef]
96. Chakraborty, S.; Liao, I.C.; Adler, A.; Leong, K.W. Electrohydrodynamics: A Facile Technique to Fabricate Drug Delivery Systems. *Adv. Drug Deliv Rev.* **2009**, *61*, 1043–1054. [CrossRef] [PubMed]
97. Maleki, H.; Gharehaghaji, A.A.; Toliyat, T.; Dijkstra, P.J. Drug Release Behavior of Electrospun Twisted Yarns as Implantable Medical Devices. *Biofabrication* **2016**, *8*, 035019. [CrossRef] [PubMed]
98. Katti, D.S.; Robinson, K.W.; Ko, F.K.; Laurencin, C.T. Bioresorbable Nanofiber-Based Systems for Wound Healing and Drug Delivery: Optimization of Fabrication Parameters. *J. Biomed. Mater. Res. Part B Appl. Biomater.* **2004**, *70*, 286–296. [CrossRef] [PubMed]
99. Charernsriwilaiwat, N.; Opanasopit, P.; Rojanarata, T.; Ngawhirunpat, T. Lysozyme-Loaded, Electrospun Chitosan-Based Nanofiber Mats for Wound Healing. *Int. J. Pharm.* **2012**, *427*, 379–384. [CrossRef]
100. Lanno, G.-M.; Ramos, C.; Preem, L.; Putrinš, M.; Laidmäe, I.; Tenson, T.; Kogermann, K. Antibacterial Porous Electrospun Fibers as Skin Scaffolds for Wound Healing Applications. *ACS Omega* **2020**, *5*, 30011–30022. [CrossRef]
101. Xing, Z.-C.; Meng, W.; Yuan, J.; Moon, S.; Jeong, Y.; Kang, I.-K. In Vitro Assessment of Antibacterial Activity and Cytocompatibility of Quercetin-Containing PLGA Nanofibrous Scaffolds for Tissue Engineering. *J. Nanomater.* **2012**, *2012*, 202608. [CrossRef]
102. Ruckh, T.T.; Oldinski, R.A.; Carroll, D.A.; Mikhova, K.; Bryers, J.D.; Popat, K.C. Antimicrobial Effects of Nanofiber Poly(Caprolactone) Tissue Scaffolds Releasing Rifampicin. *J. Mater. Sci Mater. Med.* **2012**, *23*, 1411–1420. [CrossRef]
103. Bhattarai, D.P.; Kim, M.H.; Park, H.; Park, W.H.; Kim, B.S.; Kim, C.S. Coaxially Fabricated Polylactic Acid Electrospun Nanofibrous Scaffold for Sequential Release of Tauroursodeoxycholic Acid and Bone Morphogenic Protein2 to Stimulate Angiogenesis and Bone Regeneration. *Chem. Eng. J.* **2020**, *389*, 123470. [CrossRef]
104. Chen, S.; Liu, B.; Carlson, M.A.; Gombart, A.F.; Reilly, D.A.; Xie, J. Recent Advances in Electrospun Nanofibers for Wound Healing. *Nanomed.* **2017**, *12*, 1335–1352. [CrossRef] [PubMed]
105. Rho, K.S.; Jeong, L.; Lee, G.; Seo, B.-M.; Park, Y.J.; Hong, S.-D.; Roh, S.; Cho, J.J.; Park, W.H.; Min, B.-M. Electrospinning of Collagen Nanofibers: Effects on the Behavior of Normal Human Keratinocytes and Early-Stage Wound Healing. *Biomaterials* **2006**, *27*, 1452–1461. [CrossRef] [PubMed]
106. Venugopal, J.R.; Zhang, Y.; Ramakrishna, S. In Vitro Culture of Human Dermal Fibroblasts on Electrospun Polycaprolactone Collagen Nanofibrous Membrane. *Artif. Organs* **2006**, *30*, 440–446. [CrossRef] [PubMed]
107. Simman, R.; Phavixay, L. Split-Thickness Skin Grafts Remain the Gold Standard for the Closure of Large Acute and Chronic Wounds. *J. Am. Coll. Certif. Wound Spec.* **2011**, *3*, 55–59. [CrossRef] [PubMed]
108. Ma, B.; Xie, J.; Jiang, J.; Wu, J. Sandwich-Type Fiber Scaffolds with Square Arrayed Microwells and Nanostructured Cues as Microskin Grafts for Skin Regeneration. *Biomaterials* **2014**, *35*, 630–641. [CrossRef] [PubMed]
109. Zafar, M.; Najeeb, S.; Khurshid, Z.; Vazirzadeh, M.; Zohaib, S.; Najeeb, B.; Sefat, F. Potential of Electrospun Nanofibers for Biomedical and Dental Applications. *Materials* **2016**, *9*, 73. [CrossRef]

110. Damle, S.G.; Bhattal, H.; Loomba, A. Apexification of Anterior Teeth: A Comparative Evaluation of Mineral Trioxide Aggregate and Calcium Hydroxide Paste. *J. Clin. Pediatric Dent.* **2012**, *36*, 263–268. [CrossRef]
111. Odabaş, M.E.; Alaçam, A.; Sillelioğlu, H.; Deveci, C. Clinical and Radiographic Success Rates of Mineral Trioxide Aggregate and Ferric Sulphate Pulpotomies Performed by Dental Students. *Eur. J. Paediatr. Dent.* **2012**, *13*, 118–122.
112. Kim, G.M.; Asran, A.S.; Michler, G.H.; Simon, P.; Kim, J.S. Electrospun PVA/HAp Nanocomposite Nanofibers: Biomimetics of Mineralized Hard Tissues at a Lower Level of Complexity. *Bioinspiration Biomim.* **2008**, *3*, 046003. [CrossRef]
113. Kim, J.J.; Bae, W.J.; Kim, J.M.; Kim, J.J.; Lee, E.J.; Kim, H.W.; Kim, E.C. Mineralized Polycaprolactone Nanofibrous Matrix for Odontogenesis of Human Dental Pulp Cells. *J. Biomater. Appl.* **2014**, *28*, 1069–1078. [CrossRef] [PubMed]
114. Nygaard, I.; Barber, M.D.; Burgio, K.L.; Kenton, K.; Meikle, S.; Schaffer, J.; Spino, C.; Whitehead, W.E.; Wu, J.; Brody, D.J. Prevalence of Symptomatic Pelvic Floor Disorders in US Women. *JAMA* **2008**, *300*, 1311–1316. [CrossRef]
115. Mukherjee, S.; Darzi, S.; Paul, K.; Cousins, F.L.; Werkmeister, J.A.; Gargett, C.E. Electrospun Nanofiber Meshes with Endometrial MSCs Modulate Foreign Body Response by Increased Angiogenesis, Matrix Synthesis, and Anti-Inflammatory Gene Expression in Mice: Implication in Pelvic Floor. *Front. Pharmacol.* **2020**, *11*, 353. [CrossRef] [PubMed]
116. Albouaini, K.; Egred, M.; Rao, A.; Alahmar, A.; Wright, D.J. Cardiac Resynchronisation Therapy: Evidence Based Benefits and Patient Selection. *Eur. J. Intern. Med.* **2008**, *19*, 165–172. [CrossRef] [PubMed]
117. Liu, J.; Wu, P.; Wang, H.; Wang, Y.; Du, Y.; Cheng, W.; Xu, Z.; Zhou, N.; Wang, L.; Yang, Z. Necroptosis Induced by Ad-HGF Activates Endogenous C-Kit+ Cardiac Stem Cells and Promotes Cardiomyocyte Proliferation and Angiogenesis in the Infarcted Aged Heart. *Cell. Physiol. Biochem.* **2016**, *40*, 847–860. [CrossRef]
118. Ho, C.M.B.; Mishra, A.; Lin, P.T.P.; Ng, S.H.; Yeong, W.Y.; Kim, Y.; Yoon, Y. 3D Printed Polycaprolactone Carbon Nanotube Composite Scaffolds for Cardiac Tissue Engineering. *Macromol. Biosci.* **2017**, *17*, 1600250. [CrossRef] [PubMed]
119. Kitsara, M.; Agbulut, O.; Kontziampasis, D.; Chen, Y.; Menasché, P. Fibers for Hearts: A Critical Review on Electrospinning for Cardiac Tissue Engineering. *Acta Biomater.* **2017**, *48*, 20–40. [CrossRef]
120. Hanas, T.; Sampath Kumar, T.S.; Perumal, G.; Doble, M. Tailoring Degradation of AZ31 Alloy by Surface Pre-Treatment and Electrospun PCL Fibrous Coating. *Mater. Sci. Eng. C Mater. Biol. Appl.* **2016**, *65*, 43–50. [CrossRef]
121. Yasin, A.S.; Mohamed, I.M.A.; Mousa, H.M.; Park, C.H.; Kim, C.S. Facile Synthesis of TiO_2/ZrO_2 Nanofibers/Nitrogen Co-Doped Activated Carbon to Enhance the Desalination and Bacterial Inactivation via Capacitive Deionization. *Sci. Rep.* **2018**, *8*, 541. [CrossRef]
122. Kim, H.-W.; Lee, H.-H.; Chun, G.-S. Bioactivity and Osteoblast Responses of Novel Biomedical Nanocomposites of Bioactive Glass Nanofiber Filled Poly(Lactic Acid). *J. Biomed. Mater. Res. Part A* **2008**, *85A*, 651–663. [CrossRef]
123. Bianco, A.; di Federico, E.; Moscatelli, I.; Camaioni, A.; Armentano, I.; Campagnolo, L.; Dottori, M.; Kenny, J.M.; Siracusa, G.; Gusmano, G. Electrospun Poly(ε-Caprolactone)/Ca-Deficient Hydroxyapatite Nanohybrids: Microstructure, Mechanical Properties and Cell Response by Murine Embryonic Stem Cells. *Mater. Sci. Eng. C Biomim. Mater. Sens. Syst.* **2009**, *29*, 2063–2071. [CrossRef]
124. Fabbri, P.; Bondioli, F.; Messori, M.; Bartoli, C.; Dinucci, D.; Chiellini, F. Porous Scaffolds of Polycaprolactone Reinforced with in Situ Generated Hydroxyapatite for Bone Tissue Engineering. *J. Mater. Sci. Mater. Med.* **2010**, *21*, 343–351. [CrossRef] [PubMed]
125. Mousa, H.M.; Abdal-Hay, A.; Bartnikowski, M.; Mohamed, I.M.A.; Yasin, A.S.; Ivanovski, S.; Park, C.H.; Kim, C.S. A Multifunctional Zinc Oxide/Poly(Lactic Acid) Nanocomposite Layer Coated on Magnesium Alloys for Controlled Degradation and Antibacterial Function. *ACS Biomater. Sci. Eng.* **2018**, *4*, 2169–2180. [CrossRef] [PubMed]
126. Tian, P.; Xu, D.; Liu, X. Mussel-Inspired Functionalization of PEO/PCL Composite Coating on a Biodegradable AZ31 Magnesium Alloy. *Colloids Surf. B Biointerfaces* **2016**, *141*, 327–337. [CrossRef]
127. Bakhsheshi-Rad, H.R.; Hamzah, E.; Kasiri-Asgarani, M.; Jabbarzare, S.; Iqbal, N.; Abdul Kadir, M.R. Deposition of Nanostructured Fluorine-Doped Hydroxyapatite-Polycaprolactone Duplex Coating to Enhance the Mechanical Properties and Corrosion Resistance of Mg Alloy for Biomedical Applications. *Mater. Sci. Eng. C Mater. Biol. Appl.* **2016**, *60*, 526–537. [CrossRef]
128. Cavalu, S.; Simon, V. Microstructure and Bioactivity of Acrylic Bone Cements for Prosthetic Surgery. *J. Optoelectron. Adv. Mater.* **2006**, *8*, 1520.
129. Mordike, B.L.; Ebert, T. Magnesium: Properties—Applications—Potential. *Mater. Sci. Eng. A* **2001**, *302*, 37–45. [CrossRef]
130. Rezk, A.I.; Mousa, H.M.; Lee, J.; Park, C.H.; Kim, C.S. Composite PCL/HA/Simvastatin Electrospun Nanofiber Coating on Biodegradable Mg Alloy for Orthopedic Implant Application. *J. Coat. Technol. Res.* **2019**, *16*, 477–489. [CrossRef]
131. Romanò, C.L.; Tsuchiya, H.; Morelli, I.; Battaglia, A.G.; Drago, L. Antibacterial Coating of Implants: Are We Missing Something? *Bone Jt. Res.* **2019**, *8*, 199–206. [CrossRef]
132. Zhang, L.; Yan, J.; Yin, Z.; Tang, C.; Guo, Y.; Li, D.; Wei, B.; Xu, Y.; Gu, Q.; Wang, L. Electrospun Vancomycin-Loaded Coating on Titanium Implants for the Prevention of Implant-Associated Infections. *Int. J. Nanomed.* **2014**, *9*, 3027–3036. [CrossRef]
133. Kranthi Kiran, A.S.; Kizhakeyil, A.; Ramalingam, R.; Verma, N.K.; Lakshminarayanan, R.; Kumar, T.S.S.; Doble, M.; Ramakrishna, S. Drug Loaded Electrospun Polymer/Ceramic Composite Nanofibrous Coatings on Titanium for Implant Related Infections. *Ceram. Int.* **2019**, *45*, 18710–18720. [CrossRef]
134. Wu, S.; Liu, X.; Yeung, K.W.K.; Liu, C.; Yang, X. Biomimetic Porous Scaffolds for Bone Tissue Engineering. *Mater. Sci. Eng. R Rep.* **2014**, *80*, 1–36. [CrossRef]
135. Diefenbeck, M.; Mückley, T.; Hofmann, G.O. Prophylaxis and Treatment of Implant-Related Infections by Local Application of Antibiotics. *Injury* **2006**, *37*, S95–S104. [CrossRef]

136. Hake, M.E.; Young, H.; Hak, D.J.; Stahel, P.F.; Hammerberg, E.M.; Mauffrey, C. Local Antibiotic Therapy Strategies in Orthopaedic Trauma: Practical Tips and Tricks and Review of the Literature. *Injury* **2015**, *46*, 1447–1456. [CrossRef] [PubMed]
137. Jahanmard, F.; Croes, M.; Castilho, M.; Majed, A.; Steenbergen, M.J.; Lietaert, K.; Vogely, H.C.; van der Wal, B.C.H.; Stapels, D.A.C.; Malda, J.; et al. Bactericidal Coating to Prevent Early and Delayed Implant-Related Infections. *J. Control. Release* **2020**, *326*, 38–52. [CrossRef] [PubMed]
138. Ashbaugh, A.G.; Jiang, X.; Zheng, J.; Tsai, A.S.; Kim, W.-S.; Thompson, J.M.; Miller, R.J.; Shahbazian, J.H.; Wang, Y.; Dillen, C.A.; et al. Polymeric Nanofiber Coating with Tunable Combinatorial Antibiotic Delivery Prevents Biofilm-Associated Infection in Vivo. *Proc. Natl. Acad. Sci. USA* **2016**, *113*, E6919. [CrossRef]
139. Osmon, D.R.; Berbari, E.F.; Berendt, A.R.; Lew, D.; Zimmerli, W.; Steckelberg, J.M.; Rao, N.; Hanssen, A.; Wilson, W.R. Executive Summary: Diagnosis and Management of Prosthetic Joint Infection: Clinical Practice Guidelines by the Infectious Diseases Society of Americaa. *Clin. Infect. Dis.* **2013**, *56*, 1–10. [CrossRef]
140. Sandoe, J.A.T.; Barlow, G.; Chambers, J.B.; Gammage, M.; Guleri, A.; Howard, P.; Olson, E.; Perry, J.D.; Prendergast, B.D.; Spry, M.J.; et al. Guidelines for the Diagnosis, Prevention and Management of Implantable Cardiac Electronic Device Infection. Report of a Joint Working Party Project on Behalf of the British Society for Antimicrobial Chemotherapy (BSAC, Host Organization), British Heart Rhythm Society (BHRS), British Cardiovascular Society (BCS), British Heart Valve Society (BHVS) and British Society for Echocardiography (BSE). *J. Antimicrob. Chemother.* **2015**, *70*, 325–359. [CrossRef]
141. Shahi, R.G.; Albuquerque, M.T.P.; Münchow, E.A.; Blanchard, S.B.; Gregory, R.L.; Bottino, M.C. Novel Bioactive Tetracycline-Containing Electrospun Polymer Fibers as a Potential Antibacterial Dental Implant Coating. *Odontology* **2017**, *105*, 354–363. [CrossRef]
142. Sadri, M.; Pashmfroosh, N.; Samadieh, S. Implants Modified with Polymeric Nanofibers Coating Containing the Antibiotic Vancomycin. *Nanomed. Res. J.* **2017**, *2*, 208–215. [CrossRef]
143. Cochis, A.; Ferraris, S.; Sorrentino, R.; Azzimonti, B.; Novara, C.; Geobaldo, F.; Truffa Giachet, F.; Varesano, A.; Vineis, C.; Mahmoud, A.; et al. Silver Doped Keratin Nanofibres Preserve Titanium Surface from Biofilm Contamination and Favour Soft Tissue Healing. *J. Mater. Chem. B* **2017**, *5*, 8366–8377. [CrossRef] [PubMed]
144. Song, W.; Seta, J.; Chen, L.; Bergum, C.; Zhou, Z.; Kanneganti, P.; Kast, R.E.; Auner, G.W.; Shen, M.; Markel, D.C.; et al. Doxycycline-Loaded Coaxial Nanofiber Coating of Titanium Implants Enhances Osseointegration and Inhibits Staphylococcus Aureus Infection. *Biomed. Mater.* **2017**, *12*, 045008. [CrossRef] [PubMed]
145. Maver, T.; Mastnak, T.; Mihelič, M.; Maver, U.; Finšgar, M. Clindamycin-Based 3D-Printed and Electrospun Coatings for Treatment of Implant-Related Infections. *Materials* **2021**, *14*, 1464. [CrossRef] [PubMed]
146. Li, L.L.; Wang, L.M.; Xu, Y.; Lv, L.X. Preparation of Gentamicin-Loaded Electrospun Coating on Titanium Implants and a Study of Their Properties In Vitro. *Arch. Orthop. Trauma Surg.* **2012**, *132*, 897–903. [CrossRef]
147. Chopra, H.; Kumar, S.; Singh, I. Strategies and Therapies for Wound Healing: A Review. *Curr. Drug Targets* **2022**, *23*, 87–98. [CrossRef]
148. Chopra, H.; Gandhi, S.; Gautam, R.K.; Kamal, M.A. Bacterial Nanocellulose Based Wound Dressings: Current and Future Prospects. *Curr. Pharm. Des.* **2022**, *28*, 570–580. [CrossRef]
149. Wang, X.; Ding, B.; Yu, J.; Wang, M. Engineering Biomimetic Superhydrophobic Surfaces of Electrospun Nanomaterials. *Nano Today* **2011**, *6*, 510–530. [CrossRef]
150. Yu, X.; Li, Y.; Yin, X.; Wang, X.; Han, Y.; Si, Y.; Yu, J.; Ding, B. Corncoblike, Superhydrophobic, and Phase-Changeable Nanofibers for Intelligent Thermoregulating and Water-Repellent Fabrics. *ACS Appl. Mater. Interfaces* **2019**, *11*, 39324–39333. [CrossRef]
151. Sheng, J.; Li, Y.; Wang, X.; Si, Y.; Yu, J.; Ding, B. Thermal Inter-Fiber Adhesion of the Polyacrylonitrile/Fluorinated Polyurethane Nanofibrous Membranes with Enhanced Waterproof-Breathable Performance. *Sep. Purif. Technol.* **2016**, *158*, 53–61. [CrossRef]
152. Zhao, J.; Zhu, W.; Wang, X.; Liu, L.; Yu, J.; Ding, B. Environmentally Benign Modification of Breathable Nanofibrous Membranes Exhibiting Superior Waterproof and Photocatalytic Self-Cleaning Properties. *Nanoscale Horiz.* **2019**, *4*, 867–873. [CrossRef]
153. Li, Y.; Yang, F.; Yu, J.; Ding, B. Hydrophobic Fibrous Membranes with Tunable Porous Structure for Equilibrium of Breathable and Waterproof Performance. *Adv. Mater. Interfaces* **2016**, *3*, 1600516. [CrossRef]
154. Yue, Y.; Gong, X.; Jiao, W.; Li, Y.; Yin, X.; Si, Y.; Yu, J.; Ding, B. In-Situ Electrospinning of Thymol-Loaded Polyurethane Fibrous Membranes for Waterproof, Breathable, and Antibacterial Wound Dressing Application. *J. Colloid Interface Sci.* **2021**, *592*, 310–318. [CrossRef] [PubMed]
155. Yao, Y.; Guo, Y.; Li, X.; Yu, J.; Ding, B. Asymmetric Wettable, Waterproof, and Breathable Nanofibrous Membranes for Wound Dressings. *ACS Appl. Bio Mater.* **2021**, *4*, 3287–3293. [CrossRef]
156. Guo, Y.; Zhou, W.; Wang, L.; Dong, Y.; Yu, J.; Li, X.; Ding, B. Stretchable PDMS Embedded Fibrous Membranes Based on an Ethanol Solvent System for Waterproof and Breathable Applications. *ACS Appl. Bio Mater.* **2019**, *2*, 5949–5956. [CrossRef] [PubMed]
157. Yu, Y.; Zhang, F.; Liu, Y.; Zheng, Y.; Xin, B.; Jiang, Z.; Peng, X.; Jin, S. Waterproof and Breathable Polyacrylonitrile/(Polyurethane/Fluorinated-Silica) Composite Nanofiber Membrane via Side-by-Side Electrospinning. *J. Mater. Res.* **2020**, *35*, 1173–1181. [CrossRef]
158. Xu, Y.; Sheng, J.; Yin, X.; Yu, J.; Ding, B. Functional Modification of Breathable Polyacrylonitrile/Polyurethane/TiO_2 Nanofibrous Membranes with Robust Ultraviolet Resistant and Waterproof Performance. *J. Colloid Interface Sci.* **2017**, *508*, 508–516. [CrossRef] [PubMed]

159. Mohammadi, M.R.; Rabbani, S.; Bahrami, S.H.; Joghataei, M.T.; Moayer, F. Antibacterial Performance and in Vivo Diabetic Wound Healing of Curcumin Loaded Gum Tragacanth/Poly(ε-Caprolactone) Electrospun Nanofibers. *Mater. Sci. Eng. C* **2016**, *69*, 1183–1191. [CrossRef]
160. Merrell, J.G.; McLaughlin, S.W.; Tie, L.; Laurencin, C.T.; Chen, A.F.; Nair, L.S. Curcumin-Loaded Poly(ε-Caprolactone) Nanofibres: Diabetic Wound Dressing with Anti-Oxidant and Anti-Inflammatory Properties. *Clin. Exp. Pharmacol. Physiol.* **2009**, *36*, 1149–1156. [CrossRef] [PubMed]
161. Ranjbar-Mohammadi, M.; Zamani, M.; Prabhakaran, M.P.; Bahrami, S.H.; Ramakrishna, S. Electrospinning of PLGA/Gum Tragacanth Nanofibers Containing Tetracycline Hydrochloride for Periodontal Regeneration. *Mater. Sci. Eng. C* **2016**, *58*, 521–531. [CrossRef]
162. Golchin, A.; Hosseinzadeh, S.; Jouybar, A.; Staji, M.; Soleimani, M.; Ardeshirylajimi, A.; Khojasteh, A. Wound Healing Improvement by Curcumin-Loaded Electrospun Nanofibers and BFP-MSCs as a Bioactive Dressing. *Polym. Adv. Technol.* **2020**, *31*, 1519–1531. [CrossRef]
163. GhavamiNejad, A.; Rajan Unnithan, A.; Ramachandra Kurup Sasikala, A.; Samarikhalaj, M.; Thomas, R.G.; Jeong, Y.Y.; Nasseri, S.; Murugesan, P.; Wu, D.; Hee Park, C.; et al. Mussel-Inspired Electrospun Nanofibers Functionalized with Size-Controlled Silver Nanoparticles for Wound Dressing Application. *ACS Appl. Mater. Interfaces* **2015**, *7*, 12176–12183. [CrossRef] [PubMed]
164. Zhu, C.; Cao, R.; Zhang, Y.; Chen, R. Metallic Ions Encapsulated in Electrospun Nanofiber for Antibacterial and Angiogenesis Function to Promote Wound Repair. *Front. Cell Dev. Biol.* **2021**, *9*, 707. [CrossRef] [PubMed]
165. Vargas, E.A.T.; do Vale Baracho, N.C.; de Brito, J.; de Queiroz, A.A.A. Hyperbranched Polyglycerol Electrospun Nanofibers for Wound Dressing Applications. *Acta Biomater.* **2010**, *6*, 1069–1078. [CrossRef] [PubMed]
166. Zahedi, P.; Karami, Z.; Rezaeian, I.; Jafari, S.H.; Mahdaviani, P.; Abdolghaffari, A.H.; Abdollahi, M. Preparation and Performance Evaluation of Tetracycline Hydrochloride Loaded Wound Dressing Mats Based on Electrospun Nanofibrous Poly(Lactic Acid)/Poly(Ïµ-Caprolactone) Blends. *J. Appl. Polym. Sci.* **2012**, *124*, 4174–4183. [CrossRef]
167. Kim, K.; Luu, Y.K.; Chang, C.; Fang, D.; Hsiao, B.S.; Chu, B.; Hadjiargyrou, M. Incorporation and Controlled Release of a Hydrophilic Antibiotic Using Poly(Lactide-Co-Glycolide)-Based Electrospun Nanofibrous Scaffolds. *J. Control. Release* **2004**, *98*, 47–56. [CrossRef]
168. Wang, S.; Zheng, F.; Huang, Y.; Fang, Y.; Shen, M.; Zhu, M.; Shi, X. Encapsulation of Amoxicillin within Laponite-Doped Poly(Lactic-Co-Glycolic Acid) Nanofibers: Preparation, Characterization, and Antibacterial Activity. *ACS Appl. Mater. Interfaces* **2012**, *4*, 6393–6401. [CrossRef]
169. Thakur, R.A.; Florek, C.A.; Kohn, J.; Michniak, B.B. Electrospun Nanofibrous Polymeric Scaffold with Targeted Drug Release Profiles for Potential Application as Wound Dressing. *Int. J. Pharm.* **2008**, *364*, 87–93. [CrossRef]
170. Toncheva, A.; Paneva, D.; Maximova, V.; Manolova, N.; Rashkov, I. Antibacterial Fluoroquinolone Antibiotic-Containing Fibrous Materials from Poly(l-Lactide-Co-d,l-Lactide) Prepared by Electrospinning. *Eur. J. Pharm. Sci.* **2012**, *47*, 642–651. [CrossRef]
171. Torres-Giner, S.; Martinez-Abad, A.; Gimeno-Alcañiz, J.V.; Ocio, M.J.; Lagaron, J.M. Controlled Delivery of Gentamicin Antibiotic from Bioactive Electrospun Polylactide-Based Ultrathin Fibers. *Adv. Eng. Mater.* **2012**, *14*, B112–B122. [CrossRef]
172. Su, Y.; Xiaoqiang, A.E.; Ae, L.; Wang, H.; Chuanglong, A.E.; Ae, H.; Mo, X. Fabrication and Characterization of Biodegradable Nanofibrous Mats by Mix and Coaxial Electrospinning. *J. Mater. Sci. Mater. Med.* **2009**, *20*, 2285. [CrossRef]
173. Sohrabi, A.; Shaibani, P.M.; Etayash, H.; Kaur, K.; Thundat, T. Sustained Drug Release and Antibacterial Activity of Ampicillin Incorporated Poly(Methyl Methacrylate)-Nylon6 Core/Shell Nanofibers. *Polymer* **2013**, *54*, 2699–2705. [CrossRef]
174. Khampieng, T.; Wnek, G.E.; Supaphol, P. Electrospun DOXY-h Loaded-Poly(Acrylic Acid) Nanofiber Mats: In Vitro Drug Release and Antibacterial Properties Investigation. *J. Biomater. Sci. Polym. Ed.* **2014**, *25*, 1292–1305. [CrossRef] [PubMed]
175. Celebioglu, A.; Umu, O.C.O.; Tekinay, T.; Uyar, T. Antibacterial Electrospun Nanofibers from Triclosan/Cyclodextrin Inclusion Complexes. *Colloids Surf. B Biointerfaces* **2014**, *116*, 612–619. [CrossRef] [PubMed]
176. Del Valle, L.J.; Díaz, A.; Royo, M.; Rodríguez-Galán, A.; Puiggalí, J. Biodegradable Polyesters Reinforced with Triclosan Loaded Polylactide Micro/Nanofibers: Properties, Release and Biocompatibility. *Express Polym. Lett.* **2012**, *6*, 266–282. [CrossRef]
177. Kayaci, F.; Umu, O.C.O.; Tekinay, T.; Uyar, T. Antibacterial Electrospun Poly(Lactic Acid) (PLA) Nanofibrous Webs Incorporating Triclosan/Cyclodextrin Inclusion Complexes. *J. Agric. Food Chem.* **2013**, *61*, 3901–3908. [CrossRef]
178. Ren, X.; Kocer, H.B.; Worley, S.D.; Broughton, R.M.; Huang, T.S. Biocidal Nanofibers via Electrospinning. *J. Appl. Polym. Sci.* **2013**, *127*, 3192–3197. [CrossRef]
179. Gliścińska, E.; Gutarowska, B.; Brycki, B.; Krucińska, I. Electrospun Polyacrylonitrile Nanofibers Modified by Quaternary Ammonium Salts. *J. Appl. Polym. Sci.* **2013**, *128*, 767–775. [CrossRef]
180. Chen, L.; Bromberg, L.; Hatton, T.A.; Rutledge, G.C. Electrospun Cellulose Acetate Fibers Containing Chlorhexidine as a Bactericide. *Polymer* **2008**, *49*, 1266–1275. [CrossRef]
181. Liu, X.; Lin, T.; Gao, Y.; Xu, Z.; Huang, C.; Yao, G.; Jiang, L.; Tang, Y.; Wang, X. Antimicrobial Electrospun Nanofibers of Cellulose Acetate and Polyester Urethane Composite for Wound Dressing. *J. Biomed. Mater. Res. Part B Appl. Biomater.* **2012**, *100 B*, 1556–1565. [CrossRef]
182. Spasova, M.; Manolova, N.; Paneva, D.; Rashkov, I. Preparation of Chitosan-Containing Nanofibres by Electrospinning of Chitosan/Poly(Ethylene Oxide) Blend Solutions. *E-Polymers* **2004**, *4*, 56. [CrossRef]
183. Mahapatra, A.; Garg, N.; Nayak, B.P.; Mishra, B.G.; Hota, G. Studies on the Synthesis of Electrospun PAN-Ag Composite Nanofibers for Antibacterial Application. *J. Appl. Polym. Sci.* **2012**, *124*, 1178–1185. [CrossRef]

184. Au, H.T.; Pham, L.N.; Vu, T.H.T.; Park, J.S. Fabrication of an Antibacterial Non-Woven Mat of a Poly(Lactic Acid)/Chitosan Blend by Electrospinning. *Macromol. Res.* **2012**, *20*, 51–58. [CrossRef]
185. An, J.; Zhang, H.; Zhang, J.; Zhao, Y.; Yuan, X. Preparation and Antibacterial Activity of Electrospun Chitosan/ Poly(Ethylene Oxide) Membranes Containing Silver Nanoparticles. *Colloid Polym. Sci.* **2009**, *287*, 1425–1434. [CrossRef]
186. Abdelgawad, A.M.; Hudson, S.M.; Rojas, O.J. Antimicrobial Wound Dressing Nanofiber Mats from Multicomponent (Chitosan/Silver-NPs/Polyvinyl Alcohol) Systems. *Carbohydr. Polym.* **2014**, *100*, 166–178. [CrossRef] [PubMed]
187. Heunis, T.; Bshena, O.; Klumperman, B.; Dicks, L. Release of Bacteriocins from Nanofibers Prepared with Combinations of Poly(D,L-Lactide) (PDLLA) and Poly(Ethylene Oxide) (PEO). *Int. J. Mol. Sci.* **2011**, *12*, 2158–2173. [CrossRef] [PubMed]
188. Brako, F.; Luo, C.; Craig, D.Q.M.; Edirisinghe, M. An Inexpensive, Portable Device for Point-of-Need Generation of Silver-Nanoparticle Doped Cellulose Acetate Nanofibers for Advanced Wound Dressing. *Macromol. Mater. Eng.* **2018**, *303*, 1700586. [CrossRef]
189. Liu, G.-S.; Yan, X.; Yan, F.-F.; Chen, F.-X.; Hao, L.-Y.; Chen, S.-J.; Lou, T.; Ning, X.; Long, Y.-Z. In Situ Electrospinning Iodine-Based Fibrous Meshes for Antibacterial Wound Dressing. *Nanoscale Res. Lett.* **2018**, *13*, 309. [CrossRef]
190. Haik, J.; Kornhaber, R.; Blal, B.; Harats, M. The Feasibility of a Handheld Electrospinning Device for the Application of Nanofibrous Wound Dressings. *Adv. Wound Care* **2017**, *6*, 166–174. [CrossRef]
191. Dong, W.-H.; Liu, J.-X.; Mou, X.-J.; Liu, G.-S.; Huang, X.-W.; Yan, X.; Ning, X.; Russell, S.J.; Long, Y.-Z. Performance of Polyvinyl Pyrrolidone-Isatis Root Antibacterial Wound Dressings Produced in Situ by Handheld Electrospinner. *Colloids Surf. B Biointerfaces* **2020**, *188*, 110766. [CrossRef]
192. Zhou, T.; Wang, Y.; Lei, F.; Yu, J. In-Situ Electrospinning for Intestinal Hemostasis. *Int. J. Nanomed.* **2020**, *15*, 3869. [CrossRef]
193. Zhang, J.; Zhao, Y.-T.; Hu, P.-Y.; Liu, J.-J.; Liu, X.-F.; Hu, M.; Cui, Z.; Wang, N.; Niu, Z.; Xiang, H.-F.; et al. Laparoscopic Electrospinning for in Situ Hemostasis in Minimally Invasive Operation. *Chem. Eng. J.* **2020**, *395*, 125089. [CrossRef]
194. Liu, X.-F.; Zhang, J.; Liu, J.-J.; Zhou, Q.-H.; Liu, Z.; Hu, P.-Y.; Yuan, Z.; Ramakrishna, S.; Yang, D.-P.; Long, Y.-Z. Bifunctional CuS Composite Nanofibers via in Situ Electrospinning for Outdoor Rapid Hemostasis and Simultaneous Ablating Superbug. *Chem. Eng. J.* **2020**, *401*, 126096. [CrossRef]
195. Xu, S.-C.; Qin, C.-C.; Yu, M.; Dong, R.-H.; Yan, X.; Zhao, H.; Han, W.-P.; Zhang, H.-D.; Long, Y.-Z. A Battery-Operated Portable Handheld Electrospinning Apparatus. *Nanoscale* **2015**, *7*, 12351–12355. [CrossRef] [PubMed]
196. Sofokleous, P.; Stride, E.; Bonfield, W.; Edirisinghe, M. Design, Construction and Performance of a Portable Handheld Electrohydrodynamic Multi-Needle Spray Gun for Biomedical Applications. *Mater. Sci. Eng. C* **2013**, *33*, 213–223. [CrossRef]
197. Lau, W.K.; Sofokleous, P.; Day, R.; Stride, E.; Edirisinghe, M. A Portable Device for in Situ Deposition of Bioproducts. *Bioinspired Biomim. Nanobiomater.* **2014**, *3*, 94–105. [CrossRef]
198. Perng, J.-W.; Kuo, Y.-C.; Lu, S.-P. Grounding System Cost Analysis Using Optimization Algorithms. *Energies* **2018**, *11*, 3484. [CrossRef]
199. Liu, J.-X.; Dong, W.-H.; Mou, X.-J.; Liu, G.-S.; Huang, X.-W.; Yan, X.; Zhou, C.-F.; Jiang, S.; Long, Y.-Z. In Situ Electrospun Zein/Thyme Essential Oil-Based Membranes as an Effective Antibacterial Wound Dressing. *ACS Appl. Bio Mater.* **2019**, *3*, 302–307. [CrossRef]
200. Qin, M.; Mou, X.; Dong, W.; Liu, J.; Liu, H.; Dai, Z.; Huang, X.; Wang, N.; Yan, X. In Situ Electrospinning Wound Healing Films Composed of Zein and Clove Essential Oil. *Macromol. Mater. Eng.* **2020**, *305*, 1900790. [CrossRef]
201. Qin, M.; Liu, D.; Dai, Z.; Meng, X.; Liu, G.; Liu, H.; Huang, X.; Yan, X.; Chen, S. One Step Fabrication and Application of Antibacterial Electrospun Zein/Cinnamon Oil Membrane Wound Dressing via In Situ Electrospinning Process. *Chem. Res. Chin. Univ.* **2021**, *37*, 464–469. [CrossRef]
202. Chen, H.; Zhang, H.; Shen, Y.; Dai, X.; Wang, X.; Deng, K.; Long, X.; Liu, L.; Zhang, X.; Li, Y. Instant In-Situ Tissue Repair by Biodegradable PLA/Gelatin Nanofibrous Membrane Using a 3D Printed Handheld Electrospinning Device. *Front. Bioeng. Biotechnol.* **2021**, *9*, 684105. [CrossRef]
203. Han, W.-P.; Huang, Y.-Y.; Yu, M.; Zhang, J.-C.; Yan, X.; Yu, G.-F.; Zhang, H.-D.; Yan, S.-Y.; Long, Y.-Z. Self-Powered Electrospinning Apparatus Based on a Hand-Operated Wimshurst Generator. *Nanoscale* **2015**, *7*, 5603–5606. [CrossRef] [PubMed]
204. Luo, W.-L.; Zhang, J.; Qiu, X.; Chen, L.-J.; Fu, J.; Hu, P.-Y.; Li, X.; Hu, R.-J.; Long, Y.-Z. Electric-Field-Modified in Situ Precise Deposition of Electrospun Medical Glue Fibers on the Liver for Rapid Hemostasis. *Nanoscale Res. Lett.* **2018**, *13*, 1–8. [CrossRef] [PubMed]

materials

MDPI

Article

Bioactive Properties of Composites Based on Silicate Glasses and Different Silver and Gold Structures

Zsejke-Réka Tóth [1,2], János Kiss [2], Milica Todea [1,3], Gábor Kovács [2,4], Tamás Gyulavári [2], Alina Sesarman [5], Giorgiana Negrea [6], Dan C. Vodnar [7], Anna Szabó [2], Lucian Baia [1,8,9,*] and Klára Magyari [1,2,*]

1 Nanostructured Materials and Bio-Nano-Interfaces Center, Interdisciplinary Research Institute on Bio-Nano-Sciences, Babes-Bolyai University, 400271 Cluj-Napoca, Romania; zsejke.toth@ubbcluj.ro (Z.-R.T.); milica.todea@phys.ubbcluj.ro (M.T.)
2 Department of Applied and Environmental Chemistry, University of Szeged, 6720 Szeged, Hungary; kiss.janos505@gmail.com (J.K.); k.gabor84@chem.u-szeged.hu (G.K.); gyulavarit@chem.u-szeged.hu (T.G.); szabo.anna@chem.u-szeged.hu (A.S.)
3 Faculty of Medicine, Iuliu Hațieganu University of Medicine and Pharmacy, 400012 Cluj-Napoca, Romania
4 Department of Horticulture, Faculty of Technical and Human Sciences, Sapientia Hungarian University of Transylvania, 530104 Târgu-Mureș, Romania
5 Department of Molecular Biology and Biotechnology, Center of Systems Biology, Biodiversity and Bioresources, Faculty of Biology and Geology, Babes-Bolyai University, 400006 Cluj-Napoca, Romania; alina.sesarman@ubbcluj.ro
6 Doctoral School in Integrative Biology, Faculty of Biology and Geology, Babes-Bolyai University, 400006 Cluj-Napoca, Romania; giorgiana.negrea@ubbcluj.ro
7 Faculty of Food Science and Technology, University of Agricultural Science and Veterinary Medicine, 400372 Cluj-Napoca, Romania; dan.vodnar@usamvcluj.ro
8 Faculty of Physics, Babes-Bolyai University, 400084 Cluj-Napoca, Romania
9 Institute for Research-Development-Innovation in Applied Natural Sciences, Babes-Bolyai University, 400294 Cluj-Napoca, Romania
* Correspondence: lucian.baia@ubbcluj.ro (L.B.); klara.magyari@ubbcluj.ro (K.M.); Tel.: +40-264-405300 (L.B. & K.M.)

Citation: Tóth, Z.-R.; Kiss, J.; Todea, M.; Kovács, G.; Gyulavári, T.; Sesarman, A.; Negrea, G.; Vodnar, D.C.; Szabó, A.; Baia, L.; et al. Bioactive Properties of Composites Based on Silicate Glasses and Different Silver and Gold Structures. *Materials* **2022**, *15*, 1655. https://doi.org/10.3390/ma15051655

Academic Editor: Simona Cavalu

Received: 19 January 2022
Accepted: 17 February 2022
Published: 23 February 2022

Abstract: Using an ideal biomaterial to treat injured bones can accelerate the healing process and simultaneously exhibit antibacterial properties; thus protecting the patient from bacterial infections. Therefore, the aim of this work was to synthesize composites containing silicate-based bioactive glasses and different types of noble metal structures (i.e., AgI pyramids, AgIAu composites, Au nanocages, Au nanocages with added AgI). Bioactive glass was used as an osteoconductive bone substitute and Ag was used for its antibacterial character, while Au was included to accelerate the formation of new bone. To investigate the synergistic effects in these composites, two syntheses were carried out in two ways: AgIAu composites were added in either one step or AgI pyramids and Au nanocages were added separately. All composites showed good in vitro bioactivity. Transformation of AgI in bioactive glasses into Ag nanoparticles and other silver species resulted in good antibacterial behavior. It was observed that the Ag nanoparticles remained in the Au nanocages, which was also beneficial in terms of antibacterial properties. The presence of Au nanoparticles contributed to the composites achieving high cell viability. The most outstanding result was obtained by the consecutive addition of noble metals into the bioactive glasses, resulting in both a high antibacterial effect and good cell viability.

Keywords: silver; gold nanocages; in vitro bioactivity; antibacterial activity

1. Introduction

The widespread application of bioactive glasses is not surprising since they are used to treat bone injuries, cancer metastases, and wounds [1]. Moreover, it is known that additional properties can be conferred to bioactive glasses by adding different metals to them [2–4]. Among these properties the following ones are worth highlighting: stimulation of angiogenic factors and anti-inflammatory effects, enhancement of cell viability characters, and an increase in antibacterial and osteogenic cellular activity [1,3,5].

The unique properties of gold nanoparticles are related to their (localized) surface plasmon resonance, large surface-to-volume ratio, biocompatibility, low toxicity, and stability [6]. These properties are utilized in several (biomedical) applications [7], such as drug delivery, diagnostics and therapeutics, bionanosensors, and biomedical nanodevices. It is known that gold nanoparticles with an appropriate size and morphology can simulate the proliferation of keratinocytes [8]. Liang et al. [9] reported that gold nanoparticles (with a diameter of 15 nm) loaded into mesoporous on silica nanoparticles promote the viability and proliferation of osteoblasts. Our research group successfully introduced gold nanoparticles into a bioactive glass matrix system [10] without losing their properties. In vivo assays confirmed the following results: (1) the gold nanoparticles containing bioactive glass embedded in Vaseline ointment showed faster wound regeneration in laboratory rats [11]; (2) the gold nanoparticles containing bioactive glass introduced in an alginate–pullulan composite exhibited excellent biocompatibility after in vivo subcutaneous analysis [12].

Unfortunately, surgeries sometime end with a bacterial infection that is received during hospital treatment [13]. Therefore, it is important to develop materials that exhibit antibacterial properties by involving noble metals in them, such as Ag, Pt, Pd, Cu, and Ru [12,14], or metals such as Ti, Co, Zn, and Ni [15,16]. Bioactive glass with Ag nanoparticles (AgNPs) showed high potential in antibacterial applications, as claimed by Bellantone and Hench [17]. After this finding, several studies [18–21] showed that adding different species of silver into bioactive glass matrices gave the composites antibacterial properties. It has also been stated that AgNPs are one of the best-known noble metals that possess this property but, due to their instability [22], their use still raises a lot of questions. The instability was demonstrated by Leward et al. [23] who shown that silver species in NaCl media transformed into AgNPs and AgCl, which could affect their in vivo performance.

In our previous work [24], we involved AgI particles in a bioactive glass matrix to avoid the possible formation of AgCl after its immersion in simulated body fluid (SBF). In this work, we concluded that AgI transformed into AgNPs, $Ca(IO_3)_2$, and CaI_2 but still showed acceptable antibacterial properties.

Gold nanoparticles with silver content can be obtained by synthetizing spherical gold nanocages (AuNCs), as small amounts of AgNPs remain in samples that have been used for synthesis [25,26]. By adding the latter to bioactive glass, a moderate antibacterial effect can be obtained [27].

The aim of this work was to prepare biomaterials with bactericidal effects and to maintain the beneficial properties of gold nanoparticles proven in our previous in vivo studies [11,12]. Accordingly, AuNCs and AgI microcrystals were added to bioactive glasses to obtain composites via the sol–gel the method used in our previous studies. Gold and silver particles were added into the bioactive glass using four different pathways: synthesizing two different reference materials, that is, AuNCs and AgI were added separately into the bioactive glass, or two composites were synthesized containing both noble metals. Our investigation on the human epidermal keratinocyte cell (HaCaT) line and the *Pseudomonas aeruginosa* and *Staphylococcus aureus* bacterial strain lines proved that multifunctional bioactive glasses can be successfully synthesized.

2. Materials and Methods

2.1. Materials

All chemicals were used as received without further purification. For the synthesis of AgI microcrystals and gold nanocages, ethylene glycol (EG, analytical reagent, Molar Chemical, Halasztelek, Hungary), sodium iodide (NaI, $\geq$99%, reagent grade, Honeywell, Fluka, Denmark), silver nitrate ($AgNO_3$, analytical reagent, 99.8%, Penta Industry, Prague, Czech Republic), polyvinylpyrrolidone (PVP, average mol wt. 40,000, Sigma-Aldrich, Schnelldorf, Germany), ethanol (EtOH, absolute, Molar Chemical, Halásztelek, Hungary), gold(III) chloride trihydrate ($HAuCl_4 \cdot 3\,H_2O$, trace metals basis, Sigma-Aldrich, Schnelldorf,

Germany), sodium borohydride ($NaBH_4$, 96%, Sigma-Aldrich, Schnelldorf, Germany), sodium citrate dihydrate ($Na_3C_6H_5O_7{\cdot}2H_2O$, 99.0%, Alfa-Aesar, Haverhill, MA, USA), sodium chloride (NaCl, >99.0%, Spectrum-3D, Debrecen, Hungary), and Pluronic® F-127 (Bioreagent, i.e., suitable for cell culture, Sigma-Aldrich, Schnelldorf, Germany) were used.

The precursors used for the synthesis of glass were tetraethyl orthosilicate (TEOS, ≥99%, Merck, Burlington, MA, USA), triethyl phosphate (TEP, ≥99%, Merck, Burlington, MA, USA), and calcium nitrate tetrahydrate ($Ca(NO_3)_2{\cdot}4H_2O$, ≥99%, Lach-Ner, Neratovice, Czech Republic), which were hydrolyzed in the presence of nitric acid (HNO_3, 65%, Nordic Chemical, Cluj-Napoca, Romania). Ultrapure water and absolute ethanol were used throughout the whole experimental process.

For the preparation of SBF sodium chloride (described above), sodium bicarbonate ($NaHCO_3$, Penta Industry, Prague, Czech Republic), potassium chloride (KCl, 99.5%, Nordic Chemicals, Cluj-Napoca, Romania), dipotassium hydrogen phosphate (K_2HPO_4, 99%, Penta Industry, Prague, Czech Republic), magnesium chloride hexahydrate ($MgCl_26H_2O$, 99%, Merck, Burlington, MA, USA), calcium chloride ($CaCl_2$, >97% Penta Industry, Prague, Czech Republic), sodium sulphate (Na_2SO_4, 99%, Nordic Chemicals, Cluj-Napoca, Romania), tris(hydroxymethyl)aminomethane (TRIS, 99.8%, Merck, Burlington, MA, USA), and hydrogen chloride (HCl, 37%, Nordic Chemicals, Cluj-Napoca, Romania) were used.

2.2. *Synthesis of Silver Iodide and Sperical Gold Nanocages*

2.2.1. Synthesis of Silver Iodide

AgI particles were prepared via solvothermal synthesis by the addition of silver nitrate, sodium iodide, and PVP [24]. The solutions were prepared in two separate beakers. In the first beaker, silver nitrate (0.167 M) was dissolved in 20 mL of EG. In the second beaker, sodium iodide (0.077 M) and PVP (0.0036 M; calculated to units of PVP) were dissolved in 100 mL of EG and stirred for 1 h at 60 °C. After combining the contents of the two beakers, the synthesis mixture was stirred for another hour. A yellowish-green precipitate was formed in an instantaneous reaction, indicating the formation of AgI. After stirring, the suspension was crystallized at 160 °C in an autoclave (160 mL) for 2 h. The sample was then purified with 3 × ≈35 mL Milli-Q water and 1 × ≈35 mL EtOH by centrifugation at 4400 rpm for 10 min, then dried at 40 °C for 12 h.

2.2.2. Synthesis of Spherical Gold Nanocages

The synthesis of spherical AuNCs was based on the galvanic replacement method in which AgNPs were exchanged [26,27]. In the first part of the synthesis, a sol containing AgNPs was prepared. For this purpose, 227 mL of Milli-Q water was mixed with 33 mL (2.31 mM) of sodium citrate solution. After stirring for 30 min, 2.8 mL (83.67 mM) of silver nitrate solution was added to the mixture. After stirring for another 30 min, the AgNPs were reduced with sodium borohydride (5.25 mL, 0.55 M).

The resulting solution containing AgNPs was transferred to a reflux flask, heated to 100 °C under continuous reflux, and held at that temperature for 10 min. Then, 3.15 mL of 25.39 mM $HAuCl_4{\cdot}3H_2O$ solution was added dropwise and the system was kept at 100 °C for 20 min and refluxed continuously. The solution containing AuNPs was cooled to room temperature and ≈18 g of NaCl was added to facilitate the dissolution of the resulting AgCl. Impurities were removed from the final solution by washing three times with 0.5 mM Pluronic® F-127 polymer solution.

2.2.3. Synthesis of Silver Iodide with Spherical Gold Nanocages

A composite of silver iodide with spherical gold nanocages (AgIAu) was fabricated using AgI and AuNCs. First, a sol containing AuNCs was prepared and purified by washing three times with water. Second, AgI was prepared by adding 50 mg of AgI to 100 mL of distilled water under ultrasonication for 15 min. Third, the aqueous gold sol was added to the silver-containing suspension that was washed with 3 × ≈35 mL distilled water in a centrifuge at 4400 rpm for 5 min, then dried at 40 °C for 12 h. For the thermal stability analysis, the AgIAu composite was heat treated at 500 °C in air for 3 h.

2.3. Synthesis of Bioactive Glasses with Silver and Gold Content

In the sol–gel derived bioactive glass (BG) noble metals (AgI and AuNCs) were introduced in two different ways: (i) with preliminary composite mixing, i.e, with the addition of AgIAu; and (ii) consecutively added to the system. Reactants were added consecutively after 1 h intervals under continuous stirring, as described in our previous study [27,28]. Finally, the solution of colloidal AgI, AuNCs, AgIAu, AgI, and AuNCs was added and stirred for 1 h. The solutions were stored in a closed container at 37 °C until gelation was reached (~48 h). The obtained gels were aged 3 days at 37 °C, after that they were dried at 110 °C for 24 h. Material stabilization was carried out at 500 °C in air for 3 h. Prior to characterization, the samples were milled by hand in a ceramic mortar.

Table 1 contains the abbreviations of the samples and details about the compositions. In the composites, the AuNCs:AgI ratio was 24:76. Glass composites are traditionally calculated in mol%. For comparison purposes, the amount of silver was expressed in at% in each sample, and the information is presented in Table 1.

2.4. Characterization of AgIAu Composites

The crystalline composition of the AgIAu composites was identified by using a Shimadzu X-ray diffractometer (XRD 6000, Kyoto, Japan), operating with CuKα radiation (λ = 1.54 Å) and a Ni filter. The diffraction patterns were recorded in the 2θ range from 10° to 80° with a speed of 2°/min. The specified band for AgI and the typical localized surface plasmon resonance band for Au nanocages were identified with a Jasco-V780 spectrophotometer equipped with an ILV-724 integrating sphere (UV–Vis, Jasco, Wien, Austria) with spectra resolution of 2 mm. The morphology of the samples was determined with a FEI Technai G2 20 X-TWIN transmission electron microscope (TEM, FEI, Hillsboro, OR, USA) and a Hitachi S-4700 Type II scanning electron microscope (SEM, Hitachi, Tokyo, Japan).

2.5. Structural and Morphological Characterization of Bioactive Glass Composites

Elemental composition of the glass samples was examined using an energy dispersive X-ray spectroscope (EDX, Hitachi, Tokyo, Japan) with Röntec XFlash Detector 3001 SDD system. The morpho-structural properties of the samples were analyzed by XRD and FT-IR spectroscopy. The FT-IR absorption spectra were recorded with a Jasco 6600 FT-IR spectrometer (IR, Jasco, Tokyo, Japan) within the range of 400–4000 cm^{-1} and a spectra resolution of 4 cm^{-1} by using the KBr pellet technique.

2.6. In Vitro Bioactivity Assays

The in vitro bioactivity assays were analyzed by immersion in SBF (pH = 7.4; 37 °C) for 7 days. An incubator was used to reach 37 °C and the concentration of bioactive glasses was 10 mg·mL^{-1}. After 7 days the samples were filtered, rinsed with ultrapure water, and dried. The formation of a hydroxycarbonate apatite (HCA) layer on the surface of the samples was investigate by XRD, FT-IR, UV–Vis, and SEM measurements. FT-IR spectra, XRD diffractograms, and an SEM micrograph were achieved with the same equipment described above.

Table 1. Abbreviation and composition of glasses.

Glass Samples (mol%)	Abbreviation	Silver Precursors	Silver Component Amount (mol%)	Silver Component Amount (at%)	Gold Component Amount (at%)
$60SiO_2 \cdot 32CaO \cdot 8P_2O_5$	BG	–	–	–	–
$60SiO_2 \cdot 31.25CaO \cdot 8P_2O_5 \cdot 0.75AgI$	BG-AgI	Silver iodide	0.75	0.25	–
$60SiO_2 \cdot 31.1CaO \cdot 8P_2O_5 \cdot (0.75AgI{+}0.15Au_2O)$ §	BG-AgI+AuNCs	Silver iodide **and** gold nanocages	0.75	0.249	0.089 *
$60SiO_2 \cdot 31.1CaO \cdot 8P_2O_5 \cdot (0.75AgI{-}0.15Au_2O)$ §	BG-AgIAu	Silver iodide **with** gold nanocages	0.75	0.249	0.09
$60SiO_2 \cdot 31.85CaO \cdot 8P_2O_5 \cdot 0.15Au_2O$ §	BG-AuNCs	Gold nanocages	–	0.0024	0.089 *

§ the amount of gold in the glasses is conventionally indicated as the oxidation state of gold, that is Au_2O. * the amount of silver introduced during the synthesis of gold nanocages was approximately 0.0024 at%.

To analyze surface changes before and after immersion in SBF, XPS spectra were recorded with a SPECS PHOIBOS 150 MCD system employing a monochromatic Al-K_α source (1486.6 eV), a hemispherical analyzer, and a charge neutralization device. Samples were fixed on double-sided carbon tape, and care was taken to ensure that the sample particles covered the tape. Experiments were performed by operating the X-ray source with a power of 200 W, while the pressure in the analysis chamber was in the range of 10^{-9}–10^{-10} mbar. The binding energy scale was charge referenced to that of C1s at 284.6 eV. Elemental composition was determined from the survey spectra acquired at a pass energy of 60 eV. High-resolution spectra were obtained using an analyzer pass energy of 20 eV. Analysis of the data was carried out with CasaXPS software.

2.7. Cell Viability Assay

Cell viability assessment was carried out on a human epidermal keratinocyte cell line (HaCaT, Cell Line Service, Eppelheim, Germany), via the same method that we described in our previous publications [10,27]. The cells were cultured in Dulbecco's modified Eagle's medium (Lonza) supplemented with 2 mM L-glutamine, Pen/Strep 100 U/mL, and 10% FC. Then, they were incubated in a humidified incubator under 5% CO_2 atmosphere at 37 °C. The cytotoxic effects of different samples were assayed with a WST-1 dye (Merk Millipore, Burlington, MA, USA). The HaCaT cells were seeded in a 96-well plate, at a density of 1×10^4 cells/well. The following day, different amounts of the samples were added to the test wells and cells were placed in the incubator for an additional 24 h. All glass samples were tested at three concentrations (75 $\mu g \cdot mL^{-1}$, 150 $\mu g \cdot mL^{-1}$, and 300 $\mu g \cdot mL^{-1}$). Untreated cells were used as controls. At the end of the incubation period, the medium was removed from all wells, and 100 μL of fresh medium containing 10% WST-1 solution was added to each well. The cells were incubated for another 60 min at 37 °C. Empty wells with medium containing WST-1 reagent were used as blanks. Viability of HaCaT cells after 24 h of treatments was determined by measuring the absorbance at 440 nm with a microplate reader (Fluostar Omega, BMG Labtech, Ortenberg, Germany).

2.8. Antibacterial Activity

2.8.1. Microorganism and Culture Conditions

One aerobic gram-negative *Pseudomonas aeruginosa* (ATCC 27853) and one gram-positive *Staphylococcus aureus* (ATCC 49444) bacterial strain were used for antibacterial activity analysis. The tested microorganisms were obtained from Food Biotechnology Laboratory, Life Sciences Institute, University of Agricultural Sciences and Veterinary Medicine Cluj-Napoca, Romania. Both bacteria were cultured on Mueller-Hinton Agar and cultures were stored at 4 °C and subcultured once a month.

2.8.2. Microdilution Method

A modified microdilution technique, based on the regulations of the Clinical Laboratory Standard Institute (CLSI), was used to evaluate antimicrobial activity [29]. Bacterial species were cultured overnight at 37 °C in Mueller-Hinton Broth (MHB). The bacterial cell suspensions were adjusted with a 0.9% NaCl sterile saline solution to a concentration of approximately 3×10^5 colony-forming units (CFU)/mL in a final volume of 100 μL per well. The inoculum was stored at +4 °C for further use. Dilutions of the inoculum were cultured on solid Mueller–Hinton (MH) for bacteria to verify the absence of contamination and check the inoculum's validity. Determinations of minimum inhibitory concentrations (MICs) were performed by a serial dilution technique using 96-well microtitre plates. The samples were transferred to the wells containing 100 μL MHB, and afterward, 10 μL of inoculum was added to all the wells. The microplates were incubated for 18 h at 37 °C. The MIC of the samples were detected following the addition of 20 μL (0.2 $mg \cdot mL^{-1}$) of resazurin solution to each well, and the plates were incubated for 2 h at 37 °C. A change from blue to pink indicated a reduction in resazurin and, therefore, bacterial growth. The MIC was defined as the lowest drug concentration that prevented this color change. The concentration of the samples was 10 mg·mL^{-1}. A control sample gentamicin solution of the same concentration was used for the tested samples.

2.9. Statistical Analysis

All data in the cell viability assay were reported as the mean ± SD of triplicate measurements. The values were analyzed by two-way analysis of variance ANOVA. Statistical significance was at $p < 0.05$ in all cases. Statistical values were obtained using GraphPad Prism 8.0 software. The antimicrobial analysis required no statistical validation as the method was one of the highest confidence interval methods [30].

3. Results and Discussion

3.1. Characterization of The Composites Containing AgI Microcrystals, Spherical Gold Nanocages, or AgIAu Nanostructures

After the synthesis of AgI microcrystals and AuNCs, their characterization was carried out to investigate their reproducibility. The results were compared with those published in our previous work [24,27]. The UV–Vis spectra, XRD patterns, and SEM micrographs of the AgI microcrystals confirmed the reproducibility of the samples as follows: (i) in the UV–Vis spectrum (Figure 1A) of the sample the typical absorption band for AgI (415–480 nm; [31]) was identified; (ii) in Figure 1B the corresponding reflections of AgI (β-AgI—COD 00-101-1025 and γ-AgI—COD 00-901-1693) were observed; and (iii) the SEM images confirmed the pyramidal structure (Figure 2A). It should be emphasized that AgI was semicrystalline and the two different crystal phases of AgI could not be distinguished because their reflections overlapped with each other [32].

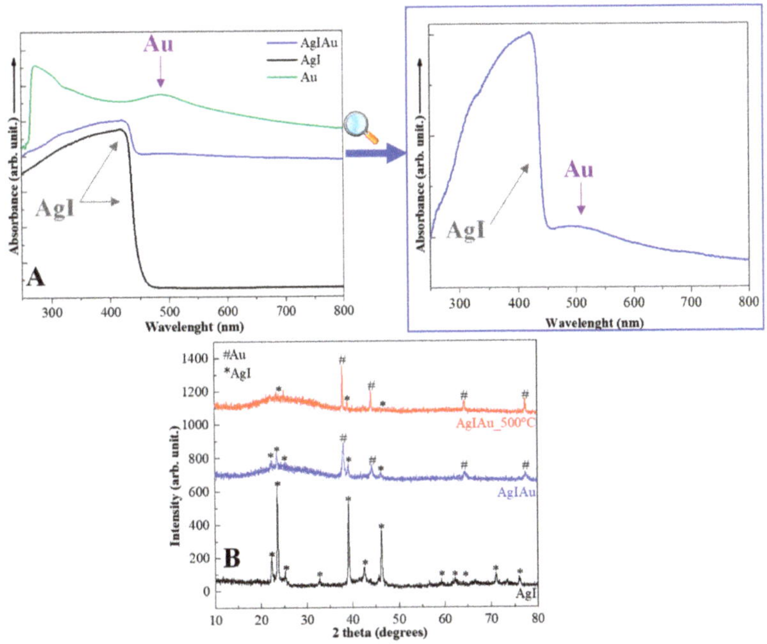

Figure 1. The characterization of the Au nanocages, AgI microcrystals, and AgIAu composite: (**A**) UV–Vis spectra and (**B**) XRD patterns.

Figure 2. SEM micrograph of pyramidal structure of AgI (**A**); TEM micrographs of Au nanocages (**B**); and AgIAu composites before (**C**) and after electron beam irradiation (**D**).

The UV–Vis spectra of the AuNCs showed the appropriate (localized) surface plasmon resonance band of AuNCs, located at ≈490 nm (Figure 1A). In the TEM micrographs the presence of Au nanostructures of about 18 ± 3 nm were also observed (Figure 2B). The first proof of successful composite mixing was obtained by UV–Vis measurements. Both the (localized) surface plasmon resonance band of AuNCs (located at 500 nm; insert graph Figure 1A) and the absorption band of AgI (between 350–470 nm) could be observed. The second proof was obtained by XRD measurements, revealing that not only could the AgI reflection be identified but the reflection of Au (COD 00-900-8463) could also be identified. Moreover, in the TEM images small particles could be observed on the surface of the AgI microparticles (Figure 2C). Unfortunately, the low stability of silver-based materials could also be noticed in Figure 2D as the surface of the particles changed with the energy of the electron beam, thereby their size could not be analyzed.

In our previous work [27] it was shown that the gold nanocages introduced into the bioactive glasses were recrystallized into spherical gold nanoparticles by heat treatment at 500 °C. It has also been demonstrated by our group [24] that AgI microcrystals remained stable after heat treatment at 500 °C. Thermal stability of the AgIAu composite was investigated to determine possible transformations during heat treatment; thus, the composite was heat treated at 500 °C. After heat treatment, the reflection for Au and AgI were still presented, the gold became crystalline, and the AgI did not have any significant changes (Figure 1B).

3.2. *Characterization and In Vitro Bioactivity of Bioactive Glass Composites*

EDX analysis was performed for elemental identification of composites. In Figure 3 the presence of all metal components can be seen. The silver and iodine values were comparable with theoretical values (Table 2). The presence of gold was also confirmed, but due to the small amount the obtained value differed to the theoretical one.

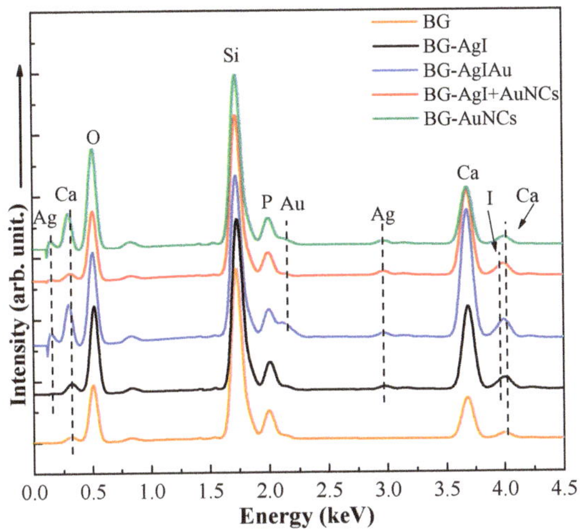

Figure 3. EDX spectra of the BG, BG-AgI, BG-AgIAu, BG-AgI+AuNCs, and BG-AuNCs samples.

Table 2. Si, Ag, I, and Au content of the heat-treated BG composites obtained from EDX measurements. For comparison, theoretical values were also calculated.

Sample		Elements (at%)			
		Si	Ag	I	Au
BG	Theoretical	20.00	-	-	-
	Heat-treated	21.76	-	-	-
BG-AgI	Theoretical	20.00	0.25	0.25	-
	Heat-treated	18.51	0.17	0.25	-
BG-AgIAu	Theoretical	20.00	0.24	0.24	0.089
	Heat-treated	14.97	0.17	0.14	0.015
BG-AgI+AuNCs	Theoretical	20.00	0.24	0.24	0.09
	Heat-treated	16.10	0.3	0.22	0.015
BG-AuNCs	Theoretical	20	0.0024	-	0.089
	Heat-treated	17.82	0.22	-	0.074

The XRD patterns (Figure 4) of gold-containing samples exhibited characteristic reflections of metallic gold (COD 00-900-8463). These signals were recorded both before and after immersion in SBF. In the BG-AgI sample an additional reflection was observed at 29.45 (2θ°). This could be attributed to the presence of $CaCO_3$ (calcite) (COD 00-101-0928), which only appeared after the heat treatment of the sample. The formation of $CaCO_3$ could have originated from the high concentration of calcium ions and the textural parameters of the bioactive glasses [33].

After immersion in SBF for 7 days, all samples showed reflections associated with the presence of an HCA layer precipitated on the surface of the samples (COD file no. 09-0432). After soaking in SBF, AgCl (COD file no. 00-901-1666) was formed on the surface of the silver containing BG composites because of the high amount of NaCl in SBF [34]. The formation of AgCl occurred in the BG–AuNCs samples when the amount of Ag was low.

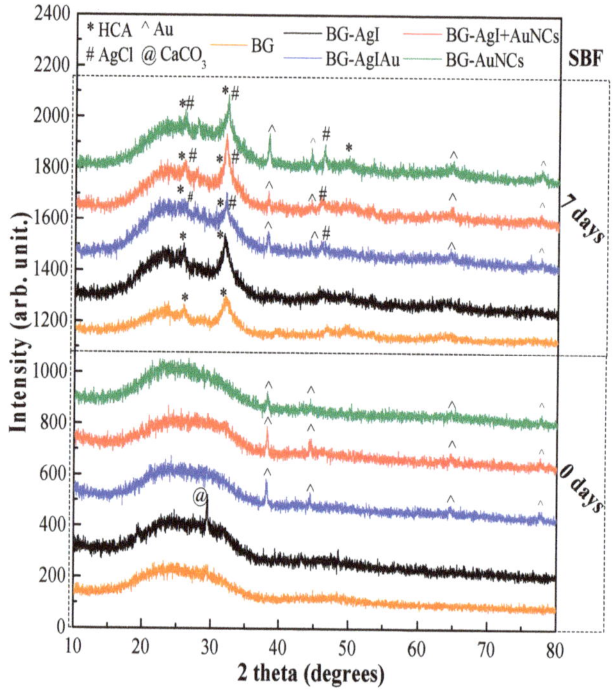

Figure 4. XRD patterns of the BG, BG-AgI, BG-AgIAu, BG-AgI+AuNCs, and BG-AuNCs samples before (0 days) and after (7 days) immersion in SBF.

The formation of BG was confirmed (Figure 5) by the presence of an absorption band at 1080 cm^{-1} and a shoulder around 950 cm^{-1}. The former could be attributed to the stretching vibration of Si-O bonds, while the latter to the vibrations of Si-O-Si groups [35]. The strong absorption band around 470 cm^{-1} was attributed to the torsional vibrations of the Si-O-Si groups, which also confirmed the presence of BG. The FT-IR spectra of the all samples immersed in SBF for seven days showed a doublet at 604 and 564 cm^{-1}. These could be attributed to the vibrations of [PO_4] units corresponding to crystalline HCA (Figure 5) [36]. In addition to the phosphate bands, a new band at 870 cm^{-1} was observed that could be ascribed to carbonate ions indicating the formation of a carbonate apatite phase.

SEM images (Figure 6) confirmed that the BG had a porous structure before immersion in SBF. It was also ascertained that samples having a cauliflower-like structure showed good bioactivity. These structures confirmed the formation of the HCA layer on the surface of the BG composites.

The plasmon resonance band of AgNPs (≈357 nm) and gold nanoparticles (≈525 nm) could be identified in the UV–Vis spectra (Figure 7A) of the BG composites before immersion in SBF. The plasmon resonance band of AgNPs (located at ≈394 nm) could also be identified in the BG-AuNCs composites. In these composites, a low amount of remaining silver was observed because their synthesis involves the addition of AgNPs as well. Ag^+ ions, with a signal around 260 nm, were visible only in the BG-AuNCs and BG-AgI+AuNCs. The similarity of these two samples was due to the fact that in both cases, AuNCs were added into the BG structure. As described before, the Ag^+ ions could originate from the silver that remained after the synthesis of AuNCs. The presence of Ag^+ ions could not be observed in the XRD patterns either because their amount was low or they were present in the amorphous phase. UV–Vis absorption bands around 290 and

250 nm of the BG samples' spectra appeared to be due to the presence of Si-O-(Si, Ca) and P-O-(P, Ca) groups, respectively [37].

After immersion in SBF for 7 days, the surface plasmon resonance of gold nanoparticles (at ≈414 nm and ≈ 425 nm) and the absorption band of Ag^+ ions (at around 260 nm) were detected (Figure 7B). In addition, a band associated with the presence of AgCl was observed at 369 nm. The surface plasmon resonance signal was observed for the sample containing AuNCs. However, most probably these bands originated from the AgNPs that remained in the system during the galvanic replacement of AuNCs. For the AgI-containing samples, an additional plasmon resonance band correspond to AgNPs could be detected, presumably due to the transformation of AgI.

XPS measurements were also carried out to analyze the chemical state of the samples' surfaces and to provide the presence of the HCA layer. The proof for the formation of a HCA layer was that the atomic percentage of Si decreased after the immersion into SBF (Table 3). Another proof was the increase in the atomic percentages of Ca and P (Table 3). It is known that XPS measurements can only provide information from the surface of a sample (≈10 nm). Thus, based on the decreasing amount of Si, and the increase in Ca and P amounts, the HCA layer was formed on the surface of the composites [24]. The obtained Ca/P ratio suggested that the surface layer contained carbonates and non-stoichiometric components (decreasing from 2.3 to 1.33) [38].

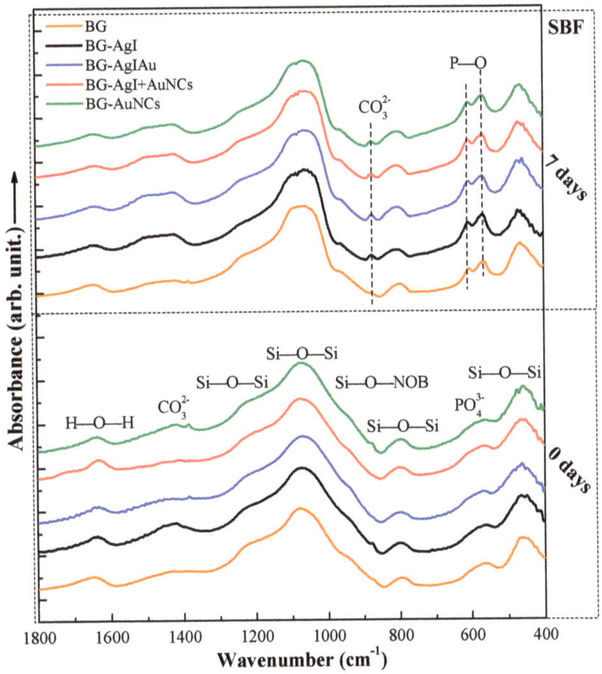

Figure 5. FT-IR spectra of the BG, BG-AgI, BG-AgIAu, BG-AgI+AuNCs, and BG-AuNCs samples before (0 days) and after (7 days) immersion in SBF.

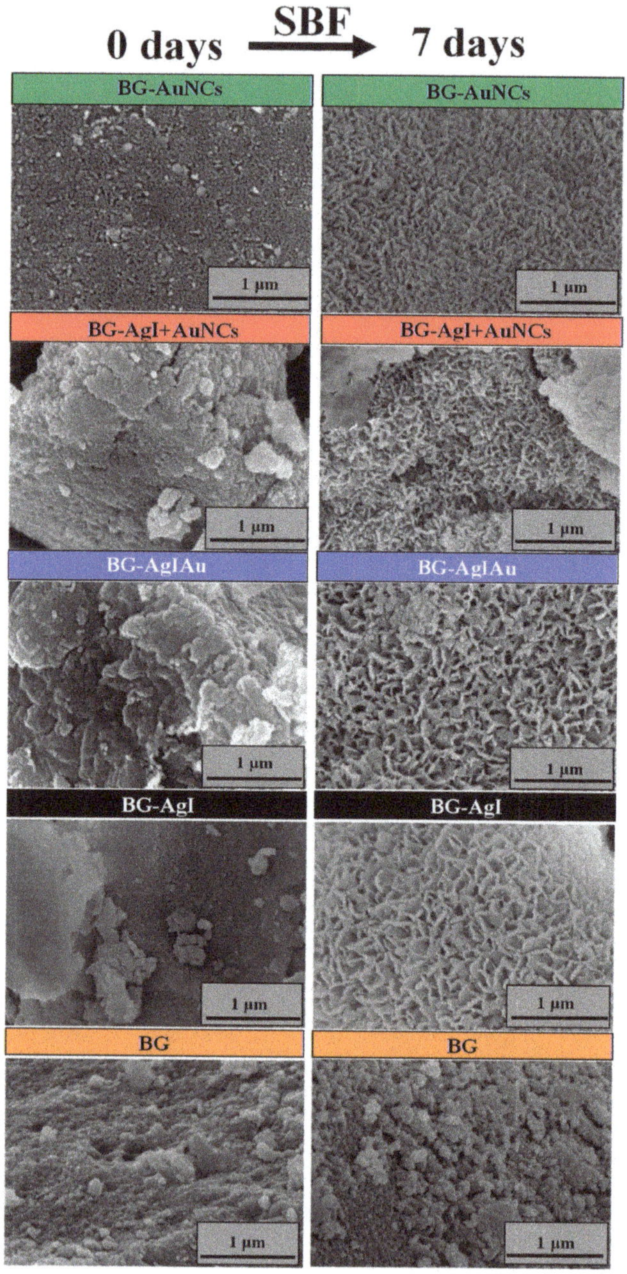

Figure 6. SEM micrographs BG, BG-AgI, BG-AgIAu, BG-AgI+AuNCs, and BG-AuNCs before (0 days) and after (7 days) immersion in SBF.

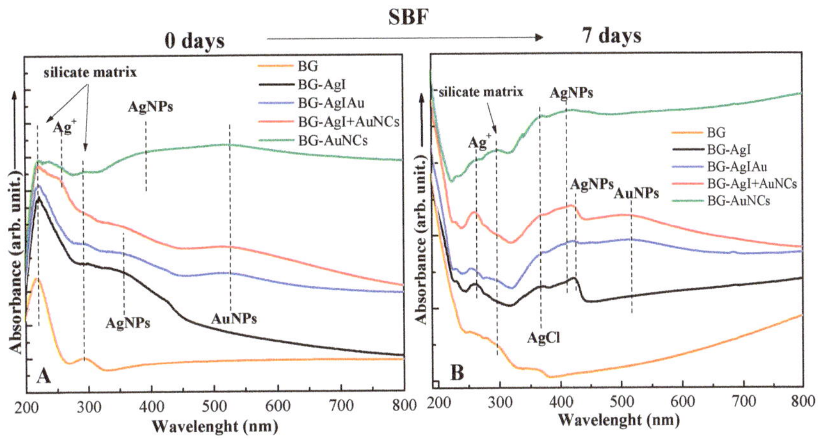

Figure 7. UV–Vis spectra of BG composites: (**A**) before (0 days) and (**B**) after (7 days) immersion in SBF.

Table 3. Si, Ca, and P content of the heat-treated and immersed BG composites obtained from XPS measurements. For comparison, theoretical values were also calculated.

Sample		Elements (at%)			Ratio
		Si	Ca	P	Ca/P
BG	Theoretical	20.00	10.66	5.33	2
	Heat-treated	21.1	9.6	7.7	1.24
	Immersed in SBF	16.4	11.8	9.4	1.25
BG-AgI	Theoretical	20.00	10.41	5.33	1.95
	Heat-treated	16.30	9.20	4.00	2.30
	Immersed in SBF	8.40	12.80	9.60	1.33
BG-AgIAu	Theoretical	20.00	10.36	5.33	1.94
	Heat-treated	15.67	8.50	2.90	2.93
	Immersed in SBF	2.30	16.30	11.70	1.39
BG-AgI+AuNCs	Theoretical	20.00	10.36	5.33	1.94
	Heat-treated	20.46	10.40	4.40	2.36
	Immersed in SBF	6.9	13.7	10.2	1.34
BG-AuNCs	Theoretical	20	10.61	5.33	1.99
	Heat-treated	15.8	12.8	6.3	2.03
	Immersed in SBF	3.9	15.8	12.1	1.31

High-resolution Ca2p and P2p spectra were also recorded (Figure 8), and the binding energy of 348 eV and 351.5 eV were attributed to Ca2p orbitals, while the one at 133 eV to P2p orbitals [24,38]. After immersion in SBF, both photoelectron peaks broadened and their position changed: for Ca2p, it changed from ≈348 eV to ≈349.3 eV, and from 351.5 eV to 352.64 eV; while for P2p, it changed from ≈133 eV to ≈135 eV. The broadening of the Ca2p peak could be due to the transformation of CaO into CaI_2 and $Ca(IO_3)_2$ as was proved in our previous study [24]. These transformations were not surprising since the presence of AgNPs was confirmed based on the UV–Vis spectra (Figure 7). This meant that the AgI in the BG transformed into AgNPs and other species of silver and into CaI_2 and $CaIO_3$. Moreover, for the BG-AuNCs the degree of these broadenings was low and could be attributed to the formation of the apatite layer (Figure 8).

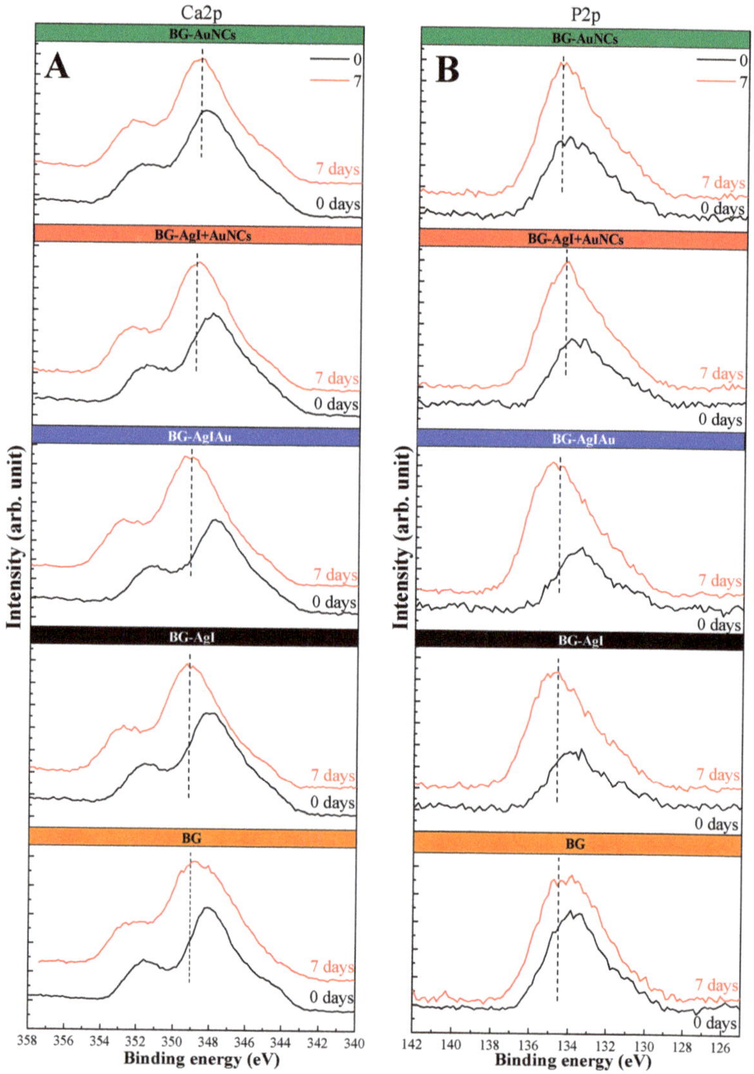

Figure 8. XPS core level spectra for Ca2p (**A**) and P2p (**B**) in BG composites.

3.3. *Influence of Silver and Gold on Cell Viability and Antibacterial Activity*

The toxicity of the silver content was evaluated via a cell viability assay following the viability of HaCaT cells in contact with the BG samples for 24 h. First, it needs to be mentioned that, independently of the BG composites used, and the different concentration of glasses, all samples showed cell viability greater than 100% (Figure 9). Therefore, it could be concluded that the composites were not cytotoxic.

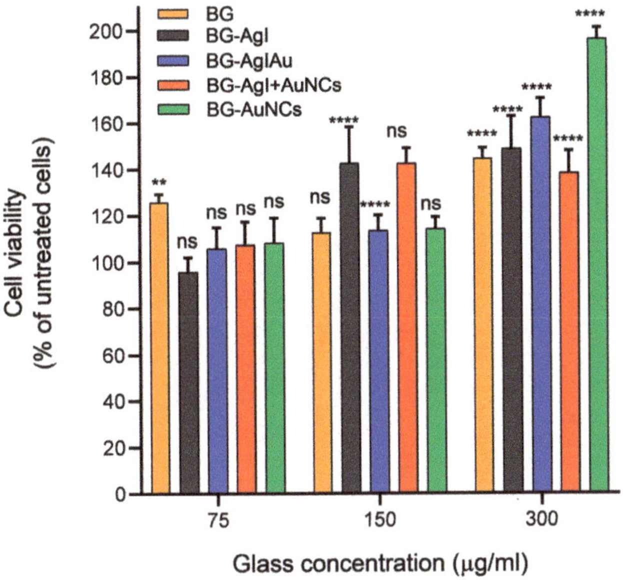

Figure 9. Viability of HaCaT cells in the presence of BG, BG-AgI, BG-AgIAu, BG-AgI+AuNCs, and BG-AuNCs in different concentrations. * $p < 0.05$.

The next step of our research was to investigate the antibacterial character of the samples using two different bacterial strains: *S. aureus* (gram-positive; Figure 10, self-colored bar) and *P. aeruginosa* (gram-negative; Figure 10, bar with pattern). Since the gram-negative bacterial strains had a thinner cell wall than that of *S. aureus,* the outstanding antibacterial character of the samples was not surprising. BG-AgI showed the highest antibacterial character, which could have originated from the transformation of AgI into AgNPs, as confirmed by the UV–Vis spectra (Figure 7A). Reasonable antibacterial activity was observed for BG-AuNCs, which could be due to the low amount of silver, as confirmed by the UV–Vis spectra (Figure 7A). Composites of BG-AgIAu and BG-AgI+AuNCs showed good antibacterial activity against *P. aeruginosa*. The composites, where the noble metal-based particles were added consecutively (BG-AgI+AuNCs) showed a higher resistance to *S. aureus*, from which it was concluded that this approach was best. The silver-free glass sample has no antibacterial effect (data not shown).

Based on the results, we could conclude that the addition of Ag and Au enabled the bioactive glasses with both antibacterial activity and with the ability to enhance the viability of HaCaT cells. Furthermore, the best combination was obtained when silver and gold content were consecutively added into the BG instead of creating a composite out of them beforehand by mixing.

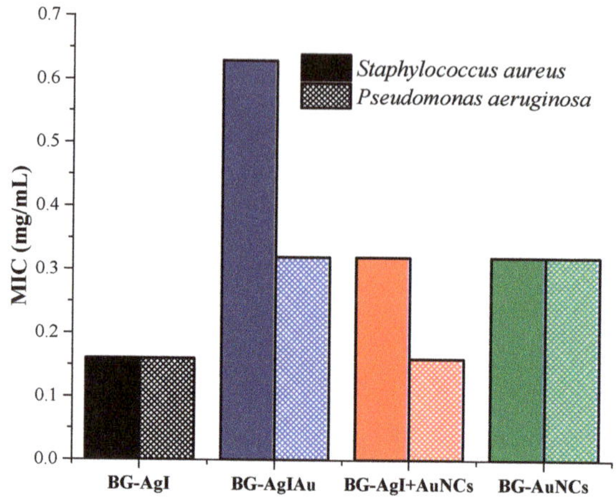

Figure 10. Minimal inhibitory concentration (MIC) of *S. aureus* and *P. aeruginosa* using BG composites. Gentamycin was used as a positive control obtaining MIC of 0.0024 mg·mL^{-1} for both bacteria.

4. Conclusions

Four different composites were prepared to investigate the synergistic effect between bioactive glass composites and two noble metals. From these four samples two were synthesized as reference materials: one containing only silver iodide and the other one containing predominantly Au nanocages. Another two samples were fabricated that contained both noble metals, either by adding them consecutively or as a AgIAu composite. In addition, noble-metal-free bioactive glass was synthetized for comparison purposes. Based on morphological and structural characterizations, bioactive glasses with amorphous structures were successfully obtained independently of the noble metal or noble-metal composites used. All composites presented excellent bioactivity and induced the proliferation of HaCaT cells. Composites with silver content possessed good antibacterial activity against the *S. aureus* and *P. aeruginosa* bacterial strains. The best results were achieved when silver iodide and gold nanocages were added consecutively to the bioactive glasses, resulting in versatile bioactive glasses.

Author Contributions: Conceptualization, K.M. and Z.-R.T.; methodology, Z.-R.T., J.K., M.T., T.G., A.S. (Alina Sesarman), D.C.V. and K.M.; validation, L.B. and K.M.; investigation, Z.-R.T., J.K., M.T., G.K., T.G., A.S. (Anna Szabó), G.N., D.C.V. and K.M.; writing—original draft preparation, Z.-R.T., T.G., A.S. (Alina Sesarman), D.C.V., L.B. and K.M.; writing—review and editing, Z.-R.T., L.B. and K.M.; visualization, T.G.; supervision, K.M.; project administration, K.M.; funding acquisition, K.M. All authors have read and agreed to the published version of the manuscript.

Funding: This work was supported by a Ministry of Research and Innovation grant CNCSIS-UEFISCDI, project number PN-III-P1-1.1-TE-2019-1138. In addition, K. Magyari and G. Kovács acknowledges the support of the János Bolyai Research Scholarship of the Hungarian Academy of Sciences.

Institutional Review Board Statement: Not applicable.

Informed Consent Statement: Not applicable.

Data Availability Statement: Not applicable.

Conflicts of Interest: The authors declare no conflict of interest.

References

1. Kargozar, S.; Mozafari, M.; Hamzehlou, S.; Baino, F. Using Bioactive Glasses in the Management of Burns. *Front. Bioeng. Biotechnol.* **2019**, *7*, 62. [CrossRef] [PubMed]
2. Craciun, A.M.; Focsan, M.; Magyari, K.; Vulpoi, A.; Pap, Z. Surface Plasmon Resonance or Biocompatibility-Key Properties for Determining the Applicability of Noble Metal Nanoparticles. *Materials* **2017**, *10*, 836. [CrossRef] [PubMed]
3. Schatkoski, V.M.; do Amaral Montanheiro, T.L.; de Menezes, B.R.C.; Pereira, R.M.; Rodrigues, K.F.; Ribas, R.G.; Morais da Silva, D.; Thim, G.P. Current Advances Concerning the Most Cited Metal Ions Doped Bioceramics and Silicate-Based Bioactive Glasses for Bone Tissue Engineering. *Ceram. Int.* **2021**, *47*, 2999–3012. [CrossRef]
4. Pantulap, U.; Arango-Ospina, M.; Boccaccini, A.R. Bioactive glasses incorporating less-common ions to improve biological and physical properties. *J. Mater. Sci. Mater. Med.* **2022**, *33*, 3. [CrossRef]
5. Cavalu, S.; Banica, F.; Gruian, C.; Vanea, E.; Goller, G.; Simon, V. Microscopic and spectroscopic investigation of bioactive glasses for antibiotic controlled release. *J. Mol. Struct.* **2013**, *1040*, 47–52. [CrossRef]
6. Yeh, Y.C.; Creran, B.; Rotello, V.M. Gold Nanoparticles: Preparation, Properties, and Applications in Bionanotechnology. *Nanoscale* **2012**, *4*, 1871–1880. [CrossRef]
7. Sau, T.K.; Goia, D.V. Biomedical Applications of Gold Nanoparticles. In *Fine Particles in Medicine and Pharmacy*; Springer: Berlin/Heidelberg, Germany, 2011; pp. 101–145, ISBN 9781461403791.
8. Lu, S.; Xia, D.; Huang, G.; Jing, H.; Wang, Y.; Gu, H. Concentration Effect of Gold Nanoparticles on Proliferation of Keratinocytes. *Colloids Surf. B Biointerfaces* **2010**, *81*, 406–411. [CrossRef]
9. Liang, H.; Jin, C.; Ma, L.; Feng, X.; Deng, X.; Wu, S.; Liu, X.; Yang, C. Accelerated Bone Regeneration by Gold-Nanoparticle-Loaded Mesoporous Silica through Stimulating Immunomodulation. *ACS Appl. Mater. Interfaces* **2019**, *11*, 41758–41769. [CrossRef]
10. Magyari, K.; Nagy-Simon, T.; Vulpoi, A.; Popescu, R.A.; Licarete, E.; Stefan, R.; Hernádi, K.; Papuc, I.; Baia, L. Novel Bioactive Glass-AuNP Composites for Biomedical Applications. *Mater. Sci. Eng. C* **2017**, *76*, 752–759. [CrossRef]
11. Mârza, S.M.; Magyari, K.; Bogdan, S.; Moldovan, M.; Peştean, C.; Nagy, A.; Tăbăran, F.; Licarete, E.; Suarasan, S.; Dreanca, A.; et al. Skin Wound Regeneration with Bioactive Glass-Gold Nanoparticles Ointment. *Biomed. Mater.* **2019**, *14*, 025011. [CrossRef]
12. Dreanca, A.; Muresan-Pop, M.; Taulescu, M.; Tóth, Z.R.; Bogdan, S.; Pestean, C.; Oren, S.; Toma, C.; Popescu, A.; Páll, E.; et al. Bioactive Glass-Biopolymers gold Nanoparticle Based Composites for Tissue Engineering Applications. *Mater. Sci. Eng. C* **2021**, *123*, 112006. [CrossRef] [PubMed]
13. Pirii, L.E.; Friedrich, A.W.; Rossen, J.W.A.; Vogels, W.; Beerthuizen, G.I.J.M.; Nieuwenhuis, M.K.; Kooistra-Smid, A.M.D.; Bathoorn, E. Extensive Colonization with Carbapenemase-Producing Microorganisms in Romanian Burn Patients: Infectious Consequences from the Colectiv Fire Disaster. *Eur. J. Clin. Microbiol. Infect. Dis.* **2018**, *37*, 175–183. [CrossRef] [PubMed]
14. Huo, L.; Li, P. *Metal Nanoparticles: Noble Metal Nanoparticles and Their Antimicrobial Properties*; Wiley Online Library: Hoboken, NJ, USA, 2018; Chapter 8, ISBN 9783527807093.
15. Yasuyuki, M.; Kunihiro, K.; Kurissery, S.; Kanavillil, N.; Sato, Y.; Kikuchi, Y. Antibacterial Properties of Nine Pure Metals: A Laboratory Study Using Staphylococcus Aureus and Escherichia Coli. *Biofouling* **2010**, *26*, 851–858. [CrossRef] [PubMed]
16. Fadeeva, I.V.; Goldberg, M.A.; Preobrazhensky, I.I.; Mamin, G.V.; Davidova, G.A.; Agafonova, N.V.; Fosca, M.; Russo, F.; Barinov, S.M.; Cavalu, S.; et al. Improved cytocompatibility and antibacterial properties of zinc-substituted brushite bone cement based on β-tricalcium phosphate. *J. Mater. Sci. Mater. Med.* **2021**, *32*, 99. [CrossRef]
17. Bellantone, M.; Hench, L.L. Bioactive Behaviour of Sol-Gel Derived Antibacterial Bioactive Glass. *Key Eng. Mater.* **2001**, *192–195*, 617–620. [CrossRef]
18. Rivadeneira, J.; Gorustovich, A. Bioactive Glasses as Delivery Systems for Antimicrobial Agents. *J. Appl. Microbiol.* **2017**, *122*, 1424–1437. [CrossRef]
19. Zheng, K.; Balasubramanian, P.; Paterson, T.E.; Stein, R.; MacNeil, S.; Fiorilli, S.; Vitale-Brovarone, C.; Shepherd, J.; Boccaccini, A.R. Ag Modified Mesoporous Bioactive Glass Nanoparticles for Enhanced Antibacterial Activity in 3D Infected Skin Model. *Mater. Sci. Eng. C* **2019**, *103*, 109764. [CrossRef]
20. Magyari, K.; Stefan, R.; Vodnar, D.C.; Vulpoi, A.; Baia, L. The Silver Influence on the Structure and Antibacterial Properties of the Bioactive 10B2O3- 30Na2O-60P2O 2 Glass. *J. Non-Cryst. Solids* **2014**, *402*, 182–186. [CrossRef]
21. Vulpoi, A.; Simon, V.; Ylänen, H.; Simon, S. Development and in Vitro Assessment of Bioactive Glass/Polymer Nanostructured Composites with Silver. *J. Compos. Mater.* **2014**, *48*, 63–70. [CrossRef]
22. Thiyagarajan, K.; Bharti, V.K.; Tyagi, S.; Tyagi, P.K.; Ahuja, A.; Kumar, K.; Raj, T.; Kumar, B. Synthesis of Non-Toxic, Biocompatible, and Colloidal Stable Silver Nanoparticle Using Egg-White Protein as Capping and Reducing Agents for Sustainable Antibacterial Application. *RSC Adv.* **2018**, *8*, 23213–23229. [CrossRef]
23. Levard, C.; Mitra, S.; Yang, T.; Jew, A.D.; Badireddy, A.R.; Lowry, G.V.; Brown, G.E. Effect of Chloride on the Dissolution Rate of Silver Nanoparticles and Toxicity to E. Coli. *Environ. Sci. Technol.* **2013**, *47*, 5738–5745. [CrossRef] [PubMed]
24. Feraru, A.; Tóth, Z.R.; Magyari, K.; Pap, Z.; Todea, M.; Mureşan-Pop, M.; Vodnar, D.C.; Licarete, E.; Hernadi, K.; Baia, L. Composites Based on Silicate Bioactive Glasses and Silver Iodide Microcrystals for Tissue Engineering Applications. *J. Non-Cryst. Solids* **2020**, *547*, 120293. [CrossRef]
25. Sun, Y.; Xia, Y. Mechanistic Study on the Replacement Reaction between Silver Nanostructures and Chloroauric Acid in Aqueous Medium. *J. Am. Chem. Soc.* **2004**, *126*, 3892–3901. [CrossRef] [PubMed]

26. Tóth, Z.R.; Kovács, G.; Hernádi, K.; Baia, L.; Pap, Z. The Investigation of the Photocatalytic Efficiency of Spherical Gold Nanocages/TiO_2 and Silver Nanospheres/TiO_2 Composites. *Sep. Purif. Technol.* **2017**, *183*, 216–225. [CrossRef]
27. Magyari, K.; Tóth, Z.R.; Pap, Z.; Licarete, E.; Vodnar, D.C.; Todea, M.; Gyulavári, T.; Hernadi, K.; Baia, L. Insights into the Effect of Gold Nanospheres, Nanotriangles and Spherical Nanocages on the Structural, Morphological and Biological Properties of Bioactive Glasses. *J. Non-Cryst. Solids* **2019**, *522*, 119552. [CrossRef]
28. Magyari, K.; Baia, L.; Vulpoi, A.; Simon, S.; Popescu, O.; Simon, V. Bioactivity Evolution of the Surface Functionalized Bioactive Glasses. *J. Biomed. Mater. Res. Part B Appl. Biomater.* **2015**, *103*, 261–272. [CrossRef] [PubMed]
29. Weinstein, M.P. *Methods for Dilution Antimicrobial Susceptibility Tests for Bacteria That Grow Aerobically*; National Committee for Clinical Laboratory Standards, Ed.; National Committee for Clinical Laboratory Standards: Wayne, PA, USA, 2018; ISBN 978-1-56238.
30. Ruparelia, J.P.; Chatterjee, A.K.; Duttagupta, S.P.; Mukherji, S. Strain Specificity in Antimicrobial Activity of Silver and Copper Nanoparticles. *Acta Biomater.* **2008**, *4*, 707–716. [CrossRef]
31. Yu, H.; Huang, B.; Wang, H.; Yuan, X.; Jiang, L.; Wu, Z.; Zhang, J.; Zeng, G. Facile Construction of Novel Direct Solid-State Z-Scheme AgI/BiOBr Photocatalysts for Highly Effective Removal of Ciprofloxacin under Visible Light Exposure: Mineralization Efficiency and Mechanisms. *J. Colloid Interface Sci.* **2018**, *522*, 82–94. [CrossRef]
32. Berry, C.R. Structure and Optical Absorytion of AgI Microcrystals. *Phys. Rev.* **1967**, *161*, 848. [CrossRef]
33. Mozafari, M.; Banijamali, S.; Baino, F.; Kargozar, S.; Hill, R.G. Calcium Carbonate: Adored and Ignored in Bioactivity Assessment. *Acta Biomater.* **2019**, *91*, 35–47. [CrossRef]
34. Feraru, A.; Tóth, Z.R.; Debreczeni, D.; Todea, M.; Popescu, R.A.; Gyulavari, T.; Sesarman, A.; Negrea, G.; Vodnar, D.C.; Hernadi, K.; et al. Influence of different silver species on the structure of bioactive silicate glasses. *J. Non-Cryst. Solids* **2022**, *583*, 121498. [CrossRef]
35. Aguiar, H.; Serra, J.; González, P.; León, B. Structural Study of Sol-Gel Silicate Glasses by IR and Raman Spectroscopies. *J. Non-Cryst. Solids* **2009**, *355*, 475–480. [CrossRef]
36. Gayathri, B.; Muthukumarasamy, N.; Velauthapillai, D.; Santhosh, S.B.; Asokan, V. Magnesium Incorporated Hydroxyapatite Nanoparticles: Preparation, Characterization, Antibacterial and Larvicidal Activity. *Arab. J. Chem.* **2018**, *11*, 645–654. [CrossRef]
37. Nicolini, V.; Varini, E.; Malavasi, G.; Menabue, L.; Menziani, M.C.; Lusvardi, G.; Pedone, A.; Benedetti, F.; Luches, P. The effect of composition on structural, thermal, redox and bioactive properties of Ce-containing glasses. *Mater. Des.* **2016**, *97*, 73–85. [CrossRef]
38. Santos, J.D.; Jha, L.J.; Monteiro, F.J. In Vitro Calcium Phosphate Formation on SiO2-Na20-CaO-P205 Glass Reinforced Hydroxyapatite Composite: A Study by XPS Analysis. *J. Mater. Sci. Mater. Med.* **1996**, *7*, 181–185. [CrossRef]

materials

MDPI

Review

Applications of Phyto-Nanotechnology for the Treatment of Neurodegenerative Disorders

Tanima Bhattacharya [1,2,*,†], Giselle Amanda Borges e Soares [3,†], Hitesh Chopra [4], Md. Mominur Rahman [5], Ziaul Hasan [6], Shasank S. Swain [7] and Simona Cavalu [8,*]

1 Innovation, Incubation & Industry (i-cube) Laboratory, Techno India NJR Institute of Technology, Udaipur 313003, India
2 Department of Science & Engineering, Novel Global Community Educational Foundation, Hebersham, NSW 2770, Australia
3 Department of Medicinal and Biological Chemistry, University of Toledo, 3000 Arlington Ave., Toledo, OH 43614, USA; giselleamanda.borgesesoares@rockets.utoledo.edu
4 Department Pharmacology, Chitkara College of Pharmacy, Chitkara University, Punjab 140401, India; chopraontheride@gmail.com
5 Department of Pharmacy, Faculty of Allied Health Sciences, Daffodil International University, Dhaka 1207, Bangladesh; mominur.ph@gmail.com
6 Department of Biosciences & Centre for Interdisciplinary Research in Basic Sciences, Jamia Millia Islamia, Jamia Nagar, New Delhi 110025, India; zhasan.biochem@gmail.com
7 Division of Microbiology and NCDs, ICMR-Regional Medical Research Center, Chandrasekharpur, Bhubaneshwar 751023, India; swain.shasanksekhar86@gmail.com
8 Faculty of Medicine and Pharmacy, University of Oradea, 410087 Oradea, Romania
* Correspondence: btanima1987@gmail.com (T.B.); simona.cavalu@gmail.com (S.C.)
† These authors contributed equally to this work.

Abstract: The strategies involved in the development of therapeutics for neurodegenerative disorders are very complex and challenging due to the existence of the blood-brain barrier (BBB), a closely spaced network of blood vessels and endothelial cells that functions to prevent the entry of unwanted substances in the brain. The emergence and advancement of nanotechnology shows favourable prospects to overcome this phenomenon. Engineered nanoparticles conjugated with drug moieties and imaging agents that have dimensions between 1 and 100 nm could potentially be used to ensure enhanced efficacy, cellular uptake, specific transport, and delivery of specific molecules to the brain, owing to their modified physico-chemical features. The conjugates of nanoparticles and medicinal plants, or their components known as nano phytomedicine, have been gaining significance lately in the development of novel neuro-therapeutics owing to their natural abundance, promising targeted delivery to the brain, and lesser potential to show adverse effects. In the present review, the promising application, and recent trends of combined nanotechnology and phytomedicine for the treatment of neurological disorders (ND) as compared to conventional therapies, have been addressed. Nanotechnology-based efforts performed in bioinformatics for early diagnosis as well as futuristic precision medicine in ND have also been discussed in the context of computational approach.

Keywords: nanomedicine; nanoparticles; phytomedicine; bioinformatics; neurodegenerative diseases

Citation: Bhattacharya, T.; Soares, G.A.B.e.; Chopra, H.; Rahman, M.M.; Hasan, Z.; Swain, S.S.; Cavalu, S. Applications of Phyto-Nanotechnology for the Treatment of Neurodegenerative Disorders. *Materials* **2022**, *15*, 804. https://doi.org/10.3390/ma15030804

Academic Editor: Alina Maria Holban

Received: 18 December 2021
Accepted: 18 January 2022
Published: 21 January 2022

1. Introduction

Nanotechnological advancements in neurological science are projected to have a significant effect on the development of novel therapeutical strategies. The capability to produce nanoparticles in the same size domain as proteins has led to a wide range of applications in the biomedical field, as they can invigorate, respond to, and affect target cells and tissues to ensure the desired physiological reactions while diminishing undesirable results. In addition, nanotechnology can permit a more exact and ideal control of complex natural frameworks than customary pharmacological methodologies can, for example, the blood-mind obstruction (BBB). "Neurodegenerative diseases" (NDs) refers to a gathering of discontinuous or potentially familial conditions portrayed by the deficiency of neuronal subtypes over a long period. Alzheimer's disease (AD) and Parkinson's disease (PD) are amongst the most decimating illnesses of the twenty-first century [1–6].

Around 24 million individuals overall experience the ill-effects of dementia, with Alzheimer's sickness representing 60% of cases [7]. Learning and memory issues occur in Alzheimer's patients. Net cerebral decay, which demonstrates the deficiency of neurons and the presence of various neuronal extracellular plaques and intracellular neurofibrillary tangles, has been found in post-mortem examinations of human brain tissue, fundamentally in the front-facing and transient flaps, including the hippocampus [8]. Notwithstanding, although the pathophysiology of AD is known, the reason for which it occurs, and its trigger, is unknown. As a result, AD treatment is only based on symptomatic treatment of the disease. The degeneration of nigrostriatal dopaminergic neurons in the midbrain causes extreme side effects including hypokinesia, bradykinesia, unbending nature, and a resting quake [9].

Although Parkinson's disease (PD) is the second most prevalent chronic and progressive non-degenerative illness (ND), it has an enormous social and economic effect on the great majority of people affected by it [10]. The ventral tegmental region and the substantia nigra of the brain produce less dopamine (DA) when dopamine neurons degenerate over time. Patients with Parkinson's disease (PD) may have a wide range of symptoms including stiffness, tremor, and slowness of movement. In certain cases, it may be difficult to tell the difference between Parkinson's disease and other conditions that have many of the same symptoms. As a result, the word "Parkinsonism" has come to describe a wide range of neurological conditions. Symptoms of Parkinsonism fall into two categories: primary and secondary. Idiopathic Parkinsonism, the most common kind of primary Parkinsonism, has no recognised aetiology. A variety of conditions may lead to secondary Parkinsonism, including progressive supranuclear palsy, multiple system atrophy, and others, the most frequent of which is drug-induced Parkinsonism following cerebrovascular illness. Vascular Parkinsonism is a distinct clinical entity that falls under the umbrella of Parkinsonism, although it has a variable natural history and a wide range of clinical manifestations. In addition, there is strong evidence for inclusion of clock genes in PD [10].

Dopamine agonists, such as levodopa, are currently utilized as first-line treatment for PD. However, as the disease advances, patients become less receptive to levodopa and disease progression continues. Profound mind incitement and foetal dopamine neuron transplantation have been examined as options in contrast to pharmacological consideration [11,12]. These medicines are currently suggestive, and do not appear to have the option to stop or counteract the progressing consumption of dopaminergic neurons [13–15]. Medication conveyance to the cerebrum remains a significant test in the treatment of AD and PD. The rise of new, useful treatment modalities for the treatment of NDs is an intriguing issue of examination at the present time [16]. Considering the various defensive hindrances that encompass the CNS, the development of effective therapeutics to treat patients with NDs is of crucial significance.

There are microvascular endothelial cells that line the cerebral capillaries of most mammals and other species with a well-developed CNS, which the BBB is composed of. Based on an average microvessel surface area of 150–200 cm^2 per g of tissue, it is anticipated that a normal adult has a total surface area of 12–18 m^2 [4]. The BBB serves as a barrier between the bloodstream and the central nervous system, protecting the brain's parenchyma from exposure to potentially harmful substances carried by the bloodstream. Specialized ion channels and transporters combine in the BBB to keep the ionic composition optimal for neuronal and synaptic signalling processes. Potassium concentrations in the CSF and ISF, for example, are 2.5–2.9 mM. The plasma potassium concentration is 4.5 mM, regardless of whether the fluctuations are induced by illness or imposed artificially. Calcium, magnesium, and pH are all controlled by the BBB [17]. Neuronal excitability and macrophage movement across the BBB are regulated by calcium and potassium homeostasis [18].

The BBB's limitation and the medication discharge energy that trigger fringe results are two significant realities about neurotherapeutics. Despite the prevalent attitude, it is presently realized that NDs can be multisystemic in nature, which represents different difficulties for future consideration. A course of a few harmful atomic and cell occasions, instead of a solitary pathogenic factor, causes the demise of specific kinds of neurons in NDs. Accordingly, nanotechnology can offer an expected answer for a significant number of these difficulties in the treatment of AD and PD by considering specific medication conveyance and improving the bioavailability as well as the viability of different medications and other bioactive specialists utilized in NDs [19].

2. Materials and Methods

Relevant studies pertaining to the applications of nanotechnology for CNS-based disorders were selected through algorithm application according to the recommendations of Page et al. [20,21]. Literature search was performed based on several scientific databases such as Google Scholar (http://www.scholar.google.co.in) (accessed on 10 November 2021), PubMed (http://www.ncbi.nlm.nih.gov/pubmed) (accessed on 10 November 2021), Elsevier (https://www.elsevier.com/en-in), ScienceDirect (http://www.sciencedirect.com) (accessed on 10 November 2021), Springer (http://www.springer.co.in) (accessed on 10 November 2021), and Scopus (http://www.scopus.com) (accessed on 10 November 2021). Only publications that had the full text available and book chapters restricted to the English language were reviewed, and figures were created using Biorender.com. (accessed on 10 November 2021).

3. Real Problem of Neurodegenerative Disease

Numerous neurodegenerative sicknesses such as AD, PD, Huntington's disease, and amyotrophic horizontal sclerosis (ALS), are connected to ageing. These neurodegenerative infections are critical social issues in numerous nations as the populace ages. The collection of misfolded and totalled proteins in the mind is an ordinary neurotic aggregate in these illnesses, as is shown in Figure 1. A developing assortment of proof demonstrates that poisonous protein conglomeration and neurodegenerative infections are connected. Revelatory endeavours for sickness-changing therapeutics have lately significantly expanded, because of a more noteworthy comprehension of the pathogenesis of these infections [22].

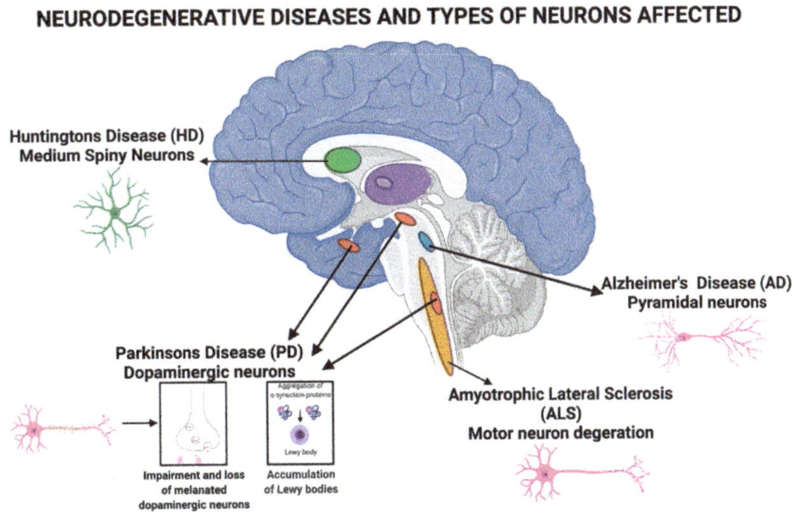

Figure 1. Neurodegenerative diseases and the types of neurons affected.

Neurodegenerative diseases are now the most common and debilitating illnesses afflicting humans. They have recently surpassed cancer, myocardial infraction (cardiovascular disorders), and stroke as the fourth leading cause of death. Neurodegenerative disorders disrupt a person's reasoning, skilled gestures, emotions, cognitive actions, and memory, causing short- and long-term disabilities.

4. Role of Nanotechnology in Neurodegenerative Disorders

Neurodegenerative disorders are a collection of neurological illnesses that damage the brain or spinal cord's neuronal structure and function. Because of the ageing population and the increasing prevalence of CNS disorders including Alzheimer's, Parkinson's, and strokes, the healthcare industry is faced with a tremendous issue. Early identification and treatment of CNS illnesses are still difficult despite advancements in our knowledge of their pathophysiology, and current medications are focused mostly on treating their symptoms. Due to unsatisfactory pharmacokinetics and unspecific targeting of new medications (such as proteins and peptides) that have been developed, many of these new treatments may not succeed because they may raise the risk of side effects. Biological barriers, such as the blood-brain barrier (BBB), are at the root of these difficulties. As a barrier between the CNS and peripheral circulation, it blocks the transfer of most chemicals from the blood to the brain [23]. In order to prevent chemicals, ions, and cells from crossing from the bloodstream into the central nervous system, BBB is made up of endothelial cells, astrocyte end-feet, and pericytes. The five primary ways by which chemicals traverse the BBB are diffusive transcellular lipoid bilayer membrane transport; transport carriers; specific receptor-mediated endocytosis and transcytosis; adsorptive transcytosis; and paracellular-tight junction transport. Toxins, infections, and inflammation are all prevented, and the CNS's homeostasis is maintained, due to the presence of BBB [24]. Drug transport to the central nervous system (CNS) was further hindered by the BBB's restrictive character. Furthermore, BBB malfunction has been reported in the majority of CNS illnesses, including Alzheimer's, Parkinson's, and strokes. Although it is unclear whether BBB failure is the primary cause of illness development, alterations in the transport system and enzymes are a significant contributor and exacerbator [24]. In CNS illnesses, the BBB impairment has major consequences for therapy. CNS illness patients may benefit from having an intact blood-brain barrier (BBB). In light of the fact that BBB malfunction plays a critical role in developing illness, this will undoubtedly be a divisive approach.

Nanotechnology, on the other hand, has advanced quickly in recent years and may offer significant benefits for the detection and treatment of CNS illnesses. Control or manipulation of nanometre-scale (one billionth of a metre) designed materials or devices is what is meant by the term "nanotechnology" [25]. Due to changes in the arrangement and spacing of surface atoms and molecules, nanomaterials exhibit markedly different characteristics than their macroscale counterparts [26].

There are several applications for biomarker discovery, treatments, and theranostics using nanomaterial-based technologies. To begin with, nanomaterials that have been changed may be utilised to identify and treat damaged cells and tissues at the molecular level. When a surface that has been modified with unique molecules, nanoengineered materials may also maintain medication release, boost bioavailability, distribute numerous agents, and preserve substances from degradation. Surface functionalization of nanomaterials may be used to target and infiltrate the BBB while also lengthening the nanomaterial's blood half-life. For CNS illness diagnosis and treatment, nanomaterials are at the top of the list because of their unique features.

There are many promising advantages to be gained from using nanoparticles in the medical field, such as a high drug-loading capacity, which reduces the risk of chemical interactions or toxicity; a high surface area-to-volume ratio, which makes it easier to administer parenterally; the ability to use active and passive drug-targeting strategies; and sustained and continuous dosing options. Nanoparticle size, targeting properties, lipid, or water solubility and their respective hydrophobicity or hydrophilicity, chemical and physical stability, surface charge, and permeability, biodegradability, biocompatibility, cytotoxicity, drug release profile and antigenicity of the final product all play a role in the selection of the nanoparticle manufacturing materials.

NPs may be used for brain medication delivery because of their multifunctional and adaptable architectures. However, several factors must be considered before they may be used, including surface chemistry, hydrophobicity, shape, size, and charge, to name a few. Biocompatibility, decreased toxicity, and the capacity to bind and transport medications or therapies are all desirable features of an ideal NP. As long as it does not degrade quickly in vivo, it can pass across the BBB and regulate the release of medicines for extended periods of time. All of these characteristics result in NPs that are very effective at penetrating the BBB [27]. An investigation into site-specific drug delivery has resulted in a non-invasive distribution of the given dosage to specified areas, allowing for a precise and concentrated administration of the therapeutic dose at the site of the pharmacological activity. There are now non-invasive approaches, such as intranasal administration using drug modification to increase BBB permeability [28] as well as invasive ones, such as intraventricular or intracerebral injection or implantation, and infusion [29]. Disruption may also be effective in certain cases to cross the BBB [30]. The rupture of the BBB is often employed to improve the efficiency of medication transport to the brain. Mannitol, for example, is used to open the BBB to treat some CNS malignancies, and ultrasound is used to create temporary holes in the BBB [31]. Small therapeutic compounds (less than 1000 Da) have recently been shown to be able to pass across the BBB in Alzheimer's disease (AD) and multiple sclerosis (MS) [32]. Physical and chemical features of NPs determine their path across the BBB and the methods by which they may cross it [33]. It is possible for NPs to penetrate the BBB and distribute medications in the sick brain when they are functionalized with a suitable ligand [34,35]. Endothelial cells may be crossed via transcytosis, allowing medications or drug-conjugated NPs to enter the central nervous system [36]; endothelial cells can be entered by endocytosis, allowing pharmaceuticals to cross the blood-brain barrier [37]. Researchers in neuropharmaceuticals are working to understand the processes of receptor-mediated and adsorptive transcytosis, as well as all the inherent physicochemical features of neuropharmeuticals. As a result, this might lead to new therapies that are more effective in crossing the blood-brain barrier (BBB) [38].

4.1. *Techniques for Preparation of Nanoparticles and Nanocapsules*

4.1.1. Nanoprecipitation

Colloidal drug delivery devices may be integrated with active substances using the nanoprecipitation technique pioneered by Fessi et al. [39]. Some of the benefits of the generated nanoparticles include controlled release, targeted distribution, and better stability in biological fluids. As a result, you may expect minimal toxicity and moderate side effects. Preparing nanoparticles by nanoprecipitation requires a good solvent (usually an organic solvent such as ethanol, isopropanol, or acetone), whereas the particle is created using a non-solvent (such as water). Nanoprecipitation relies on both an organic and non-solvent phase, often referred to as aqueous phase, to guarantee that all initial components are completely soluble. The organic phase might include a low HLB surfactant and active compounds dissolved in an organic solvent or a mixture of organic solvents. The solubility of the active molecule in the solvent is one of the factors limiting the amount of medicine that can be loaded onto the particle. Particle formation and physical stability are made easier in the non-solvent phase by water-soluble stabilizing compounds [40]. There have been reports of particles devoid of stabilizing compounds. The amphiphilic nature of isoprenoid chains allows them to be linked to the active molecule in this situation. Nanoparticles form when the organic component is abruptly dumped or mixed with the aqueous phase. Particle size and polydispersity index are unaffected by nanoprecipitation, which is a reliable process. Instead, it seems that the features of the nanosized system are determined by parameters connected with the formulation used [40]. This might be linked to the procedures proposed for the production of nanoprecipitation particles. Nano and submicron particles can only be formed when particular polymer/solvent/nonsolvent ratios are used. As a consequence of the Gibbs–Marangoni effect (interfacial turbulence and thermal inequalities produced by mutual miscibility of the solvent and non-solvent and their varying interfacial tensions), mechanical mixing breaks down the organic phase into drops inside the aqueous phase [41]. Submicron-sized droplets of organic solvent diffuse away, causing the material that causes particle precipitation to become insoluble. It has also been suggested that the "ouzo effect", which relies on the system's chemical instability, might be a mechanism. In this situation, when aqueous and organic phases are joined, particle-forming molecules are supersaturated, allowing for the generation of "protoparticles" that follow the classic nucleation-and-growth process. Work circumstances must be created in order to enable the spontaneous generation of submicron or nanoscale particle sizes with the lowest polydispersity indices. As nanoprecipitation is difficult to standardize, polymer aggregation results in a wide and asymmetric particle size ranges. Low stabilizing agent concentrations and poor phase mixing are two obvious signs of polymer aggregates: a concentrated organic phase and a high organic phase ratio [40,42].

4.1.2. Emulsification-Diffusion Method

So-called spontaneous emulsion diffusion describes a method for making biodegradable nanoparticles by dissolving a polymer in a mixture of water-immiscible and water-miscible solvents (such as dichloromethane) [43]. The quick dissolving of the miscible solvent causes the formation of a nanoemulsion when this solution is introduced to water. Immiscible liquid evaporates, forming nanoparticles. Nanoemulsions may be created by turbulence produced during solvent displacement, and there is no actual diffusion stage in the process of creating them. Solvent displacement and solvent evaporation are combined in this procedure, making it a hybrid. Due to the use of solely miscible solvents, the modified spontaneous emulsification solvent technique is obviously a method of solvent displacement [44].

The technique used to produce diffusion is a key variable in the emulsion diffusion approach. Because dilution water is added, low-solids dispersions are formed. So, armed with this information, it was suggested that the solvent (with a low boiling point) from the internal phase be extracted and then distilled directly into the exterior phase [45]. Due to the saturation of the continuous aqueous phase, its rapid displacement will prompt a free flux of the solvent globules to generate an anti-solvent medium, where the material will aggregate in the form of nanoparticles, in such a way that high solid concentrated dispersions (up to 30%) can be prepared from different materials such as polymers (e.g., poly(D,L-lactic acid), poly(!-caprolactone), and other materials. By using an emulsification-diffusion method, Yusuf et al. developed an emulsification solvent diffusion technique for coating with PS80 that reduced SOD1 levels and immobility, while raising acetylcholinesterase levels in Piperine-SLNs produced with Piperine. In addition, histological investigation revealed decreased plaques and tangles [46].

4.1.3. Double Emulsion Technique

The most popular techniques for evaporating solvent from emulsions are the single- and double-emulsion procedures. This process uses an organic phase that has been emulsified in an aqueous medium with surfactants or stabilisers before the solvent is evaporated to create the NP [47]. Double emulsion solvent technique uses a three-phase process, which is different from the single emulsion solvent method, which uses just two phases. Shear stress during the homogenization process is the fundamental downside of these two methods, which results in poor protein encapsulation effectiveness. The nanoprecipitation approach may help overcome this issue. Dropwise addition to the aqueous media of the organic solvent containing PLGA results in the fast diffusion of the miscible solvent, resulting in the creation of the NPs layer-by-layer method [48].

Sukhorukov et al. devised the approach for the production of vesicular particles. Adsorbed polymer layers may be applied to a colloidal template using either the polymer solution and washing or the addition of a miscible solvent to reduce polymer solubility. Chitosan, heparin, polylysine, protamine sulfurate, gelatine, and dextran sulphate are examples of common polymers that may have polyanionic or polycationic characteristics [49,50].

4.2. Green Method of Synthesis of Nanoparticles

Research and development in materials science and technology is entering a new phase in which "green synthesis" methodologies and technologies are becoming more popular. A major benefit to environmentally friendly synthesis of materials and nanomaterials will be improved by regulation, control, clean-up, and remediation of the production process.

'Green synthesis' is necessary in order to prevent the generation of undesirable or dangerous by-products via the development of dependable, long-lasting, and environmentally friendly synthesis methods. In order to attain this purpose, it is critical to make use of appropriate solvent systems and natural resources. The use of metallic nanoparticles in the green synthesis of biological materials has been widely accepted (e.g., bacteria, fungi, algae, and plant extracts). Using plant extracts to synthesise metal and metal oxide nanoparticles is one of the more straightforward green synthesis techniques accessible, especially when compared to bacteria- and fungus-mediated approaches. Biogenic nanoparticles are a grouping of these compounds.

Various reaction factors, such as solvent, temperature, pressure, and pH conditions, influence green synthesis methods based on biological precursors (acidic, basic, or neutral). Many plant extracts, particularly those from leaves, include potent phytochemicals, such as ketones, aldehydes, flavonoids, amides and terpenoids, carboxylic acids, phenols and ascorbic acids, that may be used to synthesize metal and metal oxide nanoparticles [51,52].

5. Role of Phyto-Nanomedicine on Neurodegenerative Diseases Treatment

Neurodegenerative Diseases (ND) are basically disorders related to age and are prevailing rapidly worldwide due to the increased elderly population in recent years. The actual problem however is not the increasing prevalence but the lack of effective treatment. As per published evidence from the past, NDs were discovered in the early 1900s, however, presently, no effective treatments for these disorders are found in modern society. Integrative medicine can sometimes delay the progression of these diseases or exhibit protective effects. It is believed by scientists that with the development of integrative medicine and modern science, the former will progressively take a more and more significant role in the treatment of ND [52,53].

A series of synthetic drugs have shown positive results for the treatment of some common NDs including Autism, PD, AD, and other chronic illnesses. The use of synthetic drugs is associated with many side effects, which make them inappropriate for regular treatment. Considering the adverse effects of these synthetic drugs, scientists have made a soft turn towards the utilization of phytochemicals, as they have minimal side effects. The antioxidative, anticholinesterase, anti-inflammatory, and anti-amyloid properties of phytochemicals makes phytochemicals a promising therapeutic agent [54,55].

In consideration that current therapeutics seem to be inadequate treatment for the above-mentioned disorders, scientists are exploring their options with plant-based drugs using nanotechnological approaches. Nanotheranostics is one such approach which is gaining wide attractions from the scientific community at a global level for the treatment of ND. It uses nanoparticles for the simultaneous diagnosis as well as for treatment. Research findings of Tripathi et al. [56] reveal that this treatment has been receiving significance in the medical field because it is extremely aggressive and specifically targets the affected area. Moreover, alterations with respect to type of disease and personalization based on the needs of the patient can be met, which further increases the applicability of the approach [57]. A novel nanotheranostic system has been developed by chemical engineers, which uses adjustable light that activates nanoparticles and opens new applicability in the field [58].

In the subsequent sections we will discuss in detail the various types of plant-based medicines that are utilized in ND treatments along with the conventional approaches and recent trends in the plant-derived nanomedicine discovered so far.

5.1. *Types of Phyto-Medicines Available for Treatment*

- Acorus *calamus*

Acorus *calamus,* also commonly referred to as sweet flag, belongs to the family Acoraceae and acts as a nervous system rejuvenator and has beneficial effects on the brain through memory enhancement, learning behaviour, and performance modification. Acorus *calamus* consists of a majority of α-and β-asarone while β-asarone brings about the suppression of beta-amyloid-induced neuronal apoptosis observed in the hippocampus through the downregulation of Bcl-w and Bcl-2, which further causes the activation of caspase-3 and phosphorylation of c-Jun N-terminal kinase (JNK). Apart from the above-mentioned benefits parts, of Acorus *calamus* also showed an inhibitory result on AChE with an IC50 value of 188 μg/mL. It has also shown positive results by improving the dopaminergic nerve function, and thus acts as neuroprotective for PD [59].

- Allium *sativum*

Allium *sativum,* commonly known as garlic, belongs to the family Amaryllidaceae and has been used since ancient times for its medicinal properties. It was used for cardiovascular disease as well as for the prevention and treatment of other metabolic diseases such as hyperlipidemia, atherosclerosis, thrombosis, dementia, hypertension, diabetes, and cancer. S-allyl cysteine (SAC), one of the major constituents found in aged garlic extract (AGE), has been extensively studied. SAC has indirect and direct antioxidant activity. Besides bringing about a decrease in lipid peroxidation and DNA fragmentation, it also causes a reduction in oxidation and nitration. In the Parkinsonian models, namely 1-methyl-4-phenyl pyridinium (MPP) and 6-hydroxydopamine (6-OHDA), SAC causes a protection of dopamine levels, prevents oxidative damage and causes peroxidation of lipids. [60].

- Bacopa monnieri

Bacopa *monnieri* (Linn), popularly called "Brahmi" (family Scrophulariaceae), is a herb found in tropical countries such as India. Steroidal bacosides (A and B) and saponins are the main active compounds present in Bacopa *monnieri.* These active compounds are responsible for enhancing memory and learning. Other constituents include bacopa saponins F, E, and D, flavonoids, phytosterols, and alkaloids. Superoxide SOD, CAT, GPx, and glutathione reductase (GSR) activity is enhanced by Bacoside A. Therefore, the levels of glutathione in the brain are upregulated significantly. Bacoside A inhibits lipid peroxidation by modifying the activity of enzymes such as Hsp 70 and cytochrome P450 in the brain. It also improves the activities of adenosine triphosphatases (ATPases), maintains ionic equilibrium, and restores level of selenium and zinc in the brain. The reduction in aggregation of alpha-synuclein protein by Bacopa *monnieri* was also found by researchers [61].

- Centella *asiatica*

Centella *asiatica* belonging to the Apiaceae (Umbelliferae) family and has been demonstrated to possess a neuroprotective property. Since ancient times, it was part of the Ayurvedic system as an alternative medicine for bringing about memory improvement. Centella *asiatica* decreases Aβ deposition in the brain, exhibits potent antioxidant activity and can scavenge free radicals, causes a reduction in ferric ions, and brings about a restoration of GSH levels through an upregulation of glutathione-S-transferase activity. Chen et al. [62] carried out a study which suggested that Centella *asiatica* ethanolic extract causes the suppression of Aβ-induced neurotoxicity through the enhancement of the antioxidative-based defence system in IMR32 and PC12 differentiated cells. Amelioration of decrease in AChE activity due to colchicine-induction, neuronal damage through induction of asiaticoside occurring through nitric oxide inhibition, also reflects the neuroprotective effect of Centella *asiatica* [63].

- Curcuma *longa*

Turmeric, derived from the plant Curcuma *longa,* belonging to the Zingiberaceae family, is a gold-coloured spice and has been used traditionally as medicine for a variety of diseases [64,65]. Curcumin, turmeric's principal constituent, has several known neuroprotective actions. In patients afflicted with AD, studies have shown that curcumin potentially binds to Aβ peptides, preventing aggregate formation of new amyloid deposits as well as promoting the de-aggregation of the existing amyloid deposits. Analogues of curcumin, namely desmethoxycurcumin and bis-desmethoxycurcumin, have also shown protective ability against Aβ-induced oxidative stress. Moreover, curcumin causes the inhibition of Aβ oligomerization and formation of fibril, causes a macrophage enhancement of Aβ uptake, and inhibits the A beta-heme complex peroxidase activity. Other components of turmeric such as curcuminoids and polyphenolic compounds attenuate mitochondrial dysfunction, inflammatory response induction, and oxidative stress in the presence of inflammatory cytokines such as iNOS and COX-2. Curcuminoids can also bind to Aβ plaques, causing inhibition of amyloid accumulation and its aggregation in the brain [63].

- Celastrus *paniculatus* Wild

Celastrus *paniculatus* Wild, commonly called as Jyotishmati, belongs to the Celastraceae family. Traditionally, it was administered as a powerful appetite stimulant, brain tonic, and emetic. Phytochemical studies show the presence of a sesquiterpene, evoninoate, alkaloids paniculatine A and B, wifornine F celapagine, celapanine, celastrine, celapanigine, polyalcohols such as malkanginnol, paniculate diolmalangunin, and malkanguniol, triterpenoids, and sterols such as β-amyrin and β-sitosterol. Scientific studies suggested that Celastrus *paniculatus* water extract modulated the glutamate receptor function that protects neurons against toxicity produced by the glutamate and resulted in the improvement in memory and learning. It also causes a noteworthy decrease in the MDA level in the brain, which is an important marker of oxidative damage, with simultaneously significantly increases in levels of glutathione and CAT; two endogenous antioxidants in the brain [63]. Jakka et al. [66] investigated research that explained the neurotrophic potential in Celastrus *paniculata* Wild whole plant methanolic extract (CPPME). A significant decrease was also seen in AChE and enhanced neurotrophic activity, which ultimately improved spatial memory formation in scopolamine-induced amnesia.

- Coriandrum *sativum L*

Coriandrum *sativum L.*, also referred to as dhanya, belongs to the family Apiaceae [30]. The major phytochemicals present include flavonoids such as quercetin 3-glucoronide, and polyphenolics such as protocatechinic acid, glycitin, and caffeic acid. In seeds, the flavonoid content was reported at a concentration of 12.6 quercetin equivalents/kg and the polyphenolic concentration was reported at 12.2 gallic acid equivalents/kg [67,68]. A study reported that the Coriandrum *sativum* extract caused an increase in total protein concentration and enzyme levels of CAT, SOD, GSH, and reduced the size of cerebral infarct, calcium levels, and lipid peroxidation (LPO) in the experimental rat. Scopolamine- and diazepam-induced memory deficits were also decreased by Coriandrum *sativum* leaves. The leaves also show an antioxidant property, having radical scavenging activity of DPPH with lipoxygenase inhibition and phospholipid peroxidation inhibition activity, which contribute to its memory enhancement effect [69].

- Galanthus *nivalis*

Galanthus *nivalis,* referred to commonly as snowdrop, belongs to the family Amaryllidaceae. Galantamine is the major constituent found in the bulbs and flowers of Galanthus *nivalis* and is a tertiary iso-quinoline alkaloid. The neuroprotective effect exerted by Galantamine is associated with a dual action. The drug is a competitive and selective AChE inhibitor and can stimulate nicotinic receptors, which further enhance cognition and memory [70].

- Ginkgo *biloba*

Gingko *biloba* belongs to the Ginkgoaceae family and is considered a 'living fossil'. Its extract majorly consists of flavonoids, whose fraction accounts to 24%. It mainly comprises of flavanols such as, isorhamnetin, quercetin, kaempferol, and terpene lactones, which account to a concentration of 6% and further comprise diterpenic lactones such as ginkgolides A, B,

C, J and M, and a sesquiterpene tri lactone-bilobalide. The extract exhibits neuroprotection by several mechanisms such as membrane lipid peroxidation inhibition, anti-inflammatory activity, as well as through direct inhibition of amyloid-b aggregation and anti-apoptotic activities. The flavonoid fraction of Ginkgo *biloba* (G. *biloba*) extract has a free-radical scavenging property as well as anti-oxidative effects, while bilobalide has been associated with a decrease in damage caused due to excitotoxicity and global brain ischemia-induced death of neuronal cells. Extract of G. *biloba* in the brain significantly inhibits the AChE activity, which indicates that the level of acetylcholine has increased considerably. Flavonoids alter several biological processes through interaction with signalling pathways, effects on protein neuronal expression, which is imperative for plasticity and repair of the synapse, variation in cerebral blood flow, and neuropathological inhibition of processes in certain cortical regions. A study by Dash SK presented that the G. *biloba* extract causes a decline in cortical Aβ levels through a reduction in cholesterol levels as free and circulating cholesterols affect amyloid genesis [71].

- Glycyrrhiza *glabra*

Glycyrrhiza *glabra,* widely referred to as Yashti-madhuh or liquorice, belongs to the Leguminosae family. Its major constituent is a flavonoid called Glabridin, which possesses multiple pharmacological activities such as anticancer, antiviral, anti-ulcer antioxidant, anti-diabetic, immunomodulatory activity, anti-inflammatory activity, antimicrobial activity, and anticonvulsant activity. Liquorice significantly increased learning and memory, however, research has indicated that its consumption improves general intelligence rather than short-term memory. In the brain, glabridin decreases the MDA level, and increases the superoxide dismutase level while reducing glutathione levels. A study suggested that G. *glabra* administration restored the decreased concentration of dopamine and glutamate in the brain and decreased activity of AChE [72].

- Hypericum *perforatum*

Hypericum *perforatum* is also known as millepertuis or hypericum. It belongs to the family Hypericaceae. Although it is found worldwide, it is mainly native to Europe, northern Africa, and western Asia. The main active component of H. *perforatum* is Hyperoside. Hypericin, kaempferol, biapigenin, and quercetin comprise the other constituents. The extract of H. *perforatum* has been reported to behave as a protector against NADPH-dependent (enzymatic) and Fe^{2+} and ascorbate-dependent (non-enzymatic) peroxidation of lipids in the cortical region of the brain. The extract also protects brain cells from cytotoxicity brought about by glutamate through the reduction of glutathione loss, overload of calcium, and cell death mediated through ROS. The ethanolic extract of H. *perforatum* may also bring about an improvement of microglial viability through the reduction of toxicity by amyloid-beta in AD. Hypericum *perforatum* inhibits AChE and MDA formation in the brain and increases the concentration of SOD, GPx and CAT. Therefore, H. *perforatum* also has the capability to bind to iron ions, has scavenging activity for hydroxyl radicals, and behaves as an antioxidant [73].

- Lycopodium *serratum*

Lycopodium is also referred to as ground pines or creeping cedar. It belongs to the family Lycopodiaceae, which comprises a family of fern-allies. Its leaves contain a single, unbranched vascular strand and are microphylls. The major constituent is huperzine A, which is a potential therapeutic agent by researchers for treating AD. The alkaloid, which has been isolated from Lycopodium *serratum,* has been used for treating inflammation, fever, and blood disorders for many decades. It acts as a highly reversible, potent, and selective inhibitor of AChE and is comparable to the potency observed by galantamine, physostigmine, tacrine, and donepezil. Huperzine A is considered to be a strong candidate for therapy in AD. It has been found that huperzine A causes a noteworthy upregulation in AChE levels in rat brains and is associated with protective effects such as amyloid precursor protein metabolism regulation, oxidative stress protection mediated by Aβ, apoptosis, dysfunction of mitochondria, and anti-inflammation [74].

- Melissa officinalis

Melissa *officinalis* L. (Lamiaceae) leaves are commonly referred to as lemon balm and have been used traditionally for spasmolytic and nerve calming effects. The leaves produce a calming and soothing effect through interaction with the $GABA_A$ benzodiazepine receptor. Its extracts contain flavonoids, namely quercitrin, apigenin, luteolin along with phenolic acids. The derivatives of these products inhibit enzymes such as monoamine oxidases (MAO) and AChE, which scavenge these free radicals and prevent apoptosis. Enzymatic inhibition of the above-mentioned enzymes leads to alleviation of depression symptoms. Research also suggests that Melissa *officinalis* produces protective effects in the PC12 cell line and protects neurons from oxidative stress [75].

- Ocimum *sanctum*

Ocimum *sanctum* is commonly referred to as 'Tulsi' in Hindi and in English, 'Holy Basil'. It belongs to the Labiatae family. The plant is reported to contain glycosides, alkaloids, saponins, tannins, vitamin C, citric acid, tartaric acid, and maleic acid. Research conducted by Kusindarta et al. [76] indicated that an ethanolic extract derived from the leaves of Ocimum *sanctum* may stimulate and restore choline acetyltransferase expression in human ageing cerebral microvascular endothelial cells and could provide nerve protection and increased production of Ach, which may enhance memory and cognitive ability. Scientific studies reveal that the hydro-alcoholic extract of Ocimum *sanctum* exhibits strong antioxidant activity against DPPH and hydroxyl radicals, which possibly occur due to the presence of a high number of polyphenols and flavonoids. It inhibits peroxidation of lipids, ROS generation, damage to DNA, and depolarization of membranes. It also decreases the enzymatic leakage of lactate dehydrogenase, preserves cellular morphology, restores superoxide dismutase, and catalyses enzyme levels, thereby preventing neuronal damage [77].

- Panax Ginseng

Ginseng belongs to the Araliaceae family and is prominently found in north-east Asia. It is used world-wide for boosting energy. Ginseng may provide neuroprotection against neuronal degradation through various mechanisms such as production of a reduction in the β-amyloid deposition or glutamate-induced excitotoxicity in a dose-dependent manner, thereby preventing apoptosis and neuronal death, improving routine in a passive-avoidance learning paradigm, and protecting neurons possibly through its potential to suppress cellular AChE activity and enhance cholinergic metabolism [78].

- Rosmarinus *officinalis*

Rosemary, commonly known as Satapatrika, belongs to the Lamiaceae family. It contains many essential oils such as eugenol, carvacrol, oleanolic acid, and ursolic acid, and thymol constituent's antioxidants such as ferulic acid and carnosic acid, which can be used against cyanide-induced damage in brain. A neuroprotective effect is also seen, which can be cultured and human-induced, as has been observed in rodents, cell-derived neurons, and pluripotent stem cells in vivo and in vitro in various parts of the brain in non-Swiss albino mouse models. It also possesses cytoprotective, anti-apoptotic, and anti-inflammatory activities that also add on to its neuroprotective mechanism [79].

- Salvia officinalis

Salvia *officinalis* belongs to the Lamiaceae family, and is well-known and reputed for improving memory, as it has been traditionally used as a memory enhancing agent. Carnosic acid and rosmarinic acid are the active ingredients found in S. *officinalis*. These components are known to have potential pharmacological activity such as antioxidant and anti-inflammatory properties as well as a low AChE inhibitory effect. It inhibits ROS formation, peroxidation of lipids, fragmentation of DNA, activation of caspase-3, and hyperphosphorylation of protein tau. Clinical evidence that has been obtained may help to prevent or reduce the symptoms of dementia. A small pilot trial involving the oral administration of S. *officinalis* essential oil to 11 patients that possessed mild-to-moderate symptoms of AD improved cognitive function significantly [80].

- Terminalia *chebula*

Terminalia *chebula* (T. *chebula*), also known as "King of Medicines" in Tibet, belongs to Combretaceae family. It has been widely used as traditional medicine in Ayurveda, Siddha, Unani, and Homeopathy. It contains compounds such as arjungenin, triterpene sarjunglucoside 1, and the tannins, chebulosides 1 and 2, chebulic acid, chebulinic acid, tannic acid, ellagic acid, 2,4-chebulyi–β-D-glucopyranose, gallic acid, ethyl gallate, punicalaginterflavin A, and terchebin. It also has presence of flavonoids such as rutins, luteolin, and quercetin. A study reported that T. *chebula* exhibits anxiolytic activity and is equivalent to standard drug diazepam. T. *chebula* has good pharmacological activities relevant to dementia therapy and possesses antioxidant activity comparable to radical scavengers such as quercetin, reflecting 95% activity, and showing an inhibitory concentration (IC50) value of 2.2 μg/mL. T. *chebula* fruit extract also shows protective effect neuronal cells against ischemia, reduces least production, and stimulates microglia cells death rate by lipopolysaccharide [81].

- Tinospora *cordifolia*

Tinospora *cordifolia* (T. *cordifolia*) belongs to the Menispermaceae family, which is commonly known as giloe. Chemical constituents extracted from the plant are alkaloids, diterpenoid lactones, steroids, glycosides, and aliphatic acids. T. *cordifolia* holds a memory increasing property, which is due to immune-stimulation and increased synthesis of acetylcholine. T. *cordifolia* exhibits the property of scavenging free radical activity against ROS and reactive species of nitrogen, which have been studied through electron paramagnetic resonance spectroscopy. It increases the concentration of glutathione and the expression of the gamma-glutamyl-cysteine ligase and superoxide dismutase of copper-zinc genes, which play a major role in neuronal injury during hypoxia and ischemia. In addition, T. *cordifolia* significantly decreases the mRNA expressions of iNOS. T. *cordifolia* also increases the dopamine level of the brain. Thus, T. *cordifolia* has shown to prevent neurodegenerative changes and enhance cognition, learning, and memory [82].

- Withania *somnifera*

Withania *somnifera* belongs to the family Solanaceae and is popularly known as Ashwagandha or Indian ginseng. The major constituents of Ashwagandha root are two withaferin A, withanolide D, and withanolides. Active glyco-withanolides of Withania *somnifera* have a significant antioxidant function, which is accomplished by elevating the activities of SOD, CAT, and GPx. It is also reported that Ashwagandha also works as a nerve tonic that boosts energy and rejuvenates the cells. According to Rajasankar et al. [83], treatment of PD mice with an oral dose of Withania *somnifera* root extract (0.1 g/kg body weight) for 1 week or 4 weeks enhanced homo-vanillic acid and dopamine, 3,4-dihydroxy phenyl acetic acid levels in the corpus striatum. Furthermore, the report suggested that Withania *somnifera* treatment enhances the anti-apoptotic proteins level such as Bcl-2 and depreciates the pro-apoptotic protein level such as Bax in the Maneb–Paraquat-induced dopaminergic neurodegeneration model of PD. Ashwagandha extract has shown to prevent lipid peroxidation and increase antioxidant activity by increasing the free-radical slinking enzymes levels in the brain [63].

- Zizyphus Jujube

Jujube fruits are used in Korean and Chinese traditional medicine to reduce anxiety and strengthen the stomach and gastrointestinal system. Jujube seeds comprise large amounts of flavonoid, phenyl glycosides, terpenoid, and alkaloid compounds mucilage, citric acid, malic acid, sugar, organic minerals, vitamin C, and protein. The herb exerts inhibition activity against the release of histamine, cyclooxygenase I and II, and AChE inhibitory activity. Flavonoids present antioxidant properties [63]. A compound known as cis-9-octadecenamide (oleamide) extracted from jujube is reported to bring about an upregulation of acetylcholine transferase to 34.1% in the in vitro models, which leads to the increase in acetylcholine level and improves mild-to-moderate cognitive functions, motor coordination, behavioural disorders, learning, and memory [84].

5.2. Conventional Approach

As a harsh reality in ND and As, the only hope is symptomatic treatment. The symptomatic treatment therapies rely on slowing down symptoms without addressing the cause and cure of the disease. Apart from symptomatic treatment, the other strategy that is being employed currently is disease modifying-based treatment. Inhibitors of anti-cholinesterase are used as symptomatic treatment, while anti-inflammatory agents and antioxidants are used for disease modifying treatment [85]. Current therapies for neurodegenerative and neurological disorders usually are involved in managing symptoms instead of providing significant activity against disease progression. For example, the symptoms of HD are controlled using 75–200 mg/day of tetrabenazine to alleviate involuntary movement (chorea). However, as it acts as a vesicular monoamine transporter inhibitor (VMAT), it causes an interference with both 5-HT (5-hydroxytryptamine) and dopamine (DA) degradation, and, as a result, patients can show neuropsychiatric-based symptoms along with various other side effects [86]. Various other first-line treatments such as L-Dopa in PD often cause a multitude of side effects and do not delay progression of the disease. Another example is that of cholinesterase inhibitors such as Donepezil, which is minimally effective in providing improvement in cognition required for AD treatment. The various conventional strategies used currently in the treatment of the two most common NDs, AD and PD, are summarized in detail in Table 1 [87].

In view of the above information, there is an essential requirement for the development of novel therapeutics that possess lesser or tolerable side effects to overcome these disease states, which can further improve the quality of life of the ageing population. Conventional drug delivery strategies, in general, fail to cross the BBB and are therefore less efficacious in terms of treatment [88].

Table 1. Summary of the various conventional strategies used for the treatment of AD and PD.

Strategy	Alzheimer's Disease	Parkinson Disease
Modulation of neurotransmitters (approved therapies)	• Acetylcholinesterase inhibitors • NMDA antagonists	• Precursors of Dopamine • MAO-B inhibitors • COMT inhibitors • Dopaminergic agonists • Anti-cholinergics
Disease modifying therapies (under investigation)	1. **Amyloid based therapy** • Secretase modulation • Amyloid aggregate prevention • Amyloid clearance promotor 2. **Tau-based therapy** • Tau hyperphosphorylation inhibition • Tau protein degradation • Tau oligomerization inhibition	1. **α-Synuclein-based therapy** • Amyloid aggregate prevention • α-synuclein fibril formation blockade • Modulation of α-synuclein related lipidome 2. **Non-dopaminergic therapy** • Antagonists of adenosine receptor • Antagonists of NMDA • Agonists of Glucagon such as peptide-1
Immunotherapy	• Passive immunization Solanezumab Crenezumab • Active immunization CAD106	• Passive immunization Solanezumab • Active immunization PD01A
Gene-based therapy	Regulation of presenilin expression	Expression of synapsin 3 modulation
Other	• ROS reduction, oxidative stress reduction • Anti-inflammatory agents • Caspase inhibitors • Metal chelators, Statins • PPAR-γ (Peroxisome proliferator-activated receptor-γ)	• Anti-inflammatory agents • Melatonin • Nicotine • Calcium channel blockers • Antioxidants • Iron chelators

6. Recent Trends of Phyto-Neuro Medicine

The latest advancements in the domains of green chemistry and nanotechnology are promising and indicate significant potential in the advancement of biomedical sciences from a theranostic perspective [89,90]. However, this potential, so far, has not been utilized for the development of therapeutics for NDs such as AD and PD. The utilization of chemically synthesized molecules poses limitations such as toxicity and cost [89,91,92]. In addition, a study reported that some chemicals that are used for the chemical synthesis of NPs have the propensity to remain attached on the NP surface and as a result, could not be applied biomedically. Therefore, focus has moved towards the development of materials using green chemistry and a green process [93]. Traditionally, the utilization of the green chemistry-based approach is based on the use of medicinal plants or phytochemicals in their pure form that possess medicinal properties, as these phytochemicals provide chelation and stability to the NPs [90].

Recently, Suganthy et al. [94] reported the neuroprotection offered by biogenic gold NPs based on Terminalia arjuna. The results obtained showed significant biocompatibility of biogenic gold NPs as well as significant neuroprotection. These particles successfully inhibited the AChE, caused a reduction in the Aβ fibrillation process, and at low concentrations caused a destabilization of mature fibrils. Trehalose functionalization of synthesized gold NPs brought about a significant improvement in the inhibition of protein aggregation along with the disintegration of mature fibrils, and possesses the potential for application in photothermal therapeutics [56]. Functionalization of gold NPs with anti-amyloidogenic molecules can be considered as a promising strategy for ensuring improvement in the neuroprotective nature. Another strategy involves the use of biogenic platinum NPs biosynthesized through utilization of Bacopa *monnieri* as a neuroprotective agent. Functionalization of NPs using phytochemicals has yielded significant results. Recently, functionalization of selenium NPs using polyphenols was achieved, where in nanoscale selenium nanoparticles were coated with EGCG, which is a polyphenol present in tea. EGCG does possess neuroprotective effects as well as inhibitory actions against a multitude of proteins that are amyloid-forming such as amyloid beta, transthyretin, α-synuclein, and huntingtin, which are involved in AD disease progression. These particles were further coated with Tet-1, which is a protein that has strong affinity to neurons. The use of curcumin and its derivatives in a similar manner has led to therapies that show promising results in patients with AD. Benzothiazolinone in conjugation with curcumin has a strong binding affinity to amyloid and therefore can be used for targeted delivery of curcumin or various other natural products that can help in the treatment of AD [95].

Because of its significant antioxidant and anti-inflammatory effects, Yusuf et al. created PLGA NPs loaded with thymoquinone (TQ) to operate on this aspect in animal models [96]. Streptozotocin (SZT)-treated male albino mice were treated with TQ-loaded PLGA NPs coated with polysorbate 80, which mimics AD oxidative stress by decreasing Streptozotocin (SOD) activity (P-80-TQN). With an average particle size of 226 nm and a zweta-potential of 45.6 mV, the single-emulsion solvent evaporation technique was used to create these nanoparticles. Initial bursts of TQ release were seen after 2 h, followed by a prolonged sustained release (stabilised dipole-dipole interactions taking place between TQ and PLGA components). Endocytosis via LDL receptors allowed the P-80-TQ NPs to traverse the BBB (mediated by the polysorbate coating). These systems had a considerable impact on SOD activity (increase) from the seventh to the 28th day after arrival on site. Further proving their positive impact, a study on animals and cognition (the "Despair test") was conducted at the same time.

Uncaria species have yielded the spirocyclic alkaloid rhynchophylline (RIN), which has been shown to have a variety of pharmacological effects, including neuroprotection. RIN suppresses soluble A-induced hyperexcitability of hippocampus neurons in the event of Alzheimer's disease. In 2020, Xu et al. released the first research on the design and development of brain-targeting treatment for Alzheimer's disease via RIN injection [97].

Methoxy Polyethylene Glycol NPs coated with Tween 80 were synthesised using the nano-precipitation process to improve RIN's pharmacological activity and target specificity. T80-coated RIN-loaded PLGA NPs showed no haemolysis, indicating that these NP solutions might be safely used. In confocal laser scanning microscopy, internalisation into bEnd.3 cells was demonstrated using DiD fluorophore-loaded PLGA NPs; T80-coated NPs showed more penetration. An in vitro BBB model using bEnd.3 cells was constructed to evaluate the crossing of such nano-systems and indicated their greater transport relative to free RIN or uncoated RIN laden NPs. Utilizing healthy C57BL/6 mice, the advantages of using T80 to target the brain were subsequently shown. Ultimately, incubation with T80 RIN NPs increased the survival rate of PC12 cells damaged by A25–35 while also reducing cell death. RIN neuroprotective effects were unaffected by the encapsulation of PLGA nanoparticles, as shown by these findings.

In totality, phyto-nanomedicines are promising as compared to currently marketed therapies for neurodegenerative disorders. Phytochemical derived nanomedicine could prospectively in future be used for neurodegenerative disorders due to their multitude of properties such as anti-inflammatory, antioxidative, and anticholinesterase activities. Further research, however, is necessary to investigate the complete neuroprotective ability of these compounds, to understand the mechanisms by which they exert protective effects, as well as determining whether combination therapy could be synergistic as neuroprotectants [85].

Undoubtedly, phyto-nanomedicines offer a great hope in developing treatment strategies against ND because of their fewer side effects and better target specificity. However, there are still a few limitations which must be considered. A limitation of the utilization of making use of the nanotheranostics for the treatment of NDs was identified by Kumar et al. in 2020. They found that due to variations in neural functioning, characteristics, and genome, a particular single approach cannot be used for all patients for the treatment of a particular disease. Another study by Kumar et al. [88] showed that this technique was proving to be ineffective as the NPs absorption time used in the technique was very low and if the injection is not performed adequately, there is a possibility of NP absorption in the blood or other body parts instead of the targeted area. Another constraint that was identified was that the treatment cannot be monitored in an effective and efficient manner, which poses a major challenge for the practitioners as they would be unable to track the progress and impacts of the treatment. In addition, the treatment and technique, besides being unproductive, is also extremely costly [88].

6.1. Protein-Based Nanoparticles

Because of their low toxicity and biodegradability, biopolymer-based nanoparticles, such as protein nanoparticles, have recently been actively employed as medicinal and functional tools [98]. Proteins are versatile building blocks for nanoparticles because of their specific roles in biology and the industrial industry. Endocytic transport of protein nanoparticles is possible because of their tiny size. As a drug delivery strategy, protein nanoparticles offer a number of benefits, including biodegradability, stability, surface modification, simplicity of particle size control, and less concerns related with toxicity issues, such as immunogenicity. Protecting the medication against enzymatic breakdown and renal clearance may increase its stability, activity, and half-life in particular. Additionally, protein nanoparticles may be employed in a number of targeted therapeutics, including cancer therapy, tumour treatment, and vaccinations, because of their non-antigenic properties. When protein nanoparticles are embedded in biodegradable polymers, they may be released over a long period of time. It is the primary goal of nanoparticle design to manage particle size and surface area so that nanoparticles containing the requisite quantity of pharmaceuticals may demonstrate desirable pharmacological activity by discharging active substances to produce part-specific action.

In silk fibres, the fibroin protein accounts for between 65 and 85 percent of the total protein content [99]. Degumming with Na_2CO_3 removes exterior sericin from silk produced by the Bombyx mori silkworm, a technique often used to extract fibroin [100]. As a result of its excellent mechanical strength, flexibility, low immunogenicity, biodegradability, and biocompatibility, fibroin has become a popular choice for the creation of nanoparticles. Fibroin nanoparticles' zeta potential is negatively charged. When a positively charged polymer such as PEI or chitosan is applied to the surface, it serves as a crosslinking agent that transforms the surface into a positive charge. The average size, size distribution, surface zeta potential, drug encapsulation, release profile, and particle formation stability of FNP may be affected by a variety of parameters, including fibroin molecular weight (MW), crystallinity, encapsulated drug characteristics, and production circumstances. Much research has been undertaken to distribute and use fibroin nanoparticles as a medication delivery method because they can overcome the drawbacks of low-molecular-weight medicines. Drug solubility and stability are enhanced, drug degradation is inhibited, and toxicity is reduced in all fibroin nanoparticles loaded with small molecule pharmaceuticals, making them better for drug therapy.

Approximately 585 amino acids make up HSA, which has a molecular weight of 66 kDa [101]. HSA is mostly present in the bloodstream. Subunits A and B may be found in each of the three major components of the HSA. Sudlow's sites I and II, found in subunits IIA and IIIA of the HSA, are the primary binding sites [102]. When HSA is used as a carrier for other compounds, it helps hydrophobic molecules become more easily soluble in blood. A variety of chemicals may be delivered to particular tissues in the body via HSA. The pH (stabilised in the pH range of 4 to 9) and temperature (may be heated at 60 °C for up to 10 h) of HSA are also highly stable, as they are organic solvents [103]. Because of its biodegradability, non-toxicity, and non-immunogenicity, as well as its high solubility, it has the added benefit of being derived from biological sources. In investigations on protein binding and targeted medication administration, bovine and HSAs are often used because of these benefits. Due to HSA's strong affinity for diverse medications, a matrix of HSA nanoparticles may efficiently incorporate these molecules.

Table 2 describes the outcomes obtained following administration of drug-entrapped NPs for neurological disorders.

Table 2. Summary of the materials used, active moiety entrapped, and outcomes obtained upon administration of NPs.

Material	Name of Active Moiety Entrapped	Size	Outcome of Study	Reference
Cholesterol	α-bisabolol	139.5 nm	NPs drastically decrease free radical generation, lower β-secretase, caspase-3, cholinesterase, and Bax expression, and increase Bcl-2 protein expression.	[104]
Cetyl palmitate miglyol-812	Quercetin	200 nm	It is non-toxic to hCMEC/D3 cells and penetrates the BBB more so than free drug. NPs also prevent A peptide fibril production.	[105]
Caprylic and capric triglycerides, sorbitan monostearate	Curcumin	247 nm	Curcumin-NPs protect against A42-induced behavioural and neurochemical alterations in AD mice model.	[106]
Cholesterol	α-bisabolol	Not reported	Inhibited A aggregation and protected Neuro-2a cells from A-induced neurotoxicity.	[104]
PLGA	Curcumin	150–200 nm	Curcumin encapsulated-PLGA nanoparticles, destroyed amyloid aggregates, exhibited an anti-oxidative property, and are non-cytotoxic.	[107]
DSPE-PEG2000-MAL	Quercetin	200 nm	After 4 h, RVG29-nanoparticles had 1.5 times the permeability across the blood-brain barrier compared to non-functionalized nanoparticles.	[108]
PEG, PLGA	Epigallocatechin-3-gallate	100 nm	GCG and AA NPs resulted in a marked increase in synapses, as judged by synaptophysin (SYP) expression, and reduction of neuroinflammation as well as amyloid β (Aβ) plaque burden and cortical levels of soluble and insoluble Aβ(1–42) peptide.	[109]
poly(ethylene glycol)-co-poly(ε-caprolactone)	Ginkgolide B	91 nm	NPs facilitated the sustained release of GB into the blood, thereby improving its ability to accumulate in the brain and to treat PD.	[110]
chitosan poly ethyleneglycol-poly lactic acid	Acteoside	100 nm	Significantly reversed dopaminergic (DA) neuron loss in the substantia nigra and striatum of sick mice.	[111]
Monomethoxy polyethylene glycol	Apomorphine (AMP)	100 nm	The encapsulation of AMP into the nanoparticles inhibits oxidization. The intranasal administration of the AMP-loaded nanoparticles transports AMP across the BBB.	[112]

6.2. Polymeric Nanoparticles

Polymeric nanoparticles are colloidal particles that have a medication encapsulated inside a biodegradable and biocompatible polymer carrier, ranging in size from 1 to 1000 nm [100,113]. Polylactide, polylactide polyglycolide copolymers, and polyacrylates are some of the most widely used polymers. Among them, Lactide glycolide copolymer has been widely studied. There are several natural polymers to choose from, including alginate, albumin, and chitosan. Various research has been ongoing for many years to establish the efficacy of curcumin (Indian solid gold) in a variety of disorders, including cancer. It was shown that curcumin was able to reduce Amyloid beta deposition and tau phosphorylation in an animal study on AD.

In addition, it increases the multiplication of neural stem cells and hippocampus neurogenesis [114]. As Curcumin has a weak water solubility, the nanoparticle format of Curcumin is thought to boost its ability to target neurons in Alzheimer's disease (AD). Curcumin nanoparticles Tet-1-targeted PLGA-coated curcumin were tested for the treatment of Alzheimer's disease by suppressing amyloid and antioxidant activities [107]. Nanoparticles having a zeta potential of 230 to 220 mV and an average size of 150–200 nm were manufactured using the solvent evaporation technique. These nanoparticles are entirely soluble in water, and they also have the ability to glow. Studies on the viability of cells have shown that these nanoparticles are not harmful to cells. According to fluorescence data, the uptake of nanoparticles targeted with Tet-1 peptide in GI-1 glioma cells in an in vitro uptake assay was significantly increased compared to the non-targeted nanoparticles.

6.2.1. Liposomes-Based Drug Delivery Systems

Liposome transport mechanisms across the blood-brain barrier are still at the budding stage. Electrostatic interactions are the method by which the negative charges on BBB and cationic liposomal drug delivery systems induce cell internalisation through absorption mechanism [115]. It is possible to traverse the BBB by attaching glucose and GSH to liposomes as nutrients. Because of the therapeutic potential of receptor-mediated transcytosis, which has the ability to connect specific ligands to a wide variety of BBB receptors [116]. Using nanoliposomes that have been double-functionalized with curcumin and HIV TAT peptide, A peptide affinity is increased and BBB bridging is improved [117]. Endocytosis and micropinocytosis were the two most often cited mechanisms for TAT uptake. A thiol-maleimide reaction covalently links TAT to nanoliposomes. TAT-CurcNL has a size range of 196.5 to 3.2 nm and was quantified using HPLC-MS/MS. Mass spectrometry, confocal microscopy, and a radioactivity assay using [3H]-sphingomyelin [42] all showed a threefold increase in nanoliposome absorption after TAT functionalization in human brain capillary endothelial cells (hCMEC and D3).

Mourtias also used the thin film hydration approach to make multifunctional nanoliposomes containing curcumin-lipid derivatives. When DSPE-PEG2000 was reacted with 4-methoxytrityl-thiol for the first time, it produced DSPEPEG2000-S-Mmt, which is the precursor to DSPE-PEG2000SH, a functionalized lipid available on the market. There were no by-products and the reaction proceeded quickly in the presence of DIPEA (Diisopropylethylamine) during the DSPE-PEG2000-S-Mmt synthesis [118]. A PEG spacer was added between curcumin and lipids to manufacture this derivative. When DSPEPEG2000-SH was deprotected with thiol, it reacted with curcumin to produce the DPSPEG2000-CURC derivative. Nanoliposome membranes were successfully included with this novel synthetic substance. For the second phenol-protons of the curcumin moiety, DIPEA was added and the resulting DIPEA salt was proved to effectively identify A deposits in post-mortem tissues of Alzheimer's disease (AD) patients' tissues. One of the best-selling herbal remedies, ginkgo *biloba,* has its origins in traditional Chinese medicine. Antioxidant activity in the CNS is observed to boost the activities of superoxide dismutase, catalase, glutathione peroxidase, and glutathione reductase. As a result of increased antioxidant activity in the hippocampi, Ginkgo *biloba* aids in memory and learning [119]. Non-ionic bilayer vesicles based on surfactants constitute niosomes. Unlike liposomes, niosomes may contain both hydrophilic and hydrophobic drugs in the same system [47,120]. It is possible that the niosomes containing medications for the treatment of CNS disorders may pass the BBB. In freeze-dried powder (661 nm) and spray-dried powder (680 nm), the size of niosomes is affected by the drying procedure. Spray-dried niosomes had a higher zeta potential than freeze-dried niosomes, on the other hand. Because the zeta potential of spray dried noisome powder is high, they are more stable, owing to the substantial electrostatic repulsion between particles [48,121].

6.2.2. Green Synthesized Nanoparticles

Green chemistry—in fact, green technology—is currently being pushed for its use in nanotechnology as an eco-friendly option. However, for herbal extracts, metal nanoparticles have been shown to be more favourable than microorganisms in the creation of nanoparticles. The organic components may decompose into biodegradable material because of the presence of polyphenols. Shankhapushpi is n antihypertensive, immunomodulatory, and anticonvulsant ayurvedic plant. The antioxidant capability of the ayurvedic medicine Convolvulus pluricaulis was discovered to improve memory in nanoparticles made using iron oxide as a precursor. Researchers are still trying to understand the molecular process of these micro particles [122]. As an anti-epileptic, insect repellent, and antioxidant, the Pulicaria undulata plant is also often used in traditional Chinese medicine. The increased concentration of silver nanoparticles (AgNPs) in the brain leads to an increase in AgNPs' interaction with proteins, which prevents fibril formation by decreasing protein conformation and self-association in Pulicaria undulata nanoparticles made with silver (AgNPs) [123].

7. Nanotechnology-Induced Bioinformatics for Early Diagnosis

In the last three decades, several computational programs have been continuously developed and applied to improve the quality of biomedicine in an interdisciplinary model. At the same time, the term bioinformatics, a combination of computational biology and mathematical algorithms, has been used for leading current biomedicine and biomedical research. Biological data-restoration (proteomics and genomics related to disease biology), biological data and text mining (analyses of massive amounts of data and text to restore biological data), and genomic-proteomic investigation (in the contest of mutation and drug resistance pathways and biomarkers) have been used to guide and solve several complex problems in biomedical areas through the wide range of cost and resource-saving technology. For example, the Human Genome Project and other OMICS (genomics, proteomics, metabolomics, and glycomics) projects have revolutionized bioinformatics support to biomedical research. Overall, integrated bioinformatics tools can gather and analyse massive amounts of biological and laboratory data to help understand diseases and select ideal therapies [124,125].

Simultaneously, nanotechnology is another demanding sector in biomedicine for generating more prospective technological and technical discoveries in medication delivery. Strategically, when nanotechnology is merged with bioinformatics, the phrase used is 'nano informatics', and this discipline has more potential for attaining breakthrough analyses in biomedical drug development applications [126,127]. Figure 2 illustrates the role of nano informatics for the development of therapies for dementia. Currently, nano informatics have become an emerging field and have been accepted by health and regulatory bodies such as US NSF, NIH, the National Cancer Institute, and the European Commission, etc., for analysing and processing the structure and physiochemical characteristics of nanoparticles, their formulation, as well as for their application in the treatment of various diseases such as neurodegenerative disorders and cancer. The experimentation for NDs is more expensive and sensitive. Bioinformatics can be used for early detection, understanding the disease, and application, while nanotechnology can be used for studying the diagnosis or treatment strategy through the use of drug delivery that is target specific [125,127].

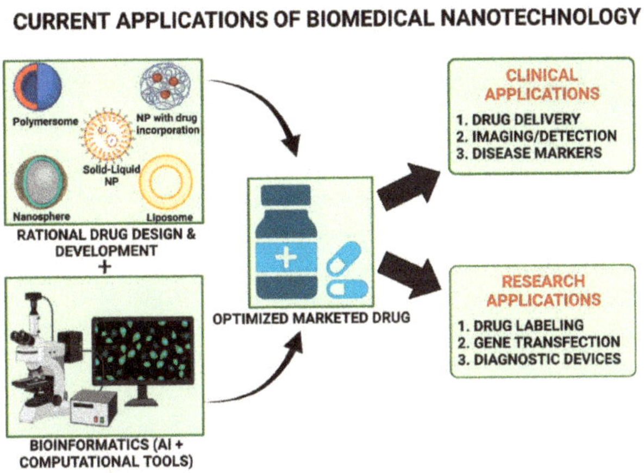

Figure 2. A schematic presentation of nano-informatic (nanotechnology and bioinformatics) in present dementia or neuro-disorder research.

Dementia, a brain disorder which is characterized by memory loss, is a common disorder; AD is its most predominant form, accounting to 60–70% of dementia cases, and about 10 million cases have been recorded per year. Mostly, in brain disorders, the synaptic transmission is affected, and its target is the ionotropic glutamate receptor (iGluR), which belongs to the most poorly understood pharmacological target, the G-coupled receptor (GPCR). However, the current updated research involves analysing brain physiology and targeting amyloid plaques amyloid-beta neurofibrillary tangles, tau protein, astrocyte associated β-secretase, cholinergic neuron associated butyrylcholinesterase, cognitive function in medial temporal lobe cortex, secretase (presenilin I), central nerves system associated GPCR-target, and dopamine-2 receptor, etc., for Alzheimer and dementia treatment [128,129]. In drug development, there are more than a hundred pipeline drugs under clinical investigation [130,131].

In the history of drug discovery, most drugs have failed due to lack of consideration of pharmacokinetics. These drugs are unable to cross the blood-brain barrier (BBB) [132,133]. For a drug to be active and to show the desired effect, the drug should have the ability to penetrate the BBB. Most of these drugs, however, have been unable to cross and as a result show about <1% effectiveness as compared to the administered dose, or sometimes exert neurotoxicity. Therefore, the development of target-specific nano-drug delivery, which can pass through the BBB, can play an important role in the management and treatment of dementia or Alzheimer's diseases [134,135]. Nevertheless, so far, the nano system is insufficient, and attempts are ongoing for the optimization of nano systems and drug applications in NDs [125,136]. Through the use of nano informatic tools, the interaction-cum-efficacy of proposed therapeutic agents with the target protein can be predicted along with obtaining relevant information on the BBB profile and possible neurotoxicity side effects that can be observed. [54,137–140]. For example, tools namely, admetSAR (http://lmmd.ecust.edu.cn/admetsar2/) (accessed on 10 November 2021), SwissADME (http://www.swissadme.ch/) (accessed on 10 November 2021), LightBBB (http://bioanalysis.cau.ac.kr:7030/) (accessed on 10 November 2021), as well as databases such as VariCarta (https://varicarta.msl.ubc.ca/index) (accessed on 10 November 2021), AlzGene (http://www.alzgene.org/) (accessed on 10 November 2021), and NDDVD (http://bioinf.suda.edu.cn/NDDvarbase/LOVDv.3.0) (accessed on 10 November 2021) have been used to obtain large data sets of neurological therapeutics. Overall, nano informatics could be a commanding approach towards exploring and understanding mechanisms along with locating and developing better suited and more efficacious therapeutic agents in a cost-effective manner. Table 3 summarizes the miRNAs used for treatment of neurological disorders.

Table 3. List of miRNAs used as therapeutics (conventional and nonconventional) for some NDDs. The investigated drug, the miRNAs, their sources, the disease state as well as the references are indicated.

Drug	miRNA	Source of miRNA	Disease State Used	Reference
Donepezil	miRNA-206-3p	Mouse-Hippocampus, cortex	Dementia	[141]
Simvastatin	miRNA-106b	SH-SY5Y cells; Mice brain tissue-APP/PS1		[142]
Osthole	miRNA-9	overexpressed APP cells	Alzheimer's Disease (AD)	[143,144]
	miRNA-107	Overexpressed APP cells Mice brain tissue-APP/PS1		[143]
	miRNA-101a-3p			[143]
AGR-GRg1	miRNA-873-5p	Mouse Hippocampus		[145]
L-Dopa	miRNA-30b-5p, miRNA-30a-5p	Plasma	PD	[146]
	miRNA-29a-3p, miRNA-30b-5p, miRNA-103a-3p	Peripheral Blood mononuclear cells (PBMC's)		[147]
	miRNA-16-2-3p, miRNA-26a-2-3p, miRNA-30a	Peripheral blood		[148]
	miRNA-155	PBMCs		[149]
L-Dopa, Amantadine	miRNA-7, miRNA-9-3p, miRNA-9-5p	Peripheral blood		[150]
Interferon-β	miRNA-29	PBMCs	Multiple Sclerosis	[151]
	miRNA-145	Whole blood		[152]
	miRNA-29b-3p	PBMCs		[153]
	miRNA-326			[154]
	miRNA-26a-5p			[155]
	miRNA-146a			[156]
Natalizumab	miRNA-150 CSF,	Plasma		[157]
	miRNA-126, miRNA-17	CD4 + T cells		[158,159]
	miRNA-17~92, miRNA-106b~25	B lymphocytes		[160]
	miRNA-26a, miRNA-155	PBMCs		[153]
	miRNA-155	Monocytes		[161]
Dimethyl fumarate	miRNA-155	Monocytes		[162]
Fingolimod	miRNA-150	Plasma		[163]
	miRNA-23a	Whole Blood		[164]
Natalizumab	miRNA-320, miRNA-320b, miRNA-629	Blood	Progressive multifocal leukoencephalopathy	[163]

8. Limitations of Nanotechnology-Based Approaches for ND

Using nanotheranostic techniques to treat neurodegenerative illnesses, as discovered by Indrasekara et al. [165], have their limits because each patient has a unique genome and neurological functioning and features, and so no one strategy can be employed to treat all of them. Every person must be diagnosed and treated for these neurodegenerative conditions individually, which is a difficult undertaking and does not aid in establishing one approach or treatment methodology that works for everyone [166]. (De Lau et al., 2006).

In addition, Kim et al. found that if the injection is not performed properly, the NPs utilised in this approach may be absorbed by the blood or other body parts instead of the intended target location [167]. This is a difficulty or restriction that has been highlighted by Kim et al. If the medicine is not injected properly, it will not be absorbed or utilised to treat the intended location for treating neurodegenerative illnesses such as Alzheimer's disease and Parkinson's disease. Because just a little amount of the medication actually makes its way to where it is needed, some have contended that it has not yet been shown that the therapy is 100% successful [168].

In order to progress this study into viable therapeutic medicines, it is critical that safety concerns be addressed. As a reminder, the most successful NP formulations for the brain delivery nonetheless accumulate extensively in other parts of the body, such as the spleen, the liver, and the kidney. As a result, it is critical to create nanoformulations that are only activated when they reach the brain [169] instead of activating when they reach other parts of the body. Nanoparticles in regenerative medicine may soon benefit from advances in triggerable nanoformulation technology. The production of nanoparticles that target particular brain cells is a significant problem that warrants more study. Targeting particular brain cells such as dopaminergic neurons (the primary target in Parkinson's disease), microglia (neuroinflammation), or neural stem cells (neuronal repair) may boost its potential therapeutic usefulness in the context of neurodegenerative illnesses.

Multifunctionality is required for brain-targeting NP because of the BBB's protective role, which makes it challenging to target molecules to the brain parenchyma. As a matter of fact, the BBB's primary job is to protect brain tissue from potentially hazardous chemicals. NPs neurotoxicity must be examined for the same reasons as a more traditional medicinal system's neurotoxicity. Microglial activation is a critical consideration in NP neurotoxicity because it plays a role in the neurodegenerative pathogenic process in the vast majority of CNS illnesses. When using cultured microglial cells, an in vitro study found that TiO_2 NPs and HAP (hydroxyapatite) NPs were able to trigger inflammation-related iNOS (inducible nitric oxide synthase) and activate the NF-B signalling pathway. All the tested NPs induced, to a variable extent, the increase in pro-inflammatory molecules [170]. It was evidenced that TiO_2 and HAP NPs may activate microglial cells and lead to probable pathogenic alterations in the olfactory bulb, striatum, and hippocampal regions of mice, concluding that these inorganic NPs contributed to the dysfunction and cytotoxicity in PC12 cells.

9. Future Nano Therapeutics for ND

Without a doubt, precision-based nanomedicine will gain more attention in the coming years. However, despite advances, the use of phages in humans has yet to gain widespread acceptance. There is a gap that must be bridged between animal trials and human use. Approval from the Food and Drug Administration is required, and it is likely to follow on the heels of increasingly promising research on phage-based nanomedicine [171].

Exosomes, which are naturally produced by human cells, are emerging as a new generation of highly protective nanoplatforms for efficient drug delivery [172]. They can transport not only therapeutics but also molecular imaging agents for use in precision therapeutics and diagnostics. Exosomes are an ideal nanoplatform for loading both hydrophilic and lipophilic agents due to their unique structure, which includes an aqueous core and a membrane rich in lipid rafts.

Nonetheless, despite the current promising research on exosomes, several major challenges remain to be overcome, including a lack of characterization of exosomes derived from different sources, low exosomal yield, and encapsulation efficiency, and a lack of advanced purification techniques with high efficiency. Exosome molecular and nanoengineering will provide insights into future effective and precision medicine for devastating diseases such as neurodegenerative disorders. Because targeted exosomal delivery is a rapidly developing field, aptamer-mediated exosomal delivery is becoming appealing for the development of smart nano-delivery systems due to its ease of use, high performance at the nanoscale, enhanced efficacy, safety, and low cost [173,174].

Although preclinical studies have provided the biological and mechanistic foundation for previous clinical trials, future efforts to improve the predictive accuracy of preclinical studies will be critical to clinical trial success. To improve clinical translation, health experts hope to develop research guidelines such as the Stroke Therapy Academic Industry Roundtable (STAIR) recommendations. The Federal Interagency Traumatic Brain Injury Research (FITBIR) system was designed to share TBI-related research across the research community [175]. Future efforts for the development of nanomaterials to treat acute brain injuries include accurate measurement of nanomaterial pharmacokinetics in living organisms, careful selection of animal models that recapitulate specific human pathology relevant to nanomaterial design and payload, designing experiments that consider multiple biological variables, and the development of technology to quantitatively measure biomarkers that can accurately predict outcomes in humans.

10. Conclusions

Nanotechnology can alter neuroscience-based information and restorative methodologies, and can be used to possibly make significant commitments for the development of nano-empowered medication for the treatment of NDs. Equal advancements in neurophysiology and neuropathology exploration would help in the advancement in nanotechnology, which can be used to provide CNS recovery and neuroprotection. Accordingly, for utilization of nanotechnology in neural system science and neurosurgery, key factors that require consideration include: (1) breakthrough discoveries and developments in drug science and material science, which can help in the manufacturing of the described methodologies; (2) development and advancement of sub-atomic science, sensory system-based neurophysiology, and neuropathology; and (3) planning and combination of explicit nano-empowered therapies to the CNS, which exploit the initial two factors. As a result, nanotechnology could provide the solution and can offer breakthrough therapies for the management and treatment of NDs and can also be used to bypass the current problem of available neurological therapies i.e., non-specific targeting and lower efficacy rates of drug therapies. Therefore, taken together, neurosurgeons, nervous system specialists, neuroscientists, and drug researchers and architects, should take part in utilizing the power of nanotechnology for drug delivery. Consistent with the profoundly interdisciplinary nature of this space of exploration, it is additionally significant to note that nano-informatics and nanotechnology can also provide innovative headways and progressions that are related to fundamental and clinical neuroscience.

Author Contributions: Conceptualization, T.B.; data curation, T.B. and G.A.B.e.S.; writing—original draft preparation, T.B., G.A.B.e.S., H.C., M.M.R., Z.H. and S.S.S.; validation, T.B. and G.A.B.e.S.; formal analysis, T.B., G.A.B.e.S., H.C., M.M.R., Z.H., S.S.S. and S.C.; investigation, T.B., G.A.B.e.S. and S.C.; resources, T.B.; writing—reviewing and editing, T.B., G.A.B.e.S. and S.C.; visualization, T.B., G.A.B.e.S., H.C.,M.M.R., Z.H., S.S.S. and S.C.; supervision, T.B. and S.C.; project administration, T.B. and S.C. All authors have read and agreed to the published version of the manuscript.

Funding: This research received no external founds.

Conflicts of Interest: The authors declare no conflict of interest.

Abbreviations

BBB	Blood brain barrier
CNS	Central nervous system
ND	Neurological and neurodegenerative diseases
NP	Nanoparticle
AD	Alzheimer's disease
PD	Parkinson's disease

ALS	Amyotrophic horizontal sclerosis
JNK	c-Jun N-terminal kinase
AChE	Acetylcholinesterase
IC50	Half-maximal inhibitory concentration
SAC	S-allyl cysteine
MPP	Phenyl pyridinium
SOD	Superoxide dismutase
6-OHDA	6-hydroxydopamine
CAT	Chloramphenicol acetyltransferase
GPx	Gluthathione peroxidase
GPR	Glutathione reductase
ATP	Adenosine triphosphate
iNOS	Inducible nitric oxide synthase
COX-2:	Cyclo-oxygenase 2
MDA	Malondialdehyde
GSH	Glutathione
CPPME	Celastrus paniculate wild whole plant methanolic extract
LPO	Lipid peroxidation
DPPH	2,2-diphenyl-1-picrylhydrazyl
$GABA_A$	Gamma- amino butyric acid
MAO	Monoamine oxidases
DNA	Deoxyribo nucleic acid
mRNA	Messenger ribonucleic acid
5-HT	5-hydroxytrptamine/serotonin
DA	Dopamine
ROS	Reactive oxygen species
VMAT	Vesicular monoamine transporter
NMDA	N-methyl-D-aspartic acid
COMT	Catechol ortho-methyl transferase
CSF	Cerebrospinal fluid
PET	Positron emission tomography
SPECT	Single photon emission computed tomography
NIH	National Institute of Health
NSF	National Science Foundation
USFDA	United States Food and Drug Administration
STAIR	Stroke Therapy Academic Industry Roundtable
FITBIR	Federal Interagency Traumatic Brain Injury Research

References

1. Bensadoun, J.-C.; Almeida, L.P.d.; Fine, E.G.; Tseng, J.L.; Déglon, N.; Aebischer, P. Comparative study of GDNF delivery systems for the CNS: Polymer rods, encapsulated cells, and lentiviral vectors. *J. Control. Release* **2003**, *87*, 107–115. [CrossRef]
2. Kishima, H.; Poyot, T.; Bloch, J.; Dauguet, J.; Conde, F.; Dolle, F.; Hinnen, F.; Pralong, W.; Palfi, S.; Deglon, N.; et al. Encapsulated GDNF-producing C2C12 cells for Parkinson's disease: A pre-clinical study in chronic MPTP-treated baboons. *Neurobiol. Dis.* **2004**, *16*, 428–439. [CrossRef]
3. McArthur, J.C. HIV dementia: An evolving disease. *J. Neuroimmunol.* **2004**, *157*, 3–10. [CrossRef] [PubMed]
4. Menei, P.; Montero-Menei, C.; Venier, M.C.; Benoit, J.P. Drug delivery into the brain using poly(lactide-co-glycolide) microspheres. *Expert Opin. Drug Deliv.* **2005**, *2*, 363–376. [CrossRef] [PubMed]
5. Flachenecker, P. Epidemiology of neuroimmunological diseases. *J. Neurol.* **2006**, *253* (Suppl. 5), V2–V8. [CrossRef]
6. Kabanov, A.V.; Gendelman, H.E. Nanomedicine in the diagnosis and therapy of neurodegenerative disorders. *Prog. Polym. Sci.* **2007**, *32*, 1054–1082. [CrossRef]
7. Ferri, C.P.; Prince, M.; Brayne, C.; Brodaty, H.; Fratiglioni, L.; Ganguli, M.; Hall, K.; Hasegawa, K.; Hendrie, H.; Huang, Y.; et al. Global prevalence of dementia: A Delphi consensus study. *Lancet* **2005**, *366*, 2112–2117. [CrossRef]
8. Popovic, N.; Brundin, P. Therapeutic potential of controlled drug delivery systems in neurodegenerative diseases. *Int. J. Pharm.* **2006**, *314*, 120–126. [CrossRef]
9. Singh, N.; Pillay, V.; Choonara, Y.E. Advances in the treatment of Parkinson's disease. *Prog. Neurobiol.* **2007**, *81*, 29–44. [CrossRef]
10. Tysnes, O.B.; Storstein, A. Epidemiology of Parkinson's disease. *J. Neura. Transm. (Vienna)* **2017**, *124*, 901–905. [CrossRef]

11. Freed, C.R.; Greene, P.E.; Breeze, R.E.; Tsai, W.Y.; DuMouchel, W.; Kao, R.; Dillon, S.; Winfield, H.; Culver, S.; Trojanowski, J.Q.; et al. Transplantation of embryonic dopamine neurons for severe Parkinson's disease. *N. Engl. J. Med.* **2001**, *344*, 710–719. [CrossRef] [PubMed]
12. Pollak, P.; Fraix, V.; Krack, P.; Moro, E.; Mendes, A.; Chabardes, S.; Koudsie, A.; Benabid, A.-L. Treatment results: Parkinson's disease. *Mov. Disord.* **2002**, *17*, S75–S83. [CrossRef] [PubMed]
13. Johnson, W.G. Late-onset neurodegenerative diseases—The role of protein insolubility. *J. Anat.* **2000**, *196 Pt 4*, 609–616. [CrossRef] [PubMed]
14. Dickson, D.W. Misfolded, protease-resistant proteins in animal models and human neurodegenerative disease. *J. Clin. Investig.* **2002**, *110*, 1403–1405. [CrossRef]
15. Woodruff, B.K.; Graff-Radford, N.R.; Ferman, T.J.; Dickson, D.W.; DeLucia, M.W.; Crook, J.E.; Arvanitakis, Z.; Brassler, S.; Waters, C.; Barker, W.; et al. Family history of dementia is a risk factor for Lewy body disease. *Neurology* **2006**, *66*, 1949–1950. [CrossRef]
16. Walia, V.; Kaushik, D.; Mittal, V.; Kumar, K.; Verma, R.; Parashar, J.; Akter, R.; Rahman, M.H.; Bhatia, S.; Al-Harrasi, A.; et al. Delineation of Neuroprotective Effects and Possible Benefits of AntioxidantsTherapy for the Treatment of Alzheimer's Diseases by Targeting Mitochondrial-Derived Reactive Oxygen Species: Bench to Bedside. *Mol. Neurobiol.* **2021**. [CrossRef]
17. Nischwitz, V.; Berthele, A.; Michalke, B. Speciation analysis of selected metals and determination of their total contents in paired serum and cerebrospinal fluid samples: An approach to investigate the permeability of the human blood-cerebrospinal fluid-barrier. *Anal. Chim. Act.* **2008**, *627*, 258–269. [CrossRef]
18. Gendelman, H.E.; Ding, S.; Gong, N.; Liu, J.; Ramirez, S.H.; Persidsky, Y.; Mosley, R.L.; Wang, T.; Volsky, D.J.; Xiong, H. Monocyte chemotactic protein-1 regulates voltage-gated K+ channels and macrophage transmigration. *J. Neuroimmune Pharm.* **2009**, *4*, 47–59. [CrossRef]
19. Modi, G.; Pillay, V.; Choonara, Y.E.; Ndesendo, V.M.; du Toit, L.C.; Naidoo, D. Nanotechnological applications for the treatment of neurodegenerative disorders. *Prog. Neurobiol.* **2009**, *88*, 272–285. [CrossRef]
20. Page, M.J.; McKenzie, J.E.; Bossuyt, P.M.; Boutron, I.; Hoffmann, T.C.; Mulrow, C.D.; Shamseer, L.; Tetzlaff, J.M.; Moher, D. Updating guidance for reporting systematic reviews: Development of the PRISMA 2020 statement. *J. Clin. Epidemiol.* **2021**, *134*, 103–112. [CrossRef]
21. Page, M.J.; McKenzie, J.E.; Bossuyt, P.M.; Boutron, I.; Hoffmann, T.C.; Mulrow, C.D.; Shamseer, L.; Tetzlaff, J.M.; Akl, E.A.; Brennan, S.E.; et al. The PRISMA 2020 statement: An updated guideline for reporting systematic reviews. *J. Clin. Epidemiol.* **2021**, *134*, 178–189. [CrossRef]
22. Kim, J.; Mook-Jung, I. Special issue on neurodegenerative diseases and their therapeutic approaches. *Exp. Mol. Med.* **2015**, *47*, e146. [CrossRef]
23. Ballabh, P.; Braun, A.; Nedergaard, M. The blood-brain barrier: An overview: Structure, regulation, and clinical implications. *Neurobiol. Dis.* **2004**, *16*, 1–13. [CrossRef]
24. Abbott, N.J.; Patabendige, A.A.; Dolman, D.E.; Yusof, S.R.; Begley, D.J. Structure and function of the blood-brain barrier. *Neurobiol. Dis.* **2010**, *37*, 13–25. [CrossRef]
25. Silva, G.A. Neuroscience nanotechnology: Progress, opportunities and challenges. *Nat. Rev. Neurosci.* **2006**, *7*, 65–74. [CrossRef]
26. Moghimi, S.M.; Hunter, A.C.; Murray, J.C. Nanomedicine: Current status and future prospects. *FASEB J.* **2005**, *19*, 311–330. [CrossRef]
27. Goldsmith, M.; Abramovitz, L.; Peer, D. Precision nanomedicine in neurodegenerative diseases. *ACS Nano* **2014**, *8*, 1958–1965. [CrossRef] [PubMed]
28. Illum, L. Nasal drug delivery—Possibilities, problems and solutions. *J. Control. Release* **2003**, *87*, 187–198. [CrossRef]
29. Pardridge, W.M. CSF, blood-brain barrier, and brain drug delivery. *Expert Opin. Drug Deliv.* **2016**, *13*, 963–975. [CrossRef] [PubMed]
30. McDannold, N.; Arvanitis, C.D.; Vykhodtseva, N.; Livingstone, M.S. Temporary disruption of the blood-brain barrier by use of ultrasound and microbubbles: Safety and efficacy evaluation in rhesus macaques. *Cancer Res.* **2012**, *72*, 3652–3663. [CrossRef] [PubMed]
31. Dasgupta, A.; Liu, M.; Ojha, T.; Storm, G.; Kiessling, F.; Lammers, T. Ultrasound-mediated drug delivery to the brain: Principles, progress and prospects. *Drug Discov. Today Technol.* **2016**, *20*, 41–48. [CrossRef]
32. Cheng, Z.; Zhang, J.; Liu, H.; Li, Y.; Zhao, Y.; Yang, E. Central nervous system penetration for small molecule therapeutic agents does not increase in multiple sclerosis- and Alzheimer's disease-related animal models despite reported blood-brain barrier disruption. *Drug Metab. Dispos.* **2010**, *38*, 1355–1361. [CrossRef] [PubMed]
33. Roy Chowdhury, S.; Mondal, S.; Muthuraj, B.; Balaji, S.N.; Trivedi, V.; Krishnan Iyer, P. Remarkably Efficient Blood-Brain Barrier Crossing Polyfluorene-Chitosan Nanoparticle Selectively Tweaks Amyloid Oligomer in Cerebrospinal Fluid and Abeta1-40. *ACS Omega* **2018**, *3*, 8059–8066. [CrossRef] [PubMed]
34. Cavalu, S.; Damian, G. Rotational Correlation Times of 3-Carbamoyl-2,2,5,5-Tetramethyl-3-Pyrrolin-1-Yloxy Spin Label with Respect to Heme and Nonheme Proteins. *Biomacromolecules* **2003**, *4*, 1630–1635. [CrossRef] [PubMed]
35. Kaushik, A.C.; Kumar, A.; Peng, Z.; Khan, A.; Junaid, M.; Ali, A.; Bharadwaj, S.; Wei, D.-Q. Evaluation and validation of synergistic effects of amyloid-beta inhibitor–gold nanoparticles complex on Alzheimer's disease using deep neural network approach. *J. Mater. Res.* **2019**, *34*, 1845–1853. [CrossRef]
36. Choi, C.H.; Alabi, C.A.; Webster, P.; Davis, M.E. Mechanism of active targeting in solid tumors with transferrin-containing gold nanoparticles. *Proc. Natl. Acad. Sci. USA* **2010**, *107*, 1235–1240. [CrossRef] [PubMed]

37. Kong, S.D.; Lee, J.; Ramachandran, S.; Eliceiri, B.P.; Shubayev, V.I.; Lal, R.; Jin, S. Magnetic targeting of nanoparticles across the intact blood-brain barrier. *J. Control. Release* **2012**, *164*, 49–57. [CrossRef]
38. Smith, M.W.; Gumbleton, M. Endocytosis at the blood-brain barrier: From basic understanding to drug delivery strategies. *J. Drug Target.* **2006**, *14*, 191–214. [CrossRef]
39. Fessi, H.; Puisieux, F.; Devissaguet, J.P.; Ammoury, N.; Benita, S. Nanocapsule formation by interfacial polymer deposition following solvent displacement. *Int. J. Pharm.* **1989**, *55*, R1–R4. [CrossRef]
40. Mora-Huertas, C.E.; Fessi, H.; Elaissari, A. Influence of process and formulation parameters on the formation of submicron particles by solvent displacement and emulsification-diffusion methods critical comparison. *Adv. Colloid. Interface Sci.* **2011**, *163*, 90–122. [CrossRef]
41. Cavalu, S.; Bisboaca, S.; Mates, I.M.; Pasca, P.M.; Laslo, V.; Costea, T.; Fritea, L.; Vicas, S. Novel Formulation Based on Chitosan-Arabic Gum Nanoparticles Entrapping Propolis Extract Production, physico-chemical and structural characterization. *Rev. Chim.* **2018**, *69*, 3756–3760. [CrossRef]
42. Erdogar, N.; Akkin, S.; Bilensoy, E. Nanocapsules for Drug Delivery: An Updated Review of the Last Decade. *Recent Pat. Drug Deliv. Formul.* **2018**, *12*, 252–266. [CrossRef] [PubMed]
43. Niwa, T.; Takeuchi, H.; Hino, T.; Kunou, N.; Kawashima, Y. Preparations of biodegradable nanospheres of water-soluble and insoluble drugs with D,L-lactide/glycolide copolymer by a novel spontaneous emulsification solvent diffusion method, and the drug release behavior. *J. Control. Release* **1993**, *25*, 89–98. [CrossRef]
44. Murakami, H.; Kobayashi, M.; Takeuchi, H.; Kawashima, Y. Preparation of poly(dl-lactide-co-glycolide) nanoparticles by modified spontaneous emulsification solvent diffusion method. *Int. J. Pharm.* **1999**, *187*, 143–152. [CrossRef]
45. Quintanar-Guerrero, D.; Allemann, E.; Fessi, H.; Doelker, E. Pseudolatex preparation using a novel emulsion-diffusion process involving direct displacement of partially water-miscible solvents by distillation. *Int. J. Pharm.* **1999**, *188*, 155–164. [CrossRef]
46. Yusuf, M.; Khan, M.; Khan, R.A.; Ahmed, B. Preparation, characterization, in vivo and biochemical evaluation of brain targeted Piperine solid lipid nanoparticles in an experimentally induced Alzheimer's disease model. *J. Drug Target.* **2013**, *21*, 300–311. [CrossRef] [PubMed]
47. McCall, R.L.; Sirianni, R.W. PLGA nanoparticles formed by single- or double-emulsion with vitamin E-TPGS. *J. Vis. Exp.* **2013**, *82*, 51015. [CrossRef]
48. Huang, W.; Zhang, C. Tuning the Size of Poly(lactic-co-glycolic Acid) (PLGA) Nanoparticles Fabricated by Nanoprecipitation. *Biotechnol. J.* **2018**, *13*, 1700203. [CrossRef] [PubMed]
49. Cavalu, S.; Prokisch, J.; Laslo, V.; Vicas, S. Preparation, structural characterisation and release study of novel hybrid microspheres entrapping nanoselenium, produced by green synthesis. *IET Nanobiotechnol.* **2017**, *11*, 426–432. [CrossRef]
50. Tarhini, M.; Greige-Gerges, H.; Elaissari, A. Protein-based nanoparticles: From preparation to encapsulation of active molecules. *Int. J. Pharm.* **2017**, *522*, 172–197. [CrossRef]
51. Zhang, D.; Ma, X.L.; Gu, Y.; Huang, H.; Zhang, G.W. Green Synthesis of Metallic Nanoparticles and Their Potential Applications to Treat Cancer. *Front. Chem.* **2020**, *8*, 799. [CrossRef] [PubMed]
52. Cavalu, S.; Antoniac, I.V.; Mohan, A.; Bodog, F.; Doicin, C.; Mates, I.; Ulmeanu, M.; Murzac, R.; Semenescu, A. Nanoparticles and Nanostructured Surface Fabrication for Innovative Cranial and Maxillofacial Surgery. *Materials* **2020**, *13*, 5391. [CrossRef]
53. Velmurugan, B.K.; Rathinasamy, B.; Lohanathan, B.P.; Thiyagarajan, V.; Weng, C.F. Neuroprotective Role of Phytochemicals. *Molecules* **2018**, *23*, 2485. [CrossRef]
54. Xu, J.; Wold, E.A.; Ding, Y.; Shen, Q.; Zhou, J. Therapeutic Potential of Oridonin and Its Analogs: From Anticancer and Antiinflammation to Neuroprotection. *Molecules* **2018**, *23*, 474. [CrossRef]
55. Hajialyani, M.; Hosein Farzaei, M.; Echeverria, J.; Nabavi, S.M.; Uriarte, E.; Sobarzo-Sanchez, E. Hesperidin as a Neuroprotective Agent: A Review of Animal and Clinical Evidence. *Molecules* **2019**, *24*, 648. [CrossRef] [PubMed]
56. Tripathy, N.; Ahmad, R.; Khang, G. Inorganic Nanotheranostics: Strategy development and applications. *Drug Deliv. Nanosyst. Biomed. Appl.* **2018**, 377–419. [CrossRef]
57. Bar-Zeev, M.; Livney, Y.D.; Assaraf, Y.G. Targeted nanomedicine for cancer therapeutics: Towards precision medicine overcoming drug resistance. *Drug Resist. Updat.* **2017**, *31*, 15–30. [CrossRef]
58. Lausanne, E.P.F.D. On the Way to Nanotheranostics: Diagnosing and Treating Diseases Simultaneously. Available online: www.sciencedaily.com/releases/2019/07/190708131203.htm (accessed on 15 May 2021).
59. Bandelow, B.; Michaelis, S.; Wedekind, D. Treatment of anxiety disorders. *Dialogues. Clin. Neurosci.* **2017**, *19*, 93–107.
60. Powell, C.R.; Dillon, K.M.; Matson, J.B. A review of hydrogen sulfide (H2S) donors: Chemistry and potential therapeutic applications. *Biochem. Pharmacol.* **2018**, *149*, 110–123. [CrossRef] [PubMed]
61. Shang, A.; Cao, S.Y.; Xu, X.Y.; Gan, R.Y.; Tang, G.Y.; Corke, H.; Mavumengwana, V.; Li, H.B. Bioactive Compounds and Biological Functions of Garlic (*Allium sativum* L.). *Foods* **2019**, *8*, 246. [CrossRef]
62. Chen, X.; Pan, W. The treatment strategies for neurodegenerative diseases by integrative medicine. *Integr. Med. Int.* **2014**, *1*, 223–225. [CrossRef]
63. Roy, S.; Awasthi, H. Herbal medicines as neuroprotective agent: A mechanistic approach. *Int. J. Pharm. Pharm. Sci.* **2017**, *9*, 1–7. [CrossRef]
64. Chopra, H.; Dey, P.S.; Das, D.; Bhattacharya, T.; Shah, M.; Mubin, S.; Maishu, S.P.; Akter, R.; Rahman, M.H.; Karthika, C.; et al. Curcumin Nanoparticles as Promising Therapeutic Agents for Drug Targets. *Molecules* **2021**, *26*, 4998. [CrossRef] [PubMed]

65. Soares, G.A.B.e.; Bhattacharya, T.; Chakrabarti, T.; Tagde, P.; Cavalu, S. Exploring Pharmacological Mechanisms of Essential Oils on the Central Nervous System. *Plants* **2021**, *11*, 21. [CrossRef]
66. Jakka, A. A Study on Nootropic Activity of Celastrus Paniculata Willd Whole Plant Methanolic Extract in Rats. *Asian J. Pharm. Clin. Res.* **2016**, *9*, 336–341.
67. Amadi, C.N.; Offor, S.J.; Frazzoli, C.; Orisakwe, O.E. Natural antidotes and management of metal toxicity. *Environ. Sci. Pollut. Res. Int.* **2019**, *26*, 18032–18052. [CrossRef]
68. Ashraf, A.; Zafar, S.; Zahid, K.; Salahuddin Shah, M.; Al-Ghanim, K.A.; Al-Misned, F.; Mahboob, S. Synthesis, characterization, and antibacterial potential of silver nanoparticles synthesized from *Coriandrum sativum* L. *J. Infect. Public Health* **2019**, *12*, 275–281. [CrossRef]
69. Owoeye, O.; Obazie, F.I.; Atiba, F.A.; Malomo, A.O. Comparative Neuroprotective Effect of Celosia argentea Linn. and Vitamin E on Mercury-induced Oxidative and Histological Parameters of Rat Brain. *Niger. J. Physiol. Sci.* **2019**, *34*, 167–175.
70. Yang, S.; Chang, M.C. Chronic Pain: Structural and Functional Changes in Brain Structures and Associated Negative Affective States. *Int. J. Mol. Sci.* **2019**, *20*, 3130. [CrossRef]
71. Yuan, Q.; Wang, C.W.; Shi, J.; Lin, Z.X. Effects of Ginkgo biloba on dementia: An overview of systematic reviews. *J. Ethnopharmacol.* **2017**, *195*, 1–9. [CrossRef] [PubMed]
72. Abdolmaleki, A.; Akram, M.; Saeed, M.M.; Asadi, A.; Kajkolah, M. Herbal medicine as neuroprotective potential agent in human and animal models: A historical overview. *J. Pharm. Care* **2020**, *8*, 75–82. [CrossRef]
73. Bridi, H.; Meirelles, G.C.; von Poser, G.L. Structural diversity and biological activities of phloroglucinol derivatives from Hypericum species. *Phytochemistry* **2018**, *155*, 203–232. [CrossRef]
74. Agatonovic-Kustrin, S.; Kettle, C.; Morton, D.W. A molecular approach in drug development for Alzheimer's disease. *Biomed. Pharm.* **2018**, *106*, 553–565. [CrossRef]
75. Eudes, J.F.; Silveria, D.; Soares, A.C.; Carneiro, F.P.; de Assis, M.S.; Leite, F.B.; Niraldo, P.; Greice, M.S.; Mônica, V.d.S.; Stéphanie, M.; et al. Effects of lemon balm (Melissa officinalis) on behavioral deficits and memory impairment of rats surviving sepsis. *J. Med. Plants Res.* **2017**, *11*, 153–160. [CrossRef]
76. Kusindarta, D.L.; Wihadmadyatami, H.; Jadi, A.R.; Karnati, S.; Lochnit, G.; Hening, P.; Haryanto, A.; Auriva, M.B.; Purwaningrum, M. Ethanolic extract Ocimum sanctum. Enhances cognitive ability from young adulthood to middle aged mediated by increasing choline acetyl transferase activity in rat model. *Res. Vet. Sci.* **2018**, *118*, 431–438. [CrossRef] [PubMed]
77. Antonescu Mintas, A.I.; Miere Groza, F.; Fritea, L.; Ganea, M.; Zdrinca, M.; Dobjanschi, L.; Antonescu, A.; Vicas, S.I.; Bodog, F.; Sindhu, R.K.; et al. Perspectives on the Combined Effects of Ocimum basilicum and Trifolium pratense Extracts in Terms of Phytochemical Profile and Pharmacological Effects. *Plants* **2021**, *10*, 1390. [CrossRef]
78. Ahuja, A.; Kim, J.H.; Kim, J.H.; Yi, Y.S.; Cho, J.Y. Functional role of ginseng-derived compounds in cancer. *J. Ginseng Res.* **2018**, *42*, 248–254. [CrossRef]
79. Nieto, G.; Ros, G.; Castillo, J. Antioxidant and Antimicrobial Properties of Rosemary (*Rosmarinus officinalis*, L.): A Review. *Medicines* **2018**, *5*, 98. [CrossRef]
80. Lopresti, A.L. Salvia (Sage): A Review of its Potential Cognitive-Enhancing and Protective Effects. *Drugs R D* **2017**, *17*, 53–64. [CrossRef]
81. Lee, D.Y.; Kim, H.W.; Yang, H.; Sung, S.H. Hydrolyzable tannins from the fruits of Terminalia chebula Retz and their alpha-glucosidase inhibitory activities. *Phytochemistry* **2017**, *137*, 109–116. [CrossRef] [PubMed]
82. Tiwari, P.; Nayak, P.; Prusty, S.K.; Sahu, P.K. Phytochemistry and pharmacology of Tinospora cordifolia: A review. *Syst. Rev. Pharm.* **2018**, *9*, 70–78. [CrossRef]
83. RajaSankar, S.; Manivasagam, T.; Sankar, V.; Prakash, S.; Muthusamy, R.; Krishnamurti, A.; Surendran, S. Withania somnifera root extract improves catecholamines and physiological abnormalities seen in a Parkinson's disease model mouse. *J. Ethnopharmacol.* **2009**, *125*, 369–373. [CrossRef]
84. Olasehinde, T.A.; Olaniran, A.O.; Okoh, A.I. Therapeutic Potentials of Microalgae in the Treatment of Alzheimer's Disease. *Molecules* **2017**, *22*, 480. [CrossRef]
85. Stone, N.L.; Murphy, A.J.; England, T.J.; O'Sullivan, S.E. A systematic review of minor phytocannabinoids with promising neuroprotective potential. *Br. J. Pharmacol.* **2020**, *177*, 4330–4352. [CrossRef]
86. Wyant, K.J.; Ridder, A.J.; Dayalu, P. Huntington's Disease-Update on Treatments. *Curr. Neurol. Neurosci. Rep.* **2017**, *17*, 33. [CrossRef] [PubMed]
87. Hussain, G.; Rasul, A.; Anwar, H.; Aziz, N.; Razzaq, A.; Wei, W.; Ali, M.; Li, J.; Li, X. Role of Plant Derived Alkaloids and Their Mechanism in Neurodegenerative Disorders. *Int. J. Biol. Sci.* **2018**, *14*, 341–357. [CrossRef] [PubMed]
88. Kumar, A.; Chaudhary, R.K.; Singh, R.; Singh, S.P.; Wang, S.Y.; Hoe, Z.Y.; Pan, C.T.; Shiue, Y.L.; Wei, D.Q.; Kaushik, A.C.; et al. Nanotheranostic Applications for Detection and Targeting Neurodegenerative Diseases. *Front. Neurosci.* **2020**, *14*, 305. [CrossRef]
89. Khalil, A.T.; Ovais, M.; Ullah, I.; Ali, M.; Shinwari, Z.K.; Khamlich, S.; Maaza, M. Sageretia thea (Osbeck.) mediated synthesis of zinc oxide nanoparticles and its biological applications. *Nanomedicine* **2017**, *12*, 1767–1789. [CrossRef] [PubMed]
90. Ovais, M.; Zia, N.; Ahmad, I.; Khalil, A.T.; Raza, A.; Ayaz, M.; Sadiq, A.; Ullah, F.; Shinwari, Z.K. Phyto-Therapeutic and Nanomedicinal Approaches to Cure Alzheimer's Disease: Present Status and Future Opportunities. *Front. Aging Neurosci* **2018**, *10*, 284. [CrossRef]

91. Ovais, M.; Nadhman, A.; Khalil, A.T.; Raza, A.; Khuda, F.; Sohail, M.F.; Islam, N.U.; Sarwar, H.S.; Shahnaz, G.; Ahmad, I.; et al. Biosynthesized colloidal silver and gold nanoparticles as emerging leishmanicidal agents: An insight. *Nanomedicine* **2017**, *12*, 2807–2819. [CrossRef] [PubMed]
92. Gaude, T.T.; Soares, G.A.B.e.; Priolkar, R.N.S.; Biradar, B.; Mamledesai, S. Synthesis of 4-hydroxy-1-(phenyl/methyl)-3-[3-(substituted amino)-2-nitropropanoyl] quinolin-2(1H)-ones as an antimicrobial andantitubercular agents. *Indian J. Heterocycl. Chem.* **2017**, *27*, 223–228.
93. Emmanuel, R.; Saravanan, M.; Ovais, M.; Padmavathy, S.; Shinwari, Z.K.; Prakash, P. Antimicrobial efficacy of drug blended biosynthesized colloidal gold nanoparticles from Justicia glauca against oral pathogens: A nanoantibiotic approach. *Microb. Pathog.* **2017**, *113*, 295–302. [CrossRef]
94. Suganthy, N.; Sri Ramkumar, V.; Pugazhendhi, A.; Benelli, G.; Archunan, G. Biogenic synthesis of gold nanoparticles from Terminalia arjuna bark extract: Assessment of safety aspects and neuroprotective potential via antioxidant, anticholinesterase, and antiamyloidogenic effects. *Environ. Sci. Pollut. Res. Int.* **2018**, *25*, 10418–10433. [CrossRef] [PubMed]
95. Reddy, P.H.; Manczak, M.; Yin, X.; Grady, M.C.; Mitchell, A.; Tonk, S.; Kuruva, C.S.; Bhatti, J.S.; Kandimalla, R.; Vijayan, M.; et al. Protective Effects of Indian Spice Curcumin against Amyloid-beta in Alzheimer's Disease. *J. Alzheimers. Dis.* **2018**, *61*, 843–866. [CrossRef] [PubMed]
96. Yusuf, M.; Khan, M.; Alrobaian, M.M.; Alghamdi, S.A.; Warsi, M.H.; Sultana, S.; Khan, R.A. Brain targeted Polysorbate-80 coated PLGA thymoquinone nanoparticles for the treatment of Alzheimer's disease, with biomechanistic insights. *J. Drug Deliv. Sci. Technol.* **2021**, *61*, 102214. [CrossRef]
97. Xu, R.; Wang, J.; Xu, J.; Song, X.; Huang, H.; Feng, Y.; Fu, C. Rhynchophylline Loaded-mPEG-PLGA Nanoparticles Coated with Tween-80 for Preliminary Study in Alzheimer's Disease. *Int. J. Nanomed.* **2020**, *15*, 1149–1160. [CrossRef]
98. Jacob, J.; Haponiuk, J.T.; Thomas, S.; Gopi, S. Biopolymer based nanomaterials in drug delivery systems: A Review. *Mater. Today Chem.* **2018**, *9*, 43–55. [CrossRef]
99. Numata, K.; Kaplan, D.L. Silk-based delivery systems of bioactive molecules. *Adv. Drug Deliv. Rev.* **2010**, *62*, 1497–1508. [CrossRef]
100. Chopra, H.; Kumar, S.; Singh, I. Biopolymer-based Scaffolds for Tissue Engineering Applications. *Curr. Drug Targets* **2021**, *22*, 282–295. [CrossRef]
101. Langer, K.; Balthasar, S.; Vogel, V.; Dinauer, N.; Briesen, H.v.; Schubert, D. Optimization of the preparation process for human serum albumin (HSA) nanoparticles. *Int. J. Pharm.* **2003**, *257*, 169–180. [CrossRef]
102. Jahanban-Esfahlan, A.; Dastmalchi, S.; Davaran, S. A simple improved desolvation method for the rapid preparation of albumin nanoparticles. *Int. J. Biol. Macromol.* **2016**, *91*, 703–709. [CrossRef]
103. Kratz, F. Albumin as a drug carrier: Design of prodrugs, drug conjugates and nanoparticles. *J. Control. Release* **2008**, *132*, 171–183. [CrossRef]
104. Sathya, S.; Shanmuganathan, B.; Devi, K.P. Deciphering the anti-apoptotic potential of alpha-bisabolol loaded solid lipid nanoparticles against Abeta induced neurotoxicity in Neuro-2a cells. *Colloids Surf. B Biointerfaces* **2020**, *190*, 110948. [CrossRef]
105. Pinheiro, R.G.R.; Granja, A.; Loureiro, J.A.; Pereira, M.C.; Pinheiro, M.; Neves, A.R.; Reis, S. Quercetin lipid nanoparticles functionalized with transferrin for Alzheimer's disease. *Eur. J. Pharm. Sci.* **2020**, *148*, 105314. [CrossRef]
106. Giacomeli, R.; Izoton, J.C.; Dos Santos, R.B.; Boeira, S.P.; Jesse, C.R.; Haas, S.E. Neuroprotective effects of curcumin lipid-core nanocapsules in a model Alzheimer's disease induced by beta-amyloid 1-42 peptide in aged female mice. *Brain Res.* **2019**, *1721*, 146325. [CrossRef]
107. Mathew, A.; Fukuda, T.; Nagaoka, Y.; Hasumura, T.; Morimoto, H.; Yoshida, Y.; Maekawa, T.; Venugopal, K.; Kumar, D.S. Curcumin loaded-PLGA nanoparticles conjugated with Tet-1 peptide for potential use in Alzheimer's disease. *PLoS ONE* **2012**, *7*, e32616. [CrossRef]
108. Pinheiro, R.G.R.; Granja, A.; Loureiro, J.A.; Pereira, M.C.; Pinheiro, M.; Neves, A.R.; Reis, S. RVG29-Functionalized Lipid Nanoparticles for Quercetin Brain Delivery and Alzheimer's Disease. *Pharm. Res.* **2020**, *37*, 139. [CrossRef]
109. Cano, A.; Ettcheto, M.; Chang, J.H.; Barroso, E.; Espina, M.; Kuhne, B.A.; Barenys, M.; Auladell, C.; Folch, J.; Souto, E.B.; et al. Dual-drug loaded nanoparticles of Epigallocatechin-3-gallate (EGCG)/Ascorbic acid enhance therapeutic efficacy of EGCG in a APPswe/PS1dE9 Alzheimer's disease mice model. *J. Control. Release* **2019**, *301*, 62–75. [CrossRef] [PubMed]
110. Zhao, Y.; Xiong, S.; Liu, P.; Liu, W.; Wang, Q.; Liu, Y.; Tan, H.; Chen, X.; Shi, X.; Wang, Q.; et al. Polymeric Nanoparticles-Based Brain Delivery with Improved Therapeutic Efficacy of Ginkgolide B in Parkinson's Disease. *Int. J. Nanomed.* **2020**, *15*, 10453–10467. [CrossRef] [PubMed]
111. Xue, Y.; Wang, N.; Zeng, Z.; Huang, J.; Xiang, Z.; Guan, Y.-Q. Neuroprotective effect of chitosan nanoparticle gene delivery system grafted with acteoside (ACT) in Parkinson's disease models. *J. Mater. Sci. Technol.* **2020**, *43*, 197–207. [CrossRef]
112. Tan, J.P.K.; Voo, Z.X.; Lim, S.; Venkataraman, S.; Ng, K.M.; Gao, S.; Hedrick, J.L.; Yang, Y.Y. Effective encapsulation of apomorphine into biodegradable polymeric nanoparticles through a reversible chemical bond for delivery across the blood-brain barrier. *Nanomedicine* **2019**, *17*, 236–245. [CrossRef]
113. Chopra, H.; Kumar, S.; Singh, I. Strategies and Therapies for Wound Healing: A Review. *Curr. Drug Targets* **2021**, *23*, 87–98. [CrossRef] [PubMed]
114. Li, J.; Han, Y.; Li, M.; Nie, C. Curcumin Promotes Proliferation of Adult Neural Stem Cells and the Birth of Neurons in Alzheimer's Disease Mice via Notch Signaling Pathway. *Cell. Reprogram* **2019**, *21*, 152–161. [CrossRef] [PubMed]

115. Joshi, S.; Singh-Moon, R.P.; Ellis, J.A.; Chaudhuri, D.B.; Wang, M.; Reif, R.; Bruce, J.N.; Bigio, I.J.; Straubinger, R.M. Cerebral hypoperfusion-assisted intra-arterial deposition of liposomes in normal and glioma-bearing rats. *Neurosurgery* **2015**, *76*, 92–100. [CrossRef]
116. Noble, G.T.; Stefanick, J.F.; Ashley, J.D.; Kiziltepe, T.; Bilgicer, B. Ligand-targeted liposome design: Challenges and fundamental considerations. *Trends Biotechnol.* **2014**, *32*, 32–45. [CrossRef] [PubMed]
117. Sancini, G.; Gregori, M.; Salvati, E.; Cambianica, I.; Re, F.; Ornaghi, F.; Canovi, M.; Fracasso, C.; Cagnotto, A.; Colombo, M.; et al. Functionalization with TAT-peptide enhances blood-brain barrier crossing in vitro of nanoliposomes carrying a curcumin-derivative to bind amyloid-β peptide. *J. Nanomed. Nanotechnol.* **2013**, *4*, 171. [CrossRef]
118. Mourtas, S.; Lazar, A.N.; Markoutsa, E.; Duyckaerts, C.; Antimisiaris, S.G. Multifunctional nanoliposomes with curcumin-lipid derivative and brain targeting functionality with potential applications for Alzheimer disease. *Eur. J. Med. Chem.* **2014**, *80*, 175–183. [CrossRef]
119. Naik, S.R.; Pilgaonkar, V.W.; Panda, V.S. Evaluation of antioxidant activity of Ginkgo biloba phytosomes in rat brain. *Phytother. Res.* **2006**, *20*, 1013–1016. [CrossRef] [PubMed]
120. Uchegbu, I.F.; Vyas, S.P. Non-ionic surfactant based vesicles (niosomes) in drug delivery. *Int. J. Pharm.* **1998**, *172*, 33–70. [CrossRef]
121. Jin, Y.; Wen, J.; Garg, S.; Liu, D.; Zhou, Y.; Teng, L.; Zhang, W. Development of a novel niosomal system for oral delivery of Ginkgo biloba extract. *Int J Nanomed.* **2013**, *8*, 421–430. [CrossRef]
122. Poka, L.P.; Mohan, G.K.; Rao, K.V.; Shanker, K. Neuroprotective Effect of Green Synthesized Iron Oxide Nanoparticles Using Aqueous Extract of Convolvulus Pluricaulis Plant in The Management of Alzheimer's Disease. *Phytopathology* **2017**, *9*, 703–709. [CrossRef]
123. Dehvari, M.; Ghahghaei, A. The effect of green synthesis silver nanoparticles (AgNPs) from Pulicaria undulata on the amyloid formation in alpha-lactalbumin and the chaperon action of alpha-casein. *Int. J. Biol. Macromol.* **2018**, *108*, 1128–1139. [CrossRef]
124. Nagamune, T. Biomolecular engineering for nanobio/bionanotechnology. *Nano Converg.* **2017**, *4*, 9. [CrossRef]
125. Kaushik, A. Biomedical Nanotechnology Related Grand Challenges and Perspectives. *Front. Nanotechnol.* **2019**, *1*. [CrossRef]
126. Maojo, V.; Fritts, M.; Martin-Sanchez, F.; De la Iglesia, D.; Cachau, R.E.; Garcia-Remesal, M.; Crespo, J.; Mitchell, J.A.; Anguita, A.; Baker, N.; et al. Nanoinformatics: Developing new computing applications for nanomedicine. *Comput. Sci. Eng.* **2012**, *94*, 521–539. [CrossRef] [PubMed]
127. Heath, J.R. Nanotechnologies for biomedical science and translational medicine. *Proc. Natl. Acad. Sci. USA* **2015**, *112*, 14436–14443. [CrossRef]
128. Zlokovic, B.V. Current treatments and therapeutic targets in Alzheimer's disease. *Adv. Drug Deliv. Rev.* **2002**, *54*, 1533–1537. [CrossRef]
129. Loera-Valencia, R.; Cedazo-Minguez, A.; Kenigsberg, P.A.; Page, G.; Duarte, A.I.; Giusti, P.; Zusso, M.; Robert, P.; Frisoni, G.B.; Cattaneo, A.; et al. Current and emerging avenues for Alzheimer's disease drug targets. *J. Intern. Med.* **2019**, *286*, 398–437. [CrossRef]
130. Cummings, J.; Lee, G.; Ritter, A.; Sabbagh, M.; Zhong, K. Alzheimer's disease drug development pipeline: 2020. *Alzheimers Dement.* **2020**, *6*, e12050. [CrossRef]
131. Huang, L.K.; Chao, S.P.; Hu, C.J. Clinical trials of new drugs for Alzheimer disease. *J. Biomed. Sci.* **2020**, *27*, 18. [CrossRef]
132. Fu, L.M.; Li, J.T. A systematic review of single chinese herbs for Alzheimer's disease treatment. *Evid. Based Complement. Altern. Med.* **2011**, *2011*, 640284. [CrossRef] [PubMed]
133. Banks, W.A.; Greig, N.H. Small molecules as central nervous system therapeutics: Old challenges, new directions, and a philosophic divide. *Future Med. Chem.* **2019**, *11*, 489–493. [CrossRef] [PubMed]
134. Grabrucker, A.M.; Ruozi, B.; Belletti, D.; Pederzoli, F.; Forni, F.; Vandelli, M.A.; Tosi, G. Nanoparticle transport across the blood brain barrier. *Tissue Barriers* **2016**, *4*, e1153568. [CrossRef] [PubMed]
135. Mulvihill, J.J.; Cunnane, E.M.; Ross, A.M.; Duskey, J.T.; Tosi, G.; Grabrucker, A.M. Drug delivery across the blood-brain barrier: Recent advances in the use of nanocarriers. *Nanomedicine* **2020**, *15*, 205–214. [CrossRef]
136. Patra, J.K.; Das, G.; Fraceto, L.F.; Campos, E.V.R.; Rodriguez-Torres, M.D.P.; Acosta-Torres, L.S.; Diaz-Torres, L.A.; Grillo, R.; Swamy, M.K.; Sharma, S.; et al. Nano based drug delivery systems: Recent developments and future prospects. *J. Nanobiotechnol.* **2018**, *16*, 71. [CrossRef]
137. Ashish, N.; Bhatt, P.; Toga, A.W. Global Data Sharing in Alzheimer Disease Research. *Alzheimer. Dis. Assoc. Disord.* **2016**, *30*, 160–168. [CrossRef]
138. Wang, Z.; Yang, H.; Wu, Z.; Wang, T.; Li, W.; Tang, Y.; Liu, G. In Silico Prediction of Blood-Brain Barrier Permeability of Compounds by Machine Learning and Resampling Methods. *ChemMedChem* **2018**, *13*, 2189–2201. [CrossRef]
139. Belmadani, M.; Jacobson, M.; Holmes, N.; Phan, M.; Nguyen, T.; Pavlidis, P.; Rogic, S. VariCarta: A Comprehensive Database of Harmonized Genomic Variants Found in Autism Spectrum Disorder Sequencing Studies. *Autism. Res.* **2019**, *12*, 1728–1736. [CrossRef]
140. Shaker, B.; Yu, M.S.; Song, J.S.; Ahn, S.; Ryu, J.Y.; Oh, K.S.; Na, D. LightBBB: Computational prediction model of blood-brain-barrier penetration based on LightGBM. *Bioinformatics* **2021**, *37*, 1135–1139. [CrossRef]
141. Wang, C.N.; Wang, Y.J.; Wang, H.; Song, L.; Chen, Y.; Wang, J.L.; Ye, Y.; Jiang, B. The Anti-dementia Effects of Donepezil Involve miR-206-3p in the Hippocampus and Cortex. *Biol. Pharm. Bull.* **2017**, *40*, 465–472. [CrossRef]
142. Huang, W.; Li, Z.; Zhao, L.; Zhao, W. Simvastatin ameliorate memory deficits and inflammation in clinical and mouse model of Alzheimer's disease via modulating the expression of miR-106b. *Biomed. Pharmacother.* **2017**, *92*, 46–57. [CrossRef]

143. Li, S.H.; Gao, P.; Wang, L.T.; Yan, Y.H.; Xia, Y.; Song, J.; Li, H.Y.; Yang, J.X. Osthole Stimulated Neural Stem Cells Differentiation into Neurons in an Alzheimer's Disease Cell Model via Upregulation of MicroRNA-9 and Rescued the Functional Impairment of Hippocampal Neurons in APP/PS1 Transgenic Mice. *Front. Neurosci.* **2017**, *11*, 340. [CrossRef]
144. Jang, M.; Choi, S.H.; Choi, J.H.; Oh, J.; Lee, R.M.; Lee, N.E.; Cho, Y.J.; Rhim, H.; Kim, H.C.; Cho, I.H.; et al. Ginseng gintonin attenuates the disruptions of brain microvascular permeability and microvascular endothelium junctional proteins in an APPswe/PSEN-1 double-transgenic mouse model of Alphalzheimer's disease. *Exp. Ther. Med.* **2021**, *21*, 310. [CrossRef]
145. Shi, R.; Zhang, S.; Cheng, G.; Yang, X.; Zhao, N.; Chen, C. Ginsenoside Rg1 and Acori Graminei Rhizoma Attenuates Neuron Cell Apoptosis by Promoting the Expression of miR-873-5p in Alzheimer's Disease. *Neurochem. Res.* **2018**, *43*, 1529–1538. [CrossRef]
146. Caggiu, E.; Paulus, K.; Mameli, G.; Arru, G.; Sechi, G.P.; Sechi, L.A. Differential expression of miRNA 155 and miRNA 146a in Parkinson's disease patients. *Eneurologicalsci* **2018**, *13*, 1–4. [CrossRef] [PubMed]
147. Serafin, A.; Foco, L.; Zanigni, S.; Blankenburg, H.; Picard, A.; Zanon, A.; Giannini, G.; Pichler, I.; Facheris, M.F.; Cortelli, P.; et al. Overexpression of blood microRNAs 103a, 30b, and 29a in L-dopa-treated patients with PD. *Neurology* **2015**, *84*, 645–653. [CrossRef]
148. Schwienbacher, C.; Foco, L.; Picard, A.; Corradi, E.; Serafin, A.; Panzer, J.; Zanigni, S.; Blankenburg, H.; Facheris, M.F.; Giannini, G.; et al. Plasma and White Blood Cells Show Different miRNA Expression Profiles in Parkinson's Disease. *J. Mol. Neurosci.* **2017**, *62*, 244–254. [CrossRef] [PubMed]
149. Alieva, A.; Filatova, E.V.; Karabanov, A.V.; Illarioshkin, S.N.; Limborska, S.A.; Shadrina, M.I.; Slominsky, P.A. miRNA expression is highly sensitive to a drug therapy in Parkinson's disease. *Parkinsonism. Relat. Disord.* **2015**, *21*, 72–74. [CrossRef]
150. Waschbisch, A.; Atiya, M.; Linker, R.A.; Potapov, S.; Schwab, S.; Derfuss, T. Glatiramer acetate treatment normalizes deregulated microRNA expression in relapsing remitting multiple sclerosis. *PLoS ONE* **2011**, *6*, e24604. [CrossRef] [PubMed]
151. Singh, J.; Deshpande, M.; Suhail, H.; Rattan, R.; Giri, S. Targeted Stage-Specific Inflammatory microRNA Profiling in Urine During Disease Progression in Experimental Autoimmune Encephalomyelitis: Markers of Disease Progression and Drug Response. *J. Neuroimmune Pharmacol.* **2016**, *11*, 84–97. [CrossRef] [PubMed]
152. Bergman, P.; Piket, E.; Khademi, M.; James, T.; Brundin, L.; Olsson, T.; Piehl, F.; Jagodic, M. Circulating miR-150 in CSF is a novel candidate biomarker for multiple sclerosis. *Neurol. Neuroimmunol. Neuroinflamm.* **2016**, *3*, e219. [CrossRef] [PubMed]
153. Fattahi, M.; Rezaei, N.; Nematalahi, F.S.; Shaygannejad, V.; Fouladi, S.; Karimi, L.; Fathi, F.; Dehghani, L.; Mirmosayyeb, O.; Eskandari, N. MicroRNA-29b variants and MxA expression change during interferon beta therapy in patients with relapsing-remitting multiple sclerosis. *Mult. Scler. Relat. Disord.* **2019**, *35*, 241–245. [CrossRef]
154. Meira, M.; Sievers, C.; Hoffmann, F.; Derfuss, T.; Kuhle, J.; Kappos, L.; Lindberg, R.L. MiR-126: A novel route for natalizumab action? *Mult. Scler.* **2014**, *20*, 1363–1370. [CrossRef] [PubMed]
155. Sievers, C.; Meira, M.; Hoffmann, F.; Fontoura, P.; Kappos, L.; Lindberg, R.L. Altered microRNA expression in B lymphocytes in multiple sclerosis: Towards a better understanding of treatment effects. *Clin. Immunol.* **2012**, *144*, 70–79. [CrossRef]
156. Ingwersen, J.; Menge, T.; Wingerath, B.; Kaya, D.; Graf, J.; Prozorovski, T.; Keller, A.; Backes, C.; Beier, M.; Scheffler, M.; et al. Natalizumab restores aberrant miRNA expression profile in multiple sclerosis and reveals a critical role for miR-20b. *Ann. Clin. Transl. Neurol.* **2015**, *2*, 43–55. [CrossRef]
157. Michell-Robinson, M.A.; Moore, C.S.; Healy, L.M.; Osso, L.A.; Zorko, N.; Grouza, V.; Touil, H.; Poliquin-Lasnier, L.; Trudelle, A.M.; Giacomini, P.S.; et al. Effects of fumarates on circulating and CNS myeloid cells in multiple sclerosis. *Ann. Clin. Transl. Neurol.* **2016**, *3*, 27–41. [CrossRef] [PubMed]
158. Eftekharian, M.M.; Komaki, A.; Mazdeh, M.; Arsang-Jang, S.; Taheri, M.; Ghafouri-Fard, S. Expression Profile of Selected MicroRNAs in the Peripheral Blood of Multiple Sclerosis Patients: A Multivariate Statistical Analysis with ROC Curve to Find New Biomarkers for Fingolimod. *J. Mol. Neurosci.* **2019**, *68*, 153–161. [CrossRef] [PubMed]
159. Vaknin-Dembinsky, A.; Charbit, H.; Brill, L.; Abramsky, O.; Gur-Wahnon, D.; Ben-Dov, I.Z.; Lavon, I. Circulating microRNAs as biomarkers for rituximab therapy, in neuromyelitis optica (NMO). *J. Neuroinflamm.* **2016**, *13*, 179. [CrossRef]
160. Ebrahimkhani, S.; Beadnall, H.N.; Wang, C.; Suter, C.M.; Barnett, M.H.; Buckland, M.E.; Vafaee, F. Serum Exosome MicroRNAs Predict Multiple Sclerosis Disease Activity after Fingolimod Treatment. *Mol. Neurobiol.* **2020**, *57*, 1245–1258. [CrossRef]
161. Fattahi, M.; Eskandari, N.; Sotoodehnejadnematalahi, F.; Shaygannejad, V.; Kazemi, M. Comparison of The Expression of miR-326 between Interferon beta Responders and Non-Responders in Relapsing-Remitting Multiple Sclerosis. *Cell J.* **2020**, *22*, 92–95. [CrossRef]
162. De Felice, B.; Mondola, P.; Sasso, A.; Orefice, G.; Bresciamorra, V.; Vacca, G.; Biffali, E.; Borra, M.; Pannone, R. Small non-coding RNA signature in multiple sclerosis patients after treatment with interferon-beta. *BMC Med. Genom.* **2014**, *7*, 26. [CrossRef] [PubMed]
163. Munoz-Culla, M.; Irizar, H.; Castillo-Trivino, T.; Saenz-Cuesta, M.; Sepulveda, L.; Lopetegi, I.; Lopez de Munain, A.; Olascoaga, J.; Baranzini, S.E.; Otaegui, D. Blood miRNA expression pattern is a possible risk marker for natalizumab-associated progressive multifocal leukoencephalopathy in multiple sclerosis patients. *Mult. Scler.* **2014**, *20*, 1851–1859. [CrossRef]
164. Fenoglio, C.; De Riz, M.; Pietroboni, A.M.; Calvi, A.; Serpente, M.; Cioffi, S.M.; Arcaro, M.; Oldoni, E.; Scarpini, E.; Galimberti, D. Effect of fingolimod treatment on circulating miR-15b, miR23a and miR-223 levels in patients with multiple sclerosis. *J. Neuroimmunol.* **2016**, *299*, 81–83. [CrossRef] [PubMed]
165. Indrasekara, A.S.D.S.; Wadams, R.C.; Fabris, L. Ligand Exchange on Gold Nanorods: Going Back to the Future. *Part. Parcle Syst. Charact.* **2014**, *31*, 819–838. [CrossRef]

166. de Lau, L.M.; Koudstaal, P.J.; Witteman, J.C.; Hofman, A.; Breteler, M.M. Dietary folate, vitamin B12, and vitamin B6 and the risk of Parkinson disease. *Neurology* **2006**, *67*, 315–318. [CrossRef]
167. Kim, J.W.; Galanzha, E.I.; Zaharoff, D.A.; Griffin, R.J.; Zharov, V.P. Nanotheranostics of circulating tumor cells, infections and other pathological features in vivo. *Mol. Pharm.* **2013**, *10*, 813–830. [CrossRef]
168. Rahman, M.; Akhter, S.; Ahmad, M.Z.; Ahmad, J.; Addo, R.T.; Ahmad, F.J.; Pichon, C. Emerging advances in cancer nanotheranostics with graphene nanocomposites: Opportunities and challenges. *Nanomedicine* **2015**, *10*, 2405–2422. [CrossRef]
169. Li, W.; Luo, R.; Lin, X.; Jadhav, A.D.; Zhang, Z.; Yan, L.; Chan, C.Y.; Chen, X.; He, J.; Chen, C.H.; et al. Remote modulation of neural activities via near-infrared triggered release of biomolecules. *Biomaterials* **2015**, *65*, 76–85. [CrossRef]
170. Xue, Y.; Wu, J.; Sun, J. Four types of inorganic nanoparticles stimulate the inflammatory reaction in brain microglia and damage neurons in vitro. *Toxicol. Lett.* **2012**, *214*, 91–98. [CrossRef] [PubMed]
171. Sunderland, K.S.; Yang, M.; Mao, C. Phage-Enabled Nanomedicine: From Probes to Therapeutics in Precision Medicine. *Angew. Chem. Int. Ed.* **2017**, *56*, 1964–1992. [CrossRef] [PubMed]
172. Soliman, H.M.; Ghonaim, G.A.; Gharib, S.M.; Chopra, H.; Farag, A.K.; Hassanin, M.H.; Nagah, A.; Emad-Eldin, M.; Hashem, N.E.; Yahya, G.; et al. Exosomes in Alzheimer's Disease: From Being Pathological Players to Potential Diagnostics and Therapeutics. *Int. J. Mol. Sci.* **2021**, *22*, 794. [CrossRef] [PubMed]
173. Tran, P.H.L.; Xiang, D.; Tran, T.T.D.; Yin, W.; Zhang, Y.; Kong, L.; Chen, K.; Sun, M.; Li, Y.; Hou, Y.; et al. Exosomes and Nanoengineering: A Match Made for Precision Therapeutics. *Adv. Mater.* **2020**, *32*, e1904040. [CrossRef] [PubMed]
174. Cavalu, S.; Ratiu, C.; Ponta, O.; Simon, V.; Rugina, D.; Miclaus, V.; Akin, I.; Goller, G. Improving osseointegration of alumina/zirconia ceramic implants by fluoride surface treatment. *Dig. J. Nanomater. Bios.* **2014**, *9*, 797–808.
175. Kandell, R.M.; Waggoner, L.E.; Kwon, E.J. Nanomedicine for Acute Brain Injuries: Insight from Decades of Cancer Nanomedicine. *Mol. Pharm.* **2021**, *18*, 522–538. [CrossRef] [PubMed]

materials

MDPI

Article

Additives Imparting Antimicrobial Properties to Acrylic Bone Cements

Alina Robu [1], Aurora Antoniac [1], Elena Grosu [1], Eugeniu Vasile [2], Anca Daniela Raiciu [3,4], Florin Iordache [5], Vasile Iulian Antoniac [1,6], Julietta V. Rau [7,8,*], Viktoriya G. Yankova [8], Lia Mara Ditu [9,*] and Vicentiu Saceleanu [10]

1 Faculty of Material Science and Engineering, UniversityPolitehnica of Bucharest, 313 Splaiul Independentei Street, District 6, 060042 Bucharest, Romania; alinarobu2021@gmail.com (A.R.); antoniac.aurora@gmail.com (A.A.); elena_grosu@yahoo.com (E.G.); antoniac.iulian@gmail.com (V.I.A.)
2 Faculty of Applied Chemistry and Material Science, University Politehnica of Bucharest, 313 Splaiul Independentei Street, District 6, 060042 Bucharest, Romania; eugeniuvasile@yahoo.com
3 Faculty of Pharmacy, Titu Maiorescu University, 22 Dambovnicului Street, District 4, 040441 Bucharest, Romania; daniela_raiciu@yahoo.com
4 S.C. Hofigal S.A., 2 Intrarea Serelor Street, District 4, 042124 Bucharest, Romania
5 Faculty of Veterinary Medicine, University of Agronomic Sciences and Veterinary Medicine Bucharest, 59 Bulevardul Marasti, 050097 Bucharest, Romania; floriniordache84@yahoo.com
6 Academy of Romanian Scientists, 54 Splaiul Independentei Street, District 5, 050094 Bucharest, Romania
7 Istituto di Struttura della Materia, Consiglio Nazionale delle Ricerche (ISM-CNR), Via del Fosso del Cavaliere 100, 00133 Rome, Italy
8 Institute of Pharmacy, Department of Analytical, Physical and Colloid Chemistry, I.M. Sechenov First Moscow State Medical University, 8 Trubetskaya Street, Build. 2, 119991 Moscow, Russia; yankova_v_g@staff.sechenov.ru
9 Faculty of Biology, University of Bucharest, Intrarea Portocalelor Street, no. 1-3, 060101 Bucharest, Romania
10 Faculty of Medicine, University Lucian Blaga Sibiu, 2A Lucian Blaga Street, 550169 Sibiu, Romania; vicentiu.saceleanu@ulbsibiu.ro
* Correspondence: giulietta.rau@ism.cnr.it (J.V.R.); lia-mara.ditu@bio.unibuc.ro (L.M.D.)

Citation: Robu, A.; Antoniac, A.; Grosu, E.; Vasile, E.; Raiciu, A.D.; Iordache, F.; Antoniac, V.I.; Rau, J.V.; Yankova, V.G.; Ditu, L.M.; et al. Additives Imparting Antimicrobial Properties to Acrylic Bone Cements. *Materials* **2021**, *14*, 7031. https://doi.org/10.3390/ma14227031

Academic Editor: Elena de la Fuente González

Received: 18 October 2021
Accepted: 16 November 2021
Published: 19 November 2021

Abstract: PMMA bone cements are mainly used to fix implanted prostheses and are introduced as a fluid mixture, which hardens over time. The problem of infected prosthesis could be solved due to the development of some new antibacterial bone cements. In this paper, we show the results obtained to develop four different modified PMMA bone cements by using antimicrobial additives, such as gentamicin, peppermint oil incorporated in hydroxyapatite, and silver nanoparticles incorporated in a ceramic glass matrix (2 and 4%). The structure and morphology of the modified bone cements were investigated by SEM and EDS. We perform experimental measurements on wettability, hydration degree, and degradation degree after immersion in simulated body fluid. The cytotoxicity was evaluated by MTT assay using the human MG-63 cell line. Antimicrobial properties were checked against standard strains *Staphylococcus aureus*, *Pseudomonas aeruginosa*, and *Candida albicans*. The addition of antimicrobial agents did not significantly affect the hydration and degradation degree. In terms of biocompatibility assessed by the MTT test, all experimental PMMA bone cements are biocompatible. The performance of bone cements with peppermint essential oil and silver nanoparticles against these two pathogens suggests that these antibacterial additives look promising to be used in clinical practice against bacterial infection.

Keywords: PMMA bone cements; peppermint essential oil; silver nanoparticles; gentamicin; antimicrobial properties

1. Introduction

Bone cements are used to fix artificial joints, such as hip or knee prostheses. These materials fill the space between the implant and the joint and are introduced as a fluid mixture, which hardens over time. Bone cements can also be defined as a mixture of substances, a family of materials consisting of a powder phase and a liquid phase which, after mixing and homogenization, forms a paste which has the ability to cure, and self-stabilizing once implanted in the body. This material has a special flexibility and modelling, which ensures the fixation of the material at the implant site and a good bone-material contact, even in complex defects from a geometric point of view [1].

Acrylic bone cements are polymeric materials, which are obtained by a polymerization reaction and produce a stable, non-absorbable material [2–4]. Poly(methyl methacrylate) (PMMA) based bone cements are two-component systems, comprising a solid phase (a polymeric powder) and a liquid phase (a liquid monomer). It also contains an inhibitor (hydroquinone) to prevent premature polymerization during storage and optionally a colouring agent, such as chlorophyll [5].

An implanted prosthesis is particularly sensitive to the development of bacteria on its surface because microorganisms can multiply as an adverse reaction of the body's immune system on a foreign body or by primary contamination due to non-sterile materials or by hematogenous spread, where the infection comes from another infected area of the body and is transferred to the implant area. As the bacteria reproduce, they form a protective biofilm, which has a low sensitivity to antibiotics [6]. The release of antibiotics from the cement matrix takes place through surface diffusion, bulk diffusion, pores, and defects in the cement start. The mixing method, particle size, and uniformity of distribution significantly influence the antibiotic release kinetics [7].

Bacteria that can cause periprosthetic joint infections have a broad spectrum, most often caused by Coagulase negative staphylococci (in 30–43% of cases), followed by *Staphylococcus aureus* (12–23%) and Streptococci (9–10%). Other bacteria encountered in rare cases that can cause infections are represented by Enterococci, Gram-negative bacilli, anaerobic, polymicrobials, or others unknown microbes [8,9]. Currently, research in this field has gathered the best evidence for an adequate selection of antibiotics available especially for staphylococcal species, while, for other bacteria, such as streptococci, enterococci, or Gram-negative, the selection of antibiotics is more complex [10]. Different antibacterial agents were used by researchers for obtaining modified PMMA bone cements with antibacterial properties: antibiotics [11–21], silver nanoparticles [22–25], gold nanoparticles [26], hydroxyapatite [27–31], graphene [32,33], essential oil [34–42], or essential oils incorporated in different materials [43–46]. A schematic representation of cement preparation procedure is shown in Figure 1.

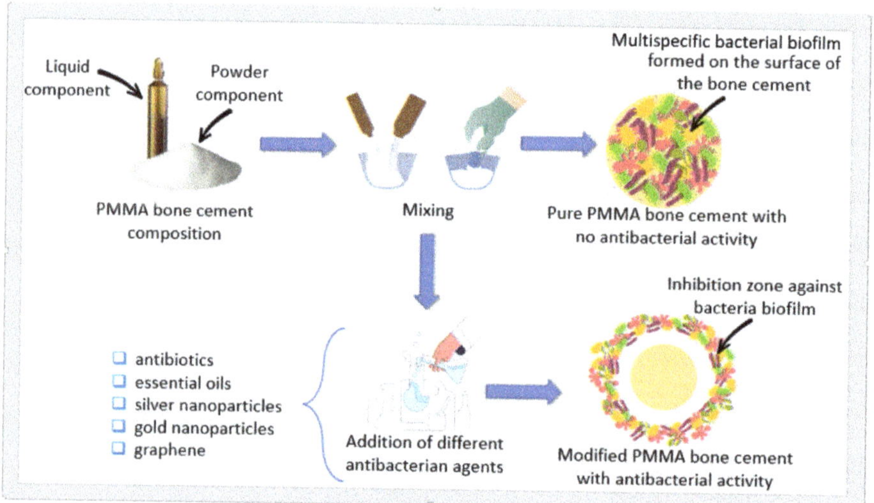

Figure 1. Schematic representation of the modified PMMA bone cements preparation procedure with antibacterial properties.

Gentamicin (GM) is the most commonly used antibiotic for treating bacterial infections that has a broad spectrum, thermal stability, and high solubility in water [11]. It is an aminoglycoside whose antibacterial activity is concentration dependent [12]. Other common antibiotics, such as tobramycin [13] and vancomycin [14], have also been used to treat periprosthetic joint infections. In order to increase the antimicrobial spectrum and, thus, the effectiveness of the treatment of periprosthetic joint infections, researchers tested the effect of acrylic bone cements with gentamicin + vancomycin + linezoid [15], gentamicin + teicoplanin [16], vancomycin + daptomycin [17], and vancomycin + cefazolin [18], which provide satisfactory results on their antibacterial activity.

However, the widespread use of antibiotics has led to the development of genetic and biochemical mechanisms that allow bacteria to survive in antibiotic environments. This is a cause for concern regarding the effectiveness of antibiotics commonly used in the composition of bone cements, especially the effectiveness of gentamicin. For example, in a study realized by Hellmark et al. [19] on joint prosthesis revision cases, a 79% resistance to gentamicin against a *Staphylococcus Epidermidis* was detected. Thornes et al. [20] confirmed these results with their research realized on rat models, which demonstrated an increased resistance of same bacteria to PMMA bone cement with gentamicin. Thus, the problem of low antibiotic efficacy in bone cement has created the need to investigate the potential to incorporate new antimicrobial agents into bone cements.

The availability of silver nanoparticles having a high surface to volume ratio and possessing which gives unique chemical and physical properties that greatly enhance the antimicrobial effects of silver. Silver ions inactivate enzymes vital to bacteria and disactivate the mechanism of bacterial DNA replication. Antimicrobial testing has also shown that cements with silver nanoparticles do not have antimicrobial activity against planktonic bacteria; however, they are able to quantitatively reduce the formation of bacterial biofilms on the surface of the cement. These results imply that cements containing silver nanoparticles have a high potential for use in arthroplasty interventions, in which it is necessary to prevent bacterial adhesion [22–25].

Other antimicrobial additives have also been studied to improve the antimicrobial activity of PMMA bone cements. For example, Russo et al. [26] used gold nanoparticles and demonstrated their antibacterial activity against Methicillin Resistant *Staphylococcus aureus* (MRSA) and *Pseudomonas aeruginosa* strains. Phakatkar et al. [27] introduced magnesium phosphate and hydroxyapatite into PMMA bone cement and obtained satisfying results on antibacterial properties and superior bioactivity related to cell viability due to the addition of hydroxyapatite. The antimicrobial effect of graphene embedded in acrylic resins used in the dental field is discussed in a review by Radhi et al. [32].

In the past couple of years, the use of plant-derived molecules in composition of some synthetic biomaterials for wound healing has started to gain widespread acceptance. Although the use of essential oils for medical applications is relatively new, it has raised interest being based on their long history of being used by humans for therapeutic purposes showing inhibition against bacteria (Gram negative and Gram positive) and fungi [34,35]. This interest in medication with straight source plants is due to the belief that natural components are safer and more dependable, compared with synthetic drugs that sometimes can have adverse effects [36]. Essential oils based on plant extracts, such as peppermint, cinnamon, lemon, and clove, have been applied to accelerate the wound healing process since ancient times due to their various therapeutic benefits, such as antibacterial activity, anti-inflammatory, and antioxidation potential [37]. However, the volatility, low stability, and high sensitivity to environmental factors have limited the usage of essential oils in wound healing applications, but, in the last decade, researchers have focused their attention on biomaterials containing essential oils to study their therapeutic properties [38].

A commonly considered essential oil is the one obtained from peppermint. Peppermint is mainly derived from the *Mentha piperita* leaf, and its antibacterial activity is due to its chemical composition of L-menthol, menthone, methyl acetate, and limonene [39,40]. Previous studies on the antimicrobial properties of organic or inorganic supports coated with peppermint essential oils reported similar results regarding their biological activity against pathogenic microorganisms [41,42]. Thus, the antibacterial activity of the hydroxyapatite coated with peppermint essential oil (HAp-P) were tested against different bacteria strains, such as *Staphylococcus aureus*, *Escherichia coli*, *Pseudomonas aeruginosa*, and *Candida albicans strains*, with the results highlighting that the antimicrobial activity of HAp-P increased significantly over that of hydroxyapatite alone [43]. Peppermint extract was also incorporated in the polyurethane (PU)-based nanofibers in order to be used in diabetic wound healing. The results showed that the release of extract from nanofibers continued for 144 h, with antibacterial efficiency of 99.9% against *Staphylococcus aureus* and *Escherichia coli* bacteria [44]. Kasiri and Fathi [45] studied the method for peppermint essential oil encapsulation with cellulose nanocrystals (CNC), loading capacity, and the *in vitro* release of the PEO, demonstrating that the main mechanism for release of peppermint oil is by diffusion transport.

The main objective of the paper was to develop novel modified PMMA bone cements with antimicrobial properties using hydroxyapatite impregnated with peppermint oil, silver nanoparticles incorporated in ceramic glass matrix, and gentamicin as antimicrobial agents. The novelty is consistent due to the antimicrobial agents, such as essential oil, incorporated in hydroxyapatite and silver nanoparticles incorporated in a ceramic glass used for bone cement. The SEM characterization, contact angle measurements, hydration, and degradation degree experiments were carried out. The *in vitro* tests were performed to demonstrate cements safety and efficacy. The biocompatibility was assessed by the MTT test using human MG-63 cell line. The antimicrobial tests of the developed new cement compositions against gram-positive *Staphylococcus aureus*, gram-negative *Pseudomonas aeruginosa*, and *Candida albicans* fungus microorganisms were performed.

2. Materials and Methods

The experimental modified PMMA bone cements were obtained by modifying the solid component of commercial cement based on PMMA, with standard viscosity.

Different bone cements-type biomaterials were developed by combining different antimicrobial additives with the powder of the commercial PMMA bone cement, in order to determine the antimicrobial activity of the cement without negatively altering its mechanical performance. As antimicrobial agents, silver nanoparticles incorporated in a ceramic glass matrix (purchased from SANITIZED AG, Burgdorf, Switzerland) were used, respectively, hydroxyapatite (HAp, >95% purity, purchased from the Plasma Biotal Limited (Tideswell, UK)) impregnated with peppermint oil and gentamicin, being obtained 4 new bone cement compositions.

The chemical composition of commercial cement used as a reference and new antimicrobial bone cements are shown in Table 1 below.

Table 1. Composition of the standard and new antimicrobial bone cements.

Samples	Composition		Antimicrobial Additive
	Powder	Liquid	
R	PMMA, BPO, $BaSO_4$	MMA, BMA, DmpT, HQ	None
GM	PMMA, BPO, $BaSO_4$	MMA, BMA, DmpT, HQ	5% gentamicin
HUM	PMMA, BPO, $BaSO_4$, HAp	MMA, BMA, DmpT, HQ	5% peppermint essential oil incorporated in hydroxyapatite
AM1	PMMA, BPO, $BaSO_4$	MMA, BMA, DmpT, HQ	2% silver nanoparticles incorporated in a ceramic glass
AM2	PMMA, BPO, $BaSO_4$	MMA, BMA, DmpT, HQ	4% silver nanoparticles incorporated in a ceramic glass

Methyl methacrylate—MMA; Butyl methacrylate—BMA; *N*,*N*-Dimethyl-*p*-toluidine—DmpT; Hydroquinone—HQ; Poly(methyl-methacrylate)—PMMA, Benzoyl peroxide—BPO, Barium sulphate—$BaSO_4$; Hydroxyapatite—HAp.

Scanning Electron Microscopy (FEI Company, Eindhoven, Netherlands)coupled with Energy Dispersive Spectroscopy analysis on bone cements samples was performed using a QUANTA INSPECT F Scanning Electron Microscope (FEI Company, Eindhoven, Netherlands) equipped with Energy Dispersive X-ray Spectrometer Detector (EDAX) (FEI Company, Eindhoven, Netherlands) with a 132 eV resolution at MnKα. Analyses were performed in order to evaluate the morphology and antimicrobial additives distribution in the polymerized matrix.

The ***contact angle*** measurements were performed on the KRÜSS DSA30 Drop Shape Analysis System, in order to determine the wetting degree of solid and liquid interaction. The contact angle measurements were based on the sessile drop method, each measurement being repeated five times.

The ***hydration degree (Ha)*** was evaluated by immersing the experimental bone cement samples in distilled water at 37 °C for a period of 7, 14, and 21 days, respectively. At each time interval, the samples were removed from the distilled water, weighed and kept in a desiccator until they reached a constant mass. The degree of hydration (Ha) was determined using the following formula:

$$H_a\ (\%) = \left(\frac{m_w - m_f}{m_0}\right) \times 100, \tag{1}$$

where:

m_w—mass of the wet bone cement samples,
m_f—mass of the bone cement samples after drying, and
m_0—initial mass of the bone cement samples.

The ***degradation degree (DR)*** test involved the immersion of the bone cement samples in solution of simulated body fluid (SBF) at 37 °C for 1, 3, 5, 7, 14, 21, and 28 days, respectively. The SBF with an ionic composition similar to human blood plasma was prepared in the laboratory (SBF chemical composition: 8.035 g/L NaCl; 0.355 g/L $NaHCO_3$; 0.225 g/L KCl; 0.231 g/L K_2HPO_4•$3H_2O$; 0.311 g/L $MgCl_2$•$6H_2O$; 39 mL 1M HCl; 0.292 g/L $CaCl_2$; 0.072 g/L Na_2SO_4; 6.118 g/L Tris-$NH_2C(CH_2OH)_3$; substances were purchased from Sigma Aldrich, Taufkirchen, Germany). In addition, at each time interval, the samples were removed from the medium, weighed, and kept in a desiccator until they reached a constant mass. Degradation rates (DR) of the experimental samples were determined with the formula:

$$DR\ (\%) = \left(\frac{m_0 - m_f}{m_o}\right) \times 100, \tag{2}$$

where:

m_f—mass of the bone cement samples after drying; and
m_0—initial mass of the bone cement samples.

Antimicrobial tests. Microbial strains used in the experiments were represented by standard strains of *Staphylococcus aureus* ATCC 25923, *Pseudomonas aeruginosa* ATCC 27853, and *Candida albicans* ATCC 10231. To avoid the impact of contaminants on the experiment, the samples were previously sterilized by UV radiation (Benchmark Scientific, Sayreville, NJ, USA), for 30 min.

The qualitative screening was performed using an adapted spot diffusion method as a standardized method recommended to investigate the antimicrobial activity of different compounds, including antibiotics (according with Performance Standards for Antimicrobial Susceptibility Testing, Clinical and Laboratory Standard Institute 2021). Microbial suspensions of 1.5 × 108 CFU/mL (corresponding to 0.5 McFarland density), obtained from 24 h microbial cultures developed on corresponding agar media, were used in the experiments. Petri dishes with Muller Hinton agar (for bacterial strains) and Sabouraud agar (for yeast strain) were seeded with microbial inoculums. Subsequently, samples of bone cements with a diameter of 1 cm were placed on the surface of the medium and gently pressed to be in direct contact with the medium. After their free diffusion, the plates were incubated for 16–18 h at room temperature. The sensitivity of microbial strains was assessed by measuring the diameters of the inhibition zones around the bone cements samples and expressing them in "mm".

Bacterial adherence assay using viable cell count method. Quantitative assessment of the capacity of the selected strains to adhere on the surface of the tested samples was performed using the 24 multi-well plates. Overnight bacterial cultures of *Staphylococcus aureus* ATCC 25923, *Pseudomonas aeruginosa* ATCC 27853, and *Candida albicans* ATCC 10231 were diluted in fresh nutrient broth media at 1.5 × 105 CFU/mL final density. Two millilitres of the obtained suspension were seeded in 24 multi-well plates containing the treated materials, previously sterilized by UV irradiation. The plates were incubated at 37 °C, for 24, 48, and 72 h, the period during which bacterial cells are multiplied and, after reaching a threshold density, begin to adhere to the surface of the treated material and generate monospecific biofilm. For the adherence assay, after the incubation period, the materials were gently washed with sterile phosphate buffered saline (PBS, Sigma-Aldrich, St. Louis, MO, USA) in order to remove the non-adherent microbial cells and placed in 14 mL centrifuge tubes containing 1 mL of sterile PBS. The samples were vigorously mixed by vortexing for 1 min and sonicated for 10 s. Serial dilutions obtained from each sample were inoculated on LB agar plates in triplicates, and viable cell counts (VCCs) were assessed after incubation at 37 °C, for 24 h [34].

Evaluation of biocompatibility properties of experimental bone cements samples. The human MG-63 cell line (ATCC CRL-1427, Manassas, VA, USA, Sigma-Aldrich) were used to evaluate the biocompatibility of 5 types of bone cement samples. Human MG-63 cell line is a stable Human osteosarcoma cell line and provides representative models to study the cytotoxicity of acrylic bone cement extracts to fabricate three-dimensional (3-D) niches and to study the cell viability, adhesion, and proliferation (product recommendations) [47]. The cells were cultured in Dulbecco's Modified Eagle's Medium (DMEM) medium (Sigma-Aldrich, MO, USA) supplemented with 10% fetal bovine serum, 1% penicillin, and 1% streptomycin antibiotics (Sigma-Aldrich, St. Louis, MO, USA). To maintain optimal culture conditions, medium was changed twice a week. The biocompatibility was assessed using MTT assay (Vybrant® MTT Cell Proliferation Assay Kit, Thermo Fischer Scientific, Waltham, MA, USA). The viable cells reduce yellow tetrazolium salt MTT (3-(4,5 dimethylthiazole)-2,5-diphenyltetrazolium bromide) to a dark blue formazan via mitochondrial enzymes. Briefly, MG-63 cells were grown in 24-well plates, with a seeding density of 10.000 cells/well in the presence of bone cements for 72 h. Then, 15 mL of Solution I was added and incubated at 37 °C for 4 h. Solution II was added and pipettes vigorously to solubilise formazan crystals. After 1 h, the absorbance was read using spectrophotometer at 570 nm (TECAN Infinite M200, Männedorf, Switzerland).

Fluorescent microscopy for tracing of living cells. The biocompatibility between the bones cements samples and human MG-63 cell line was also evaluated based on fluorescent microscopy using RED CMTPX fluorophore (Thermo Fischer Scientific, MA, USA). The CMTPX tracker was added in cell culture in the presence of bone cements, and the viability and morphology of the cells was evaluated after 5 days. The CMTPX fluorophore was added in the culture medium at a final concentration of 5 μM and incubated for 30 min in order to allow the dye penetration into the cells. Next, the cells were washed with PBS and visualized by fluorescent microscopy. The photomicrographs were taken with Olympus CKX 41 digital camera driven by Cell Sense Entry software (Olympus, Tokyo, Japan).

Alizarin Red assay. To quantify the osteogenic potential of bone cements samples, MG-63 cells were cultivated for 21 days in the presence of bone cement samples and stained with Alizarin Red dye. The cells were fixed with paraformaldehyde (4%, 2 h), washed three times with PBS, and then incubated with 40 nmol/L Alizarin Red S. This dye binds to calcium crystals in cells or matrix fibres, revealing a red colour. Unbound solution was washed out under gentle rocking. Bound dye was then released from the cells using 10% acetic acid incubation (1h), followed by neutralization with 10% ammonium hydroxide.

The concentration of dye in solution was then quantified using absorbance spectroscopy at 405 nm wavelength (TECAN Infinite M200, Männedorf, Switzerland).

3. Results and Discussion

3.1. SEM Analysis

In Figure 2, the morphologies of the obtained bone cements sample revealed by SEM microscopy images and corresponding EDS spectra are presented. The SEM analysis revealed that, after hardening, all the bone cements materials displayed the well-known microstructure typical for PMMA-based cements, in which four phases are highlighted: beads from the polymer powder; pores; the polymerized monomer; and the radiopacifying element, in this case, barium sulphate ($BaSO_4$) [48–51]. SEM images show a good dispersion of the barium sulphate and antimicrobial additives into the PMMA-PBMA matrix (poly(methyl methacrylate)-poly(butyl methacrylate) matrix). A slight tendency of antimicrobial agents to form agglomerates can be observed, mostly in the bone cements samples containing 4% silver nanoparticles (sample AM2). The EDS analysis supports this observation and highlights the presence of the majority elements, C and O, from the two polymer phases; Ba and S from the composition of the radiopacifying element, respectively; and Ag, P, Ca, and Mg from HAp and silver nanoparticles incorporated in a ceramic glass matrix.

3.2. Contact Angle

The behaviour regarding the wettability of the surface is influenced by the type of antimicrobial additives used, as can be seen in Figure 3. For testing, 3 types of liquids were used, namely: water, di-iodomethane (DIM), and ethylene glycol (EG). The behaviour regarding the wettability of the surface is influenced by both the type of sample and the type of liquid used, as can be seen in Table 2. The most hydrophobic character can be observed in the case of the GM sample, which is more accentuated than in the case of the reference sample. The HUM, AM1, and AM2 samples have a more hydrophilic character, which is caused by the ceramic components (hydroxyapatite and ceramic glass) that have been added with the respective antimicrobial agents. It is known as the optimal contact angle value for cell adhesion 55° and for bone regeneration values between 35–80° [52,53]. All additives used kept the contact angle values within the desired limits so that they remain hydrophilic.

3.3. Hydration Degree

Water absorption can influence the mechanical properties of bone cement because the fluid acts as a plasticizer inside the polymer matrix. In addition, water absorption can lead to the dissolution and release of water-soluble materials from the bone cement composition, such as antimicrobial agents or residual monomer. The absorbed water can lighten internal strains and allow the extraction of residual monomer and added additives. Moreover, water can penetrate more easily into a porous, heterogeneous structure with the interface area between the polymer and filler well-defined. The more heterogeneous areas in the structure of the material are the many spaces through which water can be accumulated. Figure 4 graphically represents the evolution of the hydration degree (Ha) evaluated in distilled water at a temperature of 37 °C over a period of 21 days, for all experimental samples obtained from bone cements based on PMMA.

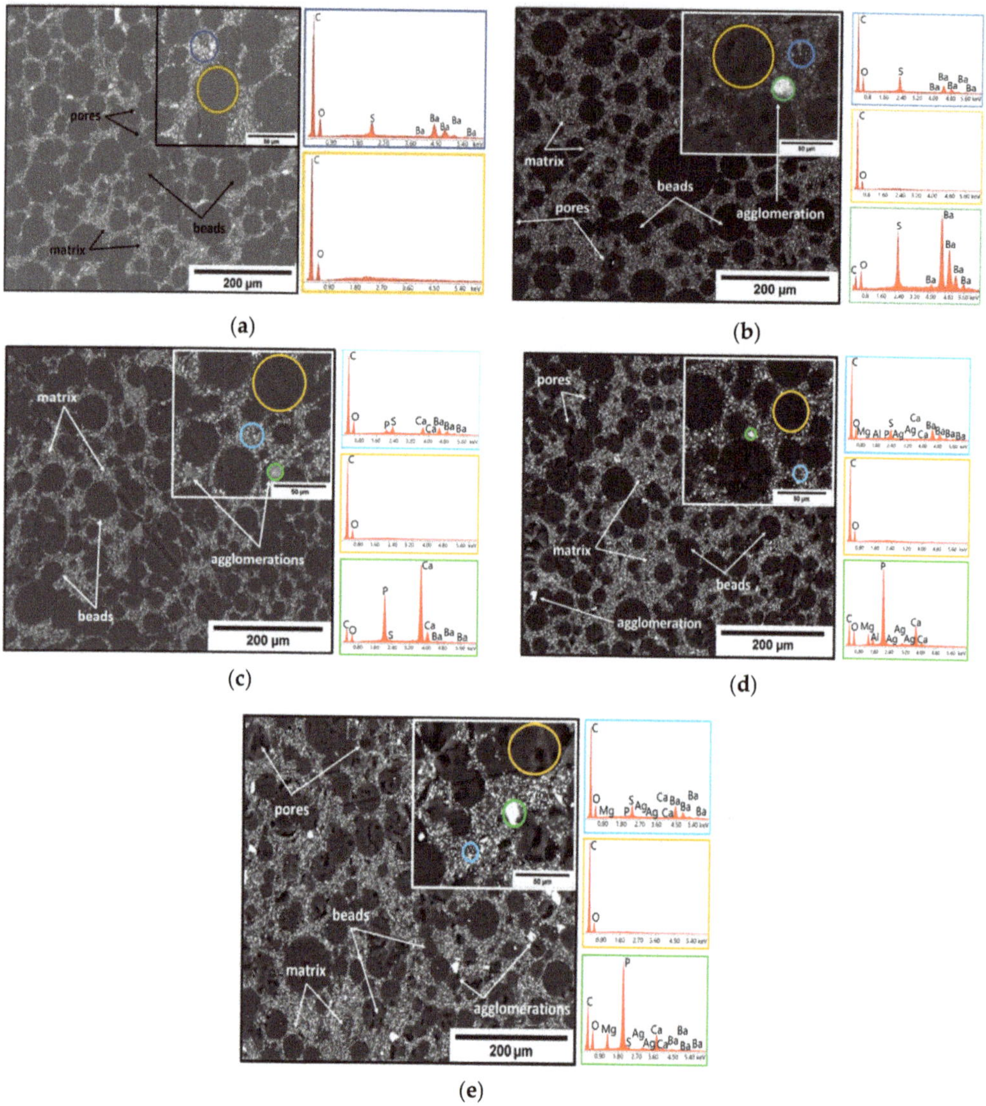

Figure 2. Representative SEM images and corresponding EDS spectra obtained from the highlighted areas of the experimental samples: (**a**) R sample; (**b**) GM sample; (**c**) HUM sample; (**d**) AM1 sample; (**e**) AM2 sample.

Table 2. Contact angle values for experimental bone cement samples.

	R	GM	HUM	AM1	AM2
Contact angle, ϴ[°], water	71	85	58	55	67
Contact angle, ϴ[°], DIM	27	40	31	51	41
Contact angle, ϴ[°], EG	57	50	50	53	46

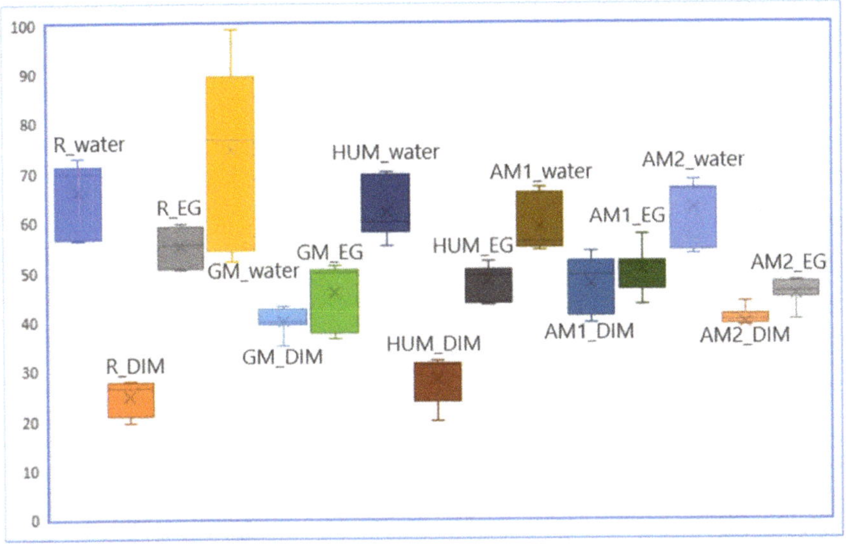

Figure 3. Surface wettability of the investigated bone cement samples determined by the measurements of the contact angle with distilled water, DIM, and EG.

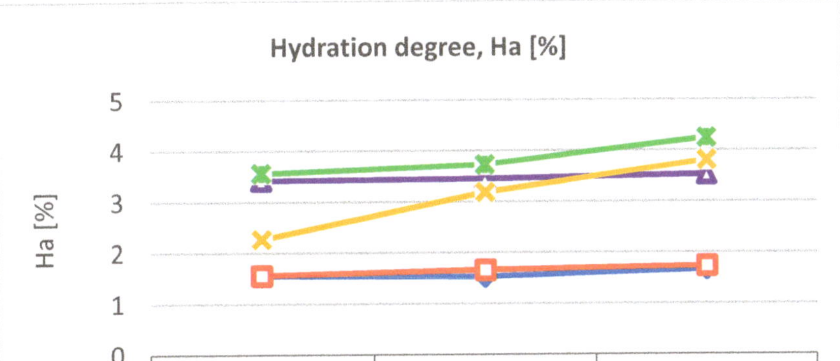

	7	14	21
R	1.5429	1.5228	1.6589
GM	1.5525	1.6544	1.7226
HUM	3.4164	3.4456	3.5205
AM1	2.2648	3.1823	3.8019
AM2	3.5545	3.7249	4.2373

Time [days]

Figure 4. Variation of hydration degree over a period of 21 days in distilled water, at 37 °C, for experimental bone cements samples.

The results show an increase in the hydration degree (Ha) of bone cements with the addition of antimicrobial agent and the amount of antimicrobial agent in AM1 and AM2 samples. Compared to the standard sample R, the values of the hydration degree (Ha) increased less in the case of the sample with gentamicin (GM sample). The largest amount of water is subsequently absorbed during the first 7 days after immersion; the values remain almost constant in the case of the reference R, GM, and HUM samples. A variation is observed in the case of samples containing silver nanoparticles in a dose-dependent manner. After exposure of the investigated bone cement samples to distilled water, no significant change in their structure was observed (Figure 5).

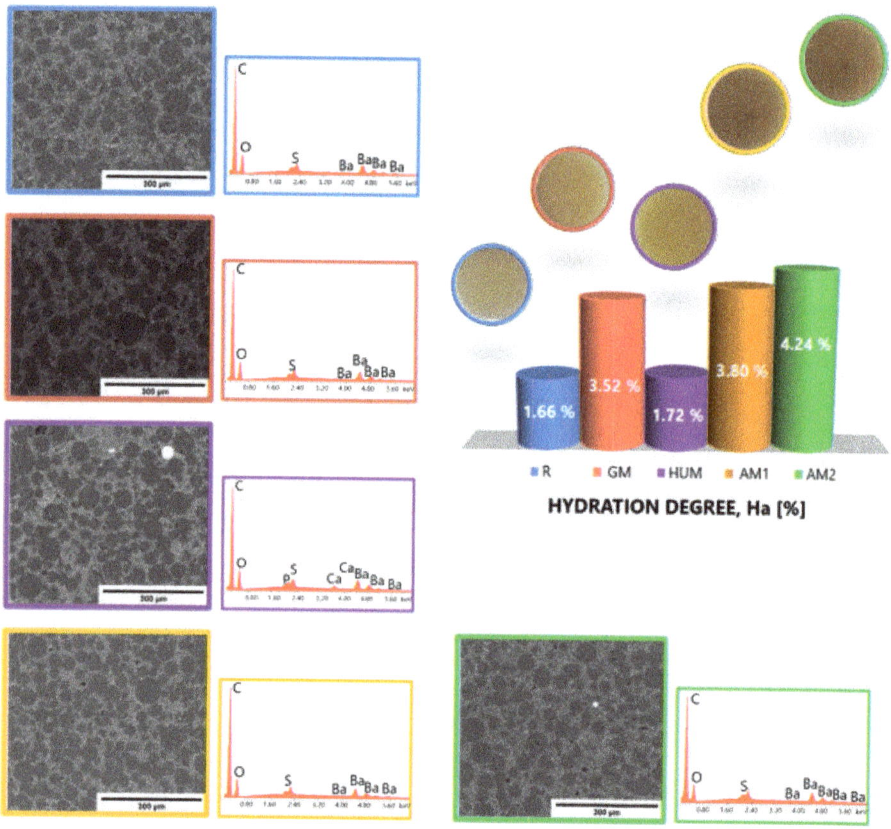

Figure 5. Morphological aspects revealed by SEM and elemental surface analysis by EDS of the samples after 21 days in distilled water.

3.4. Degradation Degree

Figure 6 shows the variation of the degradation rate (DR) over a period of 28 days in simulated body fluid (SBF) at a temperature of 37 °C, for the investigated bone cements samples. After 28 days of exposure to SBF solution, the slight weight losses were observed for GM samples, while the reference sample (R) has a constant degradation rate (DR). An increase in the degradation rate was observed for samples containing peppermint essential oil as additive. Following the exposure to the SBF solution, some particles could have been removed from the matrix of bone cements or dissolved, which led to a reduction in weight. In the case of samples containing silver nanoparticles, up to 28 days, no degradation process is observed, but only a continuous absorption process.

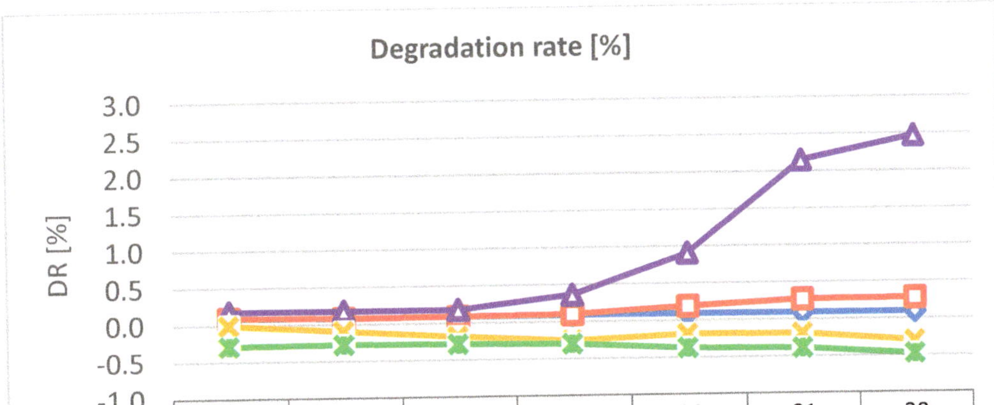

	1	3	5	7	14	21	28
R	0.0954	0.0838	0.0895	0.0918	0.0940	0.0918	0.0929
GM	0.0969	0.0864	0.0887	0.1054	0.1929	0.2712	0.2814
HUM	0.1835	0.1826	0.1830	0.3660	0.9107	2.1478	2.4713
AM1	0.0000	-0.0915	-0.1791	-0.2503	-0.1807	-0.1885	-0.2967
AM2	-0.2846	-0.2712	-0.2809	-0.2959	-0.3693	-0.3828	-0.4726

Figure 6. Variation of degradation rate over a period of 28 days in SBF, at 37 °C, for experimental PMMA bone cements.

Exposure of the investigated bone cement samples to the SBF solution did not significantly affect their structure. EDS analysis showed the presence of C, O, S, and Ba (Figure 7).

The results recorded for the hydration degree (Ha) and the degradation rate (DR) are consistent with the results obtained by evaluation of the wettability of surfaces by determining the contact angle.

3.5. Antimicrobial Tests Results

Placing the samples in contact with the culture medium inoculated with the cell suspensions allowed the qualitative evaluation of the antimicrobial effect. The experimental tests showed that bone cement sample loaded with gentamicin (GM) expressed a clear inhibition zone only towards bacterial strains, being used as positive control for bone cement samples loaded with other different antimicrobial agents. In the same time, the HUM sample (loaded with 5% peppermint essential oils active components) and AM2 sample (loaded with 4% antimicrobial additive) showed their inhibitory effect, generating clear inhibition zones with diameters compared to those of the GM control sample: 25 mm for *Staphylococcus aureus* ATCC 25923 and 20 mm for *Pseudomonas aeruginosa* ATCC 27853 (Figure 8). The lowest sensitivity was expressed by the *Candida albicans* ATCC 10231 fungus strains; in this case, all the tested samples, including GM control sample, do not show any inhibition zone (Figure 9).

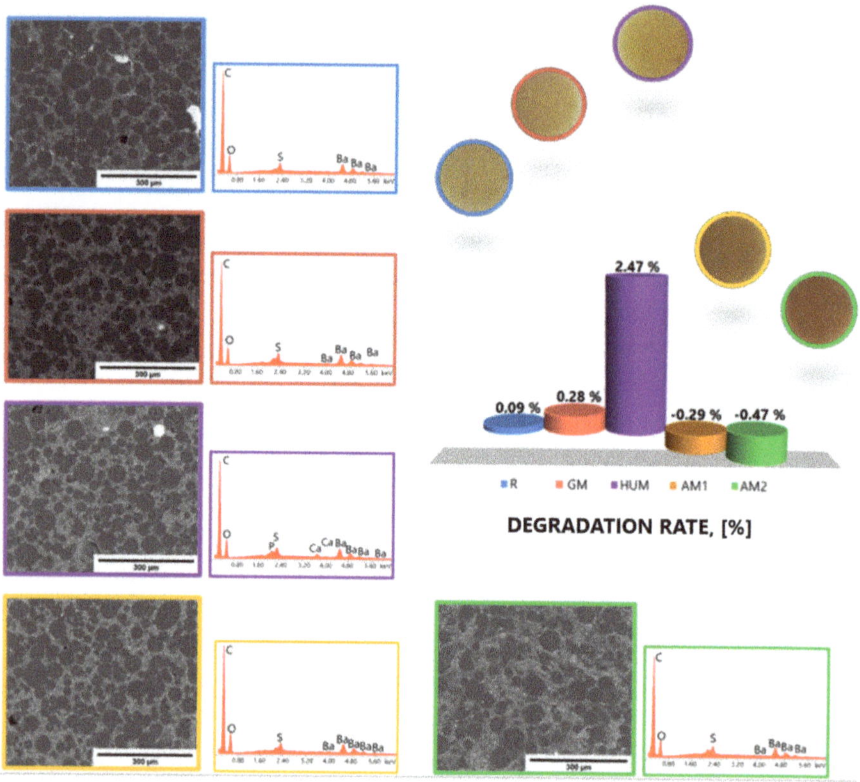

Figure 7. Morphological aspects revealed by SEM and elemental surface analysis by EDS of the experimental PMMA bone cements after 28 days in SBF.

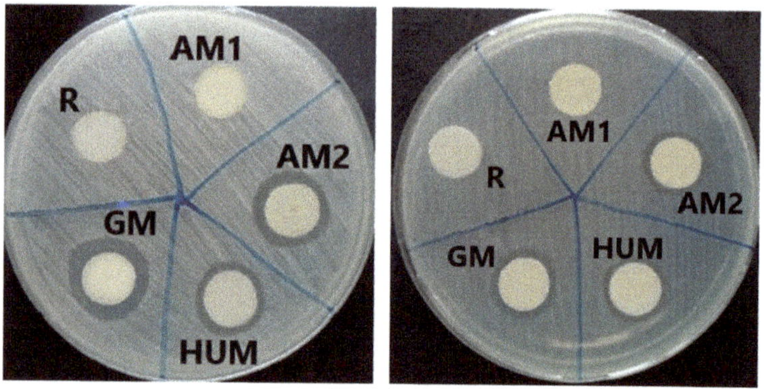

(**a**) (**b**)

Figure 8. Antibacterial property evaluation for the experimental PMMA bone cements: (**a**) growth inhibition zone for Gram-positive strain *Staphylococcus aureus* ATCC 2592; (**b**) growth inhibition zone for Gram-negative strain *Pseudomonas aeruginosa* ATCC 27853, after 72 h of incubation.

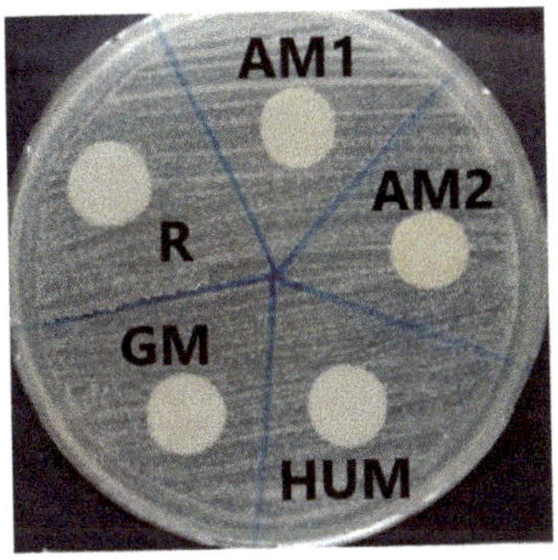

Figure 9. Contact of experimental PMMA bone cements towards *Candida albicans* ATCC 10231.

Quantitative assessment of the capacity of the selected strains to adhere and generate biofilm on the surface of the tested samples was performed using viable cell count method. The CFU/mL values were determined for the biofilms developed in 24, 48, and 72 h, on the inert substrate represented by the bone cement samples. Analysing the data presented in Figure 10, it can be observed that the MIC values corresponding to the AM2 and HUM samples decreased significantly, by more than 4 logarithmic units, compared to the R control sample, for all selected bacterial strains and at all 3 incubation times, but did not reach the MIC values of GM control samples. These results show that, although the bacterial cells poorly adhered to the surface of the bone cement after 24 incubations, they could not develop a mature biofilm, and the decrease in CFU/mL values after 48 and 72 h of incubation was not significant compared to 24 h. Instead, the capacity of the fungus strain *Candida albicans* to adhere and generate biofilm was significantly inhibited, registering the lowest values of CFU/mL, starting with 24 h, being the most sensitive tested strain.

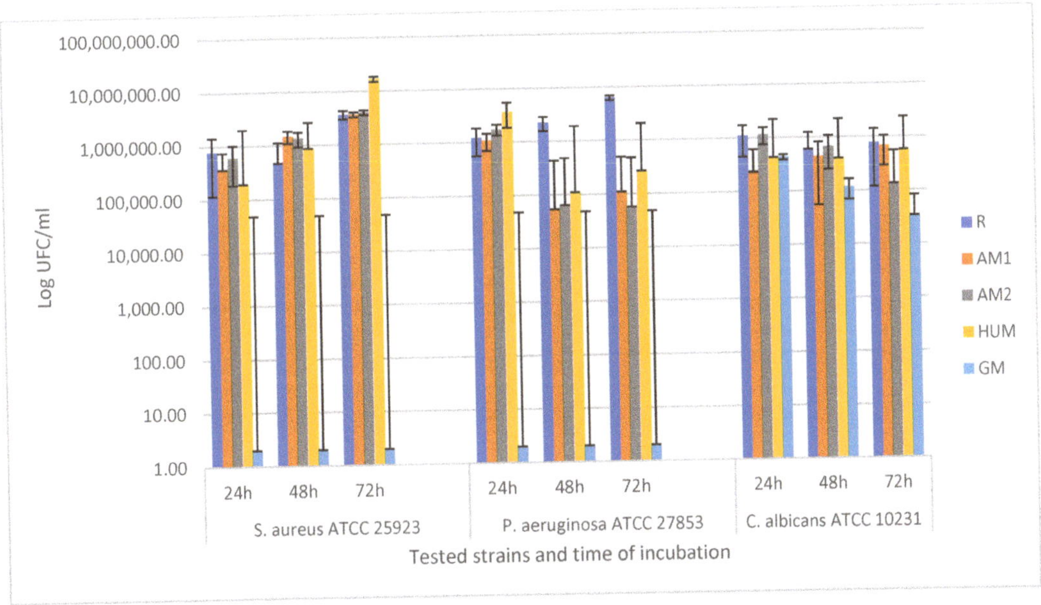

Figure 10. Graphical representation of CFU/mL values evaluating the ability of the tested strains to adhere and to develop monospecific biofilm in 24, 48, and 72h, on the surface of the experimental PMMA bone cements.

3.6. MTT Assay Results

The tested bone cement samples stimulated cellular metabolism, with a significant increase in proliferation compared to the control (CTRL). The MTT assay revealed that all bone cements tested had no cytotoxic effect, with the absorbances values being higher compared to the control sample. Moreover, GM caused the strongest increase in cell proliferation (236%), followed by the AM2 (167%), R (157%), HUM (151%), and AM1 (139%) at 24 h (Figure 11). After 48 h in the presence of bone cements, MG-63 cells remain viable, with cell proliferation being more uniform between the samples than that recorded at 24h. At 72 h, the highest proliferation rate was observed in HUM (160%), followed by AM2 (155%), R (29%), GM (129%), and AM1 (107%).

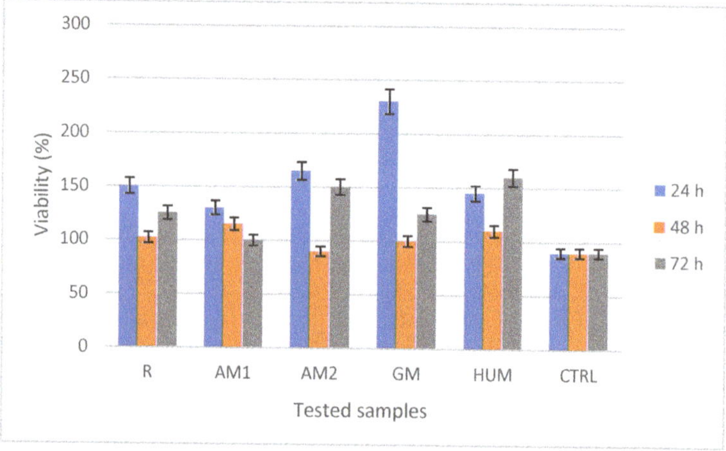

Figure 11. MTT assay showing the viability of MG-63 cells in the presence of the experimental PMMA bone cements after 24, 48, and 72 h.

After 5 days, in the presence of bone cements, the MG-63 cells showed normal morphology with a fibroblast-like characteristic appearance. Fluorescence images showed that MG-63 cells were viable, and no dead cells nor cell fragments were observed. Moreover, the cells formed phylopodia to move and establish contacts with neighboring cells, suggesting that MG-63 cells exhibited an active phenotype (Figure 12).

The osteogenic potential of bone cements onMG-63 cells was quantified using Alizarin Red assay. This test is used to stain, or locate, calcium deposits in cells and tissues, when Alizarin Red binds to the calcium to form a pigment that is orange to red in colour.

After 21 days, in the presence of bone cements, the MG-63 cells have increased their osteogenic potential. This was demonstrated by an increase in calcium deposits in all samples compared with the control. The quantification of calcium deposits ranged between 0.034 in control and 0.086 in AM2. The other cements have the following values: 0.072 for R, 0.073 for AM1, 0.082 for GM, and 0.074 HUM, respectively (Figure 13).

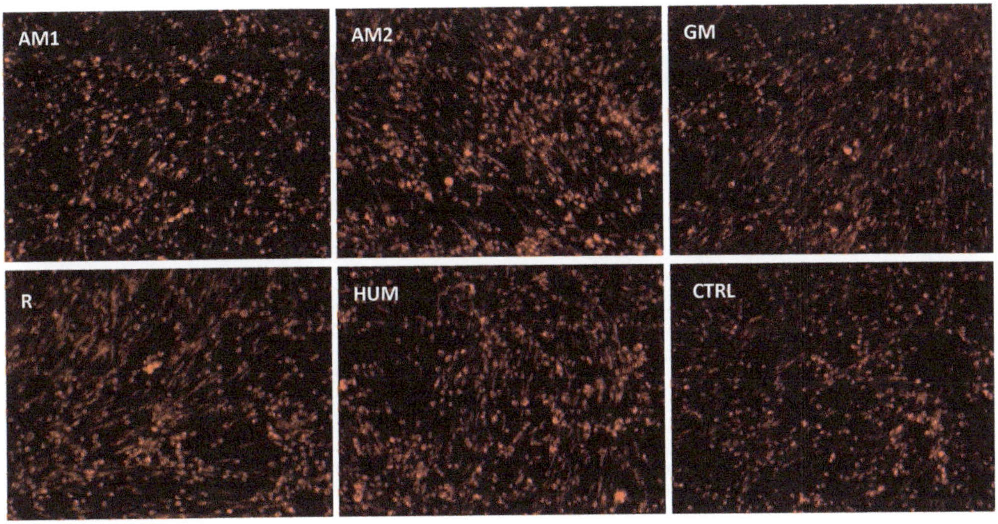

Figure 12. Fluorescence images of MG-63 cells coloured with CMTPX fluorophore in the presence of experimental PMMA bone cements.

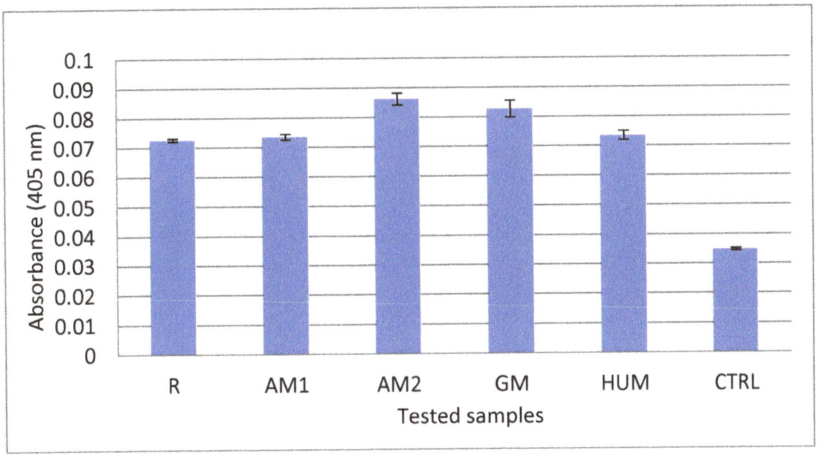

Figure 13. Alizarin Red assay showing the osteogenic potential of experimental PMMA bone cements on MG-63 cells.

4. Conclusions

The addition of antimicrobial agents did not induce structural changes in the developed novel PMMA bone cement samples. All experimental specimens possessed a typical structure and morphology for PMMA bone cements, with a slight tendency to form agglomerates, mostly in the PMMA bone cements samples containing 4% silver nanoparticles. Wettability measured by contact angle decreased by adding the antimicrobial agents, except in the case of the PMMA bone cements modified with gentamicin. All additives used kept the contact angle values within the desired limits and showed good cell adhesion. Exposure of the experimental PMMA bone cement samples to the water and SBF solution did not significantly affect their structure. The addition of antimicrobial agents did not significantly affect the hydration and degradation degree of the PMMA bone cements.

The antimicrobial properties have been demonstrated for the GM, HUM, and AM2 samples, which generated clear inhibition zones towards *Staphylococcus aureus* ATCC 25923 and *Pseudomonas aeruginosa* ATCC 27853.It was observed that not only the type of antimicrobial agent is important but also the amount used. *Staphylococcus aureus* and *Pseudomonas aeruginosa* are known to be clinically relevant pathogens associated with bone infection and intrinsic drug resistance, respectively. In terms of biocompatibility assessed by the MTT test using human MG-63 cells, all cements are biocompatible without any cytotoxicity effect. The performance of bone cements with peppermint essential oil and silver nanoparticles against these two pathogens suggests that these antibacterial additives look promising to be used in clinical practice against bacterial infection.

Author Contributions: A.R., A.A., E.G., L.M.D., and V.I.A. conceived and designed the experiments; A.R., A.A., E.G., E.V., F.I., L.M.D., and V.S. performed the experiments; V.I.A., A.D.R., J.V.R., V.G.Y. analysed the data; E.V., E.G., A.D.R., J.V.R., L.M.D., V.G.Y., and V.S. conceived the methodology; A.R., A.A., V.I.A., E.V., L.M.D., F.I. validated the obtained results; A.R., A.A., F.I., L.M.D. wrote the original draft; L.M.D. performed review & editing. All authors have read and agreed to the published version of the manuscript.

Funding: This work was supported by a grant of the Romanian Ministry of Education and Research, CCCDI-UEFISCDI, Project number PN-III-P2-2.1.-PED-2019-5236, within PNCDI III.

Institutional Review Board Statement: Not applicable.

Informed Consent Statement: Not applicable.

Data Availability Statement: Not applicable.

Conflicts of Interest: The authors declare no conflict of interest.

References

1. Deb, S. Orthopedic Bone Cement. In *Wiley Encyclopedia of Biomedical Engineering*; Metin, A., Ed.; John Wiley & Sons: Hoboken, NJ, USA, 2006; ISBN 9780471249672.
2. De Santis, R.; Russo, T.; Rau, J.V.; Papallo, I.; Martorelli, M.; Gloria, A. Design of 3D Additively Manufactured Hybrid Structures for Cranioplasty. *Materials* **2021**, *14*, 181. [CrossRef]
3. Botez, P.; Sirbu, P.; Simion, L.; Munteanu, F.; Antoniac, I. Application of a biphasic macroporous synthetic bone substitutes CERAFORM(A (R)): Clinical and histological results. *Eur. J. Orthop. Surg. Traumatol.* **2009**, *19*, 387–395. [CrossRef]
4. Rivis, M.; Pricop, M.; Talpos, S.; Ciocoiu, R.; Antoniac, I.; Gheorghita, D.; Trante, O.; Moldovan, H.; Grigorescu, G.; Seceleanu, V.; et al. Influence of the bone cements processing on the mechanical properties in cranioplasty. *Rev. Chim.* **2018**, *69*, 990–993. [CrossRef]
5. Ginebra, M.P.; Montufar, E.B. Cements as bone repair materials. In *Bone Repair Biomaterials*, 2nd ed.; Pawelec, K., Planell, J.A., Eds.; Woodhead Publishing: Sawston, UK, 2019; pp. 233–271. ISBN 9780081024515.
6. Dunne, J.N.; Wang, C.S. Acrylic cements for bone fixation in joint replacement. In *Joint Replacement Technology*; Revel, P.A., Ed.; Woodhead Publishing: Sawston, UK, 2014; pp. 212–256. ISBN 978-1-84569-245-2.
7. Zhu, W.; Liu, F.; Yu, B.; He, J. Preparation of antibacterial acrylic bone cement with methacrylate derived from benzothiazole. *J. Mech. Behav. Biomed. Mater.* **2021**, *117*, 1–7. [CrossRef]
8. Alonso, L.M.; Torres, I.F.; Zayas Tamayo, Á.M.; Ledea Lozano, O.E.; Ramos, I.D.; García-Menocal, J.D.; Rios-Donato, N.; Mendizábal, E. Antibacterial effect of acrylic bone cements loaded with drugs of different action's mechanism. *J. Infect. Dev. Ctries.* **2019**, *13*, 487–495. [CrossRef] [PubMed]
9. Bozic, K.J.; Kurtz, S.M.; Lau, E.; Ong, K.; Chiu, V.; Vail, T.P.; Rubash, H.E.; Berry, D.J. The epidemiology of revision total knee arthroplasty in the United States. *Clin. Orthop. Relat. Res.* **2010**, *468*, 45–51. [CrossRef] [PubMed]
10. Moriarty, T.F.; Kuehl, R.; Coenye, T.; Metsemakers, W.J.; Morgenstern, M.; Schwarz, E.M.; Riool, M.; Zaat, S.A.J.; Khana, N.; Kates, S.L.; et al. Orthopaedic device-related infection: Current and future interventions for improved prevention and treatment. *EFORT Open Rev.* **2016**, *1*, 89–99. [CrossRef]
11. Chen, L.; Tang, Y.; Zhao, K.; Zha, X.; Liu, J.; Bai, H.; Wu, Z. Fabrication Of The Antibiotic-Releasing Gelatin/PMMA Bone Cement. *Colloids Surf. B Biointerfaces* **2019**, *183*, 1–11. [CrossRef]
12. Lacy, M.K.; Nicolau, D.P.; Nightingale, C.H.; Quintiliani, R. The pharmacodynamics of aminoglycosides. *Clin. Infect. Dis.* **2007**, *27*, 23–27. [CrossRef]
13. Slane, J.; Gietman, B.; Squire, M. Antibiotic elution from acrylic bone cement loaded with high doses of tobramycin and vancomycin. *J. Orthop. Res.* **2018**, *36*, 1078–1085. [CrossRef]

14. Neut, D.; De Groot, E.P.; Kowalski, R.S.Z.; Van Horn, J.R.; Van Der Mei, H.C.; Busscher, H.J. Gentamicin-loaded bone cement with clindamycin or fusidic acid added: Biofilm formation and antibiotic release. *J. Biomed. Mater. Res. A* **2005**, *73*, 165–170. [CrossRef]
15. Snir, N.; Meron-Sudai, S.; Deshmukh, A.J.; Dekel, S.; Ofek, I. Antimicrobial Properties and Elution Kinetics of Linezolid from Polymethylmethacrylate. *Orthopedics* **2013**, *36*, 1412–1417. [CrossRef] [PubMed]
16. Anagnostakos, K.; Kelm, J.; Regitz, T.; Schmitt, E.; Jung, W. *In vitro* evaluation of antibiotic release from and bacteria growth inhibition by antibiotic-loaded acrylic bone cement spacers. *J. Biomed. Mater. Res. B Appl. Biomater.* **2005**, *72*, 373–378. [CrossRef] [PubMed]
17. Parra-Ruíz, F.J.; González-Gómez, A.; Fernández-Gutiérrez, M.; Parra, J.; García-García, J.; Azuara, G.; De la Torre, B.; Buján, J.; Ibarra, B.; Duocastella-Codina, L.; et al. Development of advanced antibiotic loaded bone cement spacers for arthroplasty associated infections. *Int. J. Pharm.* **2017**, *522*, 11–20. [CrossRef]
18. Paz, E.; Sanz-Ruiz, P.; Abenojar, J.; Vaquero-Martín, J.; Forriol, F.; Del Real, J.C. Evaluation of elution and mechanical properties of high-dose antibiotic-loaded bone cement: Comparative "in vitro" study of the influence of vancomycin and cephazolin. *J. Arthroplast.* **2015**, *30*, 1423–1429. [CrossRef]
19. Hellmark, B.; Unemo, M.; Nilsdotter-Augustinsson, A.; Soderquist, B. Antibiotic susceptibility among *Staphylococcus epidermidis* isolated from prosthetic joint infections with special focus on rifampicin and variability of the rpoB gene. *Clin. Microbiol. Infect.* **2009**, *15*, 238–244. [CrossRef] [PubMed]
20. Thornes, B.; Murray, P.; Bouchier-Hayes, D. Development of resistant strains of *Staphylococcus epidermidis* on gentamicin-loaded bone cement in vivo. *J. Bone Jt. Surg.* **2002**, *84*, 758–760. [CrossRef]
21. Cavalu, S.; Banica, F.; Gruian, C.; Vanea, E.; Goller, G.; Simon, V. Microscopic and spectroscopic investigation of bioactive glasses for antibiotic controlled release. *J. Mol. Struct.* **2013**, *1040*, 47–52. [CrossRef]
22. Khaled, S.M.; Charpentier, P.A.; Rizkalla, A.S. Physical and mechanical properties of PMMA bone cement reinforced with nano-sized titania fibers. *J. Biomater. Appl.* **2011**, *25*, 515–537. [CrossRef]
23. Slane, J.; Vivanco, J.; Rose, W.; Ploeg, H.L.; Squire, M. Mechanical, material, and antimicrobial properties of acrylic bone cement impregnated with silver nanoparticles. *Mater. Sci. Eng. C* **2015**, *48*, 188–196. [CrossRef]
24. Oei, J.D.; Zhao, W.W.; Chu, L.; Desilva, M.N.; Ghimire, A.; Rawls, H.R.; Whang, K. Antimicrobial acrylic materials with in situ generated silver nanoparticles. *J. Biomed. Mater. Res. B Appl. Biomater.* **2012**, *100*, 409–415. [CrossRef] [PubMed]
25. Wekwejt, M.; Michalska-Sionkowska, M.; Bartmański, M.; Nadolska, M.; Łukowicz, K.; Pałubicka, A.; Osyczka, A.M.; Zieliński, A. Influence of several biodegradable components added to pure and nanosilver-doped PMMA bone cements on its biological and mechanical properties. *Mater. Sci. Eng. C* **2020**, *117*, 111286. [CrossRef] [PubMed]
26. Russo, T.; Gloria, A.; De Santis, R.; D'Amora, U.; Balato, G.; Vollaro, A.; Oliviero, O.; Improta, G.; Triassi, M.; Ambrosio, L. Preliminary focus on the mechanical and antibacterial activity of a PMMA-based bone cement loaded with gold nanoparticles. *Bioact. Mater.* **2017**, *2*, 156–161. [CrossRef]
27. Phakatkar, A.H.; Shirdar, M.R.; Qi, M.; Taheri, M.M.; Narayanan, S.; Foroozan, T.; Sharifi-Asl, S.; Huang, Z.; Agrawal, M.; Lu, Y.; et al. Novel PMMA bone cement nanocomposites containing magnesium phosphate nanosheets and hydroxyapatite nanofibers. *Mater. Sci. Eng. C* **2020**, *109*, 110497. [CrossRef]
28. Russo, T.; De Santis, R.; Gloria, A.; Barbaro, K.; Altigeri, A.; Fadeeva, I.V.; Rau, J. Modification of PMMA Cements for Cranioplasty with Bioactive Glass and Copper Doped Tricalcium Phosphate Particles. *Polymers* **2020**, *12*, 37. [CrossRef]
29. Cojocaru, F.D.; Balan, V.; Popa, M.I.; Lobiuc, A.; Antoniac, A.; Antoniac, I.V.; Verestiuc, L. Biopolymers—Calcium phosphates composites with inclusions of magnetic nanoparticles for bone tissue engineering. *Int. J. Biol. Macromol.* **2019**, *125*, 612–620. [CrossRef]
30. Dascalu, C.A.; Maidaniuc, A.; Pandele, A.M.; Voicu, S.I.; Machedon-Pisu, T.; Stan, G.E.; Cimpean, A.; Mitran, V.; Antoniac, I.V.; Miculescu, F. Synthesis and characterization of biocompatible polymer-ceramic film structures as favorable interface in guided bone regeneration. *Appl. Surf. Sci.* **2019**, *494*, 335–352. [CrossRef]
31. Kumar, A.; Kargozar, S.; Baino, F.; Han, S.S. Additive manufacturing methods for producing hydroxyapatite and hydroxyapatite-based composite scaffolds: A review. *Front. Mater.* **2019**, *6*, 313. [CrossRef]
32. Radhi, A.; Mohamad, D.; Rahman, F.S.; Abdullah, A.M.; Hasan, H. Mechanism and factors influence of graphene based nanomaterials antimicrobial activities and application in dentistry. *J. Mater. Res. Technol.* **2021**, *11*, 1290–1307. [CrossRef]
33. Sarosi, C.; Biris, A.R.; Antoniac, A.; Boboia, S.; Alb, C.; Antoniac, I.; Moldovan, M. The nanofiller effect on properties of experimental graphene dental nanocomposites. *J. Adhes. Sci. Technol.* **2016**, *30*, 1779–1794. [CrossRef]
34. Sartoratto, A.; Machado, A.L.M.; Delarmelina, C.; Figueira, G.M.; Duarte, M.C.T.; Rehder, V.L.G. Composition and antimicrobial activity of essential oils from aromatic plants used in Brazil. *Braz. J. Microbiol.* **2004**, *34*, 1517–1520. [CrossRef]
35. Saeed, S.; Naim, A.; Tariq, P. *In vitro* antibacterial activity of peppermint. *Pak. J. Bot.* **2006**, *38*, 869–872.
36. Gordon, M.C.; David, J.N. Natural product drug discovery in the next millennium. *Pharm. Biol.* **2001**, *139*, 8–17.
37. Bakkali, F.; Averbeck, S.; Averbeck, D.; Idaomar, M. Biological effects of essential oils—A review. *Food Chem. Toxicol.* **2008**, *46*, 446–475. [CrossRef] [PubMed]
38. El Asbahani, A.; Miladi, K.; Badri, W.; Sala, M.; Addi, E.A.; Casabianca, H.; El Mousadik, A.; Hartmann, D.; Jilale, A.; Renaud, F.N.R.; et al. Essential oils: From extraction to encapsulation. *Int. J. Pharm.* **2015**, *483*, 220–243. [CrossRef]
39. Kligler, B.; Chaudhary, S. Peppermint Oil. *Am. Fam. Physician* **2007**, *75*, 1027–1030. [PubMed]
40. Briggs, C. Peppermint: Medicinal herb and flavoring agent. *Can. Pharm. J.* **1993**, *126*, 89–92.

41. Anghel, I.; Holban, A.M.; Grumezescu, A.M.; Andronescu, E.; Ficai, A.; Anghel, A.G.; Maganu, M.; Lazar, V.; Chifiriuc, M.C. Modified wound dressing with phyto-nanostructured coating to prevent Staphylococcal and Pseudomonal biofilm development. *Nanoscale Res. Lett.* **2012**, *7*, 1–8. [CrossRef]
42. Singh, R.; Shushni Muftah, A.M.; Belkheir, A. Antibacterial and antioxidant activities of Mentha piperita L. *Arab. J. Chem.* **2015**, *8*, 322–328. [CrossRef]
43. Sokovic, M.D.; Vukojevic, J.; Marin, P.D.; Brkic, D.D.; Vajs, V.; Van Griensvenu, L.J.L.D. Chemical composition of essential oils of Thymus and Mentha species and their antifungal activities. *Molecules* **2009**, *14*, 238–249. [CrossRef]
44. Almasian, A.; Najafi, F.; Eftekhari, M.; Shams Ardekani, M.R.; Sharifzadeh, M.; Khanavi, M. Preparation of Polyurethane/Pluronic F127 Nanofibers Containing Peppermint Extract Loaded Gelatin Nanoparticles for Diabetic Wounds Healing: Characterization, In Vitro, and In Vivo Studies. *Evid. Based Complement. Altern. Med.* **2021**, *2021*, 6646702. [CrossRef] [PubMed]
45. Kasiri, N.; Fathi, M. Entrapment of peppermint oil using cellulose nanocrystals. *Cellulose* **2018**, *25*, 319–329. [CrossRef]
46. Carbonell-Blasco, P.; Martín-Martínez, J.M.; Antoniac, I.V. Synthesis and characterization of polyurethane sealants containing rosin intended for sealing defect in annulus for disc regeneration. *Int. J. Adhes. Adhes.* **2013**, *42*, 11–20. [CrossRef]
47. Attik, G.N.; Villat, C.; Hallay, F.; Pradelle-Plasse, N.; Bonnet, H.; Moreau, K.; Colon, P.; Grosgogeat, B. *In vitro* biocompatibility of a dentine substitute cement on human MG63 osteoblasts cells: Biodentine™ versus MTA®. *Int. Endod. J.* **2014**, *47*, 1133–1141. [CrossRef]
48. Cavalu, S.; Antoniac, I.V.; Fritea, L.; Mates, I.M.; Milea, C.; Laslo, V.; Vicas, S.; Mohan, A. Surface modifications of the titanium mesh for cranioplasty using selenium nanoparticles coating. *J. Adhes. Sci. Technol.* **2018**, *32*, 2509–2522. [CrossRef]
49. Cavalu, S.; Ratiu, C.; Ponta, O.; Simon, V.; Rugina, D.; Miclaus, V.; Akin, I.; Goller, G. Improving Osseointegration of Alumina/Zirconia Ceramic Implants by Fluoride Surface Treatment. *Dig. J. Nanomater. Biostruct.* **2014**, *9*, 797–808.
50. Rajamony, L.L.; Gurupatham, B.G.A.; Roy, K.; Lim, J.B.P. Effect of super absorbent polymer on microstructural and mechanical properties of concrete blends using granite pulver. *Struct. Concr.* **2020**, *22*, E898–E915. [CrossRef]
51. Rajamony, L.L.; Gurupatham, B.G.A.; Roy, K.; Lim, J.B.P. Influence of super absorbent polymer on mechanical, rheological, durability, and microstructural properties of self-compacting concrete using non-biodegradable granite pulver. *Struct. Concr.* **2020**, *22*, E1093–E1116. [CrossRef]
52. Heise, S.; Forster, C.; Heer, S.; Qi, H.; Zhou, J.; Virtanen, S.; Lu, T.; Boccaccini, A.R. Electrophoretic deposition of gelatine nanoparticle/chitosan coatings. *Electrochim. Acta* **2019**, *307*, 318–325. [CrossRef]
53. Chen, S.; Guo, Y.; Liu, R.; Wu, S.; Fang, J.; Huang, B.; Li, Z.; Chen, Z.; Chen, Z. Tuning surface properties of bone biomaterials to manipulate osteoblastic cell adhesion and the signaling pathways for the enhancement of early osseointegration. *Colloids Surf. B* **2018**, *164*, 58–69. [CrossRef] [PubMed]

materials

Review

Metal Nanoparticles and Carbon-Based Nanomaterials for Improved Performances of Electrochemical (Bio)Sensors with Biomedical Applications

Luminita Fritea [1,†], Florin Banica [1,†], Traian Octavian Costea [2], Liviu Moldovan [3,*], Luciana Dobjanschi [1,*], Mariana Muresan [1] and Simona Cavalu [1]

1 Faculty of Medicine and Pharmacy, University of Oradea, 10 P-ta 1 Decembrie, 410087 Oradea, Romania; fritea_luminita@yahoo.com (L.F.); florinbanica1@gmail.com (F.B.); marianamur2002@yahoo.com (M.M.); simona.cavalu@gmail.com (S.C.)
2 Advanced Materials Research Infrastructure—SMARTMAT, University of Oradea, 1 Universitatii Street, 410087 Oradea, Romania; tcostea@uoradea.ro
3 Faculty of Electrical Engineering and Information Technology, University of Oradea, 1 Universitatii Street, 410087 Oradea, Romania
* Correspondence: liviu@uoradea.ro (L.M.); dobjanschil@yahoo.com (L.D.)
† First authors with equal contribution.

Citation: Fritea, L.; Banica, F.; Costea, T.O.; Moldovan, L.; Dobjanschi, L.; Muresan, M.; Cavalu, S. Metal Nanoparticles and Carbon-Based Nanomaterials for Improved Performances of Electrochemical (Bio)Sensors with Biomedical Applications. *Materials* **2021**, *14*, 6319. https://doi.org/10.3390/ma14216319

Academic Editor: Marta Miola

Received: 21 September 2021
Accepted: 20 October 2021
Published: 22 October 2021

Abstract: Monitoring human health for early detection of disease conditions or health disorders is of major clinical importance for maintaining a healthy life. Sensors are small devices employed for qualitative and quantitative determination of various analytes by monitoring their properties using a certain transduction method. A "real-time" biosensor includes a biological recognition receptor (such as an antibody, enzyme, nucleic acid or whole cell) and a transducer to convert the biological binding event to a detectable signal, which is read out indicating both the presence and concentration of the analyte molecule. A wide range of specific analytes with biomedical significance at ultralow concentration can be sensitively detected. In nano(bio)sensors, nanoparticles (NPs) are incorporated into the (bio)sensor design by attachment to the suitably modified platforms. For this purpose, metal nanoparticles have many advantageous properties making them useful in the transducer component of the (bio)sensors. Gold, silver and platinum NPs have been the most popular ones, each form of these metallic NPs exhibiting special surface and interface features, which significantly improve the biocompatibility and transduction of the (bio)sensor compared to the same process in the absence of these NPs. This comprehensive review is focused on the main types of NPs used for electrochemical (bio)sensors design, especially screen-printed electrodes, with their specific medical application due to their improved analytical performances and miniaturized form. Other advantages such as supporting real-time decision and rapid manipulation are pointed out. A special attention is paid to carbon-based nanomaterials (especially carbon nanotubes and graphene), used by themselves or decorated with metal nanoparticles, with excellent features such as high surface area, excellent conductivity, effective catalytic properties and biocompatibility, which confer to these hybrid nanocomposites a wide biomedical applicability.

Keywords: metal nanoparticles; carbon-based nanomaterials; screen-printed electrodes; electrochemical (bio)sensors; biomedical applications

1. Introduction

The development of sophisticated and miniaturized devices for sensing a broad range of biological molecules emerged as an imperative strategy for real-time diagnosis of different diseases and effective management of disease progression, based on rapid diagnostics, smart data analysis and statistical informatics analysis. The fabrication of smart devices based on bio-nanomaterials are considered the result of a transdisciplinary work from the materials science to the medical field, as the target analytes recognized by these devices are proteins, enzymes, antibodies, DNA/RNA probes, microorganisms, which can be detected with high sensitivity, accuracy and low detection limits [1]. The bio-detection system can be designed as bio-catalytic or bio-affinity-based system, according to the biochemical mechanism involved in the recognition: In the first case, the bioreceptor (proteins, enzymes, cells) undergoes a catalytic reaction with the analyte, while in the second case, a specific binding mechanism of the bioreceptor (aptamer, antibody) and analyte leads to an equilibrium [2]. According to the transduction pathways, the biosensors can be classified as: electrochemical, piezoelectric, optoelectronic and calorimetric.

The concept of nano(bio)sensors is related to nanostructures which are often incorporated into the (bio)sensor by attachment to a suitably modified platform. Nanoparticles and nanomaterials offer excellent properties for designing sensing systems with enhanced performances. Incorporating them in transducers, by attachment to a suitable modified platform, a high surface area can be achieved by tailoring their size and morphology, and hence, producing (bio)sensors with greater sensitivity and shorter response time. Hybrid materials and nanocomposite structures consisting of metallic NPs, combined with particular conductive polymers and a modified electrode, have been designed for electrochemical sensing, owing to their unique combination of biocompatibility, large surface area, and good conductivity [3,4]. The output of the sensing nano-platform can be connected to wireless devices for signal processing on a smart phone. Hence, these nano-devices could be easily handled and used as implantable devices in the body for health monitoring.

A nanomaterial is defined according to the European Commission as "a natural, incidental or manufactured material containing particles, in an unbound state or as an aggregate or as an agglomerate and for 50% or more of the particles in the number size distribution, one or more external dimensions is in the size range 1 nm–100 nm. In specific cases where warranted by concerns for the environment, health, safety or competitiveness, the number size distribution threshold of 50% may be replaced by a threshold between 1% to 50%" [5].

The original, innovative and complex configuration of this review combines the theoretical general aspects about metal nanoparticles, carbon-based nanomaterials and SPEs with their biomedical applications including the actual COVID-19 pandemic. The evolution from printable towards wearable sensors is also emphasized as the most recent and modern generation of sensing systems in healthcare. This review is focused on the electrochemical nano(bio)sensors based on screen-printed electrodes (SPEs), in which transduction is based on electrochemical techniques, while metal NPs and carbon-based nanomaterials are used as modifiers to improve the analytical performances. Each type of nano(bio)sensor is described in terms of the constituent nanostructures mentioning the large range of biomedical targets such as pathogens, cancer biomarkers and other relevant biomolecules, and also pharmaceutical compounds, analysis performed in the traditional laboratories or in-to-the-field (in-situ detection). Not the least, considering that intelligent diagnostics tools are urgently required to manage the COVID-19 pandemic, we present in this review recent approaches related to nanomaterials and SPEs in COVID-19 diagnosis, prevention and therapy. Some recent advances concerning the wearable sensors as non-invasive diagnostics tools for personalized healthcare management are mentioned.

2. Metal Nanoparticles

The capability to produce metal nanoparticles in the same size domain as proteins (1 to 100 nm) has led to a wide range of applications in the biomedical field. Their unique properties such as large surface to volume ratio and high percentage of atoms/molecules on the surface, can be exploited to improve the sensing and detection of several important biomolecules in healthcare-related fields. Metal NPs can be used alone and in combination with other nanostructures, with the aim of signal amplification, higher sensitivity and great improvements in the detection and quantification of different biomolecules. Actually, the role of metallic NPs in biosensing is directly related to their physico-chemical properties and changes that occur after binding the biomolecular analyte on the surface: (1) role as immobilizing platforms, (2) accelerating electron transfer, (3) catalyzing the chemiluminescent reaction with their substrates, (4) amplifying changes in mass, and (5) enhancing changes in refractive index [2].

Nanoparticles have physical and chemical properties that differ from those of the materials from which they are obtained. For example, metallic nanoparticles have a melting point significantly lower than that of the precursor metal: AuNPs melt at a temperature of around 300 °C (for 2.5 nm nanoparticles), compared to metallic Au whose melting point is 1064 °C [6]. Even the absorption of solar radiation is superior for NPs compared to thin metal sheets. AuNPs are also colored in solution from deep red to black. Moreover, metals are chemically inert in their macroscale form, while at nanoscale, their unique physicochemical features are remarkable. So far, a wide variety of metal/metal oxide nanoparticles has been synthesized for multiple applications: noble metals NPs (gold, silver and platinum NPs—used in biosensors, therapy, drug delivery, etc.), copper NPs, palladium NPs, lead NPs, selenium NPs, and metal oxide NPs (copper oxide, titanium dioxide, zinc oxide, indium oxide, iron oxide, etc.) (Figure 1) [7–20].

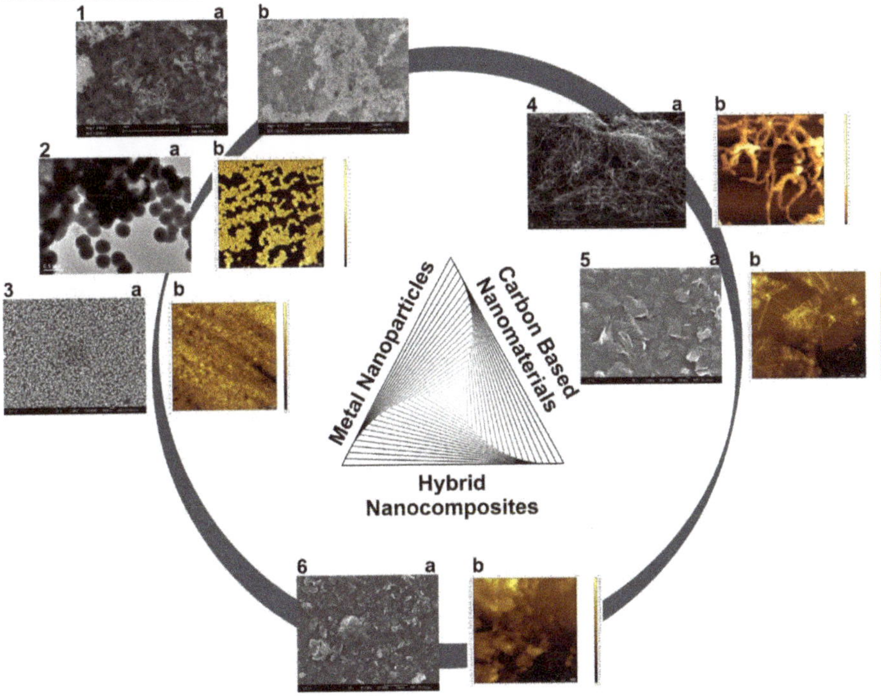

Figure 1. Microscopic images of some metal nanoparticles (1a—SEM image of TiO_2 NPs, 1b—SEM image of TiO_2 doped with SeNPs; 2a—TEM image of SeNPS; 2b—AFM image of SeNPs; 3a—SEM image of AuNPs, 3b—AFM image of AuNPs), carbon-based nanomaterials (4a—SEM image of MWCNTs, 4b—AFM image of MWCNTs; 5a—SEM image of graphene, 5b—AFM image of graphene) and hybrid nanocomposite (6a—SEM image of graphene + AuNPs, 6b—AFM image of graphene + AuNPs) (original images).

Different synthesis methods (bottom-up approach, top-down approach) have been developed in order to fabricate NPs with suitable features such as homogeneity, size and shape control, and versatile surface properties. Chemical, physical and biological fabrication route are currently employed, each of them with its own advantages and disadvantages (Figure 2).

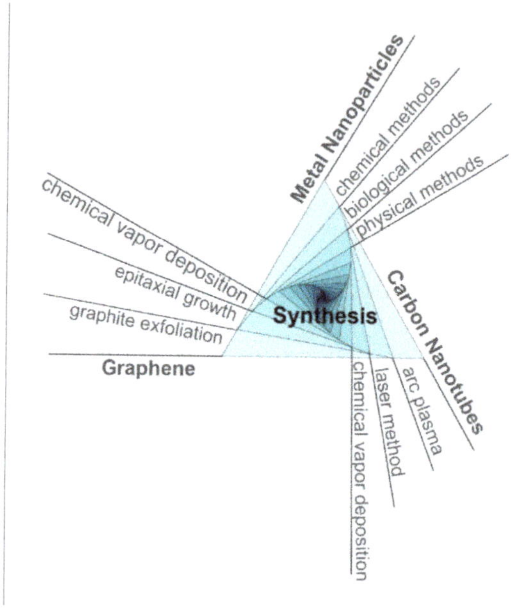

Figure 2. Schematic representation of main synthesis methods of metal NPs and carbon-based nanomaterials.

Chemical methods include chemical or photochemical reduction, co-precipitation, thermal decomposition, hydrolysis and sol-gel method, these being the simplest methods used for preparation of metal nanoparticles. Usually, a metal precursor is reduced and in some cases a stabilizer is added to avoid NPs agglomeration. As a main advantage, desired size and shape nanoparticle can be prepared by this route, but on the other hand, the processes are difficult to control and there are drawbacks related to the reducing agents in terms of toxicity, poor reducing ability, high costs of reagents and impurities [13,21–25]. In the recent years, the electrochemical methods became a powerful tool for NPs synthesis due to their advantages: Facile, cost-effective, quick, highly efficient and environmental friendliness leading to NPs with high purity, controlled size, shape and composition (multimetallic NPs have been electrodeposited) by adjusting some parameters (potential, time, current density, number of scans, etc.) [26–29].

Physical methods include mechanical milling, grinding, vapor deposition, laser ablation, sputtering, spray pyrolysis, microwave irradiation, dissolution in supercritical fluids coupled with thermal reduction, etc. These are potentially clean techniques and the advantages consist in the possibility to produce thin metal films, while the nanoparticle properties such as surface morphology and crystal structure can be controlled. The disadvantages are generally related to the expensive instruments and specific operating requirements [30–34].

Biological methods are considered green synthesis methods, an emerging trend of nanotechnology, developed to overcome safety issues with respect to health and environment. For this purpose, plants, algae, fungi and different microorganisms (bacteria, viruses) are largely employed. The NPs size and morphology are influenced by the concentration of the biological partner, the growth phase of cells, the concentration of the metal ion, the pH of the solution, the temperature and the reaction time. These techniques have multiple advantages over other physical and chemical methods, such as cost effectiveness, eco-friendliness, and are easily scaled up for large scale production. Moreover, it does not involve sophisticated instruments or use of high pressure, energy, temperature and toxic reagents [19,25,35–38].

Concerning the biocompatibility of the metal NPs, the biological (and biotechnological) methods are considered to produce non-toxic, biocompatible and well-defined NPs. The use of various biomolecules for the NPs encapsulation was also an approach for increasing their biocompatibility and determining their mechanism of action, the modified and the spherical NPs presented lower toxicity [39,40]. The cellular uptake of metal NPs by endocytosis depends on their physico-chemical properties (type, size, shape, surface properties, dose). The biocompatibility assessment is achieved by using qualitative and quantitative analysis (based on cell staining and dye-conversion assays) [39].

The NPs characterization is achieved by various techniques such as: transmission electron microscopy (TEM), scanning electron microscopy (SEM), atomic force microscopy (AFM), X-ray diffraction (XRD), energy dispersive X-ray analysis (EDAX), X-ray fluorescence (XRF), Fourier-transform infrared spectroscopy (FTIR), ultraviolet-visible spectroscopy (UV–Vis), dynamic light scattering (DLS), Zeta potential analysis, electrochemistry, etc.

Various metallic nanoparticles are currently used in (bio)sensors fabrication, noble metals NPs such as gold, silver and platinum NPs being the most popular ones. Below we will focus on these three types of NPs, but also on other metal/metal oxide NPs also applied in (bio)sensing.

2.1. Gold Nanoparticles (AuNPs)

AuNPs, or colloidal gold, can be easily synthesized in sizes ranging between 3 and 100 nm in diameter, with different shapes. In a recent paper, De Souza et al. [41] presented a comprehensive review of the methodologies used in the synthesis of gold nanoparticles by chemical reduction, including the Turkevich method, synthesis with $NaBH_4$ with/without citrate; the seeding-growth method; synthesis by ascorbic acid; green synthesis; the Brust–Schiffrin synthesis; and synthesis using other reducing agents. The most used method to synthesize quasi-spherical gold NPs is the chemical reduction of Au^{3+} and Au^{+} to Au^{0} ions using different reducing agent (sodium citrate, borohydrides, citric and oxalic acids, polyols, hydrogen peroxide, sulfites, etc.). $HAuCl_4$ is the chosen precursor salt in most of the reported studies, in which gold is in the Au^{3+} oxidation state [42]. The use of a stabilization agent is necessary, often being the same molecule that of the reduction agent. The electrochemical synthesis of AuNPs is an often-employed method in the case of electrochemical sensor development where the optimization of the electrochemical method parameters influences the size (20–40, 30–80, 90 nm, deposition by cyclic voltammetry, amperometry) and the layer thickness of the electrodeposited NPs [10,43–45].

Nowadays, a special attention is paid to green synthesis methods, and therefore numerous research groups all over the world were focus on this field. The metal NPs biosynthesis from plants can be an efficient method for the development of a fast and ecological technology [37]. As a vegetal resource, amino acids, enzymes, flavonoids, aldehydes, ketones, amines, carboxylic acids, phenols, proteins and alkaloids can be successfully used to provide electrons to reduce Au^{3+} or Au^{+} into gold nanoparticles, the reaction being dependent on plant extract concentration, metal salt, reaction pH, temperature and incubation time. It is well known that the primary and secondary metabolites of plant are consistently involved in redox reactions of metabolic pathways. *Trigonella foenum-graecum, Hibiscus extract, Elettaria cardamomum, Garcinia cambogia, Areca catechu* and *Chenopodium album* are only few examples of plant extracts successfully employed to produce small AuNPs with diameter less than 20–30 nm [41,46–50].

Different types of bacteria were reported in the synthesis of AuNPs, either by intracellular or extracellular mechanisms, the production being initiated by tetrachloroaurate salt ($AuCl_4^-$). *Bacillus subtilis* 168 strain was successfully employed to intracellularly reduce Au^{3+} ions to AuNPs with the diameter range between 5–25 nm [30], while *Rhodopseudomonas capsulata* was found to successfully produce gold nanoparticles of different sizes and shapes: Spherical gold nanoparticles with diameter in the range of 10–20 nm were observed at pH = 7, gold nanoplates were observed at pH = 4 [51]. Other examples of microorganisms assisting AuNPs production, showing pH and temperature dependent mechanisms are: *Halomonas salina* (spherical NPs 30 to 100 nm in diameter) [52], *Delftia* sp. *strain KCM-006* (spherically shaped AuNPs 11.3 nm diameter) [53], *Bacillus subtilis 168* (octahedral 5–25 nm AuNPs inside the cell wall) [54], *Stenotrophomonas* sp. (multi-shaped 10–50 nm AuNPs, extracellular) [55], etc. In other studies, AuNPs and other metal NPs have been obtained by using fungi (*Aspergillus* sp., *Fusarium* sp. and *Penicillium* sp.) [56–58], viruses (*Tabacco mosaic virus*) and yeasts (*Pichia jadinii*).

It is generally accepted that the size, shape and function of the AuNPs are highly influenced by the physical and chemical parameters of their synthesis: the temperature of reaction, the stirring rate, the ratio of gold to reducing agent. In order to control the growth of the crystal nanostructures, surfactants are often used as surface coating in nanoparticles, as they have the ability to control the growth of nanocrystals to achieve desired morphologies], and hence, surfactant coated nanoparticles will remain well dispersed in relatively dilute solutions. For example, AuNPs protected by a compact shell of organothiols, are stable for long periods of time, with the possibility to be redispersed in organic solvents [59].

The biomedical applications of AuNPs are shape and size dependent. As emphasized in several research papers, gold nanoparticles are suitable to be used in electrochemical (bio)sensors design for a sensitive and selective detection of some important biomolecules, taking advantage of their good electrical conductivity and high surface area, which provides also a stable immobilization of various biomolecules retaining their bioactivity [7,10,60–69]. Other biomedical applications of gold nanorods and gold nanoparticles include to fight against cancer cells in photothermal therapy [8,70,71] and drug delivery [12,72].

2.2. *Silver Nanoparticles (AgNPs)*

The major routes of AgNPs preparation are physical, chemical, and biological synthesis, similar to AuNPs. Conventional physical methods for AgNPs fabrication are based on the evaporation–condensation approach and laser ablation technique, which are very efficient and moreover, they allow obtaining NPs with high purity, avoiding the use of potentially toxic reagents. The size, shape and yield of the AgNPs can be tailored by changing the parameters of the tube furnace (gas temperature, pressure) in the case of the evaporation-condensation technique, or laser power, duration of irradiation and liquid media selected in the case of laser ablation [73,74].

Chemical methods or wet chemistry usually employs three main components: metal precursors (metal salts), reducing agents (organic or inorganic), and stabilizing/capping agents. By this simple route, the nucleation and subsequent growth of AgNPs can be easily achieved. Silver nitrate $AgNO_3$ is the most common salt used as precursor, while a large variety of reducing agents, such as sodium citrate, ascorbate, sodium borohydride ($NaBH_4$), elemental hydrogen, Tollens reagent, N,N-dimethylformamide, poly(ethylene glycol)-block copolymers, thyo-glycerol, hydrazine, ammonium formate, etc., are used for the reduction of the silver ions (Ag^+) in the aqueous or nonaqueous solutions [23,24]. Spherical AgNPs with diameter ranging from 10–100 nm were synthesized using ascorbic acid, sodium citrate, $NaBH_4$, thiosulfate and polyethylene glycol as the reducing agents, while the surfactants such as citrate, polyvinylpyrolidone (PVP), cetyltrimethylammonium bromide (CTAB) and polyvinyl alcohol (PVA) were employed in order to stabilize particles and avoid sedimentation and agglomeration [75–77]. Optimized spherical and hemispherical AgNPs were obtained, having a diameter less than 10 nm, by adjusting four parameters: $AgNO_3$ concentration, sodium citrate concentration, $NaBH_4$ concentration and the pH of the reaction [78] Well-dispersed silver nanorods were reported by Ojha et al. [79], obtained by mixing the precursor solution of $AgNO_3$ with citrate and adding NaOH or $NaBH_4$ while stirring, the surfactant being cetyltrimethylammonium bromide. The aspect ratio (L/d) of the obtained nanorods (estimated from TEM) were 3.0 ± 0.1, 1.8 ± 0.1 and 1.1 ± 0.1, depending on the concentration of colloidal seed solution. Similarly, by adjusting the reaction conditions, including the ratio of PVP to silver nitrate, reaction temperature and seeding conditions, silver nanowires, with diameters ranging from 30 to 40 nm, and lengths up to ∼50µm, were reported when PVP was used as stabilizing agent [80]. Zhang et al. reported the fabrication of silver triangular bipyramids by photoinduced reduction of silver nitrate in the presence of sodium citrate, bis(p-sulfonatophenyl) phenylphosphine dihydrate dipotassium salt, adjusted by NaOH, under light irradiation (wavelength range between 500–650 nm) [81]. There are many examples in literature reporting the production of different shapes of silver nanoparticles synthesized with various chemical reductants, highlighting the advantages of this method, such as ease of production, high yield (contrary to physical methods, which have low yield), low cost. The main drawback remains however, the use of chemical reducing agents is harmful to living organisms. The electrochemical method (electrolysis) has been also employed for AgNPs synthesis using sacrificial anode (Ag electrode) and $AgNO_3$ as electrolyte or precursor leading to AgNPs with size depending on the current density (10–50, 10–20, 20–80, 10–30 nm) [28,82–84].

Biologically-mediated synthesis of AgNPs emerged as a valuable option in order to overcome the shortcomings of chemical methods, and similar to AuNPs biosynthesis, bacteria, fungi, plant extracts, and small biomolecules were used as biological precursors in the context of environmentally friendly approaches. Examples of currently reported bacterial strains and fungi used for AgNPs synthesis are: *Bacillus amyloliquefaciens* [85], *Acinetobacter calcoaceticus* [86], *Pseudomonas aeruginosa* [87], *Escherichia coli* [88], *Brevibacterium casei* [89] and *Aspergillus sp., Fusarium sp., Penicillium sp.*, respectively [56,90,91]. On the other hand, plant extracts (*Aloe vera, Cocos nucifera, Ocimum tenuiflorum, Vitis vinifera, Chenopodium album*) rich in bio-compounds such as polysaccharides, tannins, saponins, phenolics, terpenoids, flavones, alkaloids, proteins, enzymes, vitamins and amino acids are largely available source of natural reagents for reduction processes involved in AgNPs production [50,92–95]. Overall, the biological methods demonstrated a controlled particle size, shape and mono-dispersity, while reducing time of preparation.

Similar to the AuNPs, AgNPs used in electrochemical (bio)sensors elaboration may improve the limit of detection of target molecules due to its high thermal, chemical stability, electrical conductivity, and catalytic activity [96–102]. Concerning other biomedical applications, AgNPs also presents a strong antimicrobial activity [103,104].

2.3. Platinum Nanoparticles (PtNPs)

PtNPs are also produced by biological or synthetic methods to be used in the biomedical field [105]. The physical methods include evaporation and condensation, laser ablation, solvothermal processes, all of them with advantages and disadvantages: high speed and no use of toxic chemicals, purity, uniform size and shape, versus low productivity, high cost, energy consuming, less thermal stability and high amount of waste. Chemical synthesis techniques include the sol–gel process, pyrolysis, microemulsion, hydrothermal, polyol synthesis and plasma chemical vapor deposition, in which metal precursors, capping or stabilizing agent, and reducing agent are the basic requirements. $PtCl_2$, H_2PtCl_6, $Pt(NH_3)_4(NO_3)_2$ are the preferred precursors, while reducing agents like ascorbate, sodium borohydride, potassium bitartrate and trisodium are often used for the reduction process in order to tailor the NPs size and shape [106]. The biological synthesis of PtNPs was performed using different plants extracts (for example *Ocimum sanctum, Pinus resinosa, Fumariae herba*), while *Desulfovibrio desulfuricans* and *Acinetobacter calcoaceticus* were reported as bacterial routes to produce PtNPs by the reduction of Pt(IV) ion into Pt(0) NPs [107]. In this case, the biosynthesis route offer advantages such as small size (2–3.5 nm), monodispersion, no toxicity, cost effectiveness, rapid synthesis and environmental friendliness, but also drawbacks, being a laborious method, with relatively high cost, and less control over the NPs size and shape. The electrochemical technique (cyclic voltammetry, amperometry) was also employed for PtNPs synthesis [43,108,109]. Despite the high cost associated with its rarity in nature, PtNPs have various analytical applications (biosensors, fuel cells etc.) [110–113].

2.4. Other Metal/Metal Oxide Nanoparticles

Other metal NPs with attractive properties are: Palladium NPs and various binary and ternary combination of metal NPs as oxides, sulfides or metallic form (Bi_2S_3, ZnO, CuO, Co_3N, Co_3O_4, Ni/ZnO, MoS_2, IrO_2, NiO, TiO_2, PbS, Pd@Pt, Ni-Co@Pt, etc.) with different morphology (nanowires, nanosheets, nanofibers, nanoflakes, nanotubes, nanorods, core-shell) [114,115]. They can act as electrocatalyst (bimetallic and trimetallic NPs have synergic effect), but also are widely used as electrode modifiers with bioanalytical applications [116–121]. A special attention is paid to TiO_2 nanomaterials due to some advantages: Ti is a biocompatible and abundant material, TiO_2 nanomaterials are chemically stable, mechanically strong, highly uniform, having large surface area, with photo-catalytic properties and multi-functionalities, therefore TiO_2 is often used as a supported material for decoration with other metal NPs with many biomedical applications [17,122–124].

Another important class is the magnetic nanoparticles, which include metal oxides, pure metals and magnetic nanocomposites, having various diagnostic and therapeutic applications in biomedical field [125]. Those containing iron oxides coated in polymer are the most used in bioanalysis due to their lower toxicity. Magnetic NPs are a powerful tool used in the design of electrochemical immunosensors due to their properties: They can be separated with an external magnet and then redispersed benefic for analytes separation and concentration, they have a large surface area leading to increased immobilization of biomolecules and facilitating the antigen-antibody reactions, controllable size, easy functionalization, superparamagnetic behavior, good conductivity and good biocompatibility [124,126]. A wide range of biomedical compounds have been analyzed by using immunosensors based on magnetic beads with increased sensitivity and reduced matrix effect [119–121,124,125].

The plenty advantages concerning the unique physical and chemical properties of metal NPs corroborated with their nanometer size are responsible for their wide range of biomedical applications (imaging, diagnostics, therapeutics). There still remains some challenges which should be addressed: Synthesis of the same size, agglomeration to be avoided, uniform distribution on the electrode surface and safe and efficient in vivo applications.

3. Carbon-Based Nanomaterials

The extensive family of nanoscale carbonaceous materials includes: Carbon nanotubes (CNTs), graphene, carbon-based quantum dots, fullerene, carbon black (CB), carbon nanowires, carbon nanofibers (CNF), carbon nanoribbons, carbon nanohorns, carbon nanocones, nanodiamonds, carbon nanoonios and mesoporous carbon [127–131]. These carbon-based nanomaterials are classified as 0D, 1D and 2D materials according to their shape with their representative members: fullerene (0D), CNTs (1D) and graphene (2D) (Figure 1). All the nanomaterials based on carbon possess some notorious inherent properties, such as high electrical conductivity, chemical stability, mechanical strength, high surface-to-volume ratio and biocompatibility [127,128,132,133]. They can be easily functionalized through covalent and non-covalent modification with functional groups or substances and also, they can be combined with other (nano)materials leading to hybrid (nano)composites with synergic effects for the envisaged applications. Their further functionalization with NPs not only drastically improves their physicochemical properties, but also prevents their agglomeration [133].

Their unique features concerning the electrical, optical, mechanical and thermal properties opened the way for a vast variety of applications, such as biosensing, bioimaging, cancer therapy, tissue engineering, drug delivery, biofuel cells, energy generation, storage, etc. [119–121,126,130,134–143].

Among all the nanostructured carbon materials, CNTs and graphene are the most widely studied, synthesized, functionalized and used for various analytical applications. The starting point for their synthesis is the graphite. As an amorphous and disordered product (concerning its structure), the graphite is stable at normal values of temperature and atmosphere. As it undergoes a controlled heating process, the disordered carbon atoms have a higher thermal energy that makes them arrange in a thermodynamically stable phase until reaching the sublimation point of 3915 K. Depending on the working temperature, graphene and CNTs are obtained [144].

Concerning the characterization of the carbon-based nanomaterials, a wide range of techniques have been employed so far such as: Raman spectroscopy, FTIR, UV–Vis, energy dispersive spectroscopy (EDS), X-ray photoelectron spectroscopy (XPS), XRD, AFM, TEM, SEM, laser scanning microscopy, electrochemistry, conductivity measurements, etc. [128,129,136,145,146]

3.1. Carbon Nanotubes

Carbon nanotubes have been discovered in 1991 by Iijima consisting in 1D cylindrical tubes of sp^2 hybridized carbon atoms in a hexagonal lattice with delocalized π electrons [129]. They are unique due to the strong intermolecular bonds between the alternating hexagonal rings leading to an agglomerated structure [147,148]. They can have different lengths, thicknesses and number of layers. Their diameter is in the nanometer scale, meanwhile their length can reach up to several millimeters, even centimeters, possessing a high aspect ratio [127]. CNTs can be classified into single-walled carbon nanotubes (SWCNTs) and multi-walled carbon nanotubes (MWCNTs) according to the number of rolled graphene layers. SWCNTs have a single wall in the form of an empty cylinder on the inside, meanwhile MWCNTs have concentric tubes inside. These nanomaterials present good mechanical strength (100 times stronger than steel), excellent conductivity (CNTs conduct heat and electricity similar to copper), excellent electrocatalytic ability (they enhance the electron transfer for proteins/enzymes) and low density (half of the aluminum density) [127,132]. Concerning their thermal properties, the thermal conductivity is higher along the nanotubes than across them, an increased number of defects negatively influences the thermal conductivity, and MWCNTs have a higher value of thermal conductivity than SWCNTs due to their multi-layers [149]. CNTs have paved the way to the discovery of graphene.

The most used methods for CNTs synthesis are arc plasma, laser method and catalyzed chemical vapor deposition (Figure 2) [134,150]. By arc plasma method, an electric current is applied between two electrodes in an inert gas atmosphere and the deposition of CNTs on an electrode is achieved by consuming the other electrode [151]. This method has the disadvantage of obtaining a complex mixture requiring an additional purification in order to separate the CNTs from the residual components [152]. The use of laser method for CNTs synthesis has led to higher yields. The CNTs were obtained by laser vaporization of graphite electrodes with a mixture of catalysts (Co:Ni = 50:50, 1200 °C, argon atmosphere). For an efficient purification of CNTs, a vacuum heat treatment was performed at 1000 °C for removing fullerenes [153]. Catalyzed chemical vapor deposition is the most used method of synthesizing CNTs by catalytic deposition of hydrocarbons (acetylene) over a metal catalyst (Co and Fe) [132,154]. The optimization of this process consisted in the choice of hydrocarbons, the control of the reaction conditions, as well as the continuous elimination of CNTs as they are formed.

In order to surpass their limitations (especially their insolubility and tendency of agglomeration) and to increase the CNTs performance in analytical applications, the CNTs functionalization is performed including some groups: covalent, defect group, non-covalent and endohedral functionalization [127]. The non-covalent functionalization (hydrophobic and π-π interactions) presents the advantage of not disturbing the conjugated system of the CNTs. They can be functionalized with different functional groups: -OH, -COOH, $-NH_2$, -F, etc. without significantly altering their properties. [155,156]. Moreover, more active binding sites can be created on the surface of nanotubes making them more easily dispersible in various solvents. Some types of functionalized CNTs are soluble in water and in other highly polar solvents [157,158]. Additionally, for biological applications different biomolecules (lipids, proteins, etc.) can be attached preserving their structural and functional integrity in order to increase the detection sensitivity [159,160]. The drop casting method widely used for CNTs deposition (especially for the electrode modification) presents a non-uniform distribution and fragility, limitations which can be overpassed by using polymers [127,161]. Recently, another approach has been employed by preparing free-standing buckypaper electrodes suitable for portable and wearable sensors and biofuel cells [135–138,162].

The various biomedical applications of CNTs requires an evaluation of their biocompatibility and toxicity. So far it was demonstrated that CNTs support the growth of neurons and osteoblastic cells; the functionalized CNTs improved the neuronal cell adhesion, mitochondrial membrane potential and concentration of acetylcholine; the CNTs coated with polymers were applied for bone tissue engineering [163]. CNTs are also promising drug delivery vehicles due to their covalent and noncovalent conjugation with drugs and biomolecules, entering the cell through cytoplasmic translocation. Nevertheless, CNTs present lung and embryonic toxicity [163].

3.2. Graphene

Graphene was discovered in 2004 by Novoselov et al. being awarded with a Nobel Prize in Physics in 2010 [164]. This 2D carbon-based nanomaterial with sp^2 hybridization consists in a single-atom-thick layer of defect-free carbon atoms in a hexagonal network with delocalized π electrons. It possesses unique properties such as: excellent electrical conductivity (six times greater than copper and 60 times greater than CNTs), huge surface area (double than SWCNTs), excellent mechanical strength (200 times higher than steel), high thermal conductivity, optical transparence and elasticity [127,134,145,146]. Its electronic properties are influenced by the number of graphene layers and of edge defects. Until now, graphene is the thinnest and the strongest material. It is considered the building block for other carbon nanomaterials: by stacking, it leads to 3D graphite, by rolling, it forms 1D CNTs and by wrapping, it results in 0D fullerenes [165,166].

The first method of obtaining graphene is mentioned by Novoselov et al. in 2004 by mechanical graphite exfoliation [134,167,168]. An improved method was presented by Stankovich et al. using graphite oxide exfoliation in water under ultrasonication for 24 h at 100 °C [169]. Chemical methods have also been used to chemically extract graphene sheets from graphite by exfoliation [170,171]. A commonly used method in the laboratory is Hummers' method of obtaining graphene oxide (GO) by adding potassium permanganate to a solution of graphite, sulfuric acid and sodium nitrate [172]. This method has been improved by removing sodium nitrate and adding sodium persulfate which ensures the complete exfoliation of graphite obtaining suspensions of individual graphite oxide sheets. Besides, it is more environmentally friendly by eliminating the formation of gases such as carbon dioxide and dinitrogen tetra oxide [173–175]. Other synthesis methods are the epitaxial growth [176] and chemical vapor deposition (Figure 2) [177,178]. Graphene can be obtained with an almost perfect structure and excellent properties by using these three methods [179].

Compared to graphene, GO can be cheaply produced by high-yielding chemical methods and is highly hydrophilic due to oxygen atoms that increase the distance between layers and can be exfoliated in water at moderate ultrasounds [179]. The presence of oxygen functional groups (hydroxyl, epoxide, carbonyl and carboxyl groups) improves the hydrophilicity, stability and the anchoring of different (bio)molecules. The chemical reduction of GO in order to obtain reduced graphene oxide (rGO) can be done by adding liquid reagents such as hydrazine, dimethylhydrazine, sodium borohydride, ascorbic acid (AA) and iodic acid to an aqueous dispersion of GO [10,180–186]. The electrochemical reduction of GO was also employed especially for electrochemical applications [127]. The difference between GO and rGO is the different percentage of oxygen content that leads to an insulating trend towards semiconductor behavior of GO, while rGO has a higher electrical conductivity [131,162,187]. Regarding the surface, rGO has a relatively larger specific surface area compared to GO [131,188]. Another modification of graphene structure is done by doping with heteroatoms (N_2, B, S, P) being used for region activation with applications in electrochemical (bio)sensors [127,128]. Graphene and its derivatives (GO, rGO, graphene quantum dots) are important nanomaterials with outstanding properties which have been employed in various fields such as: biosensors, electronics, energy storage etc. [145,146,162,164,189,190]

Due to the high biomedical potential of graphene and its composites, the biocompatibility of these nanomaterials has to be evaluated aiming towards clinical translation. According to their physical and chemical properties, it was indicated that the nano-sized and surface coated materials were more biocompatible, meanwhile the micro-sized ones presented a high inflammation response, both in vitro and in vivo [191]. Additionally, lung toxicity was also reported. The impurities (toxic reagents used in synthesis) are responsible for cytotoxicity; therefore, the production of high-quality graphene is needed. The use of polymer nanocomposites reduces the toxicity favorizing the osteoblasts adhesion and proliferation [192].

3.3. Carbon-Based Quantum Dots

The latest discovered carbon-based nanomaterials are carbon dots (CDs—defined as carbon nanoparticles with less than 10 nm diameter) and graphene quantum dots (GQDs—possessing a graphene structure with layers less than 10 nm thick and 100 nm in lateral size), containing carbon core and many functional groups [193–195]. There are two main techniques concerning the synthesis of these quasi 0D nanomaterials: "Top-down" and "bottom-up" using characteristic precursors (carbonaceous materials; ethanol, nitriles, amino-acid) by physical, chemical and electrochemical methods [194,196–198]. Due to their unique structure, they present a lot of advantages, such as biocompatibility, nontoxicity, easy functionalization, chemical stability, abundant resources, low cost, versatility, attractive optical properties (photoluminescence), excellent electronic properties, high surface area and good solubility in many solvents [193,194,198]. They are widely applied in electrochemical sensing (as signal tags or as electrode modifiers alone or in combination with other nanomaterials), electrochemical flexible devices, electrocatalysis and biofuel cells [194,198–200]. GQDs have been more often used in electrochemical (bio)sensing due to their quantum confinement and edge effect, leading to their higher electrical and thermal conductivity besides their optical properties, as compared to the CDs and traditional quantum dots [193,197,201].

The wide range of carbon-based nanomaterials advantages generated by their unique properties combined with their nano-size are responsible for their widespread use in biomedical field. There still remains some disadvantageous issues that should be solved concerning synthesis and safety, such as expensive synthesis methods, removal of metal catalyst, impurities and other chemical reagents after the synthesis step, agglomeration, determination of proper toxicity for in vivo applications.

4. Metal Nanoparticles and Carbon-Based Nanomaterials in (Bio)Sensors Design

4.1. Screen-Printed Electrodes as (Bio)Sensing Platforms

According to the International Union of Pure and Applied Chemistry (IUPAC), a chemical sensor "is a device that transforms chemical information, ranging from the concentration of a specific sample component to total composition analysis, into an analytically useful signal" [202,203]. A biosensor, according to the IUPAC, can be defined as "a device that uses specific biochemical reactions mediated by isolated enzymes, immunosystems, tissues, organelles or whole cells to detect chemical compounds usually by electrical, thermal or optical signals" [202,204]. Meanwhile, an electrochemical biosensor is "a self-contained integrated device, which is capable of providing specific quantitative or semi-quantitative analytical information using a biological recognition element (biochemical receptor) which is retained in direct spatial contact with an electrochemical transduction element" [145,202]. Therefore, two basic units are primordial for the (bio)sensor construction: a receptor (recognition element—transforms the information into a signal) and a transducer (transfers the signal to a measured result). The electrochemical methods employed for the (bio)sensor analysis are various such as: amperometry, potentiometry, conductometry, voltammetry, impedance and surface charge sensing using field-effect transistors (FETs) [202]. The transducer part of an electrochemical sensor is an electrode. There is a wide range of electrodes (ion-selective, glass, gas, metal, carbon and chemically modified electrode), among which the carbon electrodes are maybe the most used [202]. The electrochemical methods have gained a great interest due to their advantages over others analytical techniques such as: sensitivity, selectivity, low cost, simplicity, ease of use, high reproducibility, low power requirement, real-time results and possibility of miniaturization and automation, features suitable for portable sensing devices with industrial and clinical applications.

In the recent decades, SPEs have attracted an increasing interest due to their advantageous characteristics, such as miniaturized form, great variety of electrode materials, use within a wide potential range, reduced sample volume, low cost, portable, faster time response and simplicity. The recent advances in technology have enabled the miniaturization of the electrodes and also of the potentiostats, the miniaturization and portability allowing to perform on-site and real-time analysis. SPEs are produced by screen-printed technology which consists in the deposition of more layers (ink, insulating material) on a substrate offering versatility in electrode design and electrode material, reproducibility, excellent uniformity, mass production, compatibility and modifications [205,206]. The process of SPE fabrication comprises of several steps: ink manufacture, stencil formation, layer printing and sinterization (thermal, photonic, plasma, microwave, electrical and chemical agents sintering) [5,207]. The SPE contains a three-electrode configuration: one or more working electrodes -WE, the pseudoreference electrode-PRE and the counter electrode-CE, with their connections on a chemically inert supporting material (various types: plastic polymer, paper, ceramic, alumina, etc.), having small dimensions (for example 3.5–5 $\times$ 1 $\times$ 0.05 cm for a classic SPE) (Figure 3) [5,205,206]. The dimension, the thickness and the form/shape can be controlled through the screen-printed technique. There are three main printing electrode technologies: Screen printing, inkjet printing and 3D printing, each one with their own advantages and disadvantages. Various nanostructured materials are used as embedded components (dispersions of conductive nanomaterials inks) in the WE's configuration and/or as immobilized materials at the WE's surface leading to nano-based printed or modified electrodes [207]. There are few disadvantages concerning the SPEs, such as laborious procedure for fabrication, variation in stability and reproducibility for lab-made and lab-modified SPEs, the need of an electrochemical pretreatment step to increase the reproducibility and a limited life-time for SPEs containing a bio-recognition element.

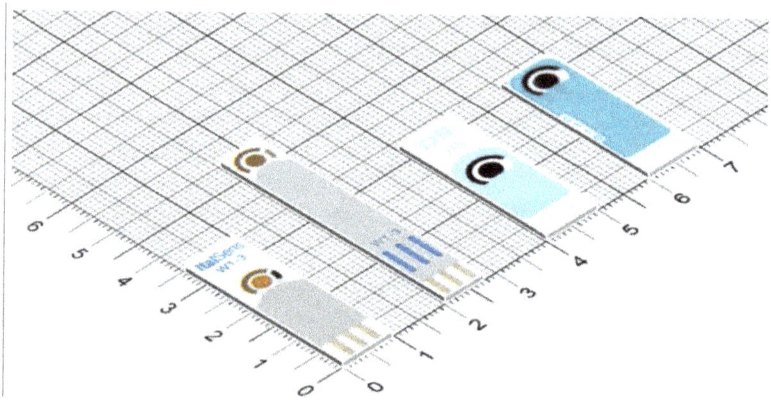

Figure 3. Some examples of commercially available screen-printed electrodes (from different manufacturers: https://www.dropsens.com/en/screen_printed_electrodes_pag.html; https://www.palmsens.com/products/sensors/screen-printed-electrodes/, accessed 10 September 2021).

The composition and/or the surface of SPEs can be easily modified with a plethora of materials/substrates/substances in order to improve their analytical properties (especially the sensitivity and selectivity) and therefore their applications. The metallic nanoparticles and carbon nanomaterials are suitable for the SPEs modification because they are biocompatible, contribute to the biomolecules' immobilization and increase the surface area, adsorption and conductivity of the electrode.

The metal nanoparticles employed in order to enhance the electrochemical signal, but also as labels in biosensors are: silver nanoparticles, gold nanoparticle, copper nanoparticles, iron nanoparticles, palladium nanoparticles, platinum nanoparticles and rhodium nanoparticles [206,208]. Their unique and excellent chemical and physical properties (high surface to volume ratio, high electron-transfer capability, high electrode conductivity) are responsible for the extensive use of NPs in (bio)sensors elaboration resulting in low limits of detection and anchoring platforms for biomolecules.

From the group of carbonaceous nanomaterials, the following can be mentioned: carbon black, carbon nanotubes, graphene, fullerene, carbon nanofibers, carbon nanohorns, carbon and graphene quantum dots with their unique electrocatalytic properties leading to sensitivity enhancement [5,206,208]. Carbon-based nanomaterials have brought uncountable benefits for electrochemical (bio)sensors due to their outstanding and attractive features such as: large specific surface area, elevated conductivity, high adsorption capability, decreased over-potentials, thermal conductivity, mechanical strength, high elasticity and functionalization possibility.

Composites based on metallic and metal oxide nanoparticles (monoatomic, polyatomic), carbon nanomaterials and polymers have been also envisaged as nanostructures in the SPE design [206,208]. The post-printing methods used for SPEs modification with nanomaterials includes: drop casting, electrodeposition, electrospraying, electrospinning, the Langmuir–Blodgett and the Langmuir–Schaefer methods [205,208–210].

The screen-printed electrochemical sensing platforms have been applied for sensing and monitoring of a wide range of target analytes in many fields, such as food and drinks, environmental analysis, pathogens, cancer biomarkers and other relevant biomolecules, pharmaceutical analysis and biological analysis, performed in the traditional laboratories and also in-to-the-field (in-situ detection) [43,119–121,205,206,208–213]. In the case of the biosensors' elaboration, the SPEs have been modified also with biological elements such as enzymes, antibodies and nucleic acids by applying various methods of immobilization (casting, physical adsorption, electrochemical coating or inclusion into the ink) [5,206,210]. In the last years, a special attention was attributed to the detection of viral pathogens, such as the Human Immunodeficiency Virus, the Hepatitis viruses, the Zika Virus, the Dengue virus and SARS-CoV-2, using the combination between SPEs and nanomaterials for the development of portable, easy-to-use and cost-effective biosensors [214].

4.2. Nano(Bio)Sensors Based on Screen-Printed Electrodes with Biomedical Applications

4.2.1. Nano(Bio)Sensors Based on Screen-Printed Electrodes with Medical Applications

Diagnostic and monitoring are essential in healthcare requiring accurate, sensitive, selective and fast results, features which can be accomplished by using screen-printed electrodes. These small size disposable electrodes can be easily integrated in point-of-care devices which are easy-to-handle, low-cost, portable and miniaturized sensors. Some examples of screen-printed electrodes modified with nanomaterials and applied for the determination of some biomolecules with great significance in the medical field are presented below and summarized in Table 1.

Table 1. Screen-printed electrodes modified with various nanomaterials for medical applications (some examples).

Type of NPs	Analyte	Method/ LOD/LR	Real Samples/ Recovery	Ref.
Reduced exfoliated graphene oxide	NADH	Amperometry 2 μM/10 μM–2.0 mM		[215]
	H_2O_2	Amperometry 0.05 μM; 0.08 μM 0.1 μM–2.0 mM; 0.15 μM–1.8 mM		
	Glucose	Amperometry 1 μM/5 μM–10.0 mM	Milk samples	
MWCNTs, AuNPs	NADH	Amperometry 3.72 μM/12.4–150 μM		[216]
Carbon black	NADH, Ascorbic acid, cysteine	Amperometry 1 μM		[217]
Ruthenium dioxide-graphene nanoribbon	NADH	Amperometry 0.52 μM/1–1300 μM		[218]
Platinum NPs, graphene sheets@cerium oxide	H_2O_2	Amperometry 0.43 μM/1 μM–10.0 mM	Contact lens clear solution/ 99.5–102%	[219]
Silver NPs, rGO@CeO_2	H_2O_2	Amperometry 0.21 μM/0.5 μM–12 mM	Contact lens clear solution/ 100–103.5%	[220]
Reduced graphene nanoribbons	Uric acid Levodopa Ascorbic acid	DPV 5 μM 10–50 μM 1–5 mM	Urine 97–101%	[221]
Reduced graphene oxide, gold NPs	Ascorbic acid Dopamine Uric acid	DPV 1.04 μM/20–375 μM 0.29 μM/1–160 μM 5.4 μM/25–200 μM	Urine 95–98.89%	[222]
Carbon black, graphene oxide	Uric acid	Flow-injection amperometry 0.01 μM/0.05–2000 μM	Urine 95%	[223]
Reduced graphene oxide, gold NPs	Dopamine	Amperometry 0.17 μM/0.57–500 μM	Blood 101.5–102.5%	[224]
MWCNTs, AuNPs	Dopamine Serotonin	DPV 0.3 μM/1–100 μM 0.8 μM/2.5–100 μM	Serum, Tears, Saliva 111.18%, 97.78%, 108.53%	[225]
Polypryrrole NPs, AuNPs	Serotonin	SWV 33.22 nM/0.1–15 μM	Serum 100.27–103.06%	[226]
Polypryrrole NPs, AuNPs	Interleukin-6	EIS 0.33 pg/mL/ 1 pg/mL–15 μg/mL	Serum 101.41–102.45%	[227]
Platinum NPs, reduced graphene oxide	Glucose H_2O_2 Cholesterol	Amperometry 44.3 μM/0.25–6.0 mM 5.24 μM/0.01–0.80 mM 40.5 μM/0.25–4.0 mM	Serum 82.2–104.1%	[228]

Table 1. *Cont.*

Type of NPs	Analyte	Method/ LOD/LR	Real Samples/ Recovery	Ref.
$Cu(OH)_2$@CoNi-LDH core–shell nanotubes	Glucose	Amperometry 6.7 μM/20 μM–8 mM	Blood 97.5%	[229]
AuNPS	Cysteine, Methionine, Glutathione, Homocysteine	Amperometry (HPLC-EC) 3.1 μM/10–80 μM 1 μM/3.3–30 μM 0.1 μM/0.3–10 μM 0.6 μM/2.2–30 μM	Dietary supplements 97.25–99%	[230]
Graphene (in the ink of SPE)	Norepinephrine	SWV 0.265 μM/1–30 μM		[231]
MWCNTs rGO	Bilirubin	Amperometry 0.3 μM/0.5–500 μM 0.1 μM/0.1–600 μM	Blood 94–106.5%	[232]
Graphene (in the ink of SPE), AuNPs	C-reactive protein	EIS 15 ng/mL/0.05–100 μg/mL	Blood 97.9–103.9%	[233]
NiO NPs, Nafion-MWCNTs	Insulin	Amperometry 6.1 nM/20–260 nM		[234]
CNTs-$NiCoO_2$ in Nafion	Insulin	Amperometry 1.06 μg/mL/ 0.17–75 μg/mL		[235]
SWCNTs (commercial SWCNTs/SPE)	Glycated hemoglobin	SWV 0.03 pg/mL/ 0.1–1000 pg/mL		[236]
Carbon nanofiber	Survival Motor Neuron Protein	SWV 0.75 pg/mL 1 pg/mL–100 ng/mL	Whole blood	[237]
PANI/AuNPs	*E. coli* DNA *E. coli* cells	CV 0.5 fM/1000–0.001 pM 4 CFU/mL 4×10^6 CFU/mL	Urine	[238]
AuNPs-CNTs, AgNPs	Hepatitis B surface antigen	DPV 0.86 ng/mL/1–40 ng/mL	Blood 80.70–91.40%	[239]
Graphene, AuNPs	Pyoverdine (Pseudomonas aeruginosa)	DPV 0.33 μM/1–100 μM	Serum, saliva, tap water 98.41–102.12%	[240]
AuNPs	Carcinoma antigen 125	EIS 6.7 U/mL/0–100 U/mL	Blood	[241]
AuNPs PtNPs	Carcinoma antigen 125	EIS 419 ng/mL 386 ng/mL 450 ng/mL–2.916 μg/mL		[242]
rGO-AuNPs	Carcinoembryonic antigen	Amperometry, CV 0.28 ng/mL/ 0.5–50 ng/mL 181.5 ng/mL/ 250–2000 ng/mL		[243]
CNTs-AuNPs (in the ink)	p53 protein	Amperometry 14 pM/20 pM–10 nM	Urine 91–132%	[244]

Table 1. *Cont.*

Type of NPs	Analyte	Method/ LOD/LR	Real Samples/ Recovery	Ref.
Graphene quantum dots—MWCNTs	Interleukin-13 receptor-α2 Cadherin-17	Amperometry 1.4 ng/mL 4.92–100 ng/mL 0.03 ng/mL 0.11–10 ng/mL	Cancer cells	[245]
Graphene (in the ink), polyaniline	Human chorionic gonadotropin	EIS 0.286 pg/mL/ 0.001–50 ng/mL	Urine	[246]
Calixarene functionalized graphene, Au@Fe_3O_4	SARS-CoV-2	DPV 200 copies/mL 3 aM/ 10^{-17}–10^{-12} M	Various biological fluids 20–100%	[247]
Carbon black	SARS-CoV-2	DPV 19 ng/mL (S protein) 8 ng/mL (N protein)	Saliva	[248]

Exfoliated graphene oxide was electrochemically reduced by the potentiostatic method (the color changed from yellow-brown to black) on SPE doped with ionic liquid, presenting a uniform surface topography. This nanomaterial promoted the oxidation of NADH on this modified electrode by increasing the electron transfer: a negative shift of the oxidation peak potential of 0.22 V, a double increase of the peak current, a good resistance to fouling induced by a high density of edge-plane-like defective sites on carbon materials and a peak separation for ascorbic acid and NADH oxidation of 220 mV, in comparison with the non-modified electrode. The same superior electrocatalytic activity was also recorded for H_2O_2 analysis: oxidation started at +0.45 V and reduction at 0 V. Glucose oxidase was immobilized onto this modified SPE by cross-linking with glutaraldehyde leading to a nanobiosensor with improved analytical performance for glucose determination at −0.2 V. A good selectivity of the 3 sensors was demonstrated in the presence of ascorbic acid, uric acid and dopamine as interferents [215]. Layer by layer method was used for the SPE modification with MWCNTs, AuNPs and electropolymerized polyneutral red by optimizing the amount of CNTs and thickness of the film, then the nanoplatform was investigated toward the electrooxidation of NADH, showing the highest anodic peak current and good analytical parameters for amperometric determination [216].

Carbon nanomaterials, such as CB, SWCNTs-COOH, GO and rGO, were employed for the modification of homemade SPEs by drop casting, being characterized by XPS, Raman spectroscopy, SEM and electrochemistry (cyclic voltammetry, amperometry, electrochemical impedance spectroscopy -EIS). These nanostructured platforms were furthermore tested for NADH, AA and cysteine sensing, exploiting their greatest advantage such as a wide potential window and a high electrochemically accessible area. In the case of NADH oxidation, the lowest potentials of 400 and 440 mV were recorded for SPE modified with CB and SWCNTs-COOH, respectively; for AA oxidation the highest current peaks and the lowest potential peaks (90, 150 mV) were detected by using SWCNTs-COOH and CB, respectively. The same two nanomaterials have been responsible for the best electrochemical behavior of cysteine (oxidation at 580 mV). Thus, the analytical performances of the CB and CNTs are similar, but CB offers some particular advantages, such as it is cost-effective, suitable to obtain homogenous and stable dispersion and mass-producible [217]. NADH quantification was also performed by using another nanocomposite based on ruthenium dioxide-graphene nanoribbon drop casted on homemade SPE. RuO_2 NPs presented an average diameter of 2 nm forming homogenous mats on the graphene matrix [218].

A hybrid nanocomposite based on platinum NPs doping into graphene sheets@cerium oxide presented good synergistic effects when tested towards the electrocatalytic reduction of H_2O_2. Various techniques, such as XRD, FTIR, SEM, EIS, cyclic voltammetry and amperometry, were applied for the characterization of the nanomaterials and of the modified SPE. The size controllable PtNPs were prepared with different electroless plating times, the best electrochemical response was obtained for PtNPs of 100 nm obtained over 200 s, indicating a larger surface area with more available active sites, but if the time and subsequently the size increases, the NPs aggregate leading to a compact surface with decreased current. The sensor also presented reliable reproducibility, long-range stability and selectivity (when adding glucose, AA, dopamine (DA), uric acid (UA)) [219]. A similar nanocomposite consisting in silver NPs, rGO@cerium oxide was synthesized and used for SPE modification being characterized by the same techniques mentioned before. Controllable in-situ synthesis of AgNPs with different sizes was achieved by a solvothermal process with a reaction time of 1–4 h, indicating that the reduction of Ag+ adsorbed onto the GO matrix is facilitated by the GO which acts as a gentle reductant. The best response for H_2O_2 reduction was registered for SPE modified with the hybrid nanomaterial with a 2 h reaction time (AgNPs having a size of 30 nm) [220].

Other important small biomolecules with high clinical importance, such as AA, UA, DA, levodopa (LD) and glucose, have been determined by using nanosensors based on SPEs. Graphene nanoribbons (obtained by chemical oxidation of CNTs) are a promising candidate for electrochemical sensors with the following advantages: Excellent electrocatalytic effect, enhanced faradaic currents and increased resistance to passivation leading to improved selectivity, sensitivity and reproducibility. SPE modified with reduced graphene nanoribbons (14% wt. oxygen content) presented the best electroanalytical performance for AA, LD and UA sensing in comparison with other related carbon nanomaterials (MWCNTs, oxidized graphene nanoribbons). This nanosensor allowed the simultaneous detection of AA, LD and UA at +0.08, +0.27 and +0.9 V, respectively, due to the reduced graphene nanoribbons containing more defects and edge sites and to the removal of oxygen functionalities [221]. In another study, the simultaneous detection of AA (−120 mV), DA (10 mV) and UA (220 mV) was achieved by employing a SPE modified with rGO and AuNPs (the simultaneous electrochemical reduction of both nanomaterials on the SPE surface is faster and more convenient), which was included in a smartphone-based integrated voltammetry system. The graphene oxide sheets containing a lot of oxygen-containing functional groups can be easier modified with NPs acting as a nanoscale building block to develop nanosensors [222]. A novel flow-injection amperometric nanosensor was developed for the accurate detection of UA based on SPE modified with a mixture of carbon black and graphene oxide by drop casting. The nanocomposite (1:1) was characterized by microscopic techniques (SEM, TEM) revealing a uniform distribution of carbon black spherical particles (30–50 nm) on the graphene oxide sheets. The electrochemical experiments indicated that the nanocomposite presented the highest current response for UA oxidation in comparison with other nanomaterials due to their synergic effects concerning the enhanced conductivity and increased surface area [223]. A SPE modified with rGO, polyneutral red and gold NPs (using commercial solutions of the nanomaterials) was elaborated for the amperometric determination of DA with good sensitivity, selectivity, reproducibility, stability and recovery rates [224]. The simultaneous detection of two neurotransmitters (dopamine and serotonin) was successfully achieved by employing SPE modified with MWCNTs and AuNPs [225]. A nanocomposite consisting in electrochemically generated polypyrrole nanoparticles and AuNPs was elaborated for SPE modification, leading to an increase of the active surface area, and this nanoplatform was used for sensitive and selective determination of serotonin [226]. In another study, the same nanocomposite elaborated through electrochemical techniques was used for immobilization of interleukin-6 aptamer and tested for the detection of the target cytokine [227].

The entrapment method for glucose oxidase immobilization onto the SPE surface has been addressed using various nanomaterials, such as SWCNTs, MWCNTs, rGO, silver NPs, platinum NPs and some polymers (poly(1-vinylimidazole), Nafion, poly(3,4-ethylenedioxythiophene, polyvinyl-alcohol and poly(3-aminobenzoic acid)) [228,249]. The nanosized platform based on platinum NPs, rGO and poly(3-aminobenzoic acid) being elaborated by one-step electrochemical deposition was also successfully applied for the determination of H_2O_2 and cholesterol [228]. Other enzymes such as glutamate dehydrogenase and lactate dehydrogenase have been immobilized onto the SPEs surface by entrapping in a mixture of chitosan and MWCNTs, drop-coating on the previously CNTs modified SPE and using a mixture of MWCNTS, glutaraldehyde and bovine serum albumin, those biosensors being applied for a sensitive detection of glutamate and lactate, respectively [249]. An enzyme free glucose sensor was elaborated by a three-step in-situ synthesis method of highly porous 3D hetero $Cu(OH)_2$@CoNi-LDH core–shell nanotubes on the SPE surface. Firstly the CuNPs were electrodeposited on SPE, then $Cu(OH)_2$ nanotubes were formed followed by the growth of CoNi-LDH nanostructures [229].

Several combinations of nanomaterials have been tested for the detection of aminothiols (cysteine, methionine, glutathione, homocysteine) including AuNPs and MWCNTs (with some polymers such as polyaniline and Nafion), deposited on a screen-printed gold electrode. AuNPs have been synthesized by a green procedure based on sonocatalysis leading to a diameter between 5 to 12 nm. The best result was obtained on the configuration consisting in AuNPs drop casted on SPE which indicates that the NPs enhance the electrochemical response of the conventional electrode [230].

The nanomaterials can be also included in the ink used for the SPE fabrication. In a study, homemade flexible screen-printed graphene electrodes have been developed and connected to a smartphone-based system. As a control, screen-printed carbon electrodes modified with graphene/graphene oxide by drop casting were also elaborated. The screen-printed graphene electrode presented on its surface a graphene-like layer structure, which led to a higher electron transfer rate in the case of norepinephrine electrochemical analysis [231]. Therefore, the sensitivity of the smartphone-based electrochemical system was improved by graphene used as a constituent material of the electrode. This novel set-up is suitable for portable and wearable point-of-care devices.

A comparison between the electrochemical performances of MWCNTs and graphene has been analyzed in the case of bilirubin oxidation. CNTs have been covalently bonded to the electrode surface, which was previously functionalized with NH_2 groups. Graphene oxide were electrochemically reduced to the SPE surface. Better electroanalytical parameters have been recorded in the case of SPE modified with graphene, explained by a more enhanced electron transfer rate and a higher surface area calculated by electrochemical measurements [232]. A paper-based sensor was fabricated using graphene and carbon ink for the WE printing. AuNPs were then electrochemically deposited resulting in a diameter of 50–70 nm and a uniform distribution. The NPs increased the electrode surface area and also the anchoring sites for the covalent antibody immobilization. Multiple steps have been employed for the electrode modification leading to the elaboration of this label-free paper-based immunosensor with improved analytical performances (good sensitivity, selectivity, stability, reproducibility, repeatability, recoveries) toward C-reactive protein detection [233].

Some biomarkers important for diabetes mellitus diagnostic and monitoring were also analyzed by using SPEs modified with nanosized materials. For insulin detection, SPE was modified with Nafion-MWCNTs, followed by the electrochemical pulse potential deposition of NiO NPs. The pulse electrodeposition method generated NPs with a diameter below 30 nm, prevented the NPs agglomeration and controlled the film thickness, leading to a good stability, excellent surface coverage and enhanced electrocatalytic activity. NiO NPs act as an active catalyst for insulin oxidation [234]. Another SPE modified with a nanocomposite of CNTs and $NiCoO_2$ (8:4.5 ratio) using Nafion as a binder was elaborated for insulin detection [235]. Glycated hemoglobin was tested by using label-free aptasensors based on six types of commercial carbon-modified SPEs (bare carbon SPE and carbon SPE modified with graphene, GO, MWCNTs, SWCNTs, CNF). The aptamer was non-covalently immobilized by π–π stacking with the nanomaterials (physical adsorption). The best results were obtained by employing SWCNTs platform, followed by MWCNTs and CNF [236]. The same 6 types of commercial carbon-based nanomaterial-modified SPEs (mentioned above) have been used as platforms for covalent immobilization of survival motor neuron protein antibody. In this case, the CNF-based immunosensor offered the best analytical performance for the protein target detection [237].

An electrochemical genosensor based on polyaniline and AuNPs electrodeposited on the SPE surface has been elaborated through multiple steps. The polymer decoration with NPs changed its redox to neutral pH suitable for biological interactions and also increased the surface area and conductivity improving the electrochemical biosensing performance. This biosensor based on a dual signal amplified strategy was applied for the detection of *E. coli* DNA and *E. coli* cells [238]. For the label-free detection of Hepatitis B surface antigen, an immunosensor was developed based on home-made SPE modified with a AuNPs-CNT nanocomposite (10–30 nm) and AgNPs. The nanocomposite was used in order to increase the surface area, conductivity and the biomolecule immobilization. AgNPs played the role of a redox probe for the direct detection, but also improved the analytical response due to its nanosize [239]. Electrochemically exfoliated graphene functionalized with AuNPs were deposited layer-by-layer on SPE and applied for pyoverdine detection (a virulence factor secreted by *Pseudomonas aeruginosa*) [240].

Tumor biomarkers are associated in patients with tumor or carcinoma; therefore, their early and accurate detection is essential. For this reason, the electrochemical immunosensors have been intensively studied combining the sensitivity of the electrochemical sensor and the specificity of the immunoreactions, paying special attention to the portable and miniaturized devices. A label free impedimetric immunosensor was elaborated based on electrodeposited AuNPs (40–100 nm) on the SPE surface, subsequently modified with self-assembled monolayers. Monoclonal antibody anti-carcinoma antigen 125 (CA125) was immobilized on the nanostructured platform which was successfully applied for the detection of CA125 (=MUC16) [241]. The detection of the same CA 125 biomarker was recorded on another immunosensor based on SPEs modified with AuNPs ("popcorn" nanostructures with 566 nm) and PtNPs (140 nm). The influence of the two nanomaterials on the electrochemical parameters have been compared, and the PtNPs presented the highest surface area leading to a lower limit of detection [242]. Carcinoembryonic antigen was determined on an immunosensor based on a SPE modified in one-step preparation with GO (electrochemically reduced) and AuNPs (140 nm—deposited by the electrochemical method). The nanocomposite increased the electron transfer kinetic and also the effective surface area, leading to a sensitive detection of the target [243]. A SPE modified with CNTs and AuNPs (the nanomaterials being incorporated in the SPE composition by the manufacturer) has been applied for an amperometric subnanomolar detection of the p53 protein (a biomarker for urinary tract carcinoma). This competitive immunosensor presented some advantages such as selectivity, precision, easy fabrication and low cost, being a valuable diagnostic tool for urologic malignancies [244]. A simultaneous determination of two metastasis-related biomarkers (interleukin-13 receptor-α2 and cadherin-17) has been realized on a dual SPE (SPE with two working electrodes) modified with graphene quantum dots (having peroxidase mimicking activity) and MWCNTs, a hybrid nanomaterial combination used for sensitivity enhancement and also as nanocarriers of antibodies for amplification purpose [245]. This dual amperometric sandwich-type immunosensor presented some relevant advantages over the traditional ELISA method such as: simplicity, rapid time response, decentralized analysis, affordability and portability. A graphene-SPE modified with electropolymerized polyaniline was tested for human chorionic gonadotropin. The impedimetric immunosensor was capable of a picogram determination [246].

4.2.2. Nanomaterials and SPEs in COVID-19 Diagnosis, Prevention and Therapy

The recent global outbreak of COVID-19 disease is a continuous threat for public health; therefore, sensitive, fast, low-cost, easy to implement, real-time response, portable and wearable point of care systems are crucial for the control and monitorization of this disease.

In addition, the nanotechnology community is able to significantly contribute and fight against coronavirus disease 2019 (COVID-19), as nanomaterials are well known to possess antiviral properties with role in the prevention, diagnosis, and treatment of virus disease. The mechanism is related to inactivation of the outer layer of the virus upon interaction with hydrophobic nanomaterials surface, which may inhibit or completely destroy the virus [250]. The spike protein located on the outer surface is easily attached by nanoparticles in the form of drugs, coatings or nano medicines in the form of vaccines. Moreover, as nanoparticles are the same size as proteins, they can enter cells to enable expression of antigens or directly target immune cells for release of antigens. In this case, lipidic formulations and polymeric nanoparticles can be used as carriers, as their size, morphology and charge are tailored according the particular situation [251–256]. In a very recent review, Ghaemi et al. [257] highlighted the most used nanomaterials/nanoparticles with high potential in therapy, prevention and detection of targeted virus proteins: carbon nanotubes, graphene and graphene oxide, fullerene, quantum dots, CuNPs, ZnNPs, AgNPs, TiO_2 and chitosan NPs. They concluded that the inclusion of nanomaterials in biosensors offer more detection capability, stability, simplicity of design, reliability and affordability. For example, TiO_2 NPs can be used for surface decontamination in combatting SARS-CoV-2, in different formulations, such as aerosol, paint, water or air treatment systems, based on the photocatalysis mechanism and the production of hydroxyl radicals due to the water molecules oxidation, which promote the disinfection activity of TiO_2 NPs [258]. In terms of prevention, different types of facemasks were designed based on graphene oxide nanoparticles as breathable barrier layers, taking into account the advantage of their hydrophobic surface to inactivate virus. Various types of nanoparticles and nanofibers have been introduced in mask production chains to improve safety performance and confer enhanced antiviral properties [259]. Graphene as a coating material for textiles is advantageous due to its mechanical properties, being also fire resistant, UV protective and conductive [260].

Lab-on-chip based strategies for smart diagnostic and personalized COVID 19 management (miniaturized SASR-CoV-2 biosensors) can be achieved via selecting a specific Anti-SARS-CoV-2 virus protein antibody for selective and sensitive detection within 30–40 min of operation time [261]. In terms of detection, optical biosensors, electrochemical biosensors, piezoelectric biosensors and thermal biosensors have been developed for respiratory virus detection. A review conducted by Samson et al. [262] summarizes the existing status of current diagnostic methods, along with their advantages and limitations, and the advantages of biosensor-based diagnostics over the conventional ones for the detection of SARS-CoV-2. A special attention was paid to novel biosensors used to detect RNA-viruses including CRISPR-Cas9-based paper strip, nucleic-acid based, aptamer-based, antigen-Au/Ag nanoparticles-based electrochemical biosensor, optical biosensor and surface plasmon resonance biosensor. As a nanotechnological approach, the gene-editing technique was modified by including a biological sensor using a CRISPR-Chip coupled with a graphene-based field effect transistor that can detect nucleic acids in a very small concentration (1.7 fM) without any amplification and in a very short time (15 min), while the detection of COVID-19 infection can be performed in less than 40 min [263].

The antigen lateral flow detection of SARS-CoV-2 as a point-of-care approach is realized by using a membrane strip with two lines: first line for the antibody-Au NPs, and second line for the captured antibodies. Biological samples such as blood or urine samples can be applied on the membrane, while, based on the capillary mechanism, proteins are drawn across the strip and an antigen/antibody AuNPs complex is formed and immobilized, a pair of red or blue lines becoming evident [257,264].

SPEs were customized in various configurations for SARS-CoV-2 detection (some components of the virus itself) [248,265–269], but also for the detection of some biomarkers useful for Covid 19 diagnosis (C-reactive protein, interleukins, tumor necrosis factor alpha, interferons, glutamate, breath pH, lymphocytes, platelets, neutrophils and D-dimer) [270]. Nanomaterials such as AuNPs, calixarene functionalized graphene, Au@Fe_3O_4 nanocomposite (400 nm) and carbon black served as catalysts, bioreceptor or labels, but, also, they improved the electrochemical performance of the SPE-based sensors applied for the virus detection [248,265–269]. In the case of biomarkers sensing, several nanomaterials have been employed for SPEs modification, such as rGO, AuNPs and Ag/Pt-graphene nanocomposite [270]. Beside AuNPs, Fe_2O_3 and Fe_3O_4 NPs docking interactions with the key amino acids in the spike protein receptor-binding domain of SARS-COV-2 were demonstrated recently, revealing that the interactions are associated with conformational changes in viral structural proteins and subsequent inactivation of the virus [271]. In a recent paper, Mahari et al. [272] described the fabrication of an in-house built biosensor based on fluorine-doped tin oxide electrode and AuNPs coupled to a nCOVID-19 antibody, which was demonstrated to be very specific in the detection of the nCOVID-19 spike antigen. In terms of sensitivity, this immunosensor could detect the nCOVID-19 antigen in concentrations from 1 fM to 1 μM within 10–30 s.

In terms of therapy, in the context of urgent treatment development and limited benefit of Dexamethasone and Remdesivir, SARS-CoV-2-specific therapies to treat coronavirus disease has emerged. In a very recent paper, Adi Idris et al. [273] reported the development and screening of two novel liposomal formulations for the delivery of small interfering RNA therapeutics to the lungs. Small interfering RNA molecules are short double-stranded RNA molecules that encodes the genome of coronaviruses, and by encapsulation of these active molecules in cationic liposomes with average size of 80 nm, an encapsulation efficiency of 97.6% was achieved, followed by in vivo injection in mice. The assessment was performed after 24 h, showing the localization of fluorescence particles in the lung (21%), liver (67%), and spleen (12%). The authors demonstrated robust repression of virus in the lungs and a pronounced survival advantage to the treated mice compared to the control. Moreover, the authors highlighted that the treatment based on the nano-liposomal approach is scalable and can be administered upon the first sign of SARS-CoV-2 infection in humans, suggesting also that this therapeutic approach is useful as an adjunctive therapy to current vaccine strategies.

The general representation of screen-printed electrodes modified with nanostructured materials for biomedical applications is presented below in Figure 4.

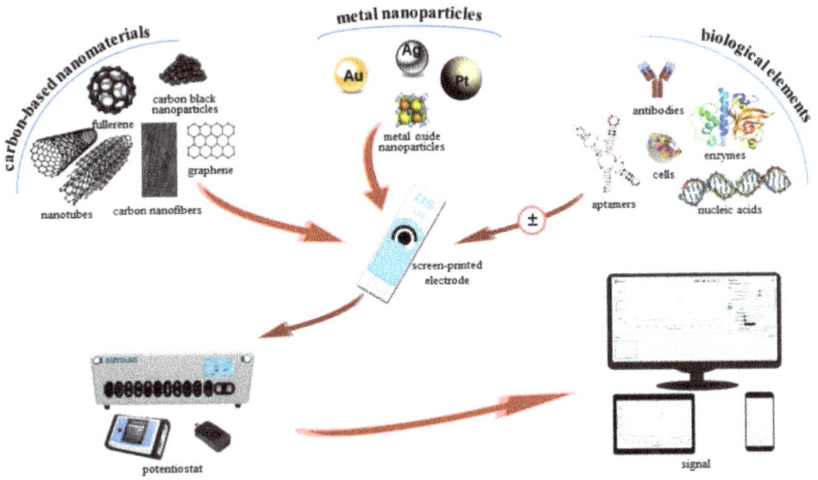

Figure 4. Schematic representation of nano(bio)sensors based on screen-printed electrodes: modification with carbon-based nanomaterials, metal nanoparticles, with/without biological elements and the electrochemical analysis.

4.2.3. Nano(Bio)Sensors Based on Screen-Printed Electrodes with Pharmaceutical Applications

The monitorization of drugs concentration in body fluids, pharmaceutical formulations and other real samples is essential in order to optimize the therapy, to monitor the treatment and to reduce the risk effects, requiring simplicity, sensitivity, selectivity, real-time response, cost effectiveness, portability and miniaturization, advantages characteristic for electrochemical nano(bio)sensors based on SPEs. Pharmaceutical substances having a wide range of pharmacological activities have been detected on screen-printed electrodes modified with various nanosized materials by exploiting various electrochemical techniques (Table 2).

Table 2. Screen-printed electrodes modified with various nanomaterials for pharmaceutical applications (some examples).

Type of NPs	Analyte	Method/ LOD/LR	Real Samples/ Recovery	Ref.
Chemically reduced graphene oxide (in the ink)	Vitamin C	DPV/ 0.95 μM/4–4500 μM	Injection formula	[249]
MWCNTs	Vitamin C	Amperometry/ 11 μM/50–400 μM	Tablet, capsule	[249]
MWCNTs	Vitamin B6	Amperometry/ 8 μM/25–300 μM DPV/ 1.5 μM/2–72 μM	Tablet, capsule, drinks, cereal	[249]
RuNPS-MWCNTs	Vitamin B6	LSW 0.8 μM/2.6–200 μM	Tablet, ampoule, drinks 92–107%	[274]
AuNPs	Vitamin B7	SWV/ 8.3 nM/0.01 nM–0.01 M Amperometry/ 14 nM/1 nM–1μM		[249]
MWCNTs	Vitamin B9	Amperometry/ 8 μM/50–400 μM	Tablet, capsule	[249]
MWCNTs	Gentamicin sulphate	Potentiometric titration 75 nM/0.1 μM–10 mM	Ampoule, ointment, cream, surface water 97.5–101.3%	[275]
Carboxylated MWCNTs-AuNPs	Amoxicillin	AdSV 0.015 μM/0.2–30 μM	Bovine milk 91.5–95.5%	[276]
AuNPs	Moxifloxacin hydrochloride	DPV 11.6 μM/8–480 μM	Urine 99.8–101.6%	[277]
Fullerene-reduced graphene oxide	Metronidazole	SWV 0.21 μM/0.25–34 μM	Serum, Urine 92–100%	[278]
BiO nanorods	Isoniazid	DPV 1.85 μM / 5–100 μM	Serum 92–104%	[279]
Fe_3O_4@polypyrrole-Pt core-shell nanoparticles	6-mercaptopurine 6-thioguanine	DPV 10 nM/0.04–330 μM 6 μM/0.1–400 μM	Urine Anticancer tablets	[280]
CeO_2 NPs	Diclofenac	SWV 0.4 μM/0.1–25.6 μM	Water samples	[281]

Table 2. *Cont.*

Type of NPs	Analyte	Method/ LOD/LR	Real Samples/ Recovery	Ref.
Carbon nanofibers	Paracetamol Ibuprofen Caffeine	DPV 0.03 mg/L 0.09–0.8 mg/L 0.6 mg/L 2.2–10.2 mg/L 0.05 mg/L 0.2–1.1 mg/L	Tap water 97.6–103.1%	[282]
Graphene nanoribbons	Melatonin	DPV 1.1 μM/0.005–3 μM	Tablet, capsule 97.8–98.3%	[283]
Manganese hexacyanoferrate/chitosan nanocomposite	Phenylalanine Chlorpheniramine Dextromethorphan	SWV 2.18 nM/0.06–25.5 μM 3.63 nM/0.045–242 μM 9.10 nM/0.062–242 μM	Pharmaceutical formulation, serum, urine 99.24–99.90%	[284]
MWCNTs	Pioglitazone	Potentiometry 0.6 μM/1μM–10 mM	Tablet 99.72–101.12%	[285]
Graphene quantum dots	Isoproterenol	DPV 0.6 μM/1–900 μM	Ampoule, urine 98–103.4%	[286]
Gd_2O_3 NPs	Venlafaxine	DPV 0.21 μM/5–900 μM	Tablet, urine, water 98–103.3%	[287]
Molecularly imprinted polymer NPs/graphene	Sertraline	SWV 1.99 nM/5–750 nM	Tablet, serum 97.98–101.33%	[288]

Vitamins' analysis has been performed on different SPEs modified with carbon-based nanomaterials leading to an μM detection limit [249]. Super critical CO_2 medium was used as a novel approach for the decoration of MWCNTs with metal nanoparticles, the best results for vitamin B6 determination was achieved by using the SPE modified with RuNPs-MWCNTs, resulting in a minimum three-fold increase in sensitivity [274]. A nanocomposite based on MWCNTs-polyvinyl chloride and calixarene as a molecular recognition element have been used for SPE modification and applied for gentamicin potentiometric determination improving the stability, response time, lifetime, sensitivity and selectivity of the sensor [275]. Carboxylated MWCNTs decorated with AuNPs using ethylenediamine as a cross-linker were drop-casted on SPE and applied for electrochemical determination of amoxicillin combining the good electron transfer property and catalytic property of the two nanomaterials [276]. AuNPs have been electrochemically deposited on SPE (with spherical shape and size between 13–58 nm) and applied for moxifloxacin hydrochloride detection with good performance [277]. A low-cost SPE modified with fullerene, reduced graphene oxide and nafion has been elaborated in order to evaluate metronidazole with high stability, repeatability, reproducibility, fast response and low cost, performance which is due to the electrocatalytic synergic effect of the nanomaterials [278]. Among the electrodes used for isoniazid and rifampicin electroanalysis, the SPEs modified with nanomaterials recorded good analytical results [279]. Fe_3O_4@polypyrrole-Pt core-shell nanoparticles have been synthesized and then drop casted on SPE allowing a simultaneous detection of two anticancer drugs (6-mercaptopurine and 6-thioguanine) [280]. A SPE was elaborated during three steps: anodic pre-treatment, drop cast of CeO_2 NPs annealed at 900 °C and heat treatment in vacuum, then it was applied for diclofenac determination. The SEM image of the nanosensor indicated o smoother surface of the electrode because the NPs filled the graphite gaps. The NPs also increased the electrochemical active area of the working electrode [281].

Three different types of SPEs modified with CNTs, carbon nanofibers and graphene were tested for the simultaneous determination of paracetamol, ibuprofen and caffeine in water samples. The best results were obtained by using the carbon nanofiber electrode [282]. The investigation of SPEs (commercial and modified in the lab) based on carbon nanomaterials (CNTs and graphene) was achieved for another simultaneous determination of melatonin and serotonin. Concerning the electroactive area of both types of SPEs, the electrode containing SWCNTs presented a higher surface in comparison with the MWCNTs, meanwhile the graphene based SPE exhibited the highest one. The SPE modified in the lab with graphene reduced nanoribbons presented the highest electroactive surface area from all the analyzed electrodes. For the simultaneous determination of the two substances, it was applied the SPE modified with graphene oxide nanoribbons [283]. Another simultaneous detection of three antiallergic drugs (phenylephrine, chlorpheniramine, dextromethorphan) was performed by using a SPE drop-casted modified with manganese hexacyanoferrate/chitosan nanocomposite. The TEM images indicated manganese hexacyanoferrate nanocubes with a size of 162.9 nm homogenously distributed over the chitosan nanoparticles [284]. MWCNTs and crown ethers have been used for SPE modification leading to a potentiometric sensor applied for a sensitive, fast and simple determination of pioglitazone [285]. A SPE based on graphene quantum dots was elaborated by drop casting and applied for isoproterenol sensitive detection [286]. Some different SPEs modified with Gd_2O_3 NPs (size 23 nm, obtained via thermal precursor decomposition), La^{3+}/Co_3O_4 nanocubes and Fe_3O_4@cellulose nanocrystal/Cu nanocomposite and graphene have been developed for the determination of two antidepressant drugs (venlafaxine and sertraline) [121,287,288]. Various nanomaterials used for SPEs modification have been also employed for cysteine and acetaminophen analysis with considerable performances [119,213].

5. Wearable Sensors in Healthcare—Some Recent Advances, Challenges and Perspectives

Although significant achievements have been realized in the development of nano-bio-sensors for point-of-care diagnostics including cancer, diabetes, malaria, HIV, it would be greatly desirable to explore a wider category of nanomaterials with superior properties to improve sensor performance for larger applications. The integration of nanomaterials in point of care testing and the opportunity of realizing portable, easy to use, cost effective, and miniaturized analytical devices represent a continuous challenge, including for personalized medicine. The next generation of point-of-care devices have to overcome the actual limitations related to inadequate detection sensitivity to distinguish biomarkers at the different stages of the diseases, while improving the selectivity at molecular level, for monitoring patient health at anywhere and anytime [289,290].

Taking into account that rigid materials could be converted to flexible ones, when their structures are reduced to nanoscale, wearable devices have revolutionized the healthcare system by decreasing the hospitalization time and by providing more reliable and timely information. Among other advantages are the comfortability and daily care possibilities. Flexible wearables, textile-based wearables, epidermal-based wearables, biofluidic-based wearables, or wearable drug delivery systems are only few examples. Wearables biosensing devices can be employed for different body parts, monitoring different psychological and physiological parameters (in saliva, blood, urine, sweat) that are crucial for the diagnoses or treatment purpose [291]. For example, in diabetes, which is among a group of underlying conditions with increased risk of severe COVID-19 disease, continuous glucose monitor devices are now commercially available as Dexcom G6 [292], with triple action: sensor, transmitter, and receiver. The automatic applicator- the sensor wire—is inserted under the skin, the readings are transmitted to the receiver and visualized in real time. Another similar approach, aiming to help diabetes monitorization and prevention of the complications, is the development of tears-based wearable device, commercially available as Triggerfish [293] that monitors the intraocular pressure of glaucoma patients for the diagnosis of diabetes. Challenges in monitoring cardiac patients may have substantial solutions by developing tattoo-based wearable device for ECG monitoring, which consists of miniaturized electronic components built on a graphene/PMMA bilayer substrate, being effective for monitoring different biopotentials like ECG, EMG and EEG signals [294–296].

As the wearables are tailored to an individual's physiological responses, such as heart rate, electrodermal activity-responsible for the emotional status and skin temperature, these signals can be extracted using the autonomic nervous system and provide feedback to patients with neurological diseases, like epilepsy, Parkinson's and Alzheimer's disease [297]. Another future approach in sensing under routine and even sedentary activity can be used to develop sweat-based biomarker monitoring practical for daily life, in a convenient glove-based form, for rapid accumulation of natural thermoregulatory sweat without active sweat stimulation. The fingertips, palm, and back of hand possess some of the highest sweat gland densities on the body, being accessible sites for monitoring natural sweat, and hence, glove-based sensing platforms (nitrile gloves and finger cots) are attractive for in situ detection of diverse biomarkers, including electrolytes and xenobiotics [298]. In the context of new challenges related to the COVID-19 pandemic, a major new driving force should be directed toward the development of modern wearable medical devices, suitable to monitor temperature, heart rate, sleep quality, blood oxygenation, which are crucial parameters for the early detection of COVID-19.

6. Conclusions and Future Outlook

In this review, we aimed to highlight the great importance and recent development of effective diagnostic tools for early detection of clinical biomarkers, not only in terms of detecting disease, but also related to physiological signatures that are predictive of potential disease. This comprehensive review is focused on the main types of metal NPs and carbon-based nanomaterials used for electrochemical (bio)sensors design, especially screen-printed electrodes, with their specific biomedical applications, improved analytical performances and miniaturized form. A brief overview about metal NPs and carbon-based nanomaterials concerning their synthesis, unique and specific properties and their use as electrode modifiers have been summarized pointing out the medical and pharmaceutical applications of the nano(bio)sensors based on SPEs. Some recent advances in the area of two important and actual topics have been also emphasized: nanomaterials and SPEs involvement in the COVID-19 management and wearable sensors in healthcare.

Nowadays, there is a tremendous need for rapid analysis, continuous monitoring systems with high accuracy for biomolecular detection. Real-time diagnostic decision and rapid manipulation is crucial, mainly in the context of COVID-19 pandemic management. Nanotechnological approaches will extend the limits of currently employed (bio)sensors and, moreover, they will open a new window toward personalized medicine, offering new solutions to the main challenges in the diagnostic and therapeutic fields. Future research should focus on some improvements concerning the nanomaterials characteristics and the sensor design in order to enhance their performances with multi-disciplinary efforts. The real sample analysis with more enhanced sensitivity and selectivity is still a challenge for researchers aiming the validation of the electrochemical nano(bio)sensors in comparison with the traditional analytical procedures. The reproducibility is another key aspect which needs to be solved for large-scale production of electrochemical sensors and their introduction on commercial market. The miniaturized, portable or wearable sensors which can perform on-site and real-time analysis will gain tremendous importance at the commercial level, with a huge impact on the health system.

Author Contributions: Conceptualization, L.F., S.C., F.B.; methodology, T.O.C., L.M., L.D., M.M.; software, L.M., T.O.C.; data curation, L.F., F.B., S.C.; writing—original draft preparation, L.F., F.B., S.C.; writing—review and editing, L.F., S.C.; visualization, L.M., L.D.; supervision, S.C.; project administration, L.F., S.C. All authors have read and agreed to the published version of the manuscript.

Funding: This research received no external funding.

Institutional Review Board Statement: Not applicable.

Informed Consent Statement: Not applicable.

Data Availability Statement: Not applicable.

Conflicts of Interest: The authors declare no conflict of interest.

References

1. Noah, N.M.; Ndangili, P.M. Current Trends of Nanobiosensors for Point-of-Care Diagnostics. *J. Anal. Methods Chem.* **2019**, *2019*, 1–16. [CrossRef] [PubMed]
2. Malekzad, H.; Sahandi Zangabad, P.; Mirshekari, H.; Karimi, M.; Hamblin, M.R. Noble Metal Nanoparticles in Biosensors: Recent Studies and Applications. *Nanotechnol. Rev.* **2017**, *6*, 301–329. [CrossRef] [PubMed]
3. Szeitner, Z.; András, J.; Gyurcsányi, R.E.; Mészáros, T. Is Less More? Lessons from Aptamer Selection Strategies. *J. Pharm. Biomed. Anal.* **2014**, *101*, 58–65. [CrossRef] [PubMed]
4. Fritea, L.; Haddache, F.; Reuillard, B.; Le Goff, A.; Gorgy, K.; Gondran, C.; Holzinger, M.; Săndulescu, R.; Cosnier, S. Electrochemical Nanopatterning of an Electrogenerated Photosensitive Poly-[Trisbipyridinyl-Pyrrole Ruthenium(II)] Metallopolymer by Nanosphere Lithography. *Electrochem. Commun.* **2014**, *46*, 75–78. [CrossRef]
5. Bounegru, A.V.; Apetrei, C. Carbonaceous Nanomaterials Employed in the Development of Electrochemical Sensors Based on Screen-Printing Technique—A Review. *Catalysts* **2020**, *10*, 680. [CrossRef]
6. Masoud, N.; Partsch, T.; de Jong, K.P.; de Jongh, P.E. Thermal Stability of Oxide-Supported Gold Nanoparticles. *Gold Bull.* **2019**, *52*, 105–114. [CrossRef]
7. Kelly, K.L.; Coronado, E.; Zhao, L.L.; Schatz, G.C. The Optical Properties of Metal Nanoparticles: The Influence of Size, Shape, and Dielectric Environment. *J. Phys. Chem. B* **2003**, *107*, 668–677. [CrossRef]
8. Huang, X.; Jain, P.K.; El-Sayed, I.H.; El-Sayed, M.A. Determination of the Minimum Temperature Required for Selective Photothermal Destruction of Cancer Cells with the Use of Immunotargeted Gold Nanoparticles. *Photochem. Photobiol.* **2006**, *82*, 412. [CrossRef]
9. Fritea, L.; Audebert, P.; Galmiche, L.; Gorgy, K.; Le Goff, A.; Villalonga, R.; Săndulescu, R.; Cosnier, S. First Occurrence of Tetrazines in Aqueous Solution: Electrochemistry and Fluorescence. *ChemPhysChem* **2015**, *16*, 3695–3699. [CrossRef]
10. Fritea, L.; Banica, F.; Costea, T.O.; Moldovan, L.; Iovan, C.; Cavalu, S. A Gold Nanoparticles-Graphene Based Electrochemical Sensor for Sensitive Determination of Nitrazepam. *J. Electroanal. Chem.* **2018**, *830*, 63–71. [CrossRef]
11. Alavi, M.; Varma, R.S. Phytosynthesis and Modification of Metal and Metal Oxide Nanoparticles/Nanocomposites for Antibacterial and Anticancer Activities: Recent Advances. *Sustain. Chem. Pharm.* **2021**, *21*, 100412. [CrossRef]
12. Paciotti, G.F.; Myer, L.; Weinreich, D.; Goia, D.; Pavel, N.; McLaughlin, R.E.; Tamarkin, L. Colloidal Gold: A Novel Nanoparticle Vector for Tumor Directed Drug Delivery. *Drug Deliv.* **2004**, *11*, 169–183. [CrossRef]
13. Cavalu, S.; Kamel, E.; Laslo, V.; Fritea, L.; Costea, T.; Antoniac, I.V.; Vasile, E.; Antoniac, A.; Semenescu, A.; Mohan, A.; et al. Eco-Friendly, Facile and Rapid Way for Synthesis of Selenium Nanoparticles Production, Structural and Morphological Characterisation. *Rev. Chim.* **2018**, *68*, 2963–2966. [CrossRef]
14. Cavalu, S.; Antoniac, I.V.; Fritea, L.; Mates, I.M.; Milea, C.; Laslo, V.; Vicas, S.; Mohan, A. Surface Modifications of the Titanium Mesh for Cranioplasty Using Selenium Nanoparticles Coating. *J. Adhes. Sci. Technol.* **2018**, *32*, 2509–2522. [CrossRef]
15. Cavalu, S.; Ratiu, C.; Ponta, O.; Simon, V.; Rugina, D.; Miclaus, V.; Akin, I.; Goller, G. Improving Osseointegration of Alumina/Zirconia Ceramic Implants by Fluoride Surface Treatment. *Dig. J. Nanomater. Biostruct.* **2014**, *9*, 797–808.
16. Cavalu, S.; Vicas, S.; Costea, T.; Fritea, L.; Copolovici, D.; Laslo, V. Preparation, Physico-Chemical Characterization and Photocatalytic Properties of Se Doped TiO2 Nanoparticles. *Rev. Chim.* **2020**, *71*, 22–27. [CrossRef]
17. Cavalu, S.; Fritea, L.; Brocks, M.; Barbaro, K.; Murvai, G.; Costea, T.O.; Antoniac, I.; Verona, C.; Romani, M.; Latini, A.; et al. Novel Hybrid Composites Based on PVA/SeTiO_2 Nanoparticles and Natural Hydroxyapatite for Orthopedic Applications: Correlations between Structural, Morphological and Biocompatibility Properties. *Materials* **2020**, *13*, 2077. [CrossRef]
18. Ding, S.-N.; Gao, B.-H.; Shan, D.; Sun, Y.-M.; Cosnier, S. Dramatically Enhanced Solid-State Electrochemiluminescence of CdTe Quantum Dots Composed with TiO_2 Nanoparticles. *Chem. Eur. J.* **2012**, *18*, 1595–1598. [CrossRef]
19. Shah, M.; Fawcett, D.; Sharma, S.; Tripathy, S.; Poinern, G. Green Synthesis of Metallic Nanoparticles via Biological Entities. *Materials* **2015**, *8*, 7278–7308. [CrossRef]
20. Cristea, C.; Tertis, M.; Galatus, R. Magnetic Nanoparticles for Antibiotics Detection. *Nanomaterials* **2017**, *7*, 119. [CrossRef]
21. Rajput, N. Methods of Preparation of Nanoparticles-A Review. *Int. J. Adv. Eng. Technol.* **2015**, *7*, 1806–1811.
22. Usman, M.; Ibrahim, N.; Shameli, K.; Zainuddin, N.; Yunus, W. Copper Nanoparticles Mediated by Chitosan: Synthesis and Characterization via Chemical Methods. *Molecules* **2012**, *17*, 14928–14936. [CrossRef]
23. García-Barrasa, J.; López-de-Luzuriaga, J.; Monge, M. Silver Nanoparticles: Synthesis through Chemical Methods in Solution and Biomedical Applications. *Open Chem.* **2011**, *9*, 7–19. [CrossRef]
24. Swain, A.K. Review on Green Synthesis of Silver Nanoparticles by Physical, Chemical and Biological Methods. *Int. J. Sci. Eng. Res.* **2016**, *7*, 551–554. [CrossRef]
25. Mates, I.; Antoniac, I.; Laslo, V.; Vicaş, S.; Brocks, M.; Fritea, L.; Milea, C.; Mohan, A.; Cavalu, S. Selenium nanoparticles: Production, characterization and possible applications in biomedicine and food science. *UPB Sci. Bull. Ser. B Chem. Mater. Sci.* **2019**, *81*, 205–216.

26. Park, J.H.; Ahn, H.S. Electrochemical Synthesis of Multimetallic Nanoparticles and Their Application in Alkaline Oxygen Reduction Catalysis. *Appl. Surf. Sci.* **2020**, *504*, 144517. [CrossRef]
27. McDarby, S.P.; Wang, C.J.; King, M.E.; Personick, M.L. An Integrated Electrochemistry Approach to the Design and Synthesis of Polyhedral Noble Metal Nanoparticles. *J. Am. Chem. Soc.* **2020**, *142*, 21322–21335. [CrossRef]
28. Roldán, M.V.; Pellegri, N.; de Sanctis, O. Electrochemical Method for Ag-PEG Nanoparticles Synthesis. *J. Nanoparticles* **2013**, *2013*, 1–7. [CrossRef]
29. Petrii, O.A. Electrosynthesis of Nanostructures and Nanomaterials. *Russ. Chem. Rev.* **2015**, *84*, 159–193. [CrossRef]
30. Jamkhande, P.G.; Ghule, N.W.; Bamer, A.H.; Kalaskar, M.G. Metal Nanoparticles Synthesis: An Overview on Methods of Preparation, Advantages and Disadvantages, and Applications. *J. Drug Deliv. Sci. Technol.* **2019**, *53*, 101174. [CrossRef]
31. Glaspell, G.; Fuoco, L.; El-Shall, M.S. Microwave Synthesis of Supported Au and Pd Nanoparticle Catalysts for CO Oxidation. *J. Phys. Chem. B* **2005**, *109*, 17350–17355. [CrossRef] [PubMed]
32. Marquardt, D.; Vollmer, C.; Thomann, R.; Steurer, P.; Mülhaupt, R.; Redel, E.; Janiak, C. The Use of Microwave Irradiation for the Easy Synthesis of Graphene-Supported Transition Metal Nanoparticles in Ionic Liquids. *Carbon* **2011**, *49*, 1326–1332. [CrossRef]
33. Ye, X.R.; Lin, Y.; Wai, C.M. Decorating Catalytic Palladium Nanoparticles on Carbon Nanotubes in Supercritical Carbon Dioxide. *Chem. Commun.* **2003**, *5*, 642–643. [CrossRef] [PubMed]
34. Ye, X.-R.; Lin, Y.; Wang, C.; Engelhard, M.H.; Wang, Y.; Wai, C.M. Supercritical Fluid Synthesis and Characterization of Catalytic Metal Nanoparticles on Carbon Nanotubes. *J. Mater. Chem.* **2004**, *14*, 908. [CrossRef]
35. Pantidos, N. Biological Synthesis of Metallic Nanoparticles by Bacteria, Fungi and Plants. *J. Nanomed. Nanotechnol.* **2014**, *5*, 1000233. [CrossRef]
36. Thakkar, K.N.; Mhatre, S.S.; Parikh, R.Y. Biological Synthesis of Metallic Nanoparticles. *Nanomed. Nanotechnol. Biol. Med.* **2010**, *6*, 257–262. [CrossRef] [PubMed]
37. Akhtar, M.S.; Panwar, J.; Yun, Y.-S. Biogenic Synthesis of Metallic Nanoparticles by Plant Extracts. *ACS Sustain. Chem. Eng.* **2013**, *1*, 591–602. [CrossRef]
38. Vicas, S.I.; Laslo, V.; Timar, A.V.; Balta, C.; Herman, H.; Ciceu, A.; Gharbia, S.; Rosu, M.; Mladin, B.; Fritea, L.; et al. Functional Food Product Based on Nanoselenium-Enriched Lactobacillus Casei against Cadmium Kidney Toxicity. *Appl. Sci.* **2021**, *11*, 4220. [CrossRef]
39. Ahmad, F.; Ashraf, N.; Ashraf, T.; Zhou, R.-B.; Yin, D.-C. Biological Synthesis of Metallic Nanoparticles (MNPs) by Plants and Microbes: Their Cellular Uptake, Biocompatibility, and Biomedical Applications. *Appl. Microbiol. Biotechnol.* **2019**, *103*, 2913–2935. [CrossRef]
40. Adabi, M.; Naghibzadeh, M.; Adabi, M.; Zarrinfard, M.A.; Esnaashari, S.S.; Seifalian, A.M.; Faridi-Majidi, R.; Tanimowo Aiyelabegan, H.; Ghanbari, H. Biocompatibility and Nanostructured Materials: Applications in Nanomedicine. *Artif. Cells Nanomed. Biotechnol.* **2017**, *45*, 833–842. [CrossRef]
41. Daruich De Souza, C.; Ribeiro Nogueira, B.; Rostelato, M.E.C.M. Review of the Methodologies Used in the Synthesis Gold Nanoparticles by Chemical Reduction. *J. Alloy. Compd.* **2019**, *798*, 714–740. [CrossRef]
42. Chandra, P.; Singh, J.; Singh, A.; Srivastava, A.; Goyal, R.N.; Shim, Y.B. Gold Nanoparticles and Nanocomposites in Clinical Diagnostics Using Electrochemical Methods. *J. Nanoparticles* **2013**, *2013*, 1–12. [CrossRef]
43. Dragan, A.-M.; Truta, F.M.; Tertis, M.; Florea, A.; Schram, J.; Cernat, A.; Feier, B.; De Wael, K.; Cristea, C.; Oprean, R. Electrochemical Fingerprints of Illicit Drugs on Graphene and Multi-Walled Carbon Nanotubes. *Front. Chem.* **2021**, *9*, 641147. [CrossRef] [PubMed]
44. Tertiş, M.; Florea, A.; Adumitrăchioaie, A.; Cernat, A.; Bogdan, D.; Barbu-Tudoran, L.; Jaffrezic Renault, N.; Săndulescu, R.; Cristea, C. Detection of Dopamine by a Biomimetic Electrochemical Sensor Based on Polythioaniline-Bridged Gold Nanoparticles. *ChemPlusChem* **2017**, *82*, 561–569. [CrossRef] [PubMed]
45. Zou, C.; Yang, B.; Bin, D.; Wang, J.; Li, S.; Yang, P.; Wang, C.; Shiraishi, Y.; Du, Y. Electrochemical Synthesis of Gold Nanoparticles Decorated Flower-like Graphene for High Sensitivity Detection of Nitrite. *J. Colloid Interface Sci.* **2017**, *488*, 135–141. [CrossRef] [PubMed]
46. Philip, D. Green Synthesis of Gold and Silver Nanoparticles Using Hibiscus Rosa Sinensis. *Phys. E Low-Dimens. Syst. Nanostruct.* **2010**, *42*, 1417–1424. [CrossRef]
47. Aswathy Aromal, S.; Philip, D. Green Synthesis of Gold Nanoparticles Using Trigonella Foenum-Graecum and Its Size-Dependent Catalytic Activity. *Spectrochim. Acta Part A Mol. Biomol. Spectrosc.* **2012**, *97*, 1–5. [CrossRef]
48. Rajan, A.; Rajan, A.R.; Philip, D. Elettaria Cardamomum Seed Mediated Rapid Synthesis of Gold Nanoparticles and Its Biological Activities. *OpenNano* **2017**, *2*, 1–8. [CrossRef]
49. Philip, D. Synthesis and Spectroscopic Characterization of Gold Nanoparticles. *Spectrochim. Acta Part A Mol. Biomol. Spectrosc.* **2008**, *71*, 80–85. [CrossRef]
50. Dwivedi, A.D.; Gopal, K. Biosynthesis of Silver and Gold Nanoparticles Using Chenopodium Album Leaf Extract. *Colloids Surf. A Physicochem. Eng. Asp.* **2010**, *369*, 27–33. [CrossRef]
51. He, S.; Guo, Z.; Zhang, Y.; Zhang, S.; Wang, J.; Gu, N. Biosynthesis of Gold Nanoparticles Using the Bacteria Rhodopseudomonas Capsulata. *Mater. Lett.* **2007**, *61*, 3984–3987. [CrossRef]
52. Shah, R.; Oza, G.; Pandey, S.; Sharon, M. Biogenic fabrication of gold nanoparticles using halomonas salina. *J. Microbiol. Biotechnol. Res.* **2012**, *2*, 485–492.

53. Ganesh Kumar, C.; Poornachandra, Y.; Mamidyala, S.K. Green Synthesis of Bacterial Gold Nanoparticles Conjugated to Resveratrol as Delivery Vehicles. *Colloids Surf. B Biointerfaces* **2014**, *123*, 311–317. [CrossRef]
54. Beveridge, T.J.; Murray, R.G. Sites of Metal Deposition in the Cell Wall of Bacillus Subtilis. *J. Bacteriol.* **1980**, *141*, 876–887. [CrossRef]
55. Malhotra, A.; Dolma, K.; Kaur, N.; Rathore, Y.S.; Ashish; Mayilraj, S.; Choudhury, A.R. Biosynthesis of Gold and Silver Nanoparticles Using a Novel Marine Strain of Stenotrophomonas. *Bioresour. Technol.* **2013**, *142*, 727–731. [CrossRef]
56. Philip, D. Biosynthesis of Au, Ag and Au–Ag Nanoparticles Using Edible Mushroom Extract. *Spectrochim. Acta Part A Mol. Biomol. Spectrosc.* **2009**, *73*, 374–381. [CrossRef]
57. Love, A.J.; Makarov, V.; Yaminsky, I.; Kalinina, N.O.; Taliansky, M.E. The Use of Tobacco Mosaic Virus and Cowpea Mosaic Virus for the Production of Novel Metal Nanomaterials. *Virology* **2014**, *449*, 133–139. [CrossRef]
58. Gericke, M.; Pinches, A. Microbial Production of Gold Nanoparticles. *Gold Bull.* **2006**, *39*, 22–28. [CrossRef]
59. Bakshi, M.S. How Surfactants Control Crystal Growth of Nanomaterials. *Cryst. Growth Des.* **2016**, *16*, 1104–1133. [CrossRef]
60. Li, Y.; Schluesener, H.J.; Xu, S. Gold Nanoparticle-Based Biosensors. *Gold Bull.* **2010**, *43*, 29–41. [CrossRef]
61. Chang, G.; Shu, H.; Ji, K.; Oyama, M.; Liu, X.; He, Y. Gold Nanoparticles Directly Modified Glassy Carbon Electrode for Non-Enzymatic Detection of Glucose. *Appl. Surf. Sci.* **2014**, *288*, 524–529. [CrossRef]
62. Saha, K.; Agasti, S.S.; Kim, C.; Li, X.; Rotello, V.M. Gold Nanoparticles in Chemical and Biological Sensing. *Chem. Rev.* **2012**, *112*, 2739–2779. [CrossRef]
63. Kurniawan, F.; Tsakova, V.; Mirsky, V.M. Gold Nanoparticles in Nonenzymatic Electrochemical Detection of Sugars. *Electroanalysis* **2006**, *18*, 1937–1942. [CrossRef]
64. Li, M.; Gao, F.; Yang, P.; Wang, L.; Fang, B. Conveniently Assembling Dithiocarbamate and Gold Nanoparticles onto the Gold Electrode: A New Type of Electrochemical Sensors for Biomolecule Detection. *Surf. Sci.* **2008**, *602*, 151–155. [CrossRef]
65. Kannan, P.; John, S.A. Determination of Nanomolar Uric and Ascorbic Acids Using Enlarged Gold Nanoparticles Modified Electrode. *Anal. Biochem.* **2009**, *386*, 65–72. [CrossRef]
66. Zhang, S.; Xu, M.; Zhang, Y. Simultaneous Voltammetric Detection of Salsolinol and Uric Acid in the Presence of High Concentration of Ascorbic Acid with Gold Nanoparticles/Functionalized Multiwalled Carbon Nanotubes Composite Film Modified Electrode. *Electroanalysis* **2009**, *21*, 2607–2610. [CrossRef]
67. Xiao, Y. "Plugging into Enzymes": Nanowiring of Redox Enzymes by a Gold Nanoparticle. *Science* **2003**, *299*, 1877–1881. [CrossRef]
68. Mao, X.; Liu, G. Nanomaterial Based Electrochemical DNA Biosensors and Bioassays. *J. Biomed. Nanotechnol.* **2008**, *4*, 419–431. [CrossRef]
69. Park, S.-J.; Taton, T.A.; Mirkin, C.A. Array-Based Electrical Detection of DNA with Nanoparticle Probes. *Science* **2002**, *295*, 1503–1506. [CrossRef] [PubMed]
70. Huang, X.; El-Sayed, I.H.; Qian, W.; El-Sayed, M.A. Cancer Cell Imaging and Photothermal Therapy in the Near-Infrared Region by Using Gold Nanorods. *J. Am. Chem. Soc.* **2006**, *128*, 2115–2120. [CrossRef] [PubMed]
71. Huff, T.B.; Tong, L.; Zhao, Y.; Hansen, M.N.; Cheng, J.-X.; Wei, A. Hyperthermic Effects of Gold Nanorods on Tumor Cells. *Nanomedicine* **2007**, *2*, 125–132. [CrossRef]
72. Jain, S.; Hirst, D.G.; O'Sullivan, J.M. Gold Nanoparticles as Novel Agents for Cancer Therapy. *Br. J. Radiol.* **2012**, *85*, 101–113. [CrossRef]
73. Iravani, S.; Korbekandi, H.; Mirmohammadi, S.V.; Zolfaghari, B. Synthesis of Silver Nanoparticles: Chemical, Physical and Biological Methods. *Res. Pharm. Sci.* **2014**, *9*, 385–406.
74. Lee, S.; Jun, B.-H. Silver Nanoparticles: Synthesis and Application for Nanomedicine. *Int. J. Mol. Sci.* **2019**, *20*, 865. [CrossRef]
75. Malassis, L.; Dreyfus, R.; Murphy, R.J.; Hough, L.A.; Donnio, B.; Murray, C.B. One-Step Green Synthesis of Gold and Silver Nanoparticles with Ascorbic Acid and Their Versatile Surface Post-Functionalization. *RSC Adv.* **2016**, *6*, 33092–33100. [CrossRef]
76. Liang, H.; Wang, W.; Huang, Y.; Zhang, S.; Wei, H.; Xu, H. Controlled Synthesis of Uniform Silver Nanospheres. *J. Phys. Chem. C* **2010**, *114*, 7427–7431. [CrossRef]
77. Güzel, R.; Erdal, G. Synthesis of Silver Nanoparticles. In *Silver Nanoparticles-Fabrication, Characterization and Applications*; Maaz, K., Ed.; InTech: London, UK, 2018. [CrossRef]
78. Quintero-Quiroz, C.; Acevedo, N.; Zapata-Giraldo, J.; Botero, L.E.; Quintero, J.; Zárate-Triviño, D.; Saldarriaga, J.; Pérez, V.Z. Optimization of Silver Nanoparticle Synthesis by Chemical Reduction and Evaluation of Its Antimicrobial and Toxic Activity. *Biomater. Res.* **2019**, *23*, 27. [CrossRef]
79. Ojha, A.K.; Forster, S.; Kumar, S.; Vats, S.; Negi, S.; Fischer, I. Synthesis of Well–Dispersed Silver Nanorods of Different Aspect Ratios and Their Antimicrobial Properties against Gram Positive and Negative Bacterial Strains. *J. Nanobiotechnol.* **2013**, *11*, 42. [CrossRef]
80. Gebeyehu, M.B.; Chala, T.F.; Chang, S.-Y.; Wu, C.-M.; Lee, J.-Y. Synthesis and Highly Effective Purification of Silver Nanowires to Enhance Transmittance at Low Sheet Resistance with Simple Polyol and Scalable Selective Precipitation Method. *RSC Adv.* **2017**, *7*, 16139–16148. [CrossRef]
81. Zhang, J.; Li, S.; Wu, J.; Schatz, G.C.; Mirkin, C.A. Plasmon-Mediated Synthesis of Silver Triangular Bipyramids. *Angew. Chem. Int. Ed.* **2009**, *48*, 7787–7791. [CrossRef]

82. Singaravelan, R.; Bangaru Sudarsan Alwar, S. Electrochemical Synthesis, Characterisation and Phytogenic Properties of Silver Nanoparticles. *Appl. Nanosci.* **2015**, *5*, 983–991. [CrossRef]
83. Nasretdinova, G.R.; Fazleeva, R.R.; Mukhitova, R.K.; Nizameev, I.R.; Kadirov, M.K.; Ziganshina, A.Y.; Yanilkin, V.V. Electrochemical Synthesis of Silver Nanoparticles in Solution. *Electrochem. Commun.* **2015**, *50*, 69–72. [CrossRef]
84. Dobre, N.; Petică, A.; Buda, M.; Anicăi, L.; Vişan, T. Electrochemical Synthesis of Silver Nanoparticles in Aqueous Electrolytes. *UPB Sci. Bull.* **2014**, *76*, 127–136.
85. Fouad, H.; Hongjie, L.; Yanmei, D.; Baoting, Y.; El-Shakh, A.; Abbas, G.; Jianchu, M. Synthesis and Characterization of Silver Nanoparticles Using *Bacillus Amyloliquefaciens* and *Bacillus Subtilis* to Control Filarial Vector *Culex Pipiens Pallens* and Its Antimicrobial Activity. *Artif. Cells Nanomed. Biotechnol.* **2017**, *45*, 1369–1378. [CrossRef] [PubMed]
86. Chopade, B.A.; Singh, R.; Wagh, P.; Wadhwani, S.; Gaidhani, S.; Kumbhar, A.; Bellare, J. Synthesis, Optimization, and Characterization of Silver Nanoparticles from Acinetobacter Calcoaceticus and Their Enhanced Antibacterial Activity When Combined with Antibiotics. *Int. J. Nanosci.* **2013**, *8*, 4277. [CrossRef] [PubMed]
87. Kumar, C.G.; Mamidyala, S.K. Extracellular Synthesis of Silver Nanoparticles Using Culture Supernatant of Pseudomonas Aeruginosa. *Colloids Surf. B Biointerfaces* **2011**, *84*, 462–466. [CrossRef] [PubMed]
88. Divya, K.; Kurian, L.C.; Vijayan, S.; Manakulam Shaikmoideen, J. Green Synthesis of Silver Nanoparticles by Escherichia Coli: Analysis of Antibacterial Activity. *J. Water Environ. Nanotechnol.* **2016**, *1*, 63–74. [CrossRef]
89. Kalishwaralal, K.; Deepak, V.; Ram Kumar Pandian, S.; Kottaisamy, M.; BarathManiKanth, S.; Kartikeyan, B.; Gurunathan, S. Biosynthesis of Silver and Gold Nanoparticles Using Brevibacterium Casei. *Colloids Surf. B Biointerfaces* **2010**, *77*, 257–262. [CrossRef]
90. Vigneshwaran, N.; Ashtaputre, N.M.; Varadarajan, P.V.; Nachane, R.P.; Paralikar, K.M.; Balasubramanya, R.H. Biological Synthesis of Silver Nanoparticles Using the Fungus Aspergillus Flavus. *Mater. Lett.* **2007**, *61*, 1413–1418. [CrossRef]
91. Kathiresan, K.; Manivannan, S.; Nabeel, M.A.; Dhivya, B. Studies on Silver Nanoparticles Synthesized by a Marine Fungus, Penicillium Fellutanum Isolated from Coastal Mangrove Sediment. *Colloids Surf. B Biointerfaces* **2009**, *71*, 133–137. [CrossRef]
92. Tippayawat, P.; Phromviyo, N.; Boueroy, P.; Chompoosor, A. Green Synthesis of Silver Nanoparticles in Aloe Vera Plant Extract Prepared by a Hydrothermal Method and Their Synergistic Antibacterial Activity. *PeerJ* **2016**, *4*, e2589. [CrossRef]
93. Gnanajobitha, G.; Paulkumar, K.; Vanaja, M.; Rajeshkumar, S.; Malarkodi, C.; Annadurai, G.; Kannan, C. Fruit-Mediated Synthesis of Silver Nanoparticles Using Vitis Vinifera and Evaluation of Their Antimicrobial Efficacy. *J. Nanostruct. Chem.* **2013**, *3*, 67. [CrossRef]
94. Patil, R.S.; Kokate, M.R.; Kolekar, S.S. Bioinspired Synthesis of Highly Stabilized Silver Nanoparticles Using Ocimum Tenuiflorum Leaf Extract and Their Antibacterial Activity. *Spectrochim. Acta Part A Mol. Biomol. Spectrosc.* **2012**, *91*, 234–238. [CrossRef]
95. Mariselvam, R.; Ranjitsingh, A.J.A.; Usha Raja Nanthini, A.; Kalirajan, K.; Padmalatha, C.; Mosae Selvakumar, P. Green Synthesis of Silver Nanoparticles from the Extract of the Inflorescence of Cocos Nucifera (Family: Arecaceae) for Enhanced Antibacterial Activity. *Spectrochim. Acta Part A Mol. Biomol. Spectrosc.* **2014**, *129*, 537–541. [CrossRef]
96. Rad, A.; Mirabi, A.; Binaian, E.; Habibollah Tayebi, H. A Review on Glucose and Hydrogen Peroxide Biosensor Based on Modified Electrode Included Silver Nanoparticles. *Int. J. Electrochem. Sci.* **2011**, *6*, 3671–3683.
97. Wang, W.; Xie, Y.; Xia, C.; Du, H.; Tian, F. Titanium Dioxide Nanotube Arrays Modified with a Nanocomposite of Silver Nanoparticles and Reduced Graphene Oxide for Electrochemical Sensing. *Microchim. Acta* **2014**, *181*, 1325–1331. [CrossRef]
98. Nycz, M.; Arkusz, K.; Pijanowska, D.G. Fabrication of Electrochemical Biosensor Based on Titanium Dioxide Nanotubes and Silver Nanoparticles for Heat Shock Protein 70 Detection. *Materials* **2021**, *14*, 3767. [CrossRef]
99. Xing, Y.; Feng, X.Z.; Zhang, L.; Hou, J.; Han, G.C.; Chen, Z. A sensitive and selective electrochemical biosensor for the determination of beta-amyloid oligomer by inhibiting the peptide-triggered in situ assembly of silver nanoparticles. *Int. J. Nanomed.* **2017**, *12*, 3171–3179. [CrossRef]
100. Verma, D.; Chauhan, D.; Das Mukherjee, M.; Ranjan, K.R.; Yadav, A.K.; Solanki, P.R. Development of MWCNT Decorated with Green Synthesized AgNps-Based Electrochemical Sensor for Highly Sensitive Detection of BPA. *J. Appl. Electrochem.* **2021**, *51*, 447–462. [CrossRef]
101. Liu, J.; Siavash Moakhar, R.; Sudalaiyadum Perumal, A.; Roman, H.N.; Mahshid, S.; Wachsmann-Hogiu, S. An AgNP-Deposited Commercial Electrochemistry Test Strip as a Platform for Urea Detection. *Sci. Rep.* **2020**, *10*, 9527. [CrossRef]
102. Nantaphol, S.; Chailapakul, O.; Siangproh, W. Sensitive and Selective Electrochemical Sensor Using Silver Nanoparticles Modified Glassy Carbon Electrode for Determination of Cholesterol in Bovine Serum. *Sens. Actuators B Chem.* **2015**, *207*, 193–198. [CrossRef]
103. Sondi, I.; Salopek-Sondi, B. Silver Nanoparticles as Antimicrobial Agent: A Case Study on E. Coli as a Model for Gram-Negative Bacteria. *J. Colloid Interface Sci.* **2004**, *275*, 177–182. [CrossRef] [PubMed]
104. Marica, A.; Fritea, L.; Banica, F.; Sinescu, C.; Iovan, C.; Hulka, I.; Rusu, G.; Cavalu, S. Carbon Nanotubes for Improved Performances of Endodontic Sealer. *Materials* **2021**, *14*, 4248. [CrossRef] [PubMed]
105. Jeyaraj, M.; Gurunathan, S.; Qasim, M.; Kang, M.-H.; Kim, J.-H. A Comprehensive Review on the Synthesis, Characterization, and Biomedical Application of Platinum Nanoparticles. *Nanomaterials* **2019**, *9*, 1719. [CrossRef] [PubMed]
106. Leong, G.J.; Schulze, M.C.; Strand, M.B.; Maloney, D.; Frisco, S.L.; Dinh, H.N.; Pivovar, B.; Richards, R.M. Shape-Directed Platinum Nanoparticle Synthesis: Nanoscale Design of Novel Catalysts: A Review of Shape-Directed Platinum Nanoparticle Synthesis. *Appl. Organometal. Chem.* **2014**, *28*, 1–17. [CrossRef]

107. Jan, H.; Gul, R.; Andleeb, A.; Ullah, S.; Shah, M.; Khanum, M.; Ullah, I.; Hano, C.; Abbasi, B.H. A Detailed Review on Biosynthesis of Platinum Nanoparticles (PtNPs), Their Potential Antimicrobial and Biomedical Applications. *J. Saudi Chem. Soc.* **2021**, *25*, 101297. [CrossRef]
108. Li, M.; Bo, X.; Mu, Z.; Zhang, Y.; Guo, L. Electrodeposition of Nickel Oxide and Platinum Nanoparticles on Electrochemically Reduced Graphene Oxide Film as a Nonenzymatic Glucose Sensor. *Sens. Actuators B Chem.* **2014**, *192*, 261–268. [CrossRef]
109. Fang, Y.; Zhang, D.; Qin, X.; Miao, Z.; Takahashi, S.; Anzai, J.; Chen, Q. A Non-Enzymatic Hydrogen Peroxide Sensor Based on Poly(Vinyl Alcohol)–Multiwalled Carbon Nanotubes–Platinum Nanoparticles Hybrids Modified Glassy Carbon Electrode. *Electrochim. Acta* **2012**, *70*, 266–271. [CrossRef]
110. Boluda, A.; Casado, C.; Alonso, B.; García Armada, M. Efficient Oxidase Biosensors Based on Bioelectrocatalytic Surfaces of Electrodeposited Ferrocenyl Polycyclosiloxanes—Pt Nanoparticles. *Chemosensors* **2021**, *9*, 81. [CrossRef]
111. Yu, H.; Yu, J.; Li, L.; Zhang, Y.; Xin, S.; Ni, X.; Sun, Y.; Song, K. Recent Progress of the Practical Applications of the Platinum Nanoparticle-Based Electrochemistry Biosensors. *Front. Chem.* **2021**, *9*, 677876. [CrossRef]
112. Hossain, M.F.; Slaughter, G. PtNPs Decorated Chemically Derived Graphene and Carbon Nanotubes for Sensitive and Selective Glucose Biosensing. *J. Electroanal. Chem.* **2020**, *861*, 113990. [CrossRef]
113. Garlyyev, B.; Kratzl, K.; Rück, M.; Michalička, J.; Fichtner, J.; Macak, J.M.; Kratky, T.; Günther, S.; Cokoja, M.; Bandarenka, A.S.; et al. Optimizing the Size of Platinum Nanoparticles for Enhanced Mass Activity in the Electrochemical Oxygen Reduction Reaction. *Angew. Chem. Int. Ed.* **2019**, *58*, 9596–9600. [CrossRef] [PubMed]
114. Sawan, S.; Maalouf, R.; Errachid, A.; Jaffrezic-Renault, N. Metal and Metal Oxide Nanoparticles in the Voltammetric Detection of Heavy Metals: A Review. *TrAC Trends Anal. Chem.* **2020**, *131*, 116014. [CrossRef]
115. Yoon, J.; Shin, M.; Lee, T.; Choi, J.-W. Highly Sensitive Biosensors Based on Biomolecules and Functional Nanomaterials Depending on the Types of Nanomaterials: A Perspective Review. *Materials* **2020**, *13*, 299. [CrossRef] [PubMed]
116. Joshi, A.; Kim, K.-H. Recent Advances in Nanomaterial-Based Electrochemical Detection of Antibiotics: Challenges and Future Perspectives. *Biosens. Bioelectron.* **2020**, *153*, 112046. [CrossRef] [PubMed]
117. Maduraiveeran, G.; Sasidharan, M.; Ganesan, V. Electrochemical Sensor and Biosensor Platforms Based on Advanced Nanomaterials for Biological and Biomedical Applications. *Biosens. Bioelectron.* **2018**, *103*, 113–129. [CrossRef] [PubMed]
118. Ambaye, A.D.; Kefeni, K.K.; Mishra, S.B.; Nxumalo, E.N.; Ntsendwana, B. Recent Developments in Nanotechnology-Based Printing Electrode Systems for Electrochemical Sensors. *Talanta* **2021**, *225*, 121951. [CrossRef] [PubMed]
119. Tajik, S.; Dourandish, Z.; Jahani, P.M.; Sheikhshoaie, I.; Beitollahi, H.; Shahedi Asl, M.; Jang, H.W.; Shokouhimehr, M. Recent Developments in Voltammetric and Amperometric Sensors for Cysteine Detection. *RSC Adv.* **2021**, *11*, 5411–5425. [CrossRef]
120. Tajik, S.; Orooji, Y.; Ghazanfari, Z.; Karimi, F.; Beitollahi, H.; Varma, R.S.; Jang, H.W.; Shokouhimehr, M. Nanomaterials Modified Electrodes for Electrochemical Detection of Sudan I in Food. *Food Meas.* **2021**, *15*, 3837–3852. [CrossRef]
121. Tajik, S.; Beitollahi, H.; Dourandish, Z.; Zhang, K.; Le, Q.V.; Nguyen, T.P.; Kim, S.Y.; Shokouhimehr, M. Recent Advances in the Electrochemical Sensing of Venlafaxine: An Antidepressant Drug and Environmental Contaminant. *Sensors* **2020**, *20*, 3675. [CrossRef] [PubMed]
122. Cavalu, S.; Antoniac, I.V.; Mohan, A.; Bodog, F.; Doicin, C.; Mates, I.; Ulmeanu, M.; Murzac, R.; Semenescu, A. Nanoparticles and Nanostructured Surface Fabrication for Innovative Cranial and Maxillofacial Surgery. *Materials* **2020**, *13*, 5391. [CrossRef] [PubMed]
123. Chen, A.; Chatterjee, S. Nanomaterials Based Electrochemical Sensors for Biomedical Applications. *Chem. Soc. Rev.* **2013**, *42*, 5425. [CrossRef] [PubMed]
124. Huo, X.; Liu, X.; Liu, J.; Sukumaran, P.; Alwarappan, S.; Wong, D.K.Y. Strategic Applications of Nanomaterials as Sensing Platforms and Signal Amplification Markers at Electrochemical Immunosensors. *Electroanalysis* **2016**, *28*, 1730–1749. [CrossRef]
125. Li, X.; Wei, J.; Aifantis, K.E.; Fan, Y.; Feng, Q.; Cui, F.-Z.; Watari, F. Current Investigations into Magnetic Nanoparticles for Biomedical Applications: Magnetic Nanoparticles for Biomedical Applications. *J. Biomed. Mater. Res.* **2016**, *104*, 1285–1296. [CrossRef] [PubMed]
126. Holzinger, M.; Le Goff, A.; Cosnier, S. Nanomaterials for Biosensing Applications: A Review. *Front. Chem.* **2014**, *2*, 63. [CrossRef]
127. Baig, N.; Sajid, M.; Saleh, T.A. Recent Trends in Nanomaterial-Modified Electrodes for Electroanalytical Applications. *Trends Anal. Chem.* **2019**, *111*, 47–61. [CrossRef]
128. Sanati, A.; Jalali, M.; Raeissi, K.; Karimzadeh, F.; Kharaziha, M.; Mahshid, S.S.; Mahshid, S. A Review on Recent Advancements in Electrochemical Biosensing Using Carbonaceous Nanomaterials. *Microchim. Acta* **2019**, *186*, 773. [CrossRef]
129. Kurbanoglu, S.; Ozkan, S.A. Electrochemical Carbon Based Nanosensors: A Promising Tool in Pharmaceutical and Biomedical Analysis. *J. Pharm. Biomed. Anal.* **2018**, *147*, 439–457. [CrossRef]
130. Pirzada, M.; Altintas, Z. Nanomaterials for Healthcare Biosensing Applications. *Sensors* **2019**, *19*, 5311. [CrossRef]
131. Sinha, A.; Dhanjai; Jain, R.; Zhao, H.; Karolia, P.; Jadon, N. Voltammetric Sensing Based on the Use of Advanced Carbonaceous Nanomaterials: A Review. *Microchim. Acta* **2018**, *185*, 89. [CrossRef]
132. Maduraiveeran, G.; Jin, W. Nanomaterials Based Electrochemical Sensor and Biosensor Platforms for Environmental Applications. *Trends Environ. Anal. Chem.* **2017**, *13*, 10–23. [CrossRef]
133. Qian, L.; Durairaj, S.; Prins, S.; Chen, A. Nanomaterial-Based Electrochemical Sensors and Biosensors for the Detection of Pharmaceutical Compounds. *Biosens. Bioelectron.* **2021**, *175*, 112836. [CrossRef]

134. Notarianni, M.; Liu, J.; Vernon, K.; Motta, N. Synthesis and Applications of Carbon Nanomaterials for Energy Generation and Storage. *Beilstein J. Nanotechnol.* **2016**, *7*, 149–196. [CrossRef]
135. Yee, M.J.; Mubarak, N.M.; Abdullah, E.C.; Khalid, M.; Walvekar, R.; Karri, R.R.; Nizamuddin, S.; Numan, A. Carbon Nanomaterials Based Films for Strain Sensing Application—A Review. *Nano Struct. Nano Objects* **2019**, *18*, 100312. [CrossRef]
136. Fritea, L.; Gross, A.J.; Gorgy, K.; O'Reilly, R.K.; Le Goff, A.; Cosnier, S. A Bifunctional Triblock Polynorbornene/Carbon Nanotube Buckypaper Bioelectrode for Low-Potential/High-Current Thionine-Mediated Glucose Oxidation by FAD-GDH. *J. Mater. Chem. A* **2019**, *7*, 1447–1450. [CrossRef]
137. Fritea, L.; Gross, A.J.; Reuillard, B.; Gorgy, K.; Cosnier, S.; Le Goff, A. A Diethyleneglycol-Pyrene-Modified Ru(II) Catalyst for the Design of Buckypaper Bioelectrodes and the Wiring of Glucose Dehydrogenases. *ChemElectroChem* **2019**, *6*, 3621–3626. [CrossRef]
138. Gross, A.J.; Holzinger, M.; Cosnier, S. Buckypaper Bioelectrodes: Emerging Materials for Implantable and Wearable Biofuel Cells. *Energy Environ. Sci.* **2018**, *11*, 1670–1687. [CrossRef]
139. Cosnier, S.; Holzinger, M.; Le Goff, A. Recent Advances in Carbon Nanotube-Based Enzymatic Fuel Cells. *Front. Bioeng. Biotechnol.* **2014**, *2*, 45. [CrossRef]
140. Holzinger, M.; Le Goff, A.; Cosnier, S. Carbon Nanotube/Enzyme Biofuel Cells. *Electrochim. Acta* **2012**, *82*, 179–190. [CrossRef]
141. Goenka, S.; Sant, V.; Sant, S. Graphene-Based Nanomaterials for Drug Delivery and Tissue Engineering. *J. Control. Release* **2014**, *173*, 75–88. [CrossRef]
142. Eivazzadeh-Keihan, R.; Maleki, A.; de la Guardia, M.; Bani, M.S.; Chenab, K.K.; Pashazadeh-Panahi, P.; Baradaran, B.; Mokhtarzadeh, A.; Hamblin, M.R. Carbon Based Nanomaterials for Tissue Engineering of Bone: Building New Bone on Small Black Scaffolds: A Review. *J. Adv. Res.* **2019**, *18*, 185–201. [CrossRef] [PubMed]
143. Maiti, D.; Tong, X.; Mou, X.; Yang, K. Carbon-Based Nanomaterials for Biomedical Applications: A Recent Study. *Front. Pharmacol.* **2019**, *9*, 1401. [CrossRef] [PubMed]
144. Proctor, J.E.; Melendrez Armada, D.A.; Vijayaraghavan, A. *An Introduction to Graphene and Carbon Nanotubes*; CRC Press, Taylor & Francis Group: Boca Raton, FL, USA, 2017.
145. Fritea, L.; Tertis, M.; Sandulescu, R.; Cristea, C. Enzymes-graphene platforms for electrochemical biosensors design with biomedical applications. In *Methods in Enzymology*; Elsevier: Amsterdam, The Netherlands, 2018; Volume 609: Enzyme Nanoarchitectures: Enzymes Armored with Graphene, pp. 293–333. [CrossRef]
146. Tertiş, M.; Fritea, L.; Săndulescu, R.; Cristea, C. Graphene and Graphene Nanocomposite-Based Electrochemical Sensors. In *Handbook of Graphene*; Palys, B., Ed.; Wiley: Hoboken, NJ, USA, 2019; Volume 6: Biosensors and Advanced Sensors, pp. 634–662.
147. Ajayan, P.M.; Ebbesen, T.W.; Ichihashi, T.; Iijima, S.; Tanigaki, K.; Hiura, H. Opening Carbon Nanotubes with Oxygen and Implications for Filling. *Nature* **1993**, *362*, 522–525. [CrossRef]
148. Khusenov, M.A.; Dushanov, E.B.; Kholmurodov, K.T.; Zaki, M.M.; Sweilam, N.H. On Correlation Effect of the Van-Der-Waals and Intramolecular Forces for the Nucleotide Chain-Metallic Nanoparticles-Carbon Nanotube Binding. *Open Bioc. J.* **2016**, *10*, 17–26. [CrossRef] [PubMed]
149. Kumanek, B.; Janas, D. Thermal Conductivity of Carbon Nanotube Networks: A Review. *J. Mater. Sci.* **2019**, *54*, 7397–7427. [CrossRef]
150. Ibrahim, K.S. Carbon Nanotubes-Properties and Applications: A Review. *Carbon Lett.* **2013**, *14*, 131–144. [CrossRef]
151. Keidar, M.; Shashurin, A.; Li, J.; Volotskova, O.; Kundrapu, M.; Zhuang, T.S. Arc Plasma Synthesis of Carbon Nanostructures: Where Is the Frontier? *J. Phys. D Appl. Phys.* **2011**, *44*, 174006. [CrossRef]
152. Volotskova, O.; Fagan, J.A.; Huh, J.Y.; Phelan, F.R., Jr.; Shashurin, A.; Keidar, M. Tailored Distribution of Single-Wall Carbon Nanotubes from Arc Plasma Synthesis Using Magnetic Fields. *ACS Nano* **2010**, *4*, 5187–5192. [CrossRef]
153. Chrzanowska, J.; Hoffman, J.; Małolepszy, A.; Mazurkiewicz, M.; Kowalewski, T.A.; Szymanski, Z.; Stobinski, L. Synthesis of Carbon Nanotubes by the Laser Ablation Method: Effect of Laser Wavelength: Synthesis of CNTs by the Laser Ablation Method. *Phys. Status Solidi B* **2015**, *252*, 1860–1867. [CrossRef]
154. Li, Y.-L. Direct Spinning of Carbon Nanotube Fibers from Chemical Vapor Deposition Synthesis. *Science* **2004**, *304*, 276–278. [CrossRef]
155. Wojtera, K.; Walczak, M.; Pietrzak, L.; Fraczyk, J.; Szymanski, L.; Sobczyk-Guzenda, A. Synthesis of Functionalized Carbon Nanotubes for Fluorescent Biosensors. *Nanotechnol. Rev.* **2020**, *9*, 1237–1244. [CrossRef]
156. Vardharajula, S.; Ali, S.Z.; Tiwari, P.M.; Eroğlu, E.; Vig, K.; Dennis, V.A.; Singh, S.R. Functionalized carbon nanotubes: Biomedical applications. *Int. J. Nanomed.* **2012**, *7*, 5361–5374.
157. Manzetti, S.; Gabriel, J.-C.P. Methods for Dispersing Carbon Nanotubes for Nanotechnology Applications: Liquid Nanocrystals, Suspensions, Polyelectrolytes, Colloids and Organization Control. *Int. Nano Lett.* **2019**, *9*, 31–49. [CrossRef]
158. Kharisov, B.I.; Kharissova, O.V.; Méndez, U.O. Methods for Dispersion of Carbon Nanotubes in Water and Common Solvents. *MRS Proc.* **2014**, *1700*, 109–114. [CrossRef]
159. Jorde, L.; Li, Z.; Pöppelwerth, A.; Piehler, J.; You, C.; Meyer, C. Biofunctionalization of Carbon Nanotubes for Reversible Site-Specific Protein Immobilization. *J. Appl. Phys.* **2021**, *129*, 094302. [CrossRef]
160. Bekyarova, E.; Ni, Y.; Malarkey, E.B.; Montana, V.; McWilliams, J.L.; Haddon, R.C.; Parpura, V. Applications of Carbon Nanotubes in Biotechnology and Biomedicine. *J. Biomed. Nanotechnol.* **2005**, *1*, 3–17. [CrossRef]

161. Banica, F.; Bungau, S.; Tit, D.M.; Behl, T.; Otrisal, P.; Nechifor, A.C.; Gitea, D.; Pavel, F.-M.; Nemeth, S. Determination of the Total Polyphenols Content and Antioxidant Activity of Echinacea Purpurea Extracts Using Newly Manufactured Glassy Carbon Electrodes Modified with Carbon Nanotubes. *Processes* **2020**, *8*, 833. [CrossRef]
162. Yang, Y.; Yang, X.; Yang, Y.; Yuan, Q. Aptamer-Functionalized Carbon Nanomaterials Electrochemical Sensors for Detecting Cancer Relevant Biomolecules. *Carbon* **2018**, *129*, 380–395. [CrossRef]
163. Li, X.; Wang, L.; Fan, Y.; Feng, Q.; Cui, F. Biocompatibility and Toxicity of Nanoparticles and Nanotubes. *J. Nanomater.* **2012**, *2012*, 1–19. [CrossRef]
164. Cernat, A.; Tertiș, M.; Fritea, L.; Cristea, C. Graphene in Sensors Design. In *Advanced 2D Materials*; Tiwari, A., Syvajarvi, M., Eds.; Scrivener Publishing: Beverly, MA, USA; John Wiley & Sons: Hoboken, NJ, USA, 2016; pp. 387–431.
165. De Bonis, A.; Curcio, M.; Santagata, A.; Rau, J.V.; Galasso, A.; Teghil, R. Fullerene-Reduced Graphene Oxide Composites Obtained by Ultrashort Laser Ablation of Fullerite in Water. *Appl. Surf. Sci.* **2015**, *336*, 67–72. [CrossRef]
166. Yang, X.; Feng, B.; He, X.; Li, F.; Ding, Y.; Fei, J. Carbon Nanomaterial Based Electrochemical Sensors for Biogenic Amines. *Microchim Acta* **2013**, *180*, 935–956. [CrossRef]
167. Choi, W.; Lahiri, I.; Seelaboyina, R.; Kang, Y.S. Synthesis of Graphene and Its Applications: A Review. *Crit. Rev. Solid State Mater. Sci.* **2010**, *35*, 52–71. [CrossRef]
168. Novoselov, K.S. Electric Field Effect in Atomically Thin Carbon Films. *Science* **2004**, *306*, 666–669. [CrossRef]
169. Stankovich, S.; Dikin, D.A.; Piner, R.D.; Kohlhaas, K.A.; Kleinhammes, A.; Jia, Y.; Wu, Y.; Nguyen, S.T.; Ruoff, R.S. Synthesis of Graphene-Based Nanosheets via Chemical Reduction of Exfoliated Graphite Oxide. *Carbon* **2007**, *45*, 1558–1565. [CrossRef]
170. Kumbar, S.; Patil, S.; Jarali, C. Graphene in Different Extraction Techniques. In *Comprehensive Analytical Chemistry*; Elsevier: Amsterdam, The Netherlands, 2020; Volume 91, pp. 49–72. [CrossRef]
171. Gambhir, S.; Jalili, R.; Officer, D.L.; Wallace, G.G. Chemically Converted Graphene: Scalable Chemistries to Enable Processing and Fabrication. *NPG Asia Mater.* **2015**, *7*, e186. [CrossRef]
172. Hummers, W.S.; Offeman, R.E. Preparation of Graphitic Oxide. *J. Am. Chem. Soc.* **1958**, *80*, 1339. [CrossRef]
173. Marcano, D.C.; Kosynkin, D.V.; Berlin, J.M.; Sinitskii, A.; Sun, Z.; Slesarev, A.; Alemany, L.B.; Lu, W.; Tour, J.M. Improved Synthesis of Graphene Oxide. *ACS Nano* **2010**, *4*, 4806–4814. [CrossRef]
174. Guerrero-Contreras, J.; Caballero-Briones, F. Graphene Oxide Powders with Different Oxidation Degree, Prepared by Synthesis Variations of the Hummers Method. *Mater. Chem. Phys.* **2015**, *153*, 209–220. [CrossRef]
175. Chen, J.; Yao, B.; Li, C.; Shi, G. An Improved Hummers Method for Eco-Friendly Synthesis of Graphene Oxide. *Carbon* **2013**, *64*, 225–229. [CrossRef]
176. Berger, C. Electronic Confinement and Coherence in Patterned Epitaxial Graphene. *Science* **2006**, *312*, 1191–1196. [CrossRef]
177. Wintterlin, J.; Bocquet, M.-L. Graphene on Metal Surfaces. *Surf. Sci.* **2009**, *603*, 1841–1852. [CrossRef]
178. Kim, K.S.; Zhao, Y.; Jang, H.; Lee, S.Y.; Kim, J.M.; Kim, K.S.; Ahn, J.-H.; Kim, P.; Choi, J.-Y.; Hong, B.H. Large-Scale Pattern Growth of Graphene Films for Stretchable Transparent Electrodes. *Nature* **2009**, *457*, 706–710. [CrossRef] [PubMed]
179. Pei, S.; Cheng, H.-M. The Reduction of Graphene Oxide. *Carbon* **2012**, *50*, 3210–3228. [CrossRef]
180. Li, D.; Müller, M.B.; Gilje, S.; Kaner, R.B.; Wallace, G.G. Processable Aqueous Dispersions of Graphene Nanosheets. *Nature Nanotech.* **2008**, *3*, 101–105. [CrossRef]
181. Shin, H.; Kim, K.K.; Benayad, A.; Yoon, S.; Park, H.K.; Jung, I.; Jin, M.H.; Jeong, H.; Kim, J.M.; Choi, J.; et al. Efficient Reduction of Graphite Oxide by Sodium Borohydride and Its Effect on Electrical Conductance. *Adv. Funct. Mater.* **2009**, *19*, 1987–1992. [CrossRef]
182. Fernández-Merino, M.J.; Guardia, L.; Paredes, J.I.; Villar-Rodil, S.; Solís-Fernández, P.; Martínez-Alonso, A.; Tascón, J.M.D. Vitamin C Is an Ideal Substitute for Hydrazine in the Reduction of Graphene Oxide Suspensions. *J. Phys. Chem. C* **2010**, *114*, 6426–6432. [CrossRef]
183. Fritea, L.; Tertis, M.; Cosnier, S.; Cristea, C.; Săndulescu, R. A novel reduced graphene oxide/β-cyclodextrin/tyrosinase biosensor for dopamine detection. *Int. J. Electrochem. Sci.* **2015**, *10*, 7292–7302.
184. Fritea, L.; Le Goff, A.; Putaux, J.-L.; Tertis, M.; Cristea, C.; Săndulescu, R.; Cosnier, S. Design of a Reduced-Graphene-Oxide Composite Electrode from an Electropolymerizable Graphene Aqueous Dispersion Using a Cyclodextrin-Pyrrole Monomer. Application to Dopamine Biosensing. *Electrochim. Acta* **2015**, *178*, 108–112. [CrossRef]
185. Pei, S.; Zhao, J.; Du, J.; Ren, W.; Cheng, H.-M. Direct Reduction of Graphene Oxide Films into Highly Conductive and Flexible Graphene Films by Hydrohalic Acids. *Carbon* **2010**, *48*, 4466–4474. [CrossRef]
186. Fritea, L.; Tertiș, M.; Le Goff, A.; Cosnier, S.; Săndulescu, R.; Cristea, C. Graphene-Based Biosensors for Dopamine Determination. *Procedia Technol.* **2017**, *27*, 106–107. [CrossRef]
187. Erickson, K.; Erni, R.; Lee, Z.; Alem, N.; Gannett, W.; Zettl, A. Determination of the Local Chemical Structure of Graphene Oxide and Reduced Graphene Oxide. *Adv. Mater.* **2010**, *22*, 4467–4472. [CrossRef]
188. Bo, Z.; Shuai, X.; Mao, S.; Yang, H.; Qian, J.; Chen, J.; Yan, J.; Cen, K. Green Preparation of Reduced Graphene Oxide for Sensing and Energy Storage Applications. *Sci. Rep.* **2015**, *4*, 4684. [CrossRef]
189. Cernat, A.; Ciui, B.; Fritea, L.; Tertis, M.; Cristea, C. New Materials for the Construction of Electrochemical Cell-Based Biosensors. In *Handbook of Cell Biosensors*; Thouand, G., Ed.; Springer Nature Switzerland AG: Cham, Switzerland, 2020; pp. 1–26. [CrossRef]
190. Tertiş, M.; Hosu, O.; Fritea, L.; Farcau, C.; Cernat, A.; Săndulescu, R.; Cristea, C. A Novel Label-Free Immunosensor Based on Activated Graphene Oxide for Acetaminophen Detection. *Electroanalysis* **2015**, *27*, 638–647. [CrossRef]

191. Bullock, C.J.; Bussy, C. Biocompatibility Considerations in the Design of Graphene Biomedical Materials. *Adv. Mater. Interfaces* **2019**, *6*, 1900229. [CrossRef]
192. Liao, C.; Li, Y.; Tjong, S. Graphene Nanomaterials: Synthesis, Biocompatibility, and Cytotoxicity. *Int. J. Mol. Sci.* **2018**, *19*, 3564. [CrossRef]
193. Campuzano, S.; Yáñez-Sedeño, P.; Pingarrón, J.M. Carbon Dots and Graphene Quantum Dots in Electrochemical Biosensing. *Nanomaterials* **2019**, *9*, 634. [CrossRef]
194. Ding, X.; Niu, Y.; Zhang, G.; Xu, Y.; Li, J. Electrochemistry in Carbon-based Quantum Dots. *Chem. Asian J.* **2020**, *15*, 1214–1224. [CrossRef]
195. Namdari, P.; Negahdari, B.; Eatemadi, A. Synthesis, Properties and Biomedical Applications of Carbon-Based Quantum Dots: An Updated Review. *Biomed. Pharmacother.* **2017**, *87*, 209–222. [CrossRef]
196. Alaghmandfard, A.; Sedighi, O.; Tabatabaei Rezaei, N.; Abedini, A.A.; Malek Khachatourian, A.; Toprak, M.S.; Seifalian, A. Recent Advances in the Modification of Carbon-Based Quantum Dots for Biomedical Applications. *Mater. Sci. Eng. C* **2021**, *120*, 111756. [CrossRef]
197. Faridbod, F.; Sanati, A.L. Graphene Quantum Dots in Electrochemical Sensors/Biosensors. *Curr. Anal. Chem.* **2019**, *15*, 103–123. [CrossRef]
198. Wang, X.; Feng, Y.; Dong, P.; Huang, J. A Mini Review on Carbon Quantum Dots: Preparation, Properties, and Electrocatalytic Application. *Front. Chem.* **2019**, *7*, 671. [CrossRef]
199. Elugoke, S.E.; Adekunle, A.S.; Fayemi, O.E.; Mamba, B.B.; Sherif, E.-S.M.; Ebenso, E.E. Carbon-Based Quantum Dots for Electrochemical Detection of Monoamine Neurotransmitters—Review. *Biosensors* **2020**, *10*, 162. [CrossRef]
200. Chen, Y.; Cao, Y.; Ma, C.; Zhu, J.-J. Carbon-Based Dots for Electrochemiluminescence Sensing. *Mater. Chem. Front.* **2020**, *4*, 369–385. [CrossRef]
201. Bhatnagar, D.; Kumar, V.; Kumar, A.; Kaur, I. Graphene Quantum Dots FRET Based Sensor for Early Detection of Heart Attack in Human. *Biosens. Bioelectron.* **2016**, *79*, 495–499. [CrossRef]
202. Thevenot, D.R.; Tóth, K.; Durst, R.A.; Wilson, G.S. Electrochemical Biosensors: Recommended Definitions and Classification. *Pure Appl. Chem.* **1999**, *71*, 2333–2348. [CrossRef]
203. Hugh Fan, Z. Chemical Sensors and Microfluidics. *J. Biosens. Bioelectron.* **2013**, *4*, 1000e117. [CrossRef]
204. Jurado-Sánchez, B. Nanoscale Biosensors Based on Self-Propelled Objects. *Biosensors* **2018**, *8*, 59. [CrossRef]
205. García-Miranda Ferrari, A.; Rowley-Neale, S.J.; Banks, C.E. Screen-Printed Electrodes: Transitioning the Laboratory in-to-the Field. *Talanta Open* **2021**, *3*, 100032. [CrossRef]
206. Couto, R.A.S.; Lima, J.L.F.C.; Quinaz, M.B. Recent Developments, Characteristics and Potential Applications of Screen-Printed Electrodes in Pharmaceutical and Biological Analysis. *Talanta* **2016**, *146*, 801–814. [CrossRef]
207. Kamyshny, A.; Magdassi, S. Conductive Nanomaterials for Printed Electronics. *Small* **2014**, *10*, 3515–3535. [CrossRef]
208. Trojanowicz, M. Impact of Nanotechnology on Design of Advanced Screen-Printed Electrodes for Different Analytical Applications. *Trends Anal. Chem.* **2016**, *84*, 22–47. [CrossRef]
209. Jubete, E.; Loaiza, O.A.; Ochoteco, E.; Pomposo, J.A.; Grande, H.; Rodríguez, J. Nanotechnology: A Tool for Improved Performance on Electrochemical Screen-Printed (Bio)Sensors. *J. Sens.* **2009**, *2009*, 1–13. [CrossRef]
210. Cinti, S.; Arduini, F. Graphene-Based Screen-Printed Electrochemical (Bio)Sensors and Their Applications: Efforts and Criticisms. *Biosens. Bioelectron.* **2017**, *89*, 107–122. [CrossRef] [PubMed]
211. Taleat, Z.; Khoshroo, A.; Mazloum-Ardakani, M. Screen-Printed Electrodes for Biosensing: A Review (2008–2013). *Microchim. Acta* **2014**, *181*, 865–891. [CrossRef]
212. Mohamed, H.M. Screen-Printed Disposable Electrodes: Pharmaceutical Applications and Recent Developments. *Trends Anal. Chem.* **2016**, *82*, 1–11. [CrossRef]
213. Kondori, T.; Tajik, S.; Akbarzadeh-T, N.; Beitollahi, H.; Graiff, C.; Jang, H.W.; Shokouhimehr, M. Synthesis and Characterization of Bipyridine Cobalt(II) Complex Modified Graphite Screen Printed Electrode: An Electrochemical Sensor for Simultaneous Detection of Acetaminophen and Naproxen. *RSC Adv.* **2021**, *11*, 3049–3057. [CrossRef]
214. Sher, M.; Faheem, A.; Asghar, W.; Cinti, S. Nano-Engineered Screen-Printed Electrodes: A Dynamic Tool for Detection of Viruses. *Trends Anal. Chem.* **2021**, *143*, 116374. [CrossRef]
215. Ping, J.; Wang, Y.; Fan, K.; Wu, J.; Ying, Y. Direct Electrochemical Reduction of Graphene Oxide on Ionic Liquid Doped Screen-Printed Electrode and Its Electrochemical Biosensing Application. *Biosens. Bioelectron.* **2011**, *28*, 204–209. [CrossRef]
216. Sahin, M.; Ayranci, E. Electrooxidation of NADH on Modified Screen-Printed Electrodes: Effects of Conducting Polymer and Nanomaterials. *Electrochim. Acta* **2015**, *166*, 261–270. [CrossRef]
217. Cinti, S.; Arduini, F.; Carbone, M.; Sansone, L.; Cacciotti, I.; Moscone, D.; Palleschi, G. Screen-Printed Electrodes Modified with Carbon Nanomaterials: A Comparison among Carbon Black, Carbon Nanotubes and Graphene. *Electroanalysis* **2015**, *27*, 2230–2238. [CrossRef]
218. Vukojević, V.; Djurdjić, S.; Ognjanović, M.; Antić, B.; Kalcher, K.; Mutić, J.; Stanković, D.M. RuO2/Graphene Nanoribbon Composite Supported on Screen Printed Electrode with Enhanced Electrocatalytic Performances toward Ethanol and NADH Biosensing. *Biosens. Bioelectron.* **2018**, *117*, 392–397. [CrossRef] [PubMed]

219. Yang, X.; Ouyang, Y.; Wu, F.; Hu, Y.; Zhang, H.; Wu, Z. In Situ & Controlled Preparation of Platinum Nanoparticles Dopping into Graphene Sheets@cerium Oxide Nanocomposites Sensitized Screen Printed Electrode for Nonenzymatic Electrochemical Sensing of Hydrogen Peroxide. *J. Electroanal. Chem.* **2016**, *777*, 85–91. [CrossRef]
220. Yao, Z.; Yang, X.; Wu, F.; Wu, W.; Wu, F. Synthesis of Differently Sized Silver Nanoparticles on a Screen-Printed Electrode Sensitized with a Nanocomposites Consisting of Reduced Graphene Oxide and Cerium(IV) Oxide for Nonenzymatic Sensing of Hydrogen Peroxide. *Microchim. Acta* **2016**, *183*, 2799–2806. [CrossRef]
221. Martín, A.; Hernández-Ferrer, J.; Martínez, M.T.; Escarpa, A. Graphene Nanoribbon-Based Electrochemical Sensors on Screen-Printed Platforms. *Electrochim. Acta* **2015**, *172*, 2–6. [CrossRef]
222. Ji, D.; Liu, Z.; Liu, L.; Low, S.S.; Lu, Y.; Yu, X.; Zhu, L.; Li, C.; Liu, Q. Smartphone-Based Integrated Voltammetry System for Simultaneous Detection of Ascorbic Acid, Dopamine, and Uric Acid with Graphene and Gold Nanoparticles Modified Screen-Printed Electrodes. *Biosens. Bioelectron.* **2018**, *119*, 55–62. [CrossRef] [PubMed]
223. Reanpang, P.; Mool-am-kha, P.; Upan, J.; Jakmunee, J. A Novel Flow Injection Amperometric Sensor Based on Carbon Black and Graphene Oxide Modified Screen-Printed Carbon Electrode for Highly Sensitive Determination of Uric Acid. *Talanta* **2021**, *232*, 122493. [CrossRef] [PubMed]
224. Altun, M.; Bilgi Kamaç, M.; Bilgi, A.; Yılmaz, M. Dopamine Biosensor Based on Screen-Printed Electrode Modified with Reduced Graphene Oxide, Polyneutral Red and Gold Nanoparticle. *Int. J. Environ. Anal. Chem.* **2020**, *100*, 451–467. [CrossRef]
225. Truţă, F.; Tertis, M.; Cristea, C.; Graur, F. Simultaneous Detection of Dopamine and Serotonin in Real Complex Matrices. *Curr. Anal. Chem.* **2021**, *17*, 374–384. [CrossRef]
226. Tertiș, M.; Cernat, A.; Lacatiș, D.; Florea, A.; Bogdan, D.; Suciu, M.; Săndulescu, R.; Cristea, C. Highly Selective Electrochemical Detection of Serotonin on Polypyrrole and Gold Nanoparticles-Based 3D Architecture. *Electrochem. Commun.* **2017**, *75*, 43–47. [CrossRef]
227. Tertiş, M.; Ciui, B.; Suciu, M.; Săndulescu, R.; Cristea, C. Label-Free Electrochemical Aptasensor Based on Gold and Polypyrrole Nanoparticles for Interleukin 6 Detection. *Electrochim. Acta* **2017**, *258*, 1208–1218. [CrossRef]
228. Phetsang, S.; Jakmunee, J.; Mungkornasawakul, P.; Laocharoensuk, R.; Ounnunkad, K. Sensitive Amperometric Biosensors for Detection of Glucose and Cholesterol Using a Platinum/Reduced Graphene Oxide/Poly(3-Aminobenzoic Acid) Film-Modified Screen-Printed Carbon Electrode. *Bioelectrochemistry* **2019**, *127*, 125–135. [CrossRef] [PubMed]
229. Shahrokhian, S.; Khaki Sanati, E.; Hosseini, H. Advanced On-Site Glucose Sensing Platform Based on a New Architecture of Free-Standing Hollow $Cu(OH)_2$ Nanotubes Decorated with CoNi-LDH Nanosheets on Graphite Screen-Printed Electrode. *Nanoscale* **2019**, *11*, 12655–12671. [CrossRef] [PubMed]
230. Bernardo-Boongaling, V.R.R.; Serrano, N.; García-Guzmán, J.J.; Palacios-Santander, J.M.; Díaz-Cruz, J.M. Screen-Printed Electrodes Modified with Green-Synthesized Gold Nanoparticles for the Electrochemical Determination of Aminothiols. *J. Electroanal. Chem.* **2019**, *847*, 113184. [CrossRef]
231. Ji, D.; Shi, Z.; Liu, Z.; Low, S.S.; Zhu, J.; Zhang, T.; Chen, Z.; Yu, X.; Lu, Y.; Lu, D.; et al. Smartphone-Based Square Wave Voltammetry System with Screen-Printed Graphene Electrodes for Norepinephrine Detection. *Smart Mater. Med.* **2020**, *1*, 1–9. [CrossRef]
232. Thangamuthu, M.; Gabriel, W.; Santschi, C.; Martin, O. Electrochemical Sensor for Bilirubin Detection Using Screen Printed Electrodes Functionalized with Carbon Nanotubes and Graphene. *Sensors* **2018**, *18*, 800. [CrossRef]
233. Boonkaew, S.; Chaiyo, S.; Jampasa, S.; Rengpipat, S.; Siangproh, W.; Chailapakul, O. An Origami Paper-Based Electrochemical Immunoassay for the C-Reactive Protein Using a Screen-Printed Carbon Electrode Modified with Graphene and Gold Nanoparticles. *Microchim. Acta* **2019**, *186*, 153. [CrossRef]
234. Rafiee, B.; Fakhari, A.R. Electrocatalytic Oxidation and Determination of Insulin at Nickel Oxide Nanoparticles-Multiwalled Carbon Nanotube Modified Screen Printed Electrode. *Biosens. Bioelectron.* **2013**, *46*, 130–135. [CrossRef]
235. Arvinte, A.; Westermann, A.C.; Sesay, A.M.; Virtanen, V. Electrocatalytic Oxidation and Determination of Insulin at CNT-Nickel–Cobalt Oxide Modified Electrode. *Sens. Actuators B Chem.* **2010**, *150*, 756–763. [CrossRef]
236. Eissa, S.; Almusharraf, A.Y.; Zourob, M. A Comparison of the Performance of Voltammetric Aptasensors for Glycated Haemoglobin on Different Carbon Nanomaterials-Modified Screen Printed Electrodes. *Mater. Sci. Eng. C* **2019**, *101*, 423–430. [CrossRef]
237. Eissa, S.; Alshehri, N.; Rahman, A.M.A.; Dasouki, M.; Abu-Salah, K.M.; Zourob, M. Electrochemical Immunosensors for the Detection of Survival Motor Neuron (SMN) Protein Using Different Carbon Nanomaterials-Modified Electrodes. *Biosens. Bioelectron.* **2018**, *101*, 282–289. [CrossRef]
238. Shoaie, N.; Forouzandeh, M.; Omidfar, K. Voltammetric Determination of the Escherichia Coli DNA Using a Screen-Printed Carbon Electrode Modified with Polyaniline and Gold Nanoparticles. *Microchim. Acta* **2018**, *185*, 217. [CrossRef]
239. Upan, J.; Banet, P.; Aubert, P.-H.; Ounnunkad, K.; Jakmunee, J. Sequential Injection-Differential Pulse Voltammetric Immunosensor for Hepatitis B Surface Antigen Using the Modified Screen-Printed Carbon Electrode. *Electrochim. Acta* **2020**, *349*, 136335. [CrossRef]
240. Gandouzi, I.; Tertis, M.; Cernat, A.; Bakhrouf, A.; Coros, M.; Pruneanu, S.; Cristea, C. Sensitive Detection of Pyoverdine with an Electrochemical Sensor Based on Electrochemically Generated Graphene Functionalized with Gold Nanoparticles. *Bioelectrochemistry* **2018**, *120*, 94–103. [CrossRef]

241. Ravalli, A.; dos Santos, G.P.; Ferroni, M.; Faglia, G.; Yamanaka, H.; Marrazza, G. New Label Free CA125 Detection Based on Gold Nanostructured Screen-Printed Electrode. *Sens. Actuators B Chem.* **2013**, *179*, 194–200. [CrossRef]
242. Baradoke, A.; Jose, B.; Pauliukaite, R.; Forster, R.J. Properties of Anti-CA125 Antibody Layers on Screen-Printed Carbon Electrodes Modified by Gold and Platinum Nanostructures. *Electrochim. Acta* **2019**, *306*, 299–306. [CrossRef]
243. Chan, K.F.; Lim, H.N.; Shams, N.; Jayabal, S.; Pandikumar, A.; Huang, N.M. Fabrication of Graphene/Gold-Modified Screen-Printed Electrode for Detection of Carcinoembryonic Antigen. *Mater. Sci. Eng. C* **2016**, *58*, 666–674. [CrossRef]
244. Giannetto, M.; Bianchi, M.V.; Mattarozzi, M.; Careri, M. Competitive Amperometric Immunosensor for Determination of P53 Protein in Urine with Carbon Nanotubes/Gold Nanoparticles Screen-Printed Electrodes: A Potential Rapid and Noninvasive Screening Tool for Early Diagnosis of Urinary Tract Carcinoma. *Anal. Chim. Acta* **2017**, *991*, 133–141. [CrossRef]
245. Serafín, V.; Valverde, A.; Garranzo-Asensio, M.; Barderas, R.; Campuzano, S.; Yáñez-Sedeño, P.; Pingarrón, J.M. Simultaneous Amperometric Immunosensing of the Metastasis-Related Biomarkers IL-13Rα2 and CDH-17 by Using Grafted Screen-Printed Electrodes and a Composite Prepared from Quantum Dots and Carbon Nanotubes for Signal Amplification. *Microchim. Acta* **2019**, *186*, 411. [CrossRef]
246. Teixeira, S.; Conlan, R.S.; Guy, O.J.; Sales, M.G.F. Label-Free Human Chorionic Gonadotropin Detection at Picogram Levels Using Oriented Antibodies Bound to Graphene Screen-Printed Electrodes. *J. Mater. Chem. B* **2014**, *2*, 1852. [CrossRef]
247. Zhao, H.; Liu, F.; Xie, W.; Zhou, T.-C.; OuYang, J.; Jin, L.; Li, H.; Zhao, C.-Y.; Zhang, L.; Wei, J.; et al. Ultrasensitive Supersandwich-Type Electrochemical Sensor for SARS-CoV-2 from the Infected COVID-19 Patients Using a Smartphone. *Sens. Actuators B Chem.* **2021**, *327*, 128899. [CrossRef]
248. Fabiani, L.; Saroglia, M.; Galatà, G.; De Santis, R.; Fillo, S.; Luca, V.; Faggioni, G.; D'Amore, N.; Regalbuto, E.; Salvatori, P.; et al. Magnetic Beads Combined with Carbon Black-Based Screen-Printed Electrodes for COVID-19: A Reliable and Miniaturized Electrochemical Immunosensor for SARS-CoV-2 Detection in Saliva. *Biosens. Bioelectron.* **2021**, *171*, 112686. [CrossRef] [PubMed]
249. Hughes, G.; Westmacott, K.; Honeychurch, K.; Crew, A.; Pemberton, R.; Hart, J. Recent Advances in the Fabrication and Application of Screen-Printed Electrochemical (Bio)Sensors Based on Carbon Materials for Biomedical, Agri-Food and Environmental Analyses. *Biosensors* **2016**, *6*, 50. [CrossRef] [PubMed]
250. Palestino, G.; García-Silva, I.; González-Ortega, O.; Rosales-Mendoza, S. Can Nanotechnology Help in the Fight against COVID-19? *Expert Rev. Anti-Infect. Ther.* **2020**, *18*, 849–864. [CrossRef] [PubMed]
251. Miere, F.; Fritea, L.; Cavalu, S.; Vicas, S.I. Formulation, characterisation and advantages of using liposomes in multiple therapies. *Pharmacophore* **2020**, *11*, 1–12.
252. Miere (Groza), F.; Vicas, S.I.; Timar, A.V.; Ganea, M.; Zdrinca, M.; Cavalu, S.; Fritea, L.; Vicas, L.; Muresan, M.; Pallag, A.; et al. Preparation and Characterization of Two Different Liposomal Formulations with Bioactive Natural Extract for Multiple Applications. *Processes* **2021**, *9*, 432. [CrossRef]
253. Abu Lila, A.S.; Ishida, T. Liposomal Delivery Systems: Design Optimization and Current Applications. *Biol. Pharm. Bull.* **2017**, *40*, 1–10. [CrossRef]
254. Miere (Groza), F.; Teusdea, A.C.; Laslo, V.; Fritea, L.; Moldovan, L.; Costea, T.; Uivarosan, D.; Vicas, S.I.; Pallag, A. Natural Polymeric Beads for Encapsulation of Stellaria Media Extract with Antioxidant Properties. *Mater. Plast.* **2019**, *56*, 671–679. [CrossRef]
255. Fritea, L.; Pasca, P.M.; Vlase, L.; Gheldiu, A.-M.; Moldovan, L.; Banica, F.; Dobjanschi, L.; Cavalu, S. Electrochemical Methods for Evaluation of Antioxidant Properties of Propolis Extract Incorporated in Chitosan Nanoparticles. *Mater. Plast.* **2021**, *57*, 96–108. [CrossRef]
256. George, A.; Shah, P.A.; Shrivastav, P.S. Natural Biodegradable Polymers Based Nano-Formulations for Drug Delivery: A Review. *Int. J. Pharm.* **2019**, *561*, 244–264. [CrossRef]
257. Ghaemi, F.; Amiri, A.; Bajuri, M.Y.; Yuhana, N.Y.; Ferrara, M. Role of Different Types of Nanomaterials against Diagnosis, Prevention and Therapy of COVID-19. *Sustain. Cities Soc.* **2021**, *72*, 103046. [CrossRef]
258. Bogdan, J.; Zarzyńska, J.; Pławińska-Czarnak, J. Comparison of Infectious Agents Susceptibility to Photocatalytic Effects of Nanosized Titanium and Zinc Oxides: A Practical Approach. *Nanoscale Res. Lett.* **2015**, *10*, 309. [CrossRef]
259. Palmieri, V.; De Maio, F.; De Spirito, M.; Papi, M. Face Masks and Nanotechnology: Keep the Blue Side Up. *Nano Today* **2021**, *37*, 101077. [CrossRef]
260. Ramaiah, G.B.; Tegegne, A.; Melese, B. Functionality of Nanomaterials and Its Technological Aspects–Used in Preventing, Diagnosing and Treating COVID-19. *Mater. Today Proc.* **2021**, *47*, 2337–2344. [CrossRef]
261. Mujawar, M.A.; Gohel, H.; Bhardwaj, S.K.; Srinivasan, S.; Hickman, N.; Kaushik, A. Nano-Enabled Biosensing Systems for Intelligent Healthcare: Towards COVID-19 Management. *Mater. Today Chem.* **2020**, *17*, 100306. [CrossRef]
262. Samson, R.; Navale, G.R.; Dharne, M.S. Biosensors: Frontiers in Rapid Detection of COVID-19. *3 Biotech* **2020**, *10*, 385. [CrossRef]
263. Broughton, J.P.; Deng, X.; Yu, G.; Fasching, C.L.; Servellita, V.; Singh, J.; Miao, X.; Streithorst, J.A.; Granados, A.; Sotomayor-Gonzalez, A.; et al. CRISPR–Cas12-Based Detection of SARS-CoV-2. *Nat. Biotechnol.* **2020**, *38*, 870–874. [CrossRef]
264. Xiang, J.; Yan, M.; Li, H.; Liu, T.; Lin, C.; Huang, S.; Shen, C. Evaluation of Enzyme-Linked Immunoassay and Colloidal Gold-Immunochromatographic Assay Kit for Detection of Novel Coronavirus (SARS-Cov-2) Causing an Outbreak of Pneumonia (COVID-19). *Epidemiology* **2020**, preprint. [CrossRef]
265. Gowri, A.; Ashwin Kumar, N.; Suresh Anand, B.S. Recent Advances in Nanomaterials Based Biosensors for Point of Care (PoC) Diagnosis of Covid-19–A Minireview. *Trends Anal. Chem.* **2021**, *137*, 116205. [CrossRef]

266. Kumar, N.; Shetti, N.P.; Jagannath, S.; Aminabhavi, T.M. Electrochemical sensors for the detection of SARS-CoV-2 virus. *Chem. Eng. J.* **2022**, *430*, 132966. [CrossRef]
267. Kudr, J.; Michalek, P.; Ilieva, L.; Adam, V.; Zitka, O. COVID-19: A Challenge for Electrochemical Biosensors. *Trends Anal. Chem.* **2021**, *136*, 116192. [CrossRef]
268. Srivastava, M.; Srivastava, N.; Mishra, P.K.; Malhotra, B.D. Prospects of Nanomaterials-Enabled Biosensors for COVID-19 Detection. *Sci. Total. Environ.* **2021**, *754*, 142363. [CrossRef]
269. Saatçi, E.; Natarajan, S. State-of-the-Art Colloidal Particles and Unique Interfaces-Based SARS-CoV-2 Detection Methods and COVID-19 Diagnosis. *Curr. Opin. Colloid Interface Sci.* **2021**, *55*, 101469. [CrossRef]
270. Kotru, S.; Klimuntowski, M.; Ridha, H.; Uddin, Z.; Askhar, A.A.; Singh, G.; Howlader, M.M.R. Electrochemical Sensing: A Prognostic Tool in the Fight against COVID-19. *Trends Anal. Chem.* **2021**, *136*, 116198. [CrossRef]
271. Abo-zeid, Y.; Ismail, N.S.M.; McLean, G.R.; Hamdy, N.M. A Molecular Docking Study Repurposes FDA Approved Iron Oxide Nanoparticles to Treat and Control COVID-19 Infection. *Eur. J. Pharm. Sci.* **2020**, *153*, 105465. [CrossRef]
272. Mahari, S.; Roberts, A.; Shahdeo, D.; Gandhi, S. ECovSens-Ultrasensitive Novel In-House Built Printed Circuit Board Based Electrochemical Device for Rapid Detection of NCovid-19 Antigen, a Spike Protein Domain 1 of SARS-CoV-2. *Bioengineering* **2020**, preprint. [CrossRef]
273. Idris, A.; Davis, A.; Supramaniam, A.; Acharya, D.; Kelly, G.; Tayyar, Y.; West, N.; Zhang, P.; McMillan, C.L.D.; Soemardy, C.; et al. A SARS-CoV-2 Targeted SiRNA-Nanoparticle Therapy for COVID-19. *Mol. Ther.* **2021**, *29*, 2219–2226. [CrossRef]
274. Moreno, V.; Llorent-Martínez, E.J.; Zougagh, M.; Ríos, A. Decoration of Multi-Walled Carbon Nanotubes with Metal Nanoparticles in Supercritical Carbon Dioxide Medium as a Novel Approach for the Modification of Screen-Printed Electrodes. *Talanta* **2016**, *161*, 775–779. [CrossRef]
275. Khaled, E.; Shoukry, E.M.; Amin, M.F.; Said, B.A.M. Novel Calixarene/Carbon Nanotubes Based Screen Printed Sensors for Flow Injection Potentiometric Determination of Naproxen. *Electroanalysis* **2018**, *30*, 2878–2887. [CrossRef]
276. Muhammad, A.; Yusof, N.; Hajian, R.; Abdullah, J. Construction of an Electrochemical Sensor Based on Carbon Nanotubes/Gold Nanoparticles for Trace Determination of Amoxicillin in Bovine Milk. *Sensors* **2016**, *16*, 56. [CrossRef]
277. Shehata, M.; Fekry, A.M.; Walcarius, A. Moxifloxacin Hydrochloride Electrochemical Detection at Gold Nanoparticles Modified Screen-Printed Electrode. *Sensors* **2020**, *20*, 2797. [CrossRef]
278. Materón, E.M.; Wong, A.; Freitas, T.A.; Faria, R.C.; Oliveira, O.N. A Sensitive Electrochemical Detection of Metronidazole in Synthetic Serum and Urine Samples Using Low-Cost Screen-Printed Electrodes Modified with Reduced Graphene Oxide and C60. *J. Pharm. Anal.* **2021**, in press. [CrossRef]
279. Farokhi-Fard, A.; Golichenari, B.; Mohammadi Ghanbarlou, M.; Zanganeh, S.; Vaziri, F. Electroanalysis of Isoniazid and Rifampicin: Role of Nanomaterial Electrode Modifiers. *Biosens. Bioelectron.* **2019**, *146*, 111731. [CrossRef] [PubMed]
280. Tajik, S.; Beitollahi, H.; Jang, H.W.; Shokouhimehr, M. A Screen Printed Electrode Modified with Fe3O4@polypyrrole-Pt Core-Shell Nanoparticles for Electrochemical Detection of 6-Mercaptopurine and 6-Thioguanine. *Talanta* **2021**, *232*, 122379. [CrossRef] [PubMed]
281. de Carvalho, R.C.; Betts, A.J.; Cassidy, J.F. Diclofenac Determination Using CeO2 Nanoparticle Modified Screen-Printed Electrodes–A Study of Background Correction. *Microchem. J.* **2020**, *158*, 105258. [CrossRef]
282. Serrano, N.; Castilla, Ò.; Ariño, C.; Diaz-Cruz, M.; Díaz-Cruz, J. Commercial Screen-Printed Electrodes Based on Carbon Nanomaterials for a Fast and Cost-Effective Voltammetric Determination of Paracetamol, Ibuprofen and Caffeine in Water Samples. *Sensors* **2019**, *19*, 4039. [CrossRef]
283. Gomez, F.J.V.; Martín, A.; Silva, M.F.; Escarpa, A. Screen-Printed Electrodes Modified with Carbon Nanotubes or Graphene for Simultaneous Determination of Melatonin and Serotonin. *Microchim. Acta* **2015**, *182*, 1925–1931. [CrossRef]
284. Rajpurohit, A.S.; Srivastava, A.K. Simultaneous Electrochemical Sensing of Three Prevalent Anti-Allergic Drugs Utilizing Nanostructured Manganese Hexacyanoferrate/Chitosan Modified Screen Printed Electrode. *Sens. Actuators B Chem.* **2019**, *294*, 231–244. [CrossRef]
285. Abou Al-Alamein, A.M.; Kamel, M.S.; Abou El-Alamin, M.M.; Khaled, E. Novel Pioglitazone Nanomaterial Based Screen Printed Sensors. *Int. J. Electrochem. Sci.* **2015**, *10*, 2400–2412.
286. Dourandish, Z.; Beitollahi, H. Electrochemical Sensing of Isoproterenol using Graphite Screen-printed Electrode Modified with Graphene Quantum Dots. *Anal. Bioanal. Electrochem.* **2018**, *10*, 192–202.
287. Beitollahi, H.; Jahani, S.; Tajik, S.; Ganjali, M.R.; Faridbod, F.; Alizadeh, T. Voltammetric Determination of Venlafaxine as an Antidepressant Drug Employing Gd2O3 Nanoparticles Graphite Screen Printed Electrode. *J. Rare Earths* **2019**, *37*, 322–328. [CrossRef]
288. Khosrokhavar, R.; Motaharian, A.; Milani Hosseini, M.R.; Mohammadsadegh, S. Screen-Printed Carbon Electrode (SPCE) Modified by Molecularly Imprinted Polymer (MIP) Nanoparticles and Graphene Nanosheets for Determination of Sertraline Antidepressant Drug. *Microchem. J.* **2020**, *159*, 105348. [CrossRef]
289. Omidfar, K.; Ahmadi, A.; Syedmoradi, L.; Khoshfetrat, S.M.; Larijani, B. Point-of-Care Biosensors in Medicine: A Brief Overview of Our Achievements in This Field Based on the Conducted Research in EMRI (Endocrinology and Metabolism Research Institute of Tehran University of Medical Sciences) over the Past Fourteen Years. *J. Diabetes Metab. Disord.* 2020. [CrossRef]
290. Syedmoradi, L.; Daneshpour, M.; Alvandipour, M.; Gomez, F.A.; Hajghassem, H.; Omidfar, K. Point of Care Testing: The Impact of Nanotechnology. *Biosens. Bioelectron.* **2017**, *87*, 373–387. [CrossRef]

291. Iqbal, S.M.A.; Mahgoub, I.; Du, E.; Leavitt, M.A.; Asghar, W. Advances in Healthcare Wearable Devices. *NPJ Flex. Electron.* **2021**, *5*, 9. [CrossRef]
292. Dexcom. The Dexcom G6 CGM System. Available online: https://www.dexcom.com/g6-cgmsystem (accessed on 10 September 2021).
293. Keum, D.H.; Kim, S.-K.; Koo, J.; Lee, G.-H.; Jeon, C.; Mok, J.W.; Mun, B.H.; Lee, K.J.; Kamrani, E.; Joo, C.-K.; et al. Wireless Smart Contact Lens for Diabetic Diagnosis and Therapy. *Sci. Adv.* **2020**, *6*, eaba3252. [CrossRef]
294. Acar, G.; Ozturk, O.; Golparvar, A.J.; Elboshra, T.A.; Böhringer, K.; Yapici, M.K. Wearable and Flexible Textile Electrodes for Biopotential Signal Monitoring: A Review. *Electronics* **2019**, *8*, 479. [CrossRef]
295. Gao, K.-P.; Shen, G.-C.; Zhao, N.; Jiang, C.-P.; Yang, B.; Liu, J.-Q. Wearable Multifunction Sensor for the Detection of Forehead EEG Signal and Sweat Rate on Skin Simultaneously. *IEEE Sens. J.* **2020**, *20*, 10393–10404. [CrossRef]
296. Clinic, M. Electromyography (EMG). Available online: https://www.mayoclinic.org/tests-procedures/emg/about/pac-20393913 (accessed on 10 September 2021).
297. Lai Kwan, C.; Mahdid, Y.; Motta Ochoa, R.; Lee, K.; Park, M.; Blain-Moraes, S. Wearable Technology for Detecting Significant Moments in Individuals with Dementia. *BioMed Res. Int.* **2019**, *2019*, 1–13. [CrossRef]
298. Bariya, M.; Li, L.; Ghattamaneni, R.; Ahn, C.H.; Nyein, H.Y.Y.; Tai, L.-C.; Javey, A. Glove-Based Sensors for Multimodal Monitoring of Natural Sweat. *Sci. Adv.* **2020**, *6*, eabb8308. [CrossRef]

materials

Article

Optimization of Hyaluronate-Based Liposomes to Augment the Oral Delivery and the Bioavailability of Berberine

Hussam I. Kutbi [1], Hani Z. Asfour [2], Ahmed K. Kammoun [3], Alaa Sirwi [4], Simona Cavalu [5,*] and Heba A. Gad [6,*]

1 Department of Pharmacy Practice, Faculty of Pharmacy, King Abdulaziz University, Jeddah 21589, Saudi Arabia; hkutbi1@kau.edu.sa
2 Department of Medical Microbiology and Parasitology, Faculty of Medicine, Princess Al-Jawhara Center of Excellence in Research of Hereditary Disorders, King Abdulaziz University, Jeddah 21589, Saudi Arabia; hasfour@hotmail.com
3 Department of Pharmaceutical Chemistry, Faculty of Pharmacy, King Abdulaziz University, Jeddah 21589, Saudi Arabia; akammoun@kau.edu.sa
4 Department of Natural Products, Faculty of Pharmacy, King Abdulaziz University, Jeddah 21589, Saudi Arabia; asirwi@kau.edu.sa
5 Faculty of Medicine and Pharmacy, University of Oradea, P-ta 1 Decembrie 10, 410087 Oradea, Romania
6 Department of Pharmaceutics and Industrial Pharmacy, Faculty of Pharmacy, Ain Shams University, Cairo 11566, Egypt
* Correspondence: simona.cavalu@gmail.com (S.C.); h.gad@pharma.asu.edu.eg (H.A.G.); Tel.: +20-100-022-6421 (H.A.G.)

Citation: Kutbi, H.I.; Asfour, H.Z.; Kammoun, A.K.; Sirwi, A.; Cavalu, S.; Gad, H.A. Optimization of Hyaluronate-Based Liposomes to Augment the Oral Delivery and the Bioavailability of Berberine. *Materials* **2021**, *14*, 5759. https://doi.org/10.3390/ma14195759

Academic Editor: Abdelwahab Omri

Received: 1 September 2021
Accepted: 28 September 2021
Published: 2 October 2021

Abstract: Various perspectives had been utilized to enhance the poor intestinal permeability and bioavailability of drugs with low water solubility. Berberine (Brb) is a unique molecule that possesses multiple therapeutic activities such as antimicrobial, anti-inflammatory, antioxidant and anti-hyperglycemic effects. To improve Brb permeability and bioavailability, this study presents a newly developed formulation, namely Brb hyaluronate-based liposomes, prepared by using film hydration method and characterized by dynamic light scattering measurements, entrapment efficiency percentage (EE%), transmission electron microscope (TEM), in vitro drug release and physical stability. The bioavailability of the selected formulations was assessed in vivo after oral administration to rats. The results revealed an enhanced effect of hyaluronic acid on the entrapment efficiency, reaching 78.1 ± 0.1% with mean size 520.7 ± 19.9 nm. Sustained release of Brb was recorded up to 24 h in comparison to Brb solution. Physical stability was maintained for three months at refrigeration temperature. Results of pharmacokinetics studies indicated the potential of the liposomal formulation to increase the oral bioavailability of Brb and to accelerate its entry into the bloodstream. The obtained results are accredited to the lipophilic nature of the prepared system, resembling the structural features of bio-membrane, in addition to their small size that enhances intestinal penetration.

Keywords: liposomes; hyaluronic acid; full-factorial; berberine; bioavailability

1. Introduction

There is an increasing interest in the utilization of naturally existing compounds with potential pharmacological effect in the management of different diseases. Berberine is a wonderful bioactive alkaloid isolated from some medicinal Chinese herbs [1]. Brb are widely applied in Chinese medicine for the management of hypertension and inflammatory conditions [2]. The valuable properties of Brb is attributed to its diverse pharmacological actions including its antimicrobial, antiprotozoal, antidiarrheal, anti-inflammatory, antioxidant [3], anticancer [4], antiviral [5], antidiabetic effects [6], in addition to its beneficial effect on the cardiovascular and central nervous systems [7,8].

All these activities could permit the use of Brb in the treatment of many diseases. However, Brb properties have limited its wide clinical applications. Brb is characterized by low aqueous solubility (1 mg/mL) and a partition coefficient of log p-value = −1.5. This makes it a class III drug in the biopharmaceutical classification system, indicating its poor membrane permeability that resulted in poor gastrointestinal absorption and low bioavailability [9]. The low aqueous solubility of Brb is attributed to its self-aggregation upon contact with GIT fluid. Moreover to Brb low solubility, the low bioavailability is assigned to the effect of drug efflux pump, which acts as an obstacle for the entrance of many drugs [10], the first hepatic pass effect and the rapid metabolism by CYP 450-dependent processes [2,11]. It has been reported that Brb has an efflux ratio of 13.8 that means that Brb is a good substrate for P-glycoprotein [12]. Encapsulation of active ingredients into lipid-based nanocarriers was a golden solution for the problems of low solubility and permeability that could enhance both drug penetration and bioavailability. Lipid-based nanocarriers could enhance intestinal absorption through different pathways including uptake through microfold cells (M-cell), transcellular and paracellular permeation pathways [13].

Liposomes are one of the most widely utilized lipid based nanocarriers for enhancing drugs bioavailability via the oral route due to their biocompatibility, safety and versatility [14]. Many studies have been carried to increase stability permeability and biodistribution of liposomes and enhance drug encapsulation [15–17]. Trials include coating liposomes using different polymers. Li et al. formulated silica coated flexible liposomes to increase liposomes structural integrity against the harshened environment of the GIT [18]. Other trials involved altering the composition of the liposomes (either lipid bilayer or core). Catalan-Latorre et al. formulated eudragit and hyaluronan liposomes to ameliorate the gastro-enteric the bioavailability and stability of curcumin [19]. Hathout et al. formulated gelatinized core liposomes to decrease hydrophilic drug leakage and to increase liposomes stability [20].

Based on the above findings, the goal of the current study was to formulate a novel formulation, namely hyaluronate-based liposomes entrapping berberine using film hydration method and to evaluate the effect of different preparation parameters on the physicochemical properties of the prepared vesicles. Optimization of the formulation variables was performed using full factorial design. Moreover, the potential of the investigated system to enhance the lipophilicity and bioavailability of Brb and to sustain its in-vivo release rate after oral administration to rats were assessed.

2. Materials and Methods

Berberine (Brb) (CAS 633-65-8) (purity > 95%) was obtained from Nanjing Zelang Medical Technology Co. Ltd., Nanjing, China. Hyaluronic acid (HA) (CAS: 9004-61-9) was purchased from Acros Organics, Fisher Scientific, England, UK. Soya bean phosphatidylcholine (PDC) CAS 8002-43-5 and cholesterol (CH) (CAS 57-88-5) were purchased from Sigma-Aldrich (Darmstadt, Germany). Organic solvent including methanol and chloroform were obtained from El-Nasr Pharmaceutical Co. (Cairo, Egypt).

2.1. Preparation and Optimization of the Berberine Loaded Liposomes

Brb loaded liposomes were formulated adapting the thin-film hydration method [21]. Full-factorial design was implemented to evaluate the impact of different variables on the properties of the prepared vesicles. The investigated factors were the total lipid amount, Brb and HA amounts all at 2 levels as follows: 100 and 200 mg, 10 and 50 mg and 0 and 30 mg for the total lipid, Brb and HA respectively. The optimized responses were the entrapment efficiency percentage (EE%) and the mean size (MS) (Table 1). Briefly, organic solvent mixture of chloroform and methanol in a ratio of 2:1 was used to dissolve the lipid mixture of PDC and CH in a ratio of 7:3. Removal of the organic solvent to form a thin film was performed using a rotary evaporator flask (IKA Laboratories, Staufen, Germany) at 40 °C. Portion-wise addition of warmed HA solution (50 °C) was used to hydrate the formed film. The formed liposomal suspension was subjected to shaking for 1 h in a water bath (Kötterman, Uetze, Germany) maintained at 50 °C.

Table 1. Composition, mean size and % EE of various berberine loaded liposomes.

Formulation	Variables (Independent Factors)		Responses (Dependent Variables)	
	Berberine:Lipid Ratio	Sodium Hyaluronate (mg)	MS (nm) ± SD	EE% ± SD
F1	1:10	0	494.9 ± 2.6	29.9 ± 0.6
F2	1:10	30	514.3 ± 9.7	55.9 ± 0.4
F3	1:20	0	207.1 ± 8.5	51.2 ± 0.5
F4	1:20	30	520.7 ± 19.9	78.1 ± 0.1
F5	1:2	0	748.5 ± 27.5	58.5 ± 0.2
F6	1:2	30	586.7 ± 36.6	77.9 ± 0.1
F7	1:4	0	825.7 ± 6.1	39.0 ± 0.1
F8	1:4	30	389.7 ± 46.8	67.0 ± 0.3

Total lipid: phosphatidylcholine and cholesterol in a ratio of 7:3. MS: mean size, % EE: percent entrapment efficiency. Each result is the mean of three determinations ± SD.

2.2. *Characterization of the Prepared Berberine Loaded Liposomes*

2.2.1. Determination of the Particle Size of the Prepared Liposomes

The particle size (PS) of Brb loaded liposomal formulations was measured at room temperature using Malvern Zetasizer Nano Series (Malvern Instruments, Malvern, UK) employing Dynamic Light Scattering [22]. Dispersions were four folds diluted with deionized water to avoid multi-scattering phenomena.

2.2.2. Measurement of the Entrapment Efficiency Percentages (EE%) of Berberine Hydrochloride

An aliquot of the prepared Brb loaded liposomes was centrifuged at 25 °C (Hermle Labortechnik, Wehingen, Germany) at 15,000 rpm for 1 h to separate the nanoparticles [23]. The free un-entrapped Brb in the clear supernatant was measured at 340 nm using a UV—visible double beam Spectrophotometer (Shimadzu, Kyoto, Japan).

The entrapment efficiency was determined as follows:

$$EE\% = [(Wt - Wf)/Wt] \times 100$$

where Wt represents the total drug concentration added to the formulation, Wf represents the amount of the free drug measured in the supernatant.

2.2.3. Transmission Electron Microscopy (TEM) of the Prepared Berberine Loaded Liposomes

Morphological examination of the selected Brb hyaluronate-based liposomes (F4) was examined using TEM at 25 °C (Jeol Electron Microscope, Tokyo, Japan). An unstained droplet of the liposomes formulation was added to a carbon film-covered copper grid and left for three minutes. Then, a filter paper was used to remove the excess liquid and the sample was dried. Observation of the prepared sample by the microscope was performed at resolution of 20,000 KV.

2.2.4. In Vitro Release Study of Berberine from the Prepared Liposomes

The in vitro drug release was performed using dissolution apparatus in phosphate buffer saline (pH 6.8) to simulate the physiological environment following oral administration [2,24]. The selected formulae (F3, F4) and Brb solution were exposed to the dissolution vessels containing 900 mL phosphate buffer saline of maintained at 37 ± 0.5 °C with a rotation of 100 rpm. At pre-determined intervals (15, 30, 45, 60, 90, and 120 min), aliquots of the medium were withdrawn, filtered with 0.45 µm filter and suitably diluted. The obtained samples were observed spectrophotometrically at a maximum wavelength of 340 nm to determine the concentration of the released Brb. All tests were conducted in triplicate.

2.2.5. Assessing the Physical Stability of Berberine Hyaluronate-Based Liposomes

The mean size and entrapment efficiency of the selected berberine hyaluronate based liposomes (F4) were re-measured after three months storage at 2–8 °C (refrigeration) using the same procedures mentioned above.

2.3. *Pharmacokinetics Study*

2.3.1. Experimental Design

Animals were obtained from the animal house of the Faculty of Pharmacy, King Abdulaziz University, Jeddah, Saudi Arabia. The in vivo study protocol was approved by the Animal Ethics Committee of the Faculty of Pharmacy, King Abdulaziz University, Jeddah, Saudi Arabia, in adherence with the Declaration of Helsinki, the Guiding Principle in Care and Use of Animals (DHEW production NIH 80-23) and the Standards of Laboratory Animal Care (NIH distribution #85-23, reconsidered in 1985). The study included three randomly divided groups of male Wistar rats (N = 6/group): group 1, 2 and 3 received Brb solution, Brb liposomes (F3), Brb hyaluronate-based liposomes (F4) respectively. Rats were fasted for 16 h before the experiment and had free access to water. All groups administered Brb dose of 50 mg/kg body weigh via intra-gastric gavage. At different time intervals, blood samples (200 μL) were obtained from the medial canthus of the eye using capillary tubes after drug administration. Blood samples were centrifuged at 3000 rpm for 10 min at 4 °C to separate plasma [25].

2.3.2. Sample Preparation and Calibration Curve

The analysis and the sample extraction technique were done applying the procedure adapted by Xue et al. 2013 [26] with certain modifications. In brief, one hundred μL plasma was transferred to test tube, then one hundred μL valsartan, 100 ng/μL as internal standard was added followed by 250 μL acetonitrile. Vortex the mixture for 1 min, followed by centrifugation for 15 min at 5300 rpm. One hundred μL of the supernatant was transferred to the total recovery vial. The injection volume was five μL. The calibration curve for Brb was assessed using blank plasma as a matrix. Stock solution of concentration 1 mg/mL of Brb was prepared using ethanol as solvent. From the stock solution a series of working solutions of Brb were prepared applying serial dilution technique. The calibration solutions were prepared by spiking separately blank plasma with Brb solutions. The calibration range was from 1.0 to 100.0 ng/mL of Brb with a fixed concentration of internal standard. All solutions were extracted and analyzed applying the developed method.

2.3.3. Chromatographic System and Instrumentation

The analysis was done using triple quad mass spectrometer (Agilent 6460, Agilent Technologies, Santa Clara, CA, USA). The instrument coupled with electrospray ionization mass spec-trometer system. The system designed with a quaternary pump, and a column compartment (Palo Alto, Santa Clara, CA, USA and controlled by MassHunter software (version B.03.01, Build 3.1.346.0). The conditions of mass spectrometric measurements were a gas temperature, 330 °C; gas flow rate, 11 L/min; nebulizer pressure; 35 psi, and capillary voltage, 4300 V. The mass spectrometric settings were optimized for berberine, including the fragmentor voltage, dwell time, and collision energy voltage. The column used for separation at positive ion mode was anAgilent Eclipse Plus C8, 5 μm, 4.6 × 100 mm column.

The mobile system composition was of acetonitrile: water containing 0.1% w/v formic acid in ratio 30: 70 v/v, with flow rate 0.5 mL/min. The TQ /MS conditions were optimized by applying different and alternative values of fragmentor voltage, dwell time, and collision energy (eV) and the optimum conditions were determined.

2.4. *Data Analysis*

Design Expert® v. 7.0 (Stat Ease, Minneapolis, MN, USA) was used to obtain the models for the entrapment efficiency percentage (%) and the mean size. Analysis of the obtained results were performed using ANOVA at $p < 0.05$ using GraphPad Prism® v.5.0

(GraphPad Software, La Jolla, CA, USA). Pharmacokinetics parameters including maximum concentration (C_{max}) and time of maximum concentration (T_{max}) were determined directly from the plasma concentration–time profiles. The area under the concentration–time curve (AUC) from time zero to test time (AUC_{0-t}) was calculated using GraphPad Prism® v.5.0 (GraphPad Software, La Jolla, CA, USA).

3. Results and Discussion

3.1. Optimization of the Berberine Loaded Liposomal Formulations

Brb loaded liposomes were successfully prepared using film hydration method that was easy to perform on small scale with the feasibility of scale up. In addition, thin film hydration method includes the hydration of the formed lipid film using preheated aqueous solution that enables decreasing the viscosity of HA. Optimization of the formulation parameters were achieved using Design Expert® v. 7.0. Table 1 shows the optimized responses (EE% and PS), while Table 2 shows the obtained models and the results of ANOVA analysis for the responses of Brb liposomes with or without HA. The obtained models were significant showing a good fitting (r-squared and adjusted r-squared values above 0.98, with high predicted r-squared values). The difference between the predicted and the adjusted values was <0.2 indicating their high agreement. In addition, the adequate precision values of the models that reflect the signal-to noise ratio were high and above the value of four showing the sensitivity and the adequacy of the models [27].

Table 2. Results of ANOVA analysis of entrapement efficiency and mean size of the berberine loaded liposomes.

	EE (%)	MS
Suggested Model	Factorial	Factorial
Equation	$\% EE = +57.22 + 1.66 \times A + 3.41 \times B + 12.53 \times C - 9.24 \times A \times B + 1.19 \times A \times C - 0.68 \times B \times C + 0.98 \times A \times B \times C$	$PS (nm) = +535.94 - 50.17 \times A + 101.68 \times B - 33.10 \times C + 20.21 \times A \times B + 2.51 \times A \times C - 116.36 \times B \times C - 71.05 \times A \times B \times C$
r^2	0.9996	0.9877
Adjusted r^2	0.9995	0.9824
Predicted r^2	0.9992	0.9724
Adequate precision	220.858	43.267

Figure 1(A) shows the 3D surface plots of EE% as a model response. The plots show the effect of factors interaction on the Brb entrapment efficiency. The results indicate the effect of the different independent variables on the EE%, as investigated by the broad range of EE% from 29.91% to 78.12%. As revealed from the magnitude and sign of the estimated coefficients of the obtained equation, the presence of HA (C) has the significant effect on EE%. In addition the higher positive coefficient value observed for HA amount (C) in comparison to lipid and Brb amounts (A and B) indicates its more enhanced effect.

The significant increase in the Brb EE% due to the presence of HA could be attributed to two main factors, firstly the ionic interaction between the quaternary ammonium group of Brb and the negatively charged carboxyl group of HA resulted in encapsulating more drug in the hyaluronated core of the prepared liposomes decreasing the drug leakage [28]. Secondly, the effect of HA presence on raising the viscosity of the inner core of the vesicles and the aqueous environment surrounding the lipid bi-layers with a subsequent decrease in the drug leakage from the vesicles.

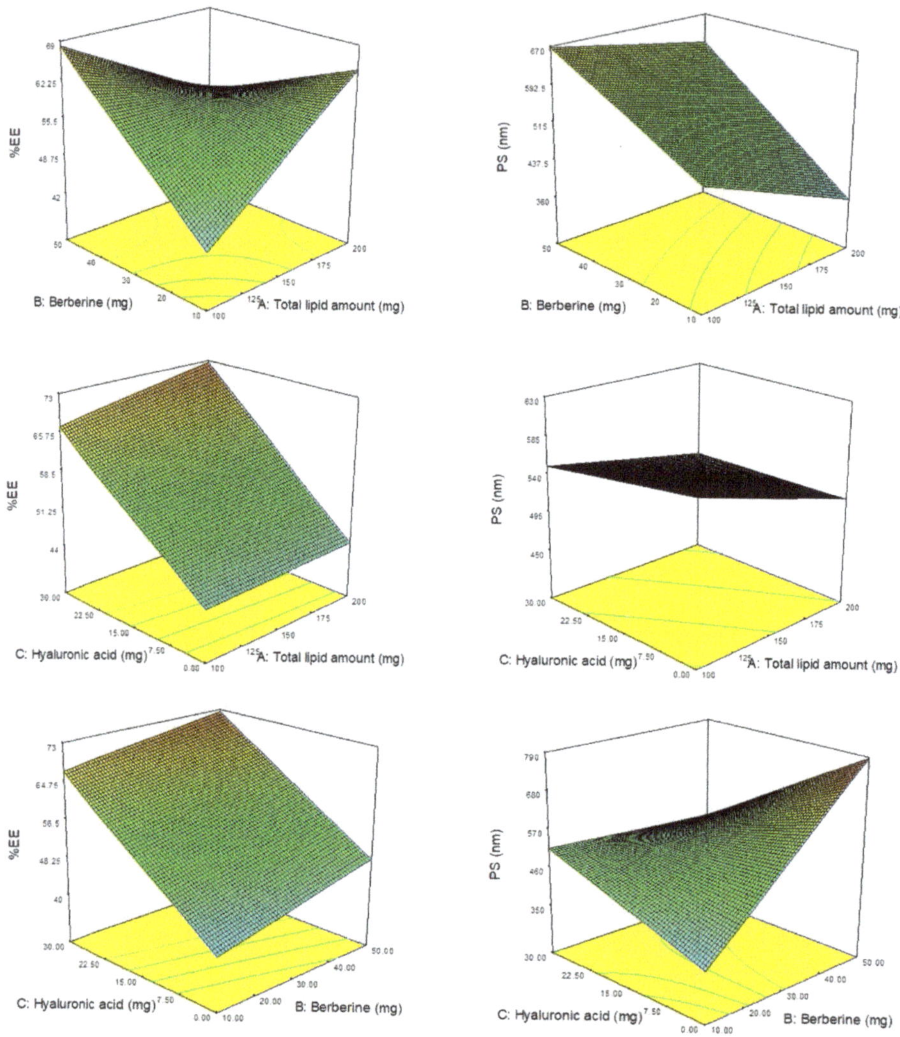

Figure 1. Three-dimensional (3D) response surface plots showing the effect of the independent variables on liposomes particle size and entrapment efficiency percent: A—total lipid amount (mg), B—berberine (mg), C—hyaluronic acid (mg).

Figure 1(B) shows the 3D surface plots related to the particle size response modeling. The plots show the effect of factors interaction on the particle size. The results indicate the enhanced effect of the different independent variables on the PS, (ranging from to 389.7 to 825.7 nm). Upon inspection of the obtained equation, it can be deduced from the magnitude and sign of the estimated coefficients that Brb amount (B) has the major effect on PS. In addition the high positive coefficient value observed for Brb amount (B) indicates its significant effect. In contrast, the negative coefficient values observed for lipid amount (A) and HA present (C), indicating that these variables are insignificant in predicting PS.

Consequently, the formulation prepared using 200 mg total lipid, 30 mg HA and 10 mg Brb (F4) with EE% of 78.12 $\pm$ 0.13% and PS equal to 520.7 $\pm$ 19.98 was selected for the further studies.

3.2. TEM

Figure 2 reveals the TEM images of selected Brb hyaluronate-based liposomes at room temperature. The figure reveals the multi-lamellar structure of the vesicles' bilayer and the obvious deformation of the prepared particles. The dark color of the interior core of the vesicles indicates the occupation of the core with HA. The irregularity of the formed vesicles is assigned to the high viscosity of HA inside the vesicles at 25 °C resulting in vesicles deformation.

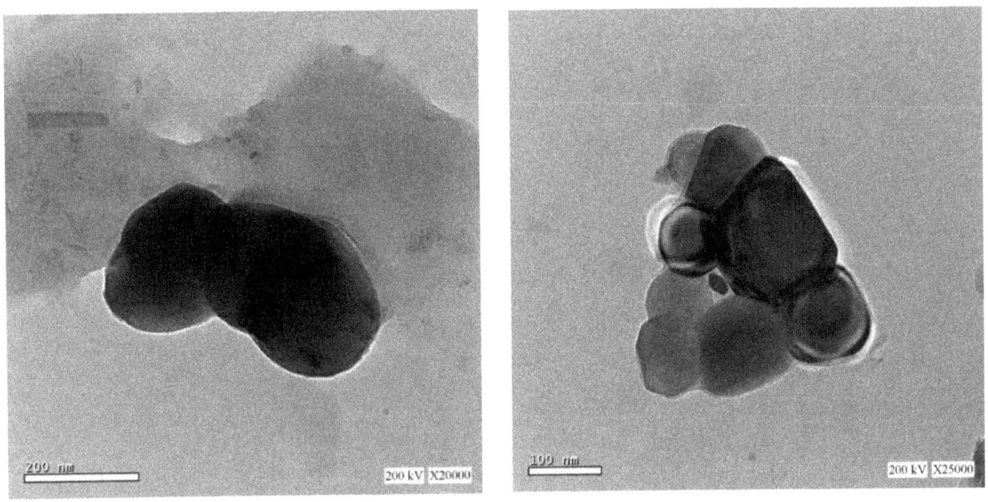

(A) (B)

Figure 2. Transmission electron microscope of berberine hyaluronate based liposomes, showing the nearly spherical shape (**A**) and the multi-lamellar structure of the vesicles bilayer (**B**) of the prepared liposomes.

3.3. *In Vitro Release Studies of Berberine from the Prepared Liposomes*

It has been reported that Brb water solubility is affected by temperature condition and buffer constitution. Brb solubility in phosphate buffer was reported to be the highest among other buffers, like borate or phthalate buffer [9], therefore phosphate buffer was used to study Brb dissolution rate from the prepared vesicles to justify its sustained behavior. Figure 3 shows the in vitro release of Brb from the selected Brb hyaluronate-based liposomes (F4) in comparison to formulation containing the same lipid and drug amounts but without HA (F3) and Brb solution in phosphate buffer saline. As revealed Brb free drug solution showed fast release rate with 100% released within 6 h. Conversely, Brb liposomes showed a biphasic slower release rates especially hyaluronate-based liposomes; the initial burst effect may be attributed to the drug near the liposomes surface. F4 demonstrated the lowest initial burst effect as compared to F3 and drug solution; in addition, it revealed the slowest and the most sustained release profile. The enhanced sustained release of Brb from hyaluronate based liposomes may be assigned to the ionic interaction between the quaternary ammonium group of Brb and the negatively charged carboxyl ion of HA. In addition, HA is a hydrophilic water soluble polymer that was solubilized in the aqueous core of the formed vesicles and its presence could increase the viscosity of the inner core due to its high molecular weight and hence decreases the entrapped drug diffusion and improves its release control. These two factors contribute to the sustained release of berberine from Brb loaded HA based liposomes. The release of Brb from HA based liposomes showed Higuchi kinetics model after fitting the release data to different models based on the highest regression coefficients. Brb is well known for its low water solubility (1 mg/mL) and log P of −1.5, which implies its ability to be incorporated within the vesicles' bilayer to a certain limit. In the present work berberine was used as its hydrochloride salt; therefore it was entrapped in the aqueous core of the liposomes. In addition, the presence of HA augment the incorporation of more drug in the hydrophilic core of the vesicles which is assured by the higher drug entrapment efficiency of hyaluronate-based liposomes. Our results are in accordance with a previous study [28,29].

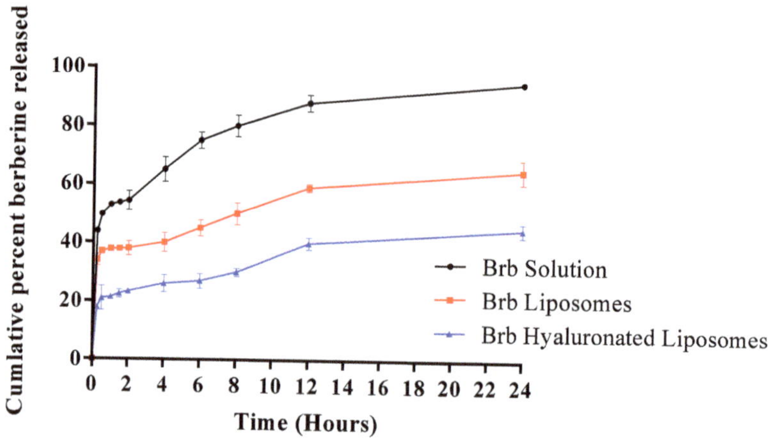

Figure 3. Cumulative in vitro release of berberine from different berberine formulations in phosphate buffer saline (pH 6.8), each result is the mean of three determinations ± SD.

3.4. Physical Stability

Results of the stability of F4 revealed acceptable stability with no statistically significant difference ($p < 0.05$) compared to initial values recording 502.57 ± 8.42 for the mean size and 77.74 ± 0.14% for the EE% after storage for 3 months. The results indicate the effect of the HA on supporting the physical stability of the prepared liposomes.

3.5. Pharmacokinetics Study

The parameters of HPLC method were validated according to ICH guidelines. The peak area ratios of Brb to internal standard were found linear with regression coefficient 0.9984 in the concentration range, 1.0 to 100.0 ng/mL of Brb. The concentrations of the Brb after extraction at different set points were calculated, referring to the obtained regression equation. The QQQ-MS optimum conditions were optimized where t_R showed 9.11 min, +MRM transition (m/z) ranged from 336.3 to 321.1, Dwell time was 200 ms at fragmentation voltage of 135 V and collision energy of 20 eV.

Figure 4 shows a representative mass spectrum of berberine of concentration 1 ng/mL after extraction from plasma applying multiple reaction monitoring mode (MRM) to increase the sensitivity and selectivity of the measurement. The MRM transition mode applied for berberine was a precursor ion of m/z coupled with main fragment ion of berberine of m/z 321.1. The comparative pharmacokinetic profiles of Brb solution, Brb liposomes and Brb hyaluronate-based liposomes after oral administration to rats at a dose of 50 mg/kg are shown in Figure 4, and pharmacokinetic parameters (C_{max}, T_{max}, and $AUC_{0–24}$) are summarized in Table 3.

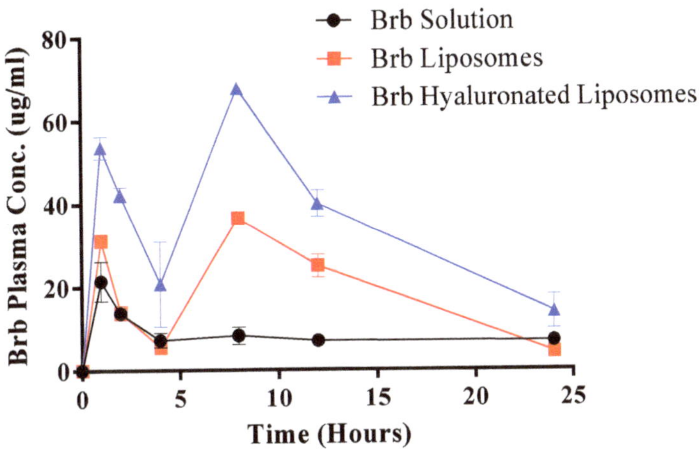

Figure 4. Plasma concentration profiles of berberine in rats after oral administrations of 50 mg/kg of berberine in various formulations. Each point represents mean ± standard deviation (N = 6).

Table 3. The calculated pharmacokinetic parameters of rats in different groups after administration of a single oral dose of berberine of 50 mg/kg body weight.

	Brb Solution	Brb Liposomes	Brb Hyaluronate Based Liposomes
C_{max} (ng/mL)	21.51 ± 4.71	36.54 ± 0.77	67.79 ± 1.12
T_{max} (h)	1	8	8
AUC	194.8 ± 6.73	439.9 ± 14.39	854 ± 23.69
Relative bioavailability	-	2.26	4.38

Data represents mean ± standard deviation.

It was obvious from Brb plasma concentration-time profile that Brb shows two absorption peaks, which may be attributed to enterohepatic circulation, which is in accordance with previous study [30]. However, other reasons may include the fast absorption of unentrapped adsorbed drug on the liposomes surface followed by the slower absorption through the lymphatic pathway [31].

C_{max} value increased in the order of Brb hyaluronate-based liposomes > Brb liposomes > Brb solution. The obtained C_{max} after oral administration of Brb solution was 21.5 ± 4.71 ng/mL, while C_{max} of Brb liposomes and Brb hyaluronate-based liposomes were 31.26 ± 0.109 and 53.61 ± respectively. Furthermore, the AUC_{0-24} of Brb solution, Brb liposomes and Brb hyaluronate-based liposomes were 194.8, 439.9 and 854 ng h/mL respectively. The higher C_{max} peaks and the greater AUC_{0-24} values of Brb liposomes and Brb hyaluronate-based liposomes indicate the greater absorption rate of Brb from liposomes. In addition, the delayed T_{max} (8 h) of Brb liposomes and Brb hyaluronate-based liposomes in comparison to Brb solution (1 h) indicates the slow sustained release of Brb, which correlates with the in vitro dissolution study.

In addition, approximately 2.26- and 4.38-fold increases in relative bioavailability were achieved within Brb liposomes and Brb hyaluronate-based liposomes, respectively, when compared with the free Brb solution. Results of pharmacokinetics studies indicate that the oral bioavailability of Brb was increased and the entrance of Brb into the bloodstream was accelerated upon its encapsulation into vesicular system. The obtained results are accredited to the lipophilic nature of the prepared system, resembling the bio-membrane, in addition to the vesicles small size that enhances intestinal penetration [32].

4. Conclusions

The present study suggests that the hyaluronate-based liposomes are suitable formulations for successful delivery of Brb. Different formulation variables (lipid, drug and hyaluronic acid amounts) have a significant effect on the physicochemical characteristics of the prepared system using film hydration method. The physicochemical characterization and the results of in vitro release displayed multi-lamellar vesicles with acceptable particle size, high Brb entrapment efficiency and sustained drug release for 24 h. The presence of hyaluronic acid as a main component in liposomes preparation was able to slow berberine diffusion from the vesicles. Moreover, observing the pharmacokinetic behavior after oral administration of Brb hyaluronate-based liposomes to rats could improve lipophilicity and bioavailability of the investigated system compared to Brb solution and Brb liposomes prepared without hyaluronic acid. Further investigations are required in order to obtain smaller liposomal formulations, with multiple applications, such as targeted and controlled delivery of the bioactive compound, in addition to pharmacodynamics studies based on the diseases.

Author Contributions: Conceptualization, methodology, software, writing—original draft preparation, H.A.G.; methodology, review and editing, A.K.K. and A.S.; funding acquisition, H.I.K.; review and editing, H.Z.A. and S.C. All authors have read and agreed to the published version of the manuscript.

Funding: This project was funded by the Deanship of Scientific Research (DSR), King Abdulaziz University, Jeddah, under grant No. G-288-166-1440. The authors, therefore, gratefully acknowledge the DSR technical and financial support.

Institutional Review Board Statement: The study was conducted according to the guidelines of the Declaration of Helsinki, the Guiding Principle in Care and Use of Animals (DHEW production NIH 80-23) and the Standards of Laboratory Animal Care (NIH distribution #85-23, reconsidered in 1985) and approved by the Animal Ethics Committee of the Faculty of Pharmacy, King Abdulaziz University, Jeddah, Saudi Arabia.

Informed Consent Statement: Not applicable.

Data Availability Statement: All data are reported in the manuscript.

Conflicts of Interest: The authors declare no conflict of interest.

Abbreviations

Brb	berberine
HA	hylauronic acid
TEM	transmission electron microscope
PDC	Soya bean phosphatidylcholine
CH	cholesterol
MS	mean size
PS	particle ssize
EE	entrapment effificency

References

1. Mirhadi, E.; Rezaee, M.; Malaekeh-Nikouei, B. Nano strategies for berberine delivery, a natural alkaloid of Berberis. *Biomed. Pharmacother.* **2018**, *104*, 465–473. [CrossRef]
2. Sahibzada, M.U.K.; Sadiq, A.; Faidah, H.S.; Khurram, M.; Amin, M.U.; Haseeb, A.; Kakar, M. Berberine nanoparticles with enhanced in vitro bioavailability: Characterization and antimicrobial activity. *Drug Des. Dev. Ther.* **2018**, *12*, 303–312. [CrossRef]
3. Vuddanda, P.R.; Chakraborty, S.; Singh, S. Berberine: A potential phytochemical with multispectrum therapeutic activities. *Expert Opin. Investig. Drugs* **2010**, *19*, 1297–1307. [CrossRef] [PubMed]
4. Kim, J.B.; Yu, J.H.; Ko, E.; Lee, K.W.; Song, A.K.; Park, S.Y.; Shin, I.; Han, W.; Noh, D.Y. The alkaloid Berberine inhibits the growth of Anoikis-resistant MCF-7 and MDA-MB-231 breast cancer cell lines by inducing cell cycle arrest. *Phytomedicine* **2010**, *17*, 436–440. [CrossRef] [PubMed]

5. Hayashi, K.; Minoda, K.; Nagaoka, Y.; Hayashi, T.; Uesato, S. Antiviral activity of berberine and related compounds against human cytomegalovirus. *Bioorg. Med. Chem. Lett.* **2007**, *17*, 1562–1564. [CrossRef] [PubMed]
6. Wei, X.-C.; Zhu, L.-Q.; Wang, C.-G. Efficacy and Safety of Berberine in Patients with Type 2 Diabetes Mellitus: A Meta-Analysis. *Chin. Herb. Med.* **2015**, *7*, 344–353. [CrossRef]
7. Huang, Z.; Han, Z.; Ye, B.; Dai, Z.; Shan, P.; Lu, Z.; Dai, K.; Wang, C.; Huang, W. Berberine alleviates cardiac ischemia/reperfusion injury by inhibiting excessive autophagy in cardiomyocytes. *Eur. J. Pharmacol.* **2015**, *762*, 1–10. [CrossRef]
8. Cai, Z.; Wang, C.; Yang, W. Role of berberine in Alzheimer's disease. *Neuropsychiatr. Dis. Treat.* **2016**, *12*, 2509–2520. [CrossRef]
9. Battu, S.K.; Repka, M.A.; Maddineni, S.; Chittiboyina, A.G.; Avery, M.A.; Majumdar, S. Physicochemical Characterization of Berberine Chloride: A Perspective in the Development of a Solution Dosage Form for Oral Delivery. *AAPS PharmSciTech* **2010**, *11*, 1466–1475. [CrossRef]
10. Raju, M.; Kulkarni, Y.A.; Wairkar, S. Therapeutic potential and recent delivery systems of berberine: A wonder molecule. *J. Funct. Foods* **2019**, *61*, 103517. [CrossRef]
11. Liu, C.S.; Zheng, Y.R.; Zhang, Y.F.; Long, X.Y. Research progress on berberine with a special focus on its oral bioavailability. *Fitoterapia* **2016**, *109*, 274–282. [CrossRef]
12. Kwon, M.; Lim, D.Y.; Lee, C.H.; Jeon, J.H.; Choi, M.K.; Song, I.S. Enhanced Intestinal Absorption and Pharmacokinetic Modulation of Berberine and Its Metabolites through the Inhibition of P-Glycoprotein and Intestinal Metabolism in Rats Using a Berberine Mixed Micelle Formulation. *Pharmaceutics* **2020**, *12*, 882. [CrossRef]
13. Pund, S.; Borade, G.; Rasve, G. Improvement of anti-inflammatory and anti-angiogenic activity of berberine by novel rapid dissolving nanoemulsifying technique. *Phytomedicine* **2014**, *21*, 307–314. [CrossRef] [PubMed]
14. Li, J.; Wang, X.; Zhang, T.; Wang, C.; Huang, Z.; Luo, X.; Deng, Y. A review on phospholipids and their main applications in drug delivery systems. *Asian J. Pharm. Sci.* **2015**, *10*, 81–98. [CrossRef]
15. Eloy, J.O.; Claro de Souza, M.; Petrilli, R.; Barcellos, J.P.A.; Lee, R.J.; Marchetti, J.M. Liposomes as carriers of hydrophilic small molecule drugs: Strategies to enhance encapsulation and delivery. *Colloids Surf. B Biointerfaces* **2014**, *123*, 345–363. [CrossRef] [PubMed]
16. Awasthi, V.D.; Garcia, D.; Goins, B.A.; Phillips, W.T. Circulation and biodistribution profiles of long-circulating PEG-liposomes of various sizes in rabbits. *Int. J. Pharm.* **2003**, *253*, 121–132. [CrossRef]
17. Yang, T.; Cui, F.D.; Choi, M.K.; Cho, J.W.; Chung, S.J.; Shim, C.K.; Kim, D.D. Enhanced solubility and stability of PEGylated liposomal paclitaxel: In vitro and in vivo evaluation. *Int. J. Pharm.* **2007**, *338*, 317–326. [CrossRef]
18. Li, C.; Zhang, Y.; Su, T.; Feng, L.; Long, Y.; Chen, Z. Silica-coated flexible liposomes as a nanohybrid delivery system for enhanced oral bioavailability of curcumin. *Int. J. Nanomed.* **2012**, *7*, 5995–6002. [CrossRef]
19. Catalan-Latorre, A.; Ravaghi, M.; Manca, M.L.; Caddeo, C.; Marongiu, F.; Ennas, G.; Escribano-Ferrer, E.; Peris, J.E.; Diez-Sales, O.; Fadda, A.M.; et al. Freeze-dried eudragit-hyaluronan multicompartment liposomes to improve the intestinal bioavailability of curcumin. *Eur. J. Pharm. Biopharm.* **2016**, *107*, 49–55. [CrossRef]
20. Hathout, R.M.; Gad, H.A.; Metwally, A.A. Gelatinized-core liposomes: Toward a more robust carrier for hydrophilic molecules. *J. Biomed. Mater. Res. Part A* **2017**, *105*, 3086–3092. [CrossRef]
21. Hathout, R.M.; Gad, H.A.; Abdel-Hafez, S.M.; Nasser, N.; Khalil, N.; Ateyya, T.; Amr, A.; Yasser, N.; Nasr, S.; Metwally, A.A. Gelatinized core liposomes: A new Trojan horse for the development of a novel timolol maleate glaucoma medication. *Int. J. Pharm.* **2019**, *556*, 192–199. [CrossRef]
22. Mansour, M.; Abo El Ezz, T.A.; Fattoh, F.N.; AbouelFadl, D.M.; Gad, H.A. Delineating the usage of Dexamethasone-loaded cubosomes as a therapeutic armamentarium for hearing loss versus its protective effect: In vitro and in vivo animal study. *J. Drug Deliv. Sci. Technol.* **2020**, *61*, 102244. [CrossRef]
23. Shamarekh, K.S.; Gad, H.A.; Soliman, M.E.; Sammour, O.A. Development and evaluation of protamine-coated PLGA nanoparticles for nose-to-brain delivery of tacrine: In-vitro and in-vivo assessment. *J. Drug Deliv. Sci. Technol.* **2020**, *57*, 101724. [CrossRef]
24. Cavalu, S.; Banica, F.; Gruian, C.; Vanea, E.; Goller, G.; Simon, V. Microscopic and spectroscopic investigation of bioactive glasses for antibiotic controlled release. *J. Mol. Struct.* **2013**, *1040*, 47–52. [CrossRef]
25. Yu, F.; Li, Y.; Chen, Q.; He, Y.; Wang, H.; Yang, L.; Guo, S.; Meng, Z.; Cui, J.; Xue, M.; et al. Monodisperse microparticles loaded with the self-assembled berberine-phospholipid complex-based phytosomes for improving oral bioavailability and enhancing hypoglycemic efficiency. *Eur. J. Pharm. Biopharm.* **2016**, *103*, 136–148. [CrossRef] [PubMed]
26. Xue, M.; Yang, M.-X.; Zhang, W.; Li, X.-M.; Gao, D.-H.; Ou, Z.-M.; Li, Z.-P.; Liu, S.-H.; Li, X.-J.; Yang, S.-Y. Characterization, pharmacokinetics, and hypoglycemic effect of berberine loaded solid lipid nanoparticles. *Int. J. Nanomed.* **2013**, *8*, 4677–4687. [CrossRef]
27. Elfaky, M.A.; Sirwi, A.; Tolba, H.H.; Shaik, R.A.; Selmi, N.M.; Alattas, A.H.; Albreki, R.S.; Alshreef, N.M.; Gad, H.A. Development, Optimization, and Antifungal Assessment of Ocular Gel Loaded With Ketoconazole Cubic Liquid Crystalline Nanoparticles. *J. Pharm. Sci.* **2021**, *110*, 2210–2220. [CrossRef]
28. Kabary, D.M.; Helmy, M.W.; Elkhodairy, K.A.; Fang, J.-Y.; Elzoghby, A.O. Hyaluronate/lactoferrin layer-by-layer-coated lipid nanocarriers for targeted co-delivery of rapamycin and berberine to lung carcinoma. *Colloids Surf. B Biointerfaces* **2018**, *169*, 183–194. [CrossRef]

29. Miere, F.; Vicas, S.I.; Timar, A.V.; Ganea, M.; Zdrinca, M.; Cavalu, S.; Fritea, L.; Vicas, L.; Muresan, M.; Pallag, A.; et al. Preparation and Characterization of Two Different Liposomal Formulations with Bioactive Natural Extract for Multiple Applications. *Processes* **2021**, *9*, 432. [CrossRef]
30. Jia, J.; Zhang, K.; Zhou, X.; Zhou, D.; Ge, F. Precise Dissolution Control and Bioavailability Evaluation for Insoluble Drug Berberine via a Polymeric Particle Prepared Using Supercritical CO_2. *Polymers* **2018**, *10*, 1198. [CrossRef] [PubMed]
31. Sheue Nee Ling, S.; Magosso, E.; Abdul Karim Khan, N.; Hay Yuen, K.; Anne Barker, S. Enhanced Oral Bioavailability and Intestinal Lymphatic Transport of a Hydrophilic Drug Using Liposomes. *Drug Dev. Ind. Pharm.* **2006**, *32*, 335–345. [CrossRef] [PubMed]
32. Yu, F.; Ao, M.; Zheng, X.; Li, N.; Xia, J.; Li, Y.; Li, D.; Hou, Z.; Qi, Z.; Chen, X.D. PEG–lipid–PLGA hybrid nanoparticles loaded with berberine–phospholipid complex to facilitate the oral delivery efficiency. *Drug Deliv.* **2017**, *24*, 825–833. [CrossRef] [PubMed]

materials

MDPI

Article

Carbon Nanotubes for Improved Performances of Endodontic Sealer

Andreea Marica [1], Luminita Fritea [1,*], Florin Banica [1,*], Cosmin Sinescu [2], Ciprian Iovan [1], Iosif Hulka [3], Gerlinde Rusu [4] and Simona Cavalu [1]

1 Faculty of Medicine and Pharmacy, University of Oradea, P-ta 1 Decembrie 10, 410087 Oradea, Romania; andreeamarica94@yahoo.com (A.M.); dr.iovan@biostandard.ro (C.I.); Simona.cavalu@gmail.com (S.C.)
2 Faculty of Dentistry, "Victor Babes" University of Medicine and Pharmacy, P-ta E. Murgu 2, 300041 Timisoara, Romania; minosinescu@yahoo.com
3 Research Institute for Renewable Energies, University Politehnica Timișoara, G. Muzicescu 138, 300501 Timișoara, Romania; iosif.hulka@upt.ro
4 Faculty of Industrial Chemistry and Environmental Engineering, University Politehnica Timișoara, C. Telbisz 6, 300001 Timișoara, Romania; gerlinde.rusu@upt.ro
* Correspondence: fritea_luminita@yahoo.com (L.F.); florinbanica1@gmail.com (F.B.)

Abstract: In order to overcome the limitations of current endodontic sealers, especially against resistant bacteria, recent developments in the field of nanotechnology have proved the necessity to reconsider the composition and physico-chemical properties of classical sealers. Nanoparticles with their unique features in terms of small size and high specific surface area, are the best choice for incorporation of antiseptic agents and effective delivery. The aim of our study is to prepare a novel platform for antibacterial drug delivery in dental adhesive systems used in endodontics. For this purpose, multi-walled carbon nanotubes (MWCNTs) encapsulating chlorhexidine (CHX) and colloidal silver nanoparticles (AgNPs) were prepared and incorporated into commercial sealer and investigated in terms of bonding performance to dentin and effectiveness against *E. faecalis*, *S. aureus* and *Candida albicans*, which are responsible for the majority of the failures in endodontic treatments. In this context, the challenges related to the long-term biological effects of CHX/AgNPs loaded MWCNTs are discussed.

Keywords: endodontic sealer; CNTs; chlorhexidine; silver nanoparticles; antimicrobial

Citation: Marica, A.; Fritea, L.; Banica, F.; Sinescu, C.; Iovan, C.; Hulka, I.; Rusu, G.; Cavalu, S. Carbon Nanotubes for Improved Performances of Endodontic Sealer. *Materials* **2021**, *14*, 4248. https://doi.org/10.3390/ma14154248

Academic Editor: Roman Perez Antoñanzas

Received: 30 June 2021
Accepted: 28 July 2021
Published: 29 July 2021

1. Introduction

The development of "smart" endodontic therapeutic agents and materials has emerged in the context of a new era of nanomaterials. Despite the reported success rate of classical endodontic treatments (96%) [1,2], failure still occurs as a result of improperly cleaning and shaping of root canal, or microleakage of sealing material [3]. A combination of adequate instrumentation, irrigation, root canal obturation and sealing are the essential steps for successful endodontic treatment. The main goal of the treatment is to eliminate the microorganisms and to prevent root canal reinfection, knowing that the primary causes of apical periodontitis are related to the microbial infection of the pulp, produced by metabolic products [4]. However, due to the complexity of anatomical and morphological architecture of root canal system and the resistant bacteria, the complete cleaning by chemo-mechanical techniques cannot be achieved. The traditional root canal irrigants are hypochlorite (NaOCl) and chlorhexidine (CHX), which are extensively used in endodontic therapy with good antimicrobial efficacy against both Gram-positive and Gram-negative bacteria and acceptable biocompatibility [5–8]. There are some limitations, however, as the direct application of NaOCl is associated with tissue destructions (mainly dissolution of collagen network, due to the braking of carbon bonds, affecting the primary structure), depending on the concentration and application time [6]. On the other hand, chlorhexidine digluconate showed a dose dependent inhibition mechanism and positive effect in stabilization of resin dentinal bonds when was applied after acid etching [9,10]. The final stage of non-surgical root canal treatment consists of application of endodontic sealers, a tree dimensionally obturating root canal system, acting for long term and being capable of filling the voids between gutta-percha and dentinal canal walls [11,12]. The adequate filling of the empty space is crucial in completing the sterility, and hence, preventing reinfection of root canal. According to the literature [12–15], the requirements of an ideal root canal sealer are: (1) excellent sealing ability when set; (2) sufficient setting time to ensure working time; (3) dimensional stability; (4) insolubility against tissue fluids; (4) good adhesion to canal walls; (5) suitable antimicrobial properties; and (6) biocompatibility.

A great variety of endodontic sealers are commercially available and classified according to the chemical composition: zinc oxide eugenol, epoxy resin-based, glass ionomer-based, calcium hydroxide-based and methacrylate-resin–based sealers [15–17]. In terms of biocompatibility, previous studies suggested that bioactive sealers may exhibit lower cytotoxic potential compared to other types of root canal sealer [17].

Nanotechnology has lately emerged as a discipline at the interface of chemistry and life sciences; nanoparticles and carbon-based nanomaterials have attracted huge interest in the recent years being widely applied in the biomedical field [18–23]. The concentrated efforts to develop "nanomodified" dental materials resulted in the incorporation of various nanoparticles to increase the surface area between the dentin and the obturating material. The smaller particles size increases the contact surface area and hence, possesses a higher antimicrobial effect than the macro sized material [24–26].

Since nanoparticles are able to penetrate into dentinal tubuli, the development of new generation of endodontic sealers is an important goal of improving comprehensive oral health. In the field of endodontics, the development of nanomaterials is focused toward overcoming the microbial challenge, and hence, new sealers are expected to possess multiple actions in terms of better sealing capacity, remineralization and enhanced antimicrobial effect [26–29]. Especially AgNPs are well known to interact with the bacterial cell membrane, and consequently, to increase permeability and prevent DNA replication. However, the increased cytotoxicity of AgNPs at elevated concentrations, due to their size and surface characteristics, may cause nonspecific oxidative damages [30,31]. Examples of organic and inorganic antimicrobial nanoparticles used in root canal obturation and adhesives are: silver nanoparticles (AgNPs), copper CuNPs, ZnO NPs, hydroxyapatite NPs, TiO_2 NPs, chitosan NPs and quaternary ammonium polyethyleneimine NPs (QAPEI), as presented in a recent review paper by Makvandi et al. [27]. Among the organic nanostructures, carbon nanotubes (CNT) have unique physico-chemical, mechanical and electrical characteristics, such as high stiffness, axial strength and excellent thermal conductivity, due to their cylindrical graphitic structure, with nano-sized diameter and larger aspect ratio [1,2]. The Young modulus value is much higher than those of typically found in stainless steel and carbon fibers [2]. CNTs also have the capacity to easily pass the biological barriers and the ability to carry macromolecules unable to pass through cellular membrane by themselves, leading to novel, biocompatible delivery systems, including in endodontic treatments [2,28].

The majority of antibiotics and antiseptic agents provide short-term antibacterial effects, with a risk of developing antibacterial resistance. In this context, the aim of our study was to prepare a platform for long-lasting antibacterial and antifungal drug delivery, as a novel endodontic sealer obtained by nanomodification, incorporating a mixture of CNTs, CHX and AgNPs. The possible synergic effect of AgNPs, CHX 2% and CNTs was also explored, without altering the structural, ultrastructural and thermal behavior of the original sealer. The phyisico-chemical, structural and morphological characterization of both the original and nanomodified sealer was performed by using FTIR spectroscopy, scanning electron microscopy (SEM), cyclic voltammetry and nanoindentation measurements, with focus on the assessment of interfacial adaptation to root canal dentine. The impact of CNTs loaded with CHX/AgNPs mixture on the nanomechanical properties of the sealer was also discussed. Complementary, thermal analysis (TGA) were performed in order to investigate the stability of modified sealer upon addition of CNTs/CHX/AgNPs mixture. The antimicrobial and antifungal efficacy of the novel endodontic sealer was tested against *Enterococcus faecalis*, *Staphylococcus aureus* and *Candida albicans*.

2. Materials and Methods

2.1. Preparation of CHX/AgNPs Loaded CNTs and Incorporation into Endodontic Sealer

The starting materials were commercial clorhexidine gluconate gel GLUCO-CheX 2% (Cerkamed, Stalowa Wola, Poland), ionic colloidal silver Argentum Special Pure Life® 77 ppm (Agnes Itara SRL, Suceava, Romania), multi-walled CNTs (NC3100, Nanocyl, Sambreville, Belgium) and ADSEAL (resin-based root canal sealer, META® BIOMED, Chungcheongbuk-do, Korea). According to the manufacturer, the formulations consist of ionic silver with purity 99.99% and particle size 4–9 nm (determined by TEM), while the size of the CNTs are 9.5 nm in diameter and 1.5 µm length. The composition of ADSEAL consists of: base (epoxy oligomer resin, ethylene glycol salicylate, calcium phosphate, zirconium oxide, bismuth subcarbonate) and catalyst (poly butanediol aminobenzoate, triethanolamine, calcium phosphate, bismuth subcarbonate, zirconium oxide).

The modified sealer was prepared as follows: 10 mL of colloidal silver (77 ppm), 12.5 mg CNTs and 0.5 mL CHX 2% were mixed by sonication and stirring during 1 h. A volume of 1 mL of this mixture was left overnight for water evaporation (leading to a gel composition) and then was added to 1 g of commercial endodontic sealer, and mixed until homogenized, following the manufacturers' instructions. The composition was applied on the endodontic canal of 20 human permanent incisors, specially prepared, as described in the Section 2.7.

2.2. FTIR Spectroscopy

The structural characterization of both commercial and nanomodified endodontic sealer was performed using an FTIR 780 spectrophotometer (PG Instruments, Leicestershire, UK) and the KBr technique, operating in the range of 400–4000 cm^{-1}, with a scanning speed of 32 cm^{-1} and spectral width of 2.0 cm^{-1}. For this purpose, after hardening, the specimens were ground using a dental milling machine (Benco Dental, Pittston, PA, USA) in order to obtain a fine powder suitable for KBr pellet preparation. FTIR spectra of CNTs as supplied from the manufacturer, were also recorded in the same conditions. CaF_2 cell accessory was used in order to acquire the FTIR spectra of colloidal mixtures containing CNTs, CHX and AgNPs.

2.3. SEM Examination (Mixture AgNPs/CHX/CNTs, Commercial and Nanomodified Sealer)

For SEM investigation, the materials were prepared according to the manufacturer instructions and applied on specially designed plastic inserts and left until hardening. The microscopic details were recorded on fracture surface of both commercial and modified material using a Leo 438VP electron microscope (SEM, Oberkochen, Germany), operating at 30 kV, with a variable vacuum level. In order to study the detailed morphology, structure and elemental composition, the polished specimens were characterized by a Quanta FEG 250 SEM instrument (FEI, Breda, The Netherlands) using a back scattered electron detector (BSD) and energy dispersive X-ray spectroscopy (EDX, using an Apolo SSD detector, EDAX Inc., Mahwah, NJ, USA). The microstructure investigations and EDX analysis were performed at about 10 mm working distance (WD) in low vacuum mode in order to avoid surface charging and damage to the analyzed samples. The details of the CNTs before and after loading with AgNPs/CHX mixture were also recorded in the same conditions.

2.4. Electrochemical Measurements

Screen printed electrodes (DRP 110, Metrohm-DropSens, Oviedo, Spain) were used for electrochemical analysis. The setup consisted of a Ag reference electrode, carbon counter electrode and another carbon electrode as a working one. The experiments were carried out using a potentiostat Autolab PGSTAT 128N (Metrohm, Utrecht, The Netherlands) equipped with Nova 2.1.2 software. In order to assess the electrochemical behavior of the colloidal mixture and of its individual components, differential pulse voltammetry (DPV) was performed employing the following parameters: start potential −1 V, end potential +1.5 V, step 0.01 V, modulation amplitude 0.05 V, modulation time 0.05 s, interval time 0.1 s. In the case of ionic Ag solution, a reduction process was previously applied using the same DPV method.

2.5. Nanoindentation Measurements

The endodontic sealer specimens (with and without modifications) were subjected to nano-mechanical tests using a depth sensing Nanoindenter G200 device (Agilent Technologies, Santa Clara, CA, USA), at room temperature and normal humidity (45–52%), applying a diamond Berkovich pyramidal shaped tip. Each indentation test (consisting of a loading and unloading phase) was repeated 12 times, the microscopic image of the selected area on the sample surface allowed precise control between the sample position and the indenter. The values of the Young modulus were obtained from load–displacement curves, by fitting parameters, using the Oliver–Pharr method [29], taking into account the tip shape function and the fact that the measured displacement may include contributions from both the specimen and the indenter.

2.6. Thermogravimetrical Analysis (TGA)

TGA thermograms were recorded using a TG 209 F1 Libra (NETZSCH-Gerätebau GmbH, Selb, Germany) thermogravimetric analyzer. The measurements were carried out in a nitrogen atmosphere, in the temperature range of 20–850 °C, with a heating rate of 10 °C/min. The data were processed with the Netzsch Proteus-Thermal Analysis program version 6.1.0. (NETZSCH-Geraetebau GmbH, Selb, Germany).

2.7. SEM Investigation of the Interfacial Adaptation to Root Canal Dentine

A total of 20 human permanent incisors were selected for this study. The selected teeth were extracted for orthodontic reasons, with patient informed consent, available from the Dental Clinic of Faculty of Medicine and Pharmacy, University of Oradea. The protocol was approved by the Institutional Research Ethics Committee of the Faculty of Medicine and Pharmacy, University of Oradea [Nr. 2/31.03.2021]. As a first step, the teeth were endodontically prepared, by making an access cavity with a round diamond bur. After the access cavity was made, each tooth was irrigated with 5% sodium hypochlorite (Cerkamed, Stalowa Wola, Poland). The next step was to determine the working length with a stainless-steel 0.10 K-file (VDW, Munich, Germany) and an endodontic ruler (Angelus, Londrina, Brazil). Having the working length determined, the next step was to enlarge the coronal one third of the canal with an endodontic preflaring file-SX file (0.19/0.04 taper) from the Protaper Universal system, at 250 rpm and 3.0 torque, with the endodontic motor X-smart plus (Dentsply Maillefer, Ballaigues, Switzerland). The glide path was established with Proglider File (0.16/0.02 variably, Dentsply Maillefer, Ballaigues, Switzerland). The next step was to prepare the teeth with the endodontic file system, ProTaper Universal (S1 0.17/0.02, S2 0.12/0.04, F1 0.12/0.07, F2 0.25/0.08, F3 0.30/0.09 Dentsply Maillefer, Ballagues, Switzerland) and cleaning with 5.25% NaOCl. A final rinse was made with 17% EDTA solution (Cerkamed, Stalowa Wola, Poland), followed by a 10 min ultrasonic treatment in distilled water. A longitudinal section was applied to each specimen using a diamond coated disk, in order to expose the root canal. Half of the sections of the prepared teeth were coated, in the root canal area, with the commercial sealer ADSEAL, respectively, the other half with the modified one, according to the manufacturer recommendations and stored at 37 °C and 100% humidity for 48 h before SEM investigation. Each half was transversely sectioned in the median zone in order to investigate the ultrastructural details of the interfacial area between the sealer and dentine. In this way, it was possible to evaluate each individual specimen in terms of interfacial adaptation toward modified and unmodified sealer, allowing a fair comparison, taking into account the structural and morphological particularities of each root canal. The SEM images at the interface between the root canal and the sealer were recorded in back-scattering mode.

2.8. Antimicrobial and Antifungal Efficacy

The following strains were considered for antimicrobial and antifungal tests:

Enterococcus faecalis ATCC® 29212, Staphylococcus aureus ATCC® 25923) and fungi (Candida albicans ATCC® 10231), being supplied by Biostandard Laboratories Oradea. Five test tubes were prepared as follows: 1 mL of bacterial inoculum suspension in concentration of 1.5×10^8 CFU/mL, which corresponds to standard 0.5 McFarland (prepared in Mueller Hinton broth) was added to 1 mL of each tested combination: (1) CNTs/AgNPs/ CXH 2%; (2) CNTs/H_2O; (3) CNTs /AgNPs; (4) CNTs /CXH 2%; (5) AgNPs; (6) CHX 2%. The test tubes were incubated for 1 h at 37 °C, and then the content was spread on the surface of the inoculated broth in Petri dishes, and incubated for 24 h at 37 °C. Similarly, an antifungal test was conducted, except that Sabouraud Dextrose agar (Oxoid, Thermo Fisher Scientific Inc., Milan, Italy) was used for the cultivation of *C. albicans* strain, in a concentration of 2.0 McFarland. All the tests were performed in triplicate and the CFU/mL was determined by manual reading.

The positive controls were well known antimicrobial and antifungal drugs, gentamicin and nystatin. Different dilutions of the control drugs were prepared in order to assess their minimum inhibition concentration and the following concentrations were considered effective: 25 µg/mL gentamicin for *S. aureus*, 100 µg/mL gentamicin for *E. faecalis* and 5 µg/mL nystatin for *C. albicans*. In order to assess the antimicrobial activity of the set sealers, both solid samples (with and without the combination CNTs/AgNPs/ CXH 2%), set and aged for 7 days, were ground in a dental mill until a fine powder was obtained and the agar diffusion test was performed in this case. A small hole of 6 mm diameter was prepared in the center of each Petri dish and filled immediately with equal quantities of the resulted powder from both modified and commercial sealer. After 24 h incubation, the diameter of the inhibition area was measured. The assay was performed in triplicate and expressed as mean value ± standard deviation.

3. Results

3.1. FTIR Spectroscopy

As presented in Figure 1, neat CNTs do not present any specific vibrational features, while AgNPs shows one single absorption peak at 572 cm^{-1}, characteristic for Ag-O stretching vibration [30,31]. Chlorhexidine gluconate presents a characteristic, intensive peak at 1590 cm^{-1}, accompanied by less intense satellites at 1650 and 1550 cm^{-1}, which can be attributed to the C=C stretching of the aromatic moiety of CHX. In the high wavenumber regions, the two peaks at 3135 cm^{-1} and 3320 cm^{-1} regions are attributed to asymmetric and symmetric –NH vibrations, which are also suggestive of CHX. The main fingerprints of the commercial sealer are 742 cm^{-1} and 914 cm^{-1}, which are due to the contribution of epoxy rings (the most important functional groups), 1020 and 1237 cm^{-1} corresponding to symmetrical and asymmetrical aromatic C-O stretch, 1510–1680 cm^{-1} corresponding to C-C stretching vibration in aromatic rings, and the peaks around 2930 cm^{-1} corresponding to symmetrical and asymmetrical C-H stretch of –CH_2 groups [32–34].

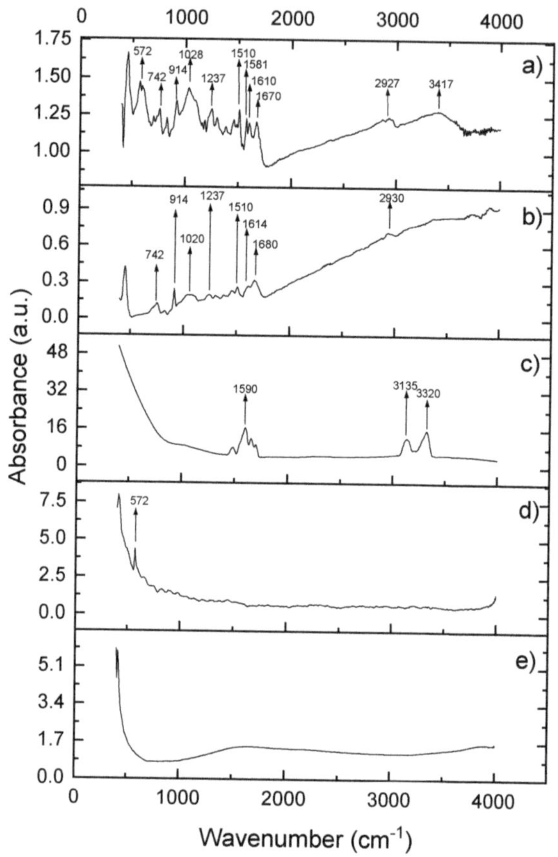

Figure 1. FTIR spectra of modified root canal sealer obtained by incorporation of (**a**) AgNPs/CHX/CNTs mixture; (**b**) commercial root canal sealer (ADSeal); (**c**) chlorhexidine gluconate gel; (**d**) colloidal AgNPs and (**e**) neat CNTs (as received from the manufacturer).

The FTIR vibrational features of modified sealer preserved all the fingerprints of the blended components, while the main peaks of the epoxide groups are visible at the same wavenumbers as in the spectrum of neat sealer and the intensity of the bands at 572 cm^{-1}, 1020 and 1237 cm^{-1} are enhanced. This behavior is due to the overlapping of C-O vibrations and those of the ether bonds. Additionally, in the high wavenumber region, the peaks between 2930–3417 cm^{-1} indicates the vibration of hydroxyl groups of epoxy resin superimposed over the asymmetric/symmetric –NH vibrations in CHX. As the main fingerprints of the neat sealer are preserved and no additional vibration bands (or disappearance of the existing ones) occurred, we can interpret that there is no chemical reaction between the original epoxy resin and the blended components (CNT, AgNPs and CHX). Hence, the sealer acts as a "reservoir" for the active ingredients.

3.2. Electrochemical Measurements

The electrochemical profile of colloidal AgNPs, CHX 2% and the mixture CNTs/AgNPs/CHX 2% investigated on screen printed electrodes, is presented in Figure 2. The electrochemical oxidation of CHX revealed four anodic peaks at 0.267 V, 0.427 V, 0.913 V and 1.305 V. The ionic Ag^+ solution indicated a cathodic peak at around −0.1 V (not visible in the figure) due to the reduction of Ag^+, which was then re-oxidized at 0.1 V [35]. This oxidation peak was also found in the DPV of the mixture at almost the same potential. The main electrochemical features of CHX in the mixture CNTs/AgNPs/CHX2% were slightly shifted concerning the potential values (0.386 V, 0.632 V, and 0.967 V), while the first two oxidation peaks of CHX alone (at 0.267 V, 0.427 V) might be merged into one single peak in the case of the mixture (at 0.386 V). According to the literature [36–38], CHX oxidation leads to two p-chloraniline and two biguanidine molecules.

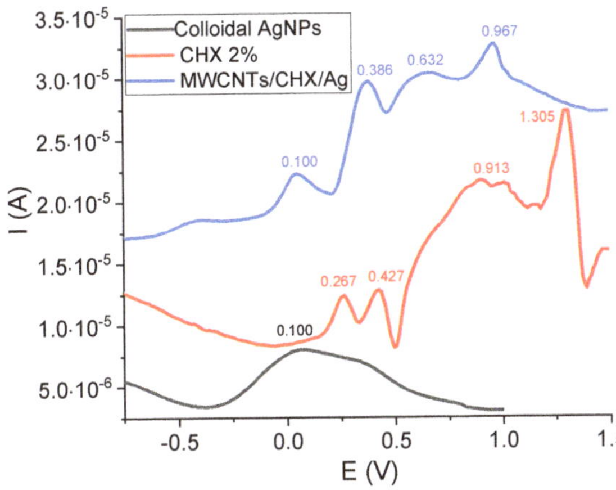

Figure 2. Differential pulse voltammetry of colloidal AgNPs, CHX 2% solution and the mixture CNTs/CHX/AgNPs.

3.3. SEM Investigation of Modified Root Canal Sealer

Prior to blending all the components, the morphology of CNTs was investigated (Figure 3). By comparing the morphological details of CNTs before and after loading with the mixture of CHX/AgNPs, an increased diameter of the nanotubes can be noticed after loading leading to a tighter matrix (Figure 3a,b). A good stability of CNTs loaded with CHX/AgNPs was also noticed compared to neat CNTs, when dispersed in distilled water, as presented in photographic images (Figure 3c), revealing the homogeneity of the colloidal mixture.

After blending the components and waiting for the setting time, according to the manufacturer's instructions, SEM images were recorded on the fractured surface of neat and modified sealer, without any other surface preparation, as presented in Figure 4.

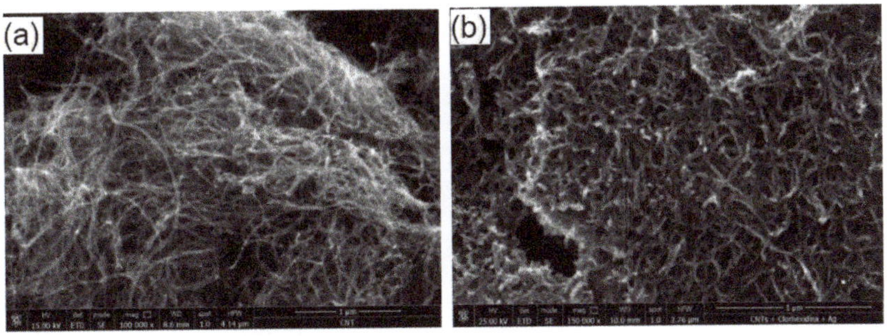

Figure 3. *Cont.*

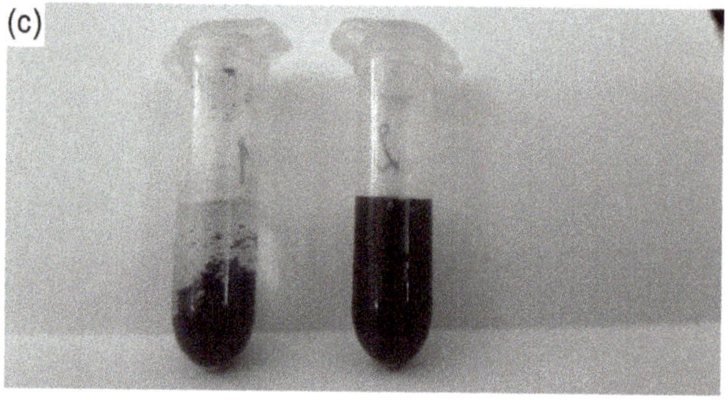

Figure 3. SEM morphological details of CNTs as received from the manufacturer, before (**a**) and after (**b**) loading with the mixture CHX/AgNPs (High magnification, 100,000×); (**c**) photographic image of colloidal CNTs in distilled water (left) compared to colloidal mixture CHX/AgNPs (right).

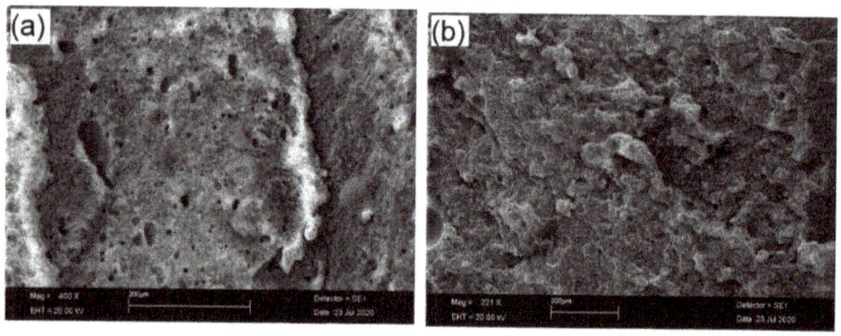

Figure 4. SEM images (low magnification, 400×) of the fracture surface of neat endodontic sealer (**a**) and modified sealer after incorporation of CNTs/CHX/AgNPs (**b**).

It can be observed that there is a good dispersion of CNTs/CHX/AgNPs in the epoxy matrix, while multiple cleavage planes can be noticed on the fracture surface of the modified sealer, compared to the neat one. According to some authors [39–43] the better the nano-fillers disperse, the greater number and the smaller size of the cleavage planes are observed on the surface fracture of epoxy nanocomposites. Furthermore, the reinforcement mechanism is more effective when the height difference between the cleavage planes increases. High magnification details of the polished specimens are presented in Figure 5, along with the corresponding EDX spectrum.

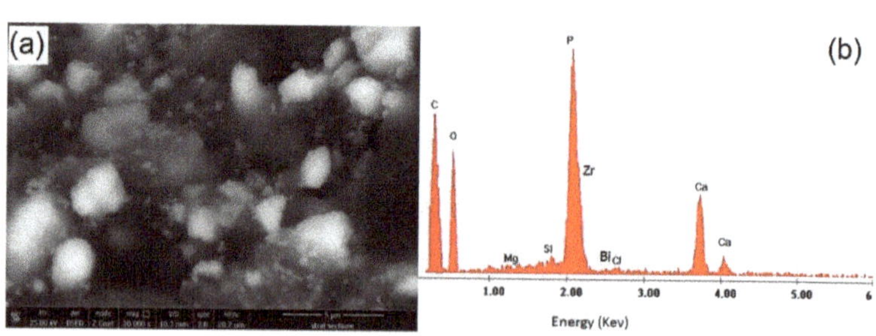

Figure 5. *Cont.*

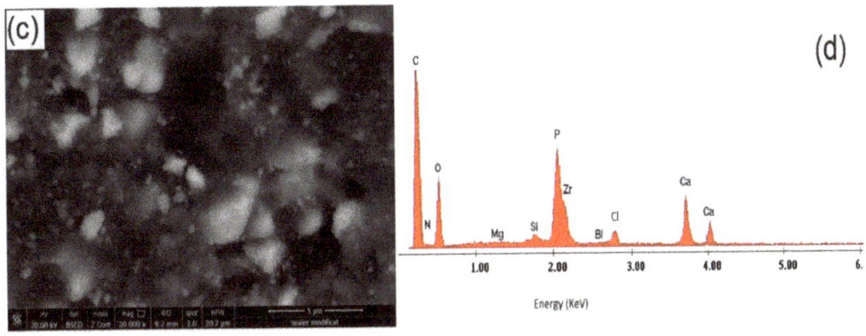

Figure 5. Ultrastructure details of commercial (**a**) and modified root canal sealer (**c**) along with the corresponding EDX spectrum (**b**,**d**).

The details presented in Figure 5 emphasize not only a good dispersion of CNTs/CHX/AgNPs in the epoxy matrix, but also the presence of inorganic components and radiopacifiers—calcium phosphate, zirconium oxide and bismuth carbonate. The relative intensity of the main elements (especially C and O) is obviously influenced by addition of CNTs/CHX/AgNPs, although the presence of Ag was not identified in the EDX spectrum, due to the detection limit of the instrument.

3.4. Nanoindentation Measurements

After setting and hardening, load–displacement curves were recorded for the two sets of prepared specimens, the commercial sealer and CNTs/CHX/AgNPs modified one, as presented in Figure 6a.

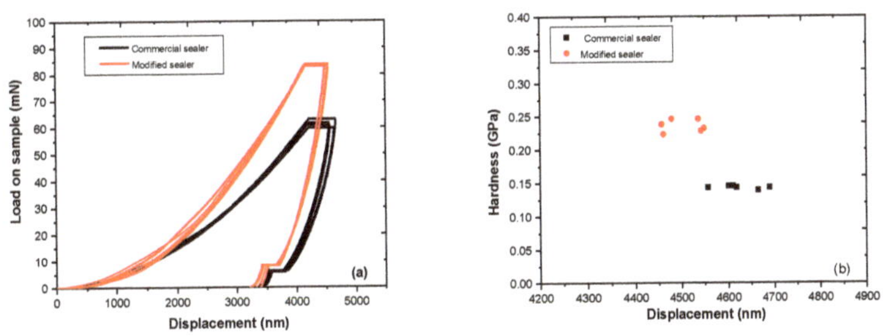

Figure 6. Load–displacement curves (**a**) recorded on the surface of commercial (black) and modified sealer (red) and the corresponding Young modulus calculation (**b**).

According to the curves' profile, it can be observed that the addition of CNTs/CHX/AgNPs to the commercial epoxy-based sealer enhanced the nanomechanical properties of the composite: a peak force of 85 mN was necessary to apply on the surface of the modified sealer, compared to 60 mN for the commercial one, in order to reach the same displacement (3300 nm). Average elastic modulus values obtained from the fitting parameters revealed a maximum value E = 0.25 GPa for the modified sealer, compared with E = 0.15 GPa for the neat sealer. The results displayed in Figure 6 demonstrate a slight reinforcement due to the addition of CNTs/CHX/AgNPs, with respect to the selected concentrations, as described in Section 2.1. An increased trend of the modulus at higher CNT content has been reported previously [44,45].

3.5. Thermal Analysis

Thermogravimetric analysis (TGA) and first derivative curves (DTG) for the modified and commercial sealer are presented in Figure 7. It can be noticed that both sealers have almost the same degradation patterns, showing one single degradation step between 200 and 450 °C. In addition, there is no mass loss until 130 °C, which denotes the absence of water content in samples [43–48]. However, DTG curves indicates that there is a small difference between the decomposition point of modified and commercial sealer (359 °C and 368 °C), while the residual mass of the modified sealer is slightly higher (59.99% toward 57.80%), indicating a good stability upon addition of CNTs/CHX/AgNPs mixture. The small peak at 293 °C in the DTG curve indicates the thermal decomposition of AgNPs [43].

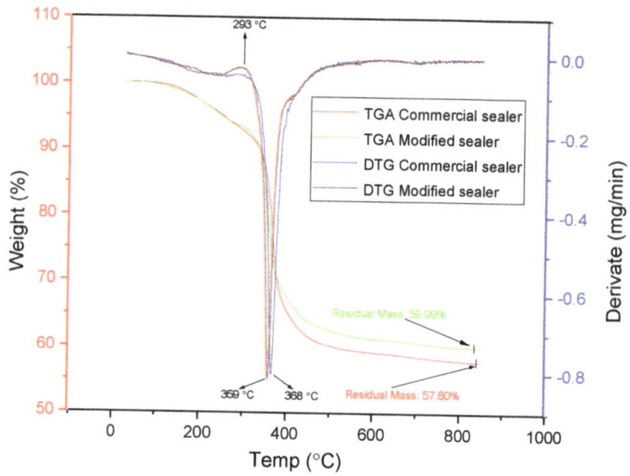

Figure 7. TGA and DTG thermograms of the modified and commercial sealers.

3.6. SEM Investigation of Interfacial Adaptation to Root Canal Dentine

It can be observed that there is a very good adaptation of both the commercial and modified sealers to the root canal walls, and no gaps were noticed across the investigated sections (Figure 8a,c). Moreover, based on the EDX spectrum recorded at the interface, it can be noticed that there is diffusion of chlorhexidine molecules from the sealer into the dentinal tissue, as evidenced by the presence of significant amount of chloride in the EDX spectrum (Figure 8b,d). We assume that AgNPs were also diffused across the interfacial area, but due to the instrument limitation, Ag was not visible in the quantitative spectrum. In addition, a more intense C peak was identified on the root canal dentine in contact with the modified sealer, suggesting the penetration of the modified resin component into dentine [44,49,50].

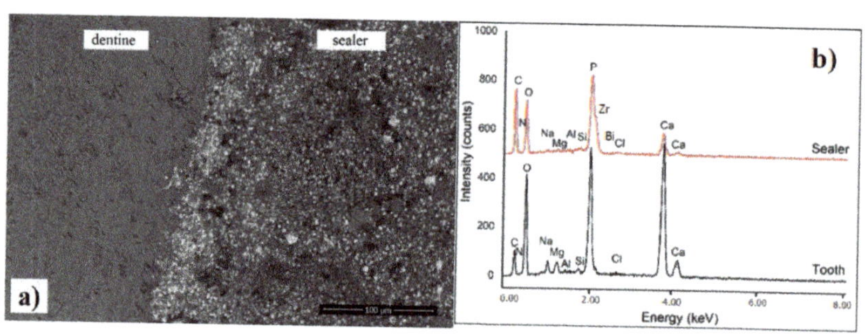

Figure 8. *Cont.*

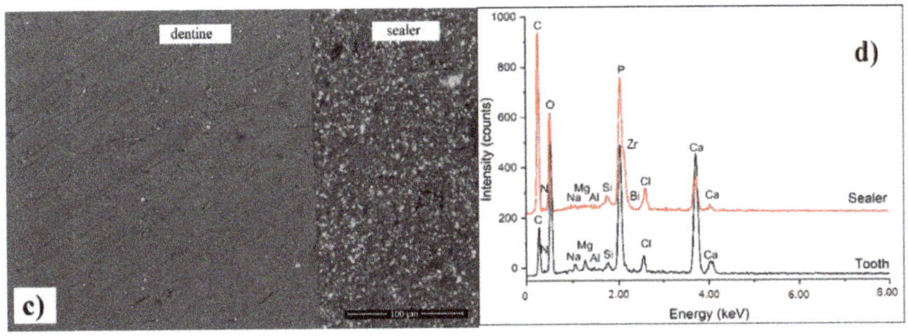

Figure 8. SEM images (**a**,**c**) of interfacial adaptation between sealer and root canal dentine (polished specimens) along with the corresponding EDX spectra (**b**,**d**): (**a**) neat sealer; (**c**) CNTs/CHX/AgNPs modified sealer. The transversal section was performed in the middle zone of the root.

The SEM details (back scatter mode) of the neat and modified sealer recorded at the interface are presented in Figure 9, emphasizing the details of inorganic compounds (white particles) and homogenous distribution within the resin matrix. Overall, it can be noticed that there is a similar microstructure of the modified sealer compared to the original one, suggesting that the addition of CNTs/CHX/AgNPs did not disturb the local distribution of the components, as there is no agglomeration of the calcium phosphate or radio opacifier particles related to the material.

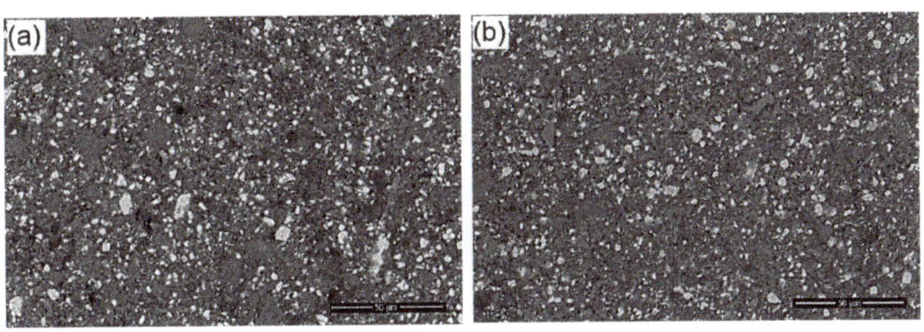

Figure 9. Back scatter SEM images with high magnification recorded on the surface of (**a**) neat sealer and (**b**) modified sealer, at the interface with root canal area.

3.7. Antimicrobial and Antifungal Effect of the Modified Sealer

In order to evaluate the antimicrobial effect of different combinations CNTs/AgNPs/CHX 2%, CNTs/CXH 2%, CNTs/AgNPs and their individual components, the prepared solutions were tested against *S. aureus*, *E. faecalis* and *C. albicans*, in concentration of 1 mg/mL. The results expressed as CFU/mL (Log_{10}) are presented in Figure 10. The antimicrobial and antifungal assay clearly demonstrated the best efficiency of CNTs/AgNPs/CHX 2% mixture against *E. faecalis*. However, the efficiency of different combinations with respect to *E. faecalis* were noticed to follow the order: CNT + H_2O < AgNPs < CNT/AgNPs < CHX 2% < CHX 2%/AgNPs < CNTs/AgNPs/CHX 2%. With respect to *S. aureus* and *C. albicans*, the effect was only moderate by comparison, however the same order being noticed. In addition, it can be observed that, if the active antimicrobial agent (AgNPs or CHX) was combined with CNTs, the effect was significantly enhanced, especially for the combination CNTs /CHX 2% against *E. faecalis* and *C. albicans*.

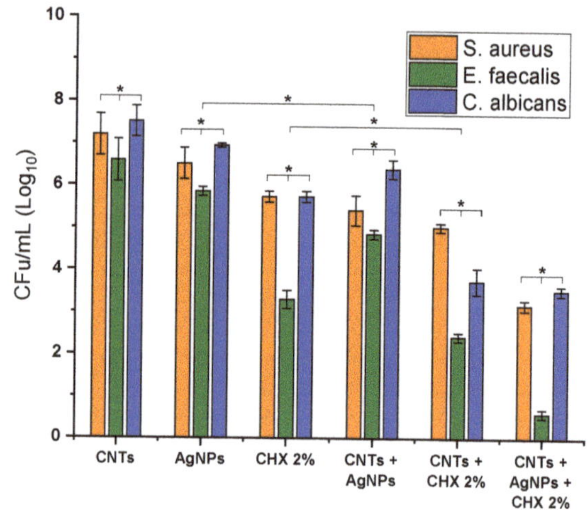

Figure 10. Antimicrobial and antifungal effect of different combinations and the mixture CNTs/AgNPs/CHX2% against the tested strains. Data are expressed as average value ± standard deviation of triplicate samples (statistical significance * $p < 0.05$).

The agar-diffusion test applied for the commercial and modified sealer, 7 days after setting, revealed that the commercial sealer had no antimicrobial and antifungal effect. The measurement of the diameter of the inhibition zone was performed for the modified sealer with respect to each strain and presented in Figure 11, compared to the positive controls. In accordance with the previous assay, the maximum efficiency was noticed against *E. faecalis* followed by *S. aureus* and *C. albicans*.

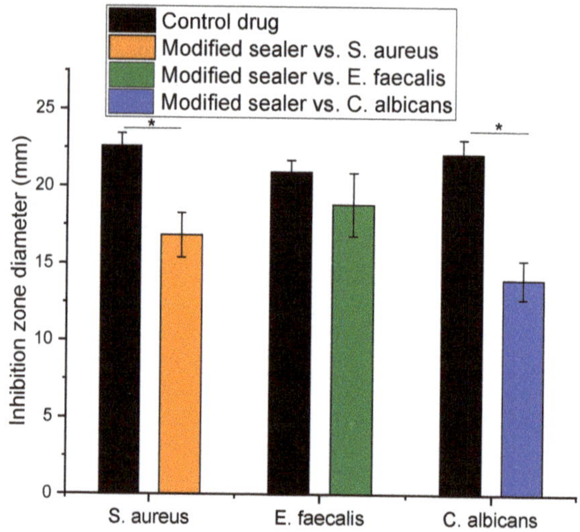

Figure 11. Diameter of inhibition zone in agar-diffusion assay with modified sealer and control drugs. Data are expressed as average value ± standard deviation of triplicate samples (* $p < 0.05$).

As compared to the controls, which are well known antibiotics and antifungal drugs, the results are considered very promising. For example, in the case of *E. faecalis* ATCC® 29212, a concentration of 100 μg/mL gentamicin was found to effectively inhibit the bacterial growth, which is about six times higher than MIC reported for most of the isolated

Enterococcus species [24,51]. In this case, no significant difference was noticed between the control and modified sealer ($p < 0.05$).

4. Discussions

The antibacterial ability of sealers is still insufficient and hence, continuous improvements are required, as the full impact of nanotechnology in endodontics is far from being fully understood. The incorporation of CNTs in epoxy-based matrix and the effective reinforcement is dependent on the aspect ratio, dispersion, alignment and interfacial stress transfer at the CNT-matrix interface [2]. It is generally considered that well-dispersed, randomly aligned CNTs are a preferred alternative for reinforcement, as aligned nanocomposites, in contrast to CNTs, possess anisotropic mechanical and electrical properties [52]. In addition to the reinforcement purpose, encouraging results have shown the ability of CNTs to be used as delivery systems for genes, peptides, oligonucleotides, cytotoxic drug molecules and antimicrobial agents [52–54]. Especially, the antimicrobial activity against Gram-positive and Gram-negative microorganisms, was evidenced due to the penetration of the sharp and narrow structure of CNTs through the cell's membrane, concomitant with the drug release, compromising wall integrity and finally, death via cell lysis.

CHX is a well-known non-specific matrix metalloproteinase (MMPs) inhibitor, having a dose-dependent inhibition mechanism, based on the interaction with the sulfhydryl groups and cysteine site of MMPs [55,56]. Some previous studies attempted to incorporate CHX into etchant materials or into primers or adhesive components because it has been suggested that CHX effect is only available immediately after application of the modified materials [56]. AgNPs have been also incorporated into bonding agents or restorative materials in endodontics, attempting to reduce *E. faecalis* adherence to dentine, to eliminate biofilms, or as endodontic irrigants and intracanal drugs [8,31,57].

Gram-positive bacteria such as *S. aureus* and *E. faecalis* have been detected in periapical infections being responsible for endodontic failure [24]. Particularly, the incidence of *E. faecalis* was reported to be in the range from 24% to 77% in the cases of apical periodontitis [58], being also reported to be resistant to several antimicrobial agents. It has a great ability to invade the dentinal tubules, and, hence, to resist during the chemo-mechanical endodontic procedures. On the other hand, the occurrence of *C. albicans* in infected root canals varies between 1% and 17%, according to previous studies [59], being also resistant to conventional root canal irrigants. In this respect, the continuous improvement of endodontic sealers with high antimicrobial activity requires new developments with the aim to prevent or to decrease the bacterial growth within the root canal, while contributing to the repair of dentinal tissue. The currently available commercial endodontic sealers, possess different physico-chemical properties in terms of sealing ability, adhesiveness, solubility to oral fluids and dimensional stability [49]. As mentioned in literature, based on in vitro studies, different categories of sealers (such as resin, zinc oxide eugenol, calcium hydroxide, glass ionomer, silicon or silicate- based sealers) demonstrated good antimicrobial activity as freshly prepared, but the antimicrobial activity was lost as the material set [59–62]. Actually, no bacterial growth inhibition was noticed for 2–7 days set samples, as evidenced in a systematic review conducted by Al Shawaimi et al. [60]. However, the in vitro studies have some limitations due to the lack of standardization in terms of inoculum density, culture medium selection, agar viscosity, condition of plate storage, or diffusion–solubility relation of the material and culture medium.

Our original approach, in the context of new generation sealers expecting to have a long-lasting antimicrobial effect, was to demonstrate that the antimicrobial effect of the mixture CNTs/AgNPs/CXH 2% incorporated in commercial sealer, was preserved long enough to efficiently inhibit Gram-positive germs, with excellent results towards *E. faecalis* in a concentration of 1 mg/mL. It is considered that epoxy resin-based root canal sealer possesses a certain degree of antibacterial properties attributed to bisphenol diglycidyl ether and formaldehyde released during the polymerization reaction [63]. Our results suggest that the modified sealer is able to release the antimicrobial agents long enough after setting and aging (7 days), which is an obvious improvement compared to the neat sealer.

The structural properties of the modified sealer, evidenced by FTIR spectroscopy, suggests that no chemical reaction between the original epoxy resin and the blended components (CNT, AgNPs and CHX) was established during the preparation. Hence, the sealer acts as a "reservoir" for the active ingredients. According to the literature, CNTs interact with organic and charged molecules via π-π^* and electrostatic interactions, while the small diameter of CNTs restricts the adsorption of larger organic molecules. Instead, most of the organic molecules can easily penetrate into the spaces between the tube bundles or at the external surface [2], being released in contact with dentine. The electrochemical profile of colloidal AgNPs, CHX 2% and the mixture CNTs/AgNPs/CHX 2% suggest that electrons transfer involving Ag^+ and CHX molecules might be responsible for the synergic antimicrobial effect.

A good dispersion of CNTs/CHX/AgNP in the epoxy matrix was demonstrated by SEM investigation and a slight reinforcement was noticed compared to the commercial sealer, as evidenced by nanoindentation measurements, which is in agreement with previous reported data [45]. The addition of CNTs/CHX/AgNPs did not disturb the local distribution of the inorganic components, as no agglomeration of the calcium phosphate or radio opacifier particles was noticed.

In terms of thermal behavior, both modified and commercial sealers have almost the same degradation patterns, while only a small difference in residual mass was noticed. Overall, a good stability of the modified sealer can be noticed. The corroborated results clearly demonstrated that the modifications induced by addition of CNTs/AgNPs/CHX 2% mixture did not impair the structural, ultrastructural and thermal behavior of the original sealer.

Moreover, a similar, very good adaptation, without gaps, of both commercial and modified sealer to the root canal walls was evidenced by SEM. The EDX spectrum recorded at the interface, demonstrated the diffusion of chlorhexidine molecules from the sealer into the dentinal tissue. It is well known that chemical or mechanical dentin pretreatments (for example sandblasting procedure) are usually made in order to increase the roughness of the treated surface, which influences the adhesion properties and the contact area between the dentinal tissue and the sealer [64]. In our work, we did not apply any mechanical pretreatment, except the routine procedure of endodontic cleaning.

The antibacterial and antifungal assay clearly demonstrated a synergic effect of AgNPs, CHX 2% and CNTs with excellent results towards *E. faecalis*, which is responsible for the primary etiologic factors in pulp and periapical lesions [7].

5. Conclusions

In this work, we report the improved antibacterial effect of a modified endodontic sealer, upon addition of CNTs incorporating CHX 2% and AgNPs. The structural, ultrastructural and thermal properties of the modified sealer were not affected by this modification, while demonstrating a perfect interfacial adaptation with the root canal dentine. As successful root canal therapy involves a combination of proper instrumentation, irrigation, obturation, and sealing, we suggest that our results are supportive of the near future potential of nanoparticles in clinical endodontics.

Author Contributions: Conceptualization, A.M. and S.C.; methodology, L.F., F.B. and C.I.; software, F.B.; validation, I.H., C.S. and G.R.; investigation, I.H., G.R., C.S.; resources, A.M. and C.I.; writing—original draft preparation, A.M. and S.C.; writing—review and editing, F.B., L.F. and S.C.; visualization, I.H. and F.B.; supervision, S.C. All authors have read and agreed to the published version of the manuscript.

Funding: This research received no external funding.

Institutional Review Board Statement: The study was conducted according to the guidelines of the Declaration of Helsinki, and the protocol was approved by the Institutional Research Ethics Committee of the Faculty of Medicine and Pharmacy, University of Oradea [Nr. 2/31.03.2021].

Informed Consent Statement: Not applicable.

Conflicts of Interest: The authors declare no conflict of interest.

References

1. Chogle, S.M.A.; Kinaia, B.M.; Goodis, H.E. Scope of Nanotechnology in Endodontics. In *Nanobiomaterials in Clinical Dentistry*, 1st ed.; Subramani, K., Ahmed, W., Hartsfield, J.K., Eds.; Elsevier: Amsterdam, The Netherlands, 2013; pp. 432–445.
2. The, S.J.; Lai, C.W. *Carbon Nanotubes for Dental Implants in Applications of Nanocomposite Materials in Dentistry*; Elsevier: Amsterdam, The Netherlands, 2019; pp. 93–105. [CrossRef]
3. Alenazy, M.S.; Mosadomi, H.A.; Al-Nazhan, S.; Rayan, M.R. Clinical considerations of nanobiomaterials in endodontics. A systematic review. *Saudi Endod. J.* **2018**, *8*, 163–169. [CrossRef]
4. Gomes, B.P.; Pinheiro, E.T.; Gade-Neto, C.R.; Sousa, E.L.R.; Ferraz, C.C.R.; Zaia, A.A.; Teixeria, F.B.; Souza-Filho, F.J. Microbiological examination of infected dental root canals. *Oral Microbiol. Immunol.* **2004**, *19*, 71–76. [CrossRef] [PubMed]
5. Mohammadi, Z.; Abbott, P.V. The properties and applications of chlorhexidine in endodontics. *Int. Endodod. J.* **2009**, *42*, 288–302. [CrossRef]
6. Sirtes, G.; Waltimo, T.; Schaetzle, M.; Zehnder, M. The effects of temperature on sodium hypochlorite short term stability, pulp dissolution capacity and antimicrobial efficacy. *J. Endod.* **2005**, *31*, 669–671. [CrossRef] [PubMed]
7. Gomes, B.; Souza, S.F.C.; Ferraz, C.C.R.; Teixeira, F.B.; Zaia, A.A.; Valdrighi, L.; Souza-Filho, F.J. Effectiveness of 2% chlorhexidine gel and calcium hydroxide against Enterococcus faecalis in bovine root dentine in vitro. *Int. Endod. J.* **2003**, *36*, 267–275. [CrossRef]
8. Kandaswamy, D.; Venkateshbabu, N. Root canal irrigants. *J. Conserv. Dent.* **2010**, *13*, 256–264. [CrossRef]
9. Chang, Y.E.; Shin, D.H. Effect of chlorhexidine application methods on microtensile bond strength to dentine in class I cavities *Oper. Dent.* **2010**, *35*, 618–623.
10. Dionysopoulos, D. Effect of digluconate chlorhexidine on bond strength between dentinal adhesive systems and dentin: A systematic review. *J. Conserv. Dent.* **2016**, *19*, 11–16. [CrossRef] [PubMed]
11. Barbizam, J.V.; Trope, M.; Tanomaru-Filho, M.; Teixeira, E.C.; Teixeira, F.B. Bond strength of different endodontic sealers to dentin: Pushout test. *J. Appl. Oral Sci.* **2011**, *19*, 644–647. [CrossRef] [PubMed]
12. Marciano, M.A.; Guimaraes, B.M.; Ordinola-Zapata, R.; Bramante, C.M.; Cavenago, B.C.; Garcia, R.B.; Bernardineli, N.; Andrade, F.B.; Moraes, I.G.; Duarte, M.A. Physical properties and interfacial adaptation of three epoxy resin-based sealers. *J. Endod.* **2011**, *37*, 1417–1421. [CrossRef] [PubMed]
13. Grossman, L.I. *Endodontic Practice*, 10th ed.; Henry Kimpton: Philadelphia, PA, USA, 1981; p. 297.
14. Al-Haddad, A.; Che Ab Aziz, Z.A. Bioceramic-based root canal sealers: A review. *Int. J. Biomater.* **2016**. [CrossRef]
15. Lee, J.K.; Kwak, S.W.; Ha, J.H.; Lee, W.; Kim, H.C. Physicochemical properties of epoxy resin-based and bioceramic-based root canal sealers. *Bioinorg. Chem. Appl.* **2017**. [CrossRef]
16. Marin-Bauza, G.A.; Silva-Sousa, Y.T.C.; da Cunha, S.A.; Rached-Junior, F.J.A.; Bonetti-Filho, I.; Sousa-Neto, M.D.; Miranda, C.E.S. Physicochemical properties of endodontic sealers of different bases, *J. Appl. Oral Sci.* **2012**, *20*, 455–461. [CrossRef] [PubMed]
17. Fonseca, D.A.; Paula, A.B.; Marto, C.M.; Coelho, A.; Paulo, S.; Martinho, J.P.; Carrilho, E.; Ferreira, M.M. Biocompatibility of root canal sealers: A systematic review of in vitro and in vivo studies. *Materials* **2019**, *12*, 4113. [CrossRef]
18. Cavalu, S.; Simon, V. Microstructure and bioactivity of acrylic bone cements for prosthetic surgery. *J. Optoelectron. Adv. Mater.* **2006**, *8*, 1520–1523.
19. Fritea, L.; Bănică, F.; Costea, T.O.; Moldovan, L.; Iovan, C.; Cavalu, S. A gold nanoparticles—Graphene based electrochemical sensor for sensitive determination of nitrazepam. *J. Electroanal. Chem.* **2018**, *830–831*, 63–71. [CrossRef]
20. Fritea, L.; Tertis, M.; Sandulescu, R.; Cristea, C. Enzymes-graphene platforms for electrochemical biosensors design with biomedical applications. *Methods Enzymol.* **2018**, *609*, 293–333.
21. Cavalu, S.; Antoniac, I.; Fritea, L.; Mohan, A.; Semenescu, A.; Laslo, V.; Vicas, S. Surface modifications of the titanium mesh for cranioplasty using selenium nanoparticles coating. *J. Adhes. Sci. Technol.* **2018**, *32*, 2509–2522. [CrossRef]
22. Pirzada, M.; Altintas, Z. Nanomaterials for healthcare biosensing applications. *Sensors* **2019**, *19*, 5311. [CrossRef] [PubMed]
23. Cavalu, S.; Ratiu, C.; Ponta, O.; Simon, V.; Rugina, D.; Miclaus, V. Improving osseointegration of alumina/zirconia ceramic implants by fluoride surface treatment. *Dig. J. Nanomater. Biostruct.* **2014**, *9*, 797–808.
24. Tabassum, S.; Khan, F.R. Failure of endodontic treatment: The usual suspects. *Eur. J. Dent.* **2016**, *10*, 144–147. [CrossRef]
25. Cavalu, S.; Fritea, L.; Brocks, M.; Barbaro, K.; Murvai, G.; Costea, T.O.; Antoniac, I.; Verona, C.; Romani, M.; Latini, A.; et al. Novel hybrid composites based on PVA/SeTiO_2 nanoparticles and natural hydroxyapatite for orthopedic applications: Correlations between structural, morphological and biocompatibility properties. *Materials* **2020**, *13*, 2077. [CrossRef]
26. Baras, B.H.; Melo, M.A.S.; Thumbigere-Math, V.; Tay, F.R.; Fouad, A.F.; Oates, T.W.; Weir, M.D.; Cheng, L.; Xu, H.H.K. Novel bioactive and therapeutic root canal sealers with antibacterial and remineralization properties. *Materials* **2020**, *13*, 1096. [CrossRef] [PubMed]

27. Makvandi, P.; Gu, J.T.; Zare, E.N.; Ashtari, K.; Moeini, A.; Tay, F.R.; Niu, L.N. Polymeric and inorganic nanoscopical antimicrobial fillers in dentistry. *Acta Biomater.* **2020**, *101*, 69–101. [CrossRef]
28. Yitzhak, R.; Noel, M.E. Carbon nanotubes in drug delivery: Focus on infectious diseases. *Expert Opin. Drug Deliv.* **2009**, *6*, 517–530.
29. Oliver, W.C.; Pharr, G.M. An improved technique for determining hardness and elastic modulus using load and displacement sensing indentation experiments. *J. Mater. Res.* **1992**, *7*, 1564–1583. [CrossRef]
30. Chen, Q.-Y.; Xiao, S.-L.; Shi, S.Q.; Cai, L.-P. A One-pot synthesis and characterization of antibacterial silver nanoparticle–cellulose film. *Polymers* **2020**, *12*, 440. [CrossRef] [PubMed]
31. Cavalu, S.; Simon, V.; Goller, G.; Akin, I. Bioactivity and antimicrobial properties of PMMA/Ag_2O acrylic bone cement collagen coated. *Dig. J. Nanomater. Biostruct.* **2011**, *6*, 779–790.
32. Rema, T.; Lawrence, J.R.; Dynes, J.J.; Hitchcock, A.P.; Korber, D.R. Microscopic and spectroscopic analyses of chlorhexidine tolerance in Delftia acidovorans biofilms. *Antimicrob. Agents Chemother.* **2014**, *58*, 5673–5686. [CrossRef] [PubMed]
33. Cortés, M.E.; Sinesterra, R.D.; Avila-Campos, M.J.; Tortamano, N.; Rocha, R.G. The chlorhexidine:β-cyclodextrin inclusion compound: Preparation, characterization and microbiological evaluation. *J. Incl. Phenom. Macrocycl. Chem.* **2001**, *40*, 297–302. [CrossRef]
34. Maity, P.; Kasisomayajula, S.V.; Parameswaran, V.; Basu, S.; Gupta, N. Improvement in surface degradation properties of polymer composites due to pre-processed nanometric alumina fillers. *IEEE Trans. Dielectr. Electr. Insul.* **2008**, *15*, 63–72. [CrossRef]
35. Choi, Y.J.; Luo, T.J.M. Electrochemical properties of silver nanoparticle doped aminosilica nanocomposite. *Int. J. Electrochem.* **2011**, *2011*, 404937. [CrossRef]
36. Khan, M.A.; Mujahid, M.; Gul, I.H.; Hussain, A. Electrochemical study of magnetic nanogel designed for controlled re-lease of chlorhexidine gluconate. *Electrochim. Acta* **2019**, *295*, 113–123. [CrossRef]
37. de Lima, A.P.; Stefano, J.S.; Montes, R.H.O.; Cunha, R.R.; Silva, L.A.J.; Richter, E.M.; Muñoz, R.A.A. Electrochemical oxidation of chlorhexidine and its amperometric determination by flow-injection analysis. *J. Braz. Chem. Soc.* **2014**, *25*, 448–452.
38. Sousa, C.P.; de Oliveira, R.C.; Freire, T.M.; Fechine, P.B.A.; Salvador, M.A.; Homem-de-Mello, P.; Morais, S.; Lima-Neto, P.; Correia, A.N. Chlorhexidine digluconate on chitosan-magnetic iron oxide nanoparticles modified electrode: Electroanalysis and mechanistic insights by computational simulations. *Sens. Actuators B Chem.* **2017**, *240*, 417–425. [CrossRef]
39. Fernandes, I.J.; Vieira Santos, R.; Araujo dos Santos, E.C.; Avila Campos Rocha, T.L.; Domingues Junior, N.S.; Mendes Moraes, C.A. Replacement of commercial silica by rice husk ash in epoxy composites: A comparative analysis. *Mater. Res.* **2018**, *21*, 0562. [CrossRef]
40. Nikafshar, S.; Zabihi, O.; Hamidi, S.; Moradi, Y.; Barzegar, S.; Ahmadi, M.; Naeb, M. A renewable bio-based epoxy resin with improved mechanical performance that can compete with DGEBA. *RSC Adv.* **2017**, *7*, 8694. [CrossRef]
41. Jen, Y.-M.; Chang, H.-H.; Lu, C.-M.; Liang, S.-Y. Temperature-dependent synergistic effect of multi-walled carbon nanotubes and graphene nanoplatelets on the tensile quasi-static and fatigue properties of epoxy nanocomposites. *Polymers* **2021**, *13*, 84. [CrossRef]
42. Cavalu, S.; Banica, F.; Gruian, C.; Vanea, E.; Goller, G.; Simon, V. Microscopic and spectroscopic investigation of bioactive glasses for antibiotic controlled release. *J. Mol. Struct.* **2013**, *1040*, 47–52. [CrossRef]
43. Majeed Khan, M.A.; Kumar, S.; Ahamed, M.; Alrokayan, S.A.; AlSalhi, M.S. Structural and thermal studies of silver nanoparticles and electrical transport study of their thin films. *Nanoscale Res. Lett.* **2011**, *6*, 434. [CrossRef]
44. Zhou, Y.; Jeelani, M.I.; Jeelani, S. Development of photo micro-graph method to characterize dispersion of CNT in epoxy. *Mater. Sci. Eng. A.* **2009**, *506*, 39–44. [CrossRef]
45. Esmizadeh, E.; Yousefi, A.A.; Naderi, G. Effect of type and aspect ratio of different carbon nanotubes on cure behavior of epoxy-based nanocomposites. *Iran. Polym. J.* **2015**, *24*, 1–12. [CrossRef]
46. Atmeh, A.R.; AlSwaimi, E. The effect of heating time and temperature on epoxy resin and calcium silicate-based endodontic sealer. *J. Endod.* **2017**, *43*, 2112–2118. [CrossRef]
47. Marian, E.; Jurca, T.; Banica, F.; Morgovan, C.; Bratu, I. Thermal analysis and raman spectrometry of some complexes of theophylline with transitional. *Rev. Chim.* **2010**, *61*, 569–574.
48. Hsieh, Y.C.; Chou, Y.C.; Lin, C.P.; Hsieh, T.F.; Shu, C.M. Thermal analysis of multi-walled carbon nanotubes by kissinger's corrected kinetic equation. *Aerosol Air Qual. Res.* **2010**, *10*, 212–218. [CrossRef]
49. Priyadarshini, B.; Selvan, S.; Narayanan, K.; Fawzy, A. Characterization of chlorhexidine-loaded calcium-hydroxide microparticles as a potential dental pulp-capping material. *Bioengineering* **2017**, *4*, 59. [CrossRef] [PubMed]
50. Viapiana, R.; Guerreiro-Tanomaru, J.; Tanomaru-Filho, M.; Camilleri, J. Interface of dentine to root canal sealers. *J. Dent.* **2014**, *42*, 336–350. [CrossRef]
51. Chow, J.W. Aminoglycoside resistance in enterococci. *Clin. Infect. Dis.* **2000**, *31*, 586–589. [CrossRef]
52. Treacy, M.M.J.; Ebbesen, T.W.; Gibson, J.M. Exceptionally high young modulus observed for individual nanotubes. *Nature* **1996**, *381*, 678–680. [CrossRef]
53. Kapralos, V.; Rukke, H.V.; Koutroulis, D.O.A.; Camilleri, J.; Sunde, P.T. Antimicrobial and physicochemical characterization of endodontic sealers after exposure to chlorhexidine digluconate. *Dent. Mater.* **2021**, *37*, 249–263. [CrossRef]
54. Coleman, J.N.; Khan, U.; Gun'ko, Y.K. Mechanical reinforcement of polymers using CNTs. *Adv. Mater.* **2006**, *18*, 689–706. [CrossRef]

55. Kalagi, S.; Feitosa, S.A.; Münchow, E.A.; Martins, V.M.; Karczewski, A.E.; Cook, N.B.; Diefenderfer, K.; Eckert, G.J.; Geraldeli, S.; Bottino, M.C. Chlorhexidine-modified nanotubes and their effects on the polymerization and bonding performance of a dental adhesive. *Dent. Mater.* **2020**, *36*, 687–697. [CrossRef]
56. Stanislawczuk, R.; Reis, A.; Loguercio, A.D. A 2-year in vitro evaluation of chlorhexidine-containing acid on the durability of resin-dentin interfaces. *J. Dent.* **2011**, *39*, 40–47. [CrossRef]
57. Rodrigues, C.T.; de Andrade, F.B.; de Vasconcelos, L.R.S.M.; Midena, R.Z.; Pereira, T.C.; Kuga, M.C.; Duarte, M.A.H.; Bernardineli, N. Antibacterial properties of silver nanoparticles as a root canal irrigant against Enterococcus faecalis biofilm and infected dentinal tubules. *Int. Endod. J.* **2018**, *51*, 901–911. [CrossRef] [PubMed]
58. Stuart, C.H.; Schwartz, S.A.; Beeson, T.J.; Owatz, C.B. "Enterococcus faecalis": Its role in root canal treatment failure and current concepts in retreatment. *J. Endod.* **2006**, *32*, 93–98. [CrossRef]
59. Kumar, J.; Sharma, R.; Sharma, M.; Prabhavathi, V.; Paul, J.; Chowdary, C.D. Presence of candida albicans in root canals of teeth with apical periodontitis and evaluation of their possible role in failure of endodontic treatment. *J. Int. Oral Health* **2015**, *7*, 42–45. [PubMed]
60. AlShwaimi, E.; Bogari, D.; Ajaj, R.; Al-Shahrani, S.; Almas, K.; Majeed, A. In vitro antimicrobial effectiveness of root canal sealers against enterococcus faecalis: A systematic review. *J. Endod.* **2016**, *42*, 1588–1597. [CrossRef] [PubMed]
61. Cavalu, S.; Simion, V.; Banica, F. In vitro study of collagen coating by electrodeposition on acrylic bone cement with antimicrobial. *Dig. J. Nanomater. Biostruct.* **2010**, *6*, 89–97.
62. Ceci, M.; Viola, M.; Rattalino, D.; Beltrami, R.; Colombo, M.; Poggio, C. Discoloration of different esthetic restorative materials: A spectrophotometric evaluation. *Eur. J. Dent.* **2017**, *11*, 149–156. [CrossRef] [PubMed]
63. Leonardo, M.R.; Bezerra da Silva, L.A.; Filho, M.T.; Santana da Silva, R. Release of formaldehyde by 4 endodontic sealers. *Oral Surg. Oral Med. Oral Pathol. Oral Radiol.* **1999**, *88*, 221–225. [CrossRef]
64. Sinjari, B.; Santilli, M.; D'Addazio, G.; Rexhepi, I.; Gigante, A.; Caputi, S.; Traini, T. Influence of dentine pre-treatment by sandblasting with aluminum oxide in adhesive restorations. an in vitro study. *Materials* **2020**, *13*, 3026. [CrossRef] [PubMed]

materials

Article

Nano Selenium—Enriched Probiotics as Functional Food Products against Cadmium Liver Toxicity

Simona Ioana Vicas [1,†], Vasile Laslo [1,†], Adrian Vasile Timar [1,†], Cornel Balta [2,†], Hildegard Herman [2,†], Alina Ciceu [2,†], Sami Gharbia [2,†], Marcel Rosu [2,†], Bianca Mladin [2,†], Laurentiu Chiana [3,†], József Prokisch [4,†], Maria Puschita [5,†], Eftimie Miutescu [5,†], Simona Cavalu [6,*,†], Coralia Cotoraci [5,†] and Anca Hermenean [2,5,*,†]

1 Faculty of Environmental Protection, University of Oradea, 24 Gen. Magheru St., 410048 Oradea, Romania; sim_vicas@yahoo.com (S.I.V.); vasilelaslo@yahoo.com (V.L.); timar.adrian@gmail.com (A.V.T.)
2 "Aurel Ardelean" Institute of Life Sciences, Vasile Goldis Western University of Arad, 86 Liviu Rebreanu St., 310414 Arad, Romania; baltacornel@gmail.com (C.B.); hildegard.i.herman@gmail.com (H.H.); alina_ciceu@yahoo.com (A.C.); samithgh2@hotmail.com (S.G.); ramrosu@gmail.com (M.R.); biancaonitamaria@gmail.com (B.M.)
3 Doctoral School of Biomedical Science, University of Oradea, 1 University St., 410087 Oradea, Romania; laurentiu.chiana@gmail.com (L.C.)
4 Institute of Animal Science, Biotechnology and Nature Conservation, Faculty of Agricultural and Food Sciences and Environmental Management, University of Debrecen, 4032 Debrecen, Hungary; jprokisch@agr.unideb.hu (J.P.)
5 Faculty of Medicine, Vasile Goldis Western University of Arad, 86 Liviu Rebreanu St., 310414 Arad, Romania; puschita.maria@uvvg.ro (M.P.); miutescu.eftimie@uvvg.ro (E.M.); ccotoraci@yahoo.com (C.C.)
6 Faculty of Medicine and Pharmacy, University of Oradea, 10 Pta 1 Decembrie St., 410073 Oradea, Romania
* Correspondence: simona.cavalu@gmail.com (S.C.); anca.hermenean@gmail.com (A.H.)
† All authors contributed equally to this work.

Citation: Vicas, S.I.; Laslo, V.; Timar, A.V.; Balta, C.; Herman, H.; Ciceu, A.; Gharbia, S.; Rosu, M.; Mladin, B.; Chiana, L.; et al. Nano Selenium—Enriched Probiotics as Functional Food Products against Cadmium Liver Toxicity. *Materials* **2021**, *14*, 2257. https://doi.org/10.3390/ma14092257

Academic Editor: Loic Hilliou

Received: 17 February 2021
Accepted: 22 April 2021
Published: 27 April 2021

Abstract: Since cadmium is a toxic metal that can cause serious health problems for humans, it is necessary to find bioremediation solutions to reduce its harmful effects. The main goal of our work was to develop a functional food based on elemental selenium nanoparticles (SeNPs) obtained by green synthesis using *Lactobacillus casei* and to validate their ability to annihilate the hepatic toxic effects induced by cadmium. The characterization of SeNPs was assessed by UV–Vis spectroscopy, FTIR, XRD, DLS and TEM. In order to investigate the dose-dependent protective effects of SeNPs on Cd liver toxicity, mice were assigned to eight experimental groups and fed by gavage, with 5 mg/kg b.w. cadmium, respectively, with co-administration with SeNPs or lacto-SeNPs (LSeNPs) in 3 doses (0.1, 0.2 and 0.4 mg/kg b.w.) for 30 days. The protective effect was demonstrated by the restoration of blood hepatic markers (AST, ALT, GGT and total bilirubin) and antioxidant enzymes, such as catalase (CAT) and glutathione peroxidase (GPx). Moreover, the antioxidant capacity of mice plasma by the FRAP assay, revealed the highest antioxidant capacity for the 0.2 mg/kg LSeNPs group. Histopathological analysis demonstrated the morphological alteration in the group that received only cadmium and was restored after the administration of SeNPs or LSeNPs, while the immunohistochemical analysis of the *bcl* family revealed anti-apoptotic effects; the Q-PCR analysis showed an upregulation of hepatic inflammatory markers for the group exposed to Cd and a decreased value for the groups receiving oral SeNPs/ LSeNPs in a dose-dependent manner. The best protective effects were obtained for LSeNPs. A functional food that includes both probiotic bacteria and elemental SeNPs could be successfully used to annihilate Cd-induced liver toxicity, and to improve both nutritional values and health benefits.

Keywords: selenium nanoparticles; *Lactobacillus casei*; cadmium; antioxidant enzymes; liver; histology; anti-apoptotic; anti-inflammatory

1. Introduction

Heavy metals are widely found in our environment and have adverse health effects on the human metabolism. Acute heavy metal intoxications may damage central nervous function, the cardiovascular and gastrointestinal systems, the liver and kidney [1,2].

Cadmium is an environmental toxicant that presents higher rates of soil-to-plant compared with other heavy metals, making foodstuffs the major source of cadmium exposure for non-smoking consumers [3,4]. EFSA's Panel on Contaminants in the Food Chain established for cadmium a tolerable weekly intake of 2.5 μg/kg b.w., a level that ensures a high level of protection for consumers [5]. The greatest dietary impact of cadmium occurs when certain foods are consumed in high quantities, such as grains or grain products (26.9%), vegetables (16.0%) and starchy roots and tubers (potatoes and potato products) (13.2%) [4].

Conventional treatment against heavy metals toxicity is based on chelation therapy using different chemical chelators that can have several adverse effects, such as kidney overload, cardiac arrest, mineral deficiency and anemia [6]. In recent years, interesting candidates for the treatment of heavy metal intoxications, including nanoparticles, probiotics, vitamins (C, E), folate, and essential amino acids have been used [7–9].

It has been widely accepted that a functional food provides both nutritional values and health benefits. Even though the concept and "functional food" term were first mentioned in Japan, in 1984 [10], it has since undergone some additional European statements: "Food products can only be considered functional if together with the basic nutritional impact it has beneficial effects on one or more functions of the human organism thus either improving the general and physical conditions or/and decreasing the risk of the evolution of diseases" [11]. Therefore, it can be accepted that functional food science emerged as a fusion between food science, nutrition, medicine, and pharmaceutics.

Selenium (Se) is a contradictory mineral, because at high levels it can become toxic for the organism, while its deficiency also produces several health problems [12]. At the same time, Se is an essential nutrient in human life as it is involved in major biochemical reactions in the body and also in the structure of many enzymes or selenoproteins that play important roles in antioxidant pathways, the endocrine and immune system, reproduction, muscle function, and tumor prevention [12,13].

Selenium is classified as a metalloid and elementally as Se, and it has different allotropic forms including rhombohedral, three deep-red monoclinic forms (α-, β-, and γ-Se), trigonal gray Se, amorphous red Se, and black vitreous Se [12]. Meat, seafood and cereals are the most important food sources of Se.

Se exists in different oxidation states including selenate (Se^{+6}), selenite (Se^{+4}), selenides (Se^{-2}) and elemental selenium (Se^0) [14]. In food products, Se occurs in combination with proteins: the main Se species includes organic Se such as Se-methyl-selenocysteine, γ-glutamyl-Se-methyl-selenocysteine and selenomethionine. Selenocysteine dominates in the products of animal origin. The main sources of Se are foods enriched in proteins, such as meat and dairy products, fish seafood, milk, and nuts. A low level of Se is found in fruits and vegetables [15].

However, the biological and toxicological effects of Se strongly depend on its chemical form, since it was accepted that its organic form is more favorable in terms of bioavailability [16]. The World Health Organization has established a value of 70 μg/day for the maximum daily intake, considering that doses above 400 μg/day may exert toxic actions [13].

On the other hand, by comparison with organic or inorganic selenium compounds, selenium nanoparticles (SeNPs) display better bioavailability, higher biological activity and lower toxicity, as demonstrated by several studies [17–19]. In addition to its antioxidant effect, its use in chemopreventive agents and anticancer drugs is also well documented [20,21], along with evidence of its antimicrobial and antifungal properties [22,23]. SeNPs can be synthesized by chemical [24], physical [25] or biological method, also known as green synthesis [26].

The biogenic synthesis of SeNPs as a green, eco-friendly approach, has attracted attention in recent years due to its low cost and simplicity, demonstrating the accumulation and biotransformation of selenium into both organic (seleno-aminoacid) and elemental forms (Se^0) by lactic bacteria [27]. Different lactic bacteria strains are able to produce SeNPs with a different size, ranging from 50–100 nm (*Streptococcus thermofilus*) and 100–200 nm (*Lactobacillus* sp.) to 400–500 nm (*Bifidobacter* sp.) [27]. Moreover, nanosized selenium in the range of 100–500 nm may help in the bio-fortification of crops (*Brassica* species), with a higher nutritional impact and health benefits, since it was demonstrated that selenium uptake by plants from the soil is strongly related to its selenium forms: elemental selenium, selenite and selenate, in association with other elements or in organic forms [28].

As the microbial transformations of selenium species with relevance to heavy metal bioremediation has been already documented [29], in this study we obtained, characterized and validated the SeNPs obtained by green synthesis using *Lactobacillus casei*. Moreover, we proposed investigating for the first time the protective effect of SeNPs and lacto-SeNPs (LSeNPs) administered orally to mice for 30 days in different concentrations (0.1, 0.2 and 0.4 mg/kg b.w.), against the toxic effects exerted by cadmium at the hepatic level. Blood biochemical parameters (transaminases, bilirubin, gamma glutamyl transferase), antioxidant enzymes (catalase and glutathione peroxidase), the antioxidant capacity of plasma along with the histology, immunohistochemistry for mitochondrial apoptosis markers (*bcl-2*, *bax*) and gene expression of hepatic inflammatory markers (NF-kB, TNFα, IL-6) were analyzed in terms of the comparative evaluation of the dose-dependent protective activity of SeNPs and LSeNPs against cadmium intoxication.

2. Materials and Methods

2.1. Biosynthesis and Characterization of SeNPs

Lactobacillus casei (Lyofast LC4P1, Sacco, Cadorago, Italy) was selected for SeNPs synthesis via a reduction route using sodium hydrogen selenite ($NaHSeO_3$) as a reducing agent, according to a protocol described by Eszenyi et al. [27]. In this study, two products were obtained: purified nanoselenium (SeNPs) and lacto-nanoselenium (LSeNPs).

MRS culture medium was inoculated with *L. casei* and sodium hydrogen selenite at a concentration of 200 mg/L in order to promote SeNPs synthesis. Instead, in order to obtain LSeNPs, the culture medium was replaced with skimmed milk. The reaction was allowed to start in a fermentation bottle during 48 h at 37 °C until the characteristic red color of the elemental nano-selenium was achieved. Then, the bacterial cells were removed from the mixture by centrifugation at 6000 rpm, for 15 min, the supernatant was discarded and the pellet was recovered in distilled water. As the mechanism of elemental Se formation is mainly intracellular for lactic acid bacteria [27], the acid digestion was performed in order to remove the bacterial cell wall. After washing, vacuum filtering and the freeze-drying procedure, the collected precipitate was characterized by TEM (transmission electron microscopy, TecnaiG2 F30 S-TWIN, FEI, Frankfurt, Germany), XRD (X-ray diffraction, Mini Flex 600, Rigaku, Tokyo, Japan, operating at 40 kV, 15 mA, with CuK α monochromatic radiation) and FT-IR spectroscopy (Fourier transform infrared spectroscopy, Spectrum BXII spectrophotometer, Perkin Elmer equipped with MIRacle ATR accessory, at scanning speed of 32 cm^{-1} and spectral width 2.0 cm^{-1}, Buckinghamshire, UK). For DLS (dynamic light scattering, ZEN 3690, Malvern Instruments, Malvern, Worcestershire, UK) and zeta potential measurements, the SeNPs powders were resuspended in distillated water and sonicated during 10 min before each measurement to prevent aggregation.

2.2. Animal and Experimental Design

Six- to eight-week-old CD1 female mice, weighing 26 $\pm$ 3 g, were used for the experiments. The mice were housed in a controlled microclimate environment, with a dark–light cycle of 12/12 h, and watering and feeding ad libitum. Mice were fed with an autoclavable standard scientific diet for rodents (Safe D40 diet, SAFE Complete Care Competence, Germany), which is certified as free of toxic substances and balanced regarding the content of amino acids, fatty acids, minerals, and vitamins.

All experimental procedures were approved by the Ethical Committee of the "Vasile Goldis" Western University of Arad and certified by the National Sanitary Veterinary and Food Safety Authority of Romania (005/02.27.2017). We chose females for the animal model, considering the larger study this work is part of which seeks to evaluate multi-organ effects, and knowing that cadmium is more retained in their body compared to males, due to estrogenic effects [30].

The mice were divided into 8 experimental groups (n = 10), as follows: Group 1 (control group), where only the vehicle was administered by gavage (water); Group 2, which received orally $CdCl_2$ (Cd group) at a dose of 5 mg/kg b.w.; Group 3 (SeNPs 0.1 group), 4 (SeNPs 0.2 group) and 5 (SeNPs 0.4 group) which were given purified SeNPs in 3 different doses: 0.1, 0.2 and 0.4 mg/kg b.w., respectively, together with 5 mg/kg b.w. of Cd for each group. Groups 6 (LSeNPs 0.1 group), 7 (LSeNPs 0.2 group) and 8 (LSeNPs 0.4 group) received LSeNPs at doses of 0.1, 0.2 and 0.4 mg/kg b.w., respectively, together with 5 mg/kg b.w. of Cd for each group. The administration of SeNPs and LSeNPs was performed one hour after the administration of Cd.

The three doses of SeNPs (0.1, 0.2, 0.4 mg/kg b.w.) and the route of administration were selected according to the results in which protection was obtained against cadmium, administered to mice at a dose of 5 mg/kg b.w. [31].

Thirty days after the first oral administration, the mice were euthanized under anesthesia with a mixture of ketamine and xylazine. Blood and liver tissues were collected for further analysis.

2.3. *Blood Biochemical Parameters*

Blood samples were collected by cardiac puncture. The samples were centrifuged at 3500 rpm for 10 min. Samples were analyzed for aspartate aminotransferase (AST), alanine aminotransferase (ALT), gamma-glutamyltransferase (GGT) and total bilirubin, levels (ChemaDiagnostica, Monsano, Italy) with a Mindray BS-120 Chemistry Analyzer (ShenzenMindray Bio-Medical Electronics Co., Ltd., Nanshan, Shenzhen, China).

2.4. *Antioxidant Enzymes Assay*

Antioxidant enzyme catalase (CAT) was determined from liver samples using a commercially available Catalase Activity Colorimetric Assay kit (Canvax Biotech, S.L., Córdoba, Spain). The absorbance of samples was measured at 570 nm, using a Tecan microplate reader. The results were expressed as nmol of H_2O_2 decomposed by catalase in 30 min reactions.

Glutathione peroxidase (GPx) was determined from liver samples using the glutathione peroxidase activity kit (Enzo, Catalog No. ADI-900-158) and the rate of decrease in the absorbance at 340 nm was directly proportional with glutathione peroxidase activity in the samples. The GPx activity was expressed as Units/mL.

2.5. *Antioxidant Capacity of Mice Plasma—FRAP Assay*

The FRAP values of mice plasma were determined according to Benzie and Strain, 1996 [32], with some modification. All reagents were prepared and used the same day. The working FRAP reagent was prepared by mixing in ratio 10:1:1 (*v/v/v*) of 300 mM acetate buffer (pH 3.6), 10 mM TPTZ (2,4,6-tripyridyl-s-triazine) in 40 mM HCl and 20 mM $FeCl_3 \cdot 6H_2O$. Briefly, a 50 µL sample was mixed with 150 µL distilled water and 1.5 mL working FRAP solution. After 30 min, the absorbance at 595 nm was measured against a reagent blank at 37 °C using The PharmaSpec Shimadzu UV–Vis 1700 (Shimadzu, Kyoto, Japan) spectrophotometer. The FRAP values were expressed as mmol $FeSO_4$/L by using different concentrations of aqueous solutions of $FeSO_4 \cdot 7H_2O$ (in the range of 50–1000 µM) for a standard curve.

2.6. Histopathology Analysis

Formalin-fixed samples of the liver were dehydrated with gradient solutions of ethanol, embedded in paraffin and then sectioned at 5 μm. Slides were stained with hematoxylin and eosin (H&E) for routine histological evaluation. Mounted sample slides were examined under an Olympus BX43 light microscope (Tokyo, Japan) and the images were captured using an XC30 CCD camera (Tokyo, Japan).

Histopathological changes were qualitatively described, and the alterations were graded (n = 10), using a modified scale from our previous studies [33]:

Grade 1: normal aspect of the hepatocytes and sinusoids; Grade 2: regularly/irregularly shaped hepatocytes with slightly dilated blood capillaries or blood congestion; Grade 3: vacuolated hepatocytes, mild dilatation of sinusoids and blood congestion, inflammatory infiltrates; Grade 4: <5% necrosis, lysis or strong inflammatory changes, large dilatations of sinusoids, increased macrophage number; Grade 5: >5% necrosis, lysis or strong inflammatory reactions.

2.7. Immunohistochemical Analysis

Immunohistochemistry analysis was performed on paraffin embedded 5 μm-thick liver sections. Liver sections were deparaffinized in Dewax (Biosystems, Nussloch, Germany) and rehydrated prior to epitope retrieval in Novocastra sol. (Leica Biosystems, Nussloch, Germany). Following the neutralization of endogenous peroxidase (3% H_2O_2), the sections were incubated at 4 °C overnight with anti *bax* and *bcl*-2 antibodies (1:100). Detection was then performed using a polymer detection system (cat. no. RE7280 K; Novolink Max Polymer Detection system) and 3,3′diaminobenzidine (DAB) as a chromogenic substrate. Nuclei were stained with hematoxylin, dehydrated in a gradient of alcohol, and mounted onto slides. The slides were examined and images were captured as described previously.

2.8. RT-PCR Analysis

Tissue samples collected for the analysis of the expression of NF-kB p65,TNF-α, IL-6 gene involved in the inflammatory process were stored in an RNA Shield solution at a temperature of −80 °C until processing. The total RNA was extracted using an SV Total RNA Isolation System extraction kit, purchased from Promega, according to the manufacturer's recommendations. The quantitative and qualitative of purified RNA evaluation was assessed spectrophotometrically, using the NanoDrop 8000 spectrophotometer produced by Thermo Scientific, Waltham, MA, USA. The conversion of the total RNA to complementary DNA was performed using 2 micrograms of total RNA and the First Strand cDNA Synthesis Kit conversion kit. To determine the quantitative expression, we used a LuminarisHiGreenqPCT Master Mix kit (Thermo Scientific, Waltham, MA, USA), low ROX, each sample being determined in triplicate. The PCR system used was Applied Biosystems 7500 Real Time PCR System (Foster City, CA, USA). Primers used were included in Table 1. The results obtained were interpreted using the 2ΔΔCT method, proposed by Livak in 2001 [34].

Table 1. Primer sequences for RT-PCR.

Target	Sense	Antisense
NF-kB 65	5′CTTGGCAACAGCACAGACC3′	5′GAGAAGTCCATGTCCGCAAT3′
TNF-α	5′CTGTAGCCCACGTCGTAGC3′	5′TTGAGATCCATGCCGTTG3′
IL-6	5′AAAGAGTTGTGCAATGGCAATTCT3′	5′AAGTGCATCATCGTTGTTCATACA3′
GAPDH	5′CGACTTCAACAGCAACTCCCACTCTTCC3′	5′TGGGTGGTCCAGGGTTTCTTACTCCTT3′

2.9. Statistical Analysis

Data were statistically processed using GraphPad Prism 3.03 software (GraphPad Software, Inc., La Jolla, CA, USA), and one-way analysis of variance, followed by a Bonferroni test. $p < 0.05$ was considered to indicate a statistically significant difference.

The data expressed as mean ± standard deviation (SD) and statistically significant differences (* $p < 0.05$ and # $p < 0.05$) were determined compared with control (group 1) and the cadmium group (group 2), respectively.

3. Results and Discussion

3.1. Physico–Chemical Characterization of SeNPs

The morphology of SeNPS was evidenced by TEM as presented in Figure 1, showing homogenous and spherical shape particles, with a diameter of a maximum 80 nm, as confirmed by DLS. Zeta potential measurement indicated −22 mV with a polydispersity index less than 0.2, indicating a good stability.

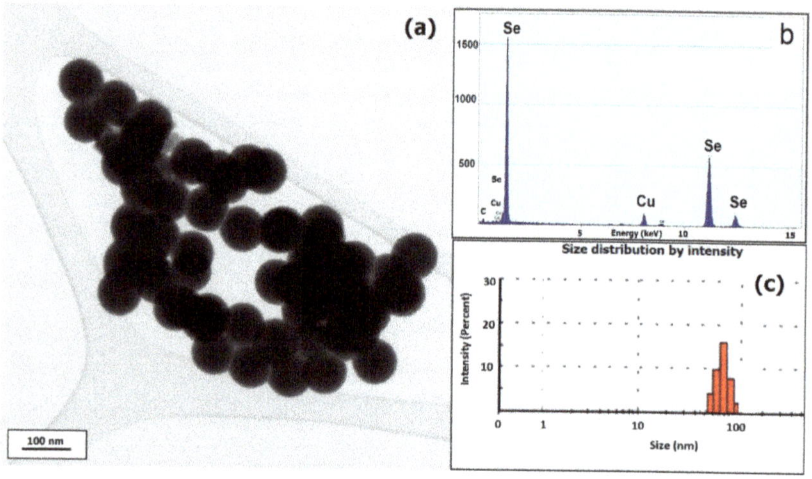

Figure 1. (**a**) TEM micrograph of SeNPs synthesized using *L. casei* and $NaHSeO_3$ as a reducing agent; (**b**) energy dispersive X-ray analysis; and (**c**) particles size distribution confirmed by DLS.

A preliminary confirmation of SeNPs formation was achieved by plasmon resonance in UV–Vis spectroscopy, observing the maximum adsorption at 270 nm (Figure 2a). This result is in agreement with some previous papers [35,36]. The SeNPs' production using different lactic acid bacteria show unique structured nanospheres with regular and uniform size, comparative with chemical synthesis [27,37,38]. To date, different microorganisms (*Lactobacillus* sp. *Bifidobacter* sp. *Streptococcus thermophilus*) are able to reduce inorganic selenium (selenite and selenite oxoanions) into SeNPs in different size, ranging from 50 to 500 nm [24]. The conversion of selenite to SeNPs by microorganism involves different mechanisms, where the reductase enzymes play an important role [29]. The biosynthesis of SeNPs by microorganisms is a cheaper and faster process, where high purity selenium spheres should be produced. Furthermore, the probiotic bacteria and SeNPs are safe for clinical administration and human consumption [24].

In order to determine the functional groups that exist on the surface of SeNPs, FTIR vibrational features were obtained and presented in Figure 2b. The main vibrational bands in the high wavenumber region are 3300 cm^{-1}, 2920 cm^{-1} and 2850 cm^{-1} corresponding to the stretching vibration of OH groups, aliphatic C–H and carboxylic acid O–H groups, respectively. In the low wavenumber region, 1720 cm^{-1} is assigned to carbonyl C=O stretch, 1658 cm^{-1} to amide I vibration, 1540 cm^{-1} to amide II and 1230 cm^{-1} to amide III vibrations. The most intense vibrational band at 1040 cm^{-1} represents the characteristic of the Se–O bond stretching, according to previously reported data in the literature [36,39]. These results indicated the presence of both proteins and carbohydrates on the surface of nanoparticles, deriving from the cell membrane, which may improve the long-term stability and protection of the core-particle. The XRD pattern of the Se nanoparticles (Figure 2c) shows a broad peak at about $2\theta = 22°$ suggesting the amorphous structure of SeNPs, in agreement with previous data [24,35,36], which demonstrated that stable amorphous forms (or even low crystallinity) are advantageous for biological applications, as they exhibit better solubility and subsequent adsorption and bioavailability.

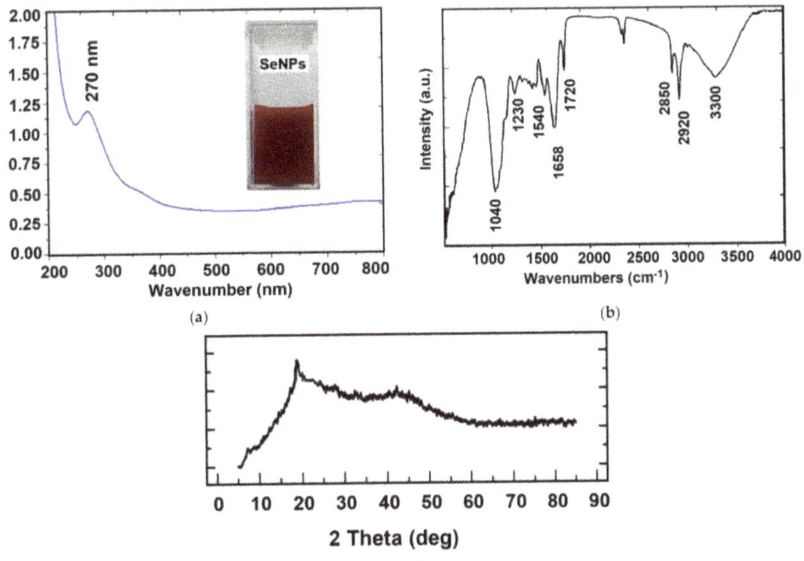

Figure 2. Physico–chemical characterization of SeNPs by UV–Vis spectroscopy (**a**); FTIR spectroscopy (**b**); and XRD pattern (**c**).

The bioconversion of the inorganic Se^{+4} into elemental selenium form by *Lactobacilus casei* was performed in this study to decrease its toxicity and to provide highly bioavailable forms, which can be further used in functional foods. There are some studies which provide data regarding the conversion of inorganic to organic selenium, such as on cabbage and yeast which have been applied for the industrial production of selenium–methionine from inorganic selenium [40]. Moreover, a previous study demonstrated that some anaerobic or semi-anaerobic bacteria can convert the Se^{+4} into elemental selenium ranging between 100 and 500 nm [27,41]. The biological activity of SeNPs depends on their size, meaning that the smaller particles have a higher activity. The SeNPs is selenium in the zero-oxidation state that presents a lower toxicity and very good bioavailability compared to other oxidation states (Se^{+4}, Se^{+6}) [42].

3.2. *Effect of SeNPs on Blood Biochemical Parameters*

The results of biochemical parameters are shown in Table 2. Blood alanine transaminase (ALT) and aspartate transaminase (AST) activity increased after Cd administration in mice (group 2) by 37.92 and 33.40%, respectively. The treatment with SeNPs or LSeNPs significantly decreased the activity of these enzymes compared to group 2, but not for group 3, where the lowest dose of SeNPs (0.1 mg/kg b.w.) was used. The GGT activity was significantly decreased only for group 3. The total bilirubin level was not affected by the treatment of both forms of SeNPs.

Table 2. Effects of cadmium and selenium on biochemical parameters after 30 days treatments [1].

Biochemical Parameters	Group 1 (Control)	Group 2 (Cd)	Group 3 (Cd + 0.1SeNPs)	Group 4 (Cd + 0.2SeNPs)	Group 5 (Cd + 0.4SeNPs)	Group 6 (Cd + 0.1LSeNPs)	Group 7 (Cd + 0.2LSeNPs)	Group 8 (Cd + 0.4LSeNPs)
AST (U/L)	106.91 ± 16.50	142.62 ± 42.39	131.19 ± 42.44	74.71 ± 4.36 $^{###}$	74.20 ± 12.19 $^{###}$	79.09 ± 10.71 $^{###}$	88.66 ± 30.46 $^{##}$	78.60 ± 16.61 $^{###}$
ALT (U/L)	57.15 ± 5.89	78.82 ± 14.17	72.32 ± 17.23	38.62 ± 8.42 $^{###}$	35.74 ± 11.97 $^{###}$	34.16 ± 7.73 $^{*,###}$	35.53 ± 10.91 $^{*,###}$	33.43 ± 7.16 $^{*,###}$
GGT (U/L)	2.37 ± 0.31	2.62 ± 0.54	1.35 ± 0.71 $^{*,##}$	2.5 ± 0.10	2.35 ± 0.42	2.02 ± 0.63	2.40 ± 0.73	2.96 ± 0.23
Total bilirubin (mg/dL)	0.19 ± 0.11	0.15 ± 0.04	0.22 ± 0.07	0.22 ± 0.06	0.19 ± 0.07	0.16 ± 0.05	0.15 ± 0.04	0.099 ± 0.07

[1] All values were expressed as the mean ± SD for 10 mice in each group. Groups: Cd + SeNPs and Cd + LSeNPs at different concentration vs. control group: * $p < 0.05$. Groups: Cd + SeNPs and Cd + LSeNPs at different concentration vs. cadmium group: ## $p < 0.01$; ### $p < 0.001$. Groups: 1—control; 2—$CdCl_2$; 3—$CdCl_2$ + 0.1 mg/kg SeNPs; 4—$CdCl_2$ + 0.2 mg/kg SeNPs; 5—$CdCl_2$ + 0.4 mg/kg SeNPs; 6—$CdCl_2$ + 0.1 mg/kg LSeNPs; 7—$CdCl_2$ + 0.2 mg/kg LSeNPs; 8—$CdCl_2$ + 0.4 mg/kg LSeNPs. In the groups 2 to 8, orally $CdCl_2$ at a dose of 5 mg/kg b.w. was administrated.

Plasma transaminases are known to be important indicators for assessing the health of liver tissue. Cd leads to an increase in blood transaminase enzymes by transferring them to the blood and destroying membrane permeability due to lipid oxidation [43,44].

As expected, elevated levels of hepatic enzymes (AST and ALT) were recorded after Cd administration compared to the control, which is in line with other similar studies [45–47]. Co-administration of Cd and both the form of SeNPs showed that liver markers decreased significantly in a dose-dependent manner compared to the metal-group and highlights the protective effect of both forms of SeNPs against the toxic injuries induced by the metal. The best effects were obtained with LSeNPs.

3.3. *Effect of SeNPs on Antioxidant Enzyme Assay*

Cadmium is absorbed from the gastrointestinal tract and is mainly accumulated in liver and kidney where it is bounded to metallothionein (MT) and provides de novo protection against Cd. When it exceeds the binding capability of MT due to a higher exposure to metal, the non-bounded Cd ions cause hepato- and nephrotoxicity [46]. Moreover, the fact that cadmium stimulates free radical production, resulting in oxidative changes of lipids, proteins and DNA, may in turn suppress hepatic and renal functions and initiate different pathologies [47].

Catalase and glutathione peroxidase are included in the enzymatic antioxidant defense system that protects cells against reactive oxygen species toxicity and lipid peroxidation. Catalase cleaves hydrogen peroxide into water and oxygen [48]. In the case of Cd intoxication, the activity of this enzyme is decreased [43]. Moreover, glutathione peroxidase (GPx) catalyzes the reduction of hydrogen peroxide (or a variety of organic hydroperoxides) with reduced gluthatione to form gluthathione disulfide (GSSG). The low GPx activity is one of the consequences of oxidative stress [49].

Our results showed that Cd alone induced a significant reduction in catalase activity (Figure 3A) compared to the control. After the combination of Cd and SeNPs (groups 3–5), an improvement in catalase activity was recorded compared with cadmium group in a dose-dependent manner. Instead, using LSeNPs at a high dose (4 mg/kg b.w.) results a significantly increased in catalase activity comparative with both the control and cadmium groups. This result demonstrates the beneficial effect of SeNPs in combination with probiotics against Cd toxicity. On the other hand, the treatment with Cd (group 2) showed a significant decrease in GPx activity ($p < 0.001$) compared with the control group (Figure 3B). The treatments using a combination of Cd and SeNPs or LSeNPs at different concentrations also gave a significant decrease in enzyme activity.

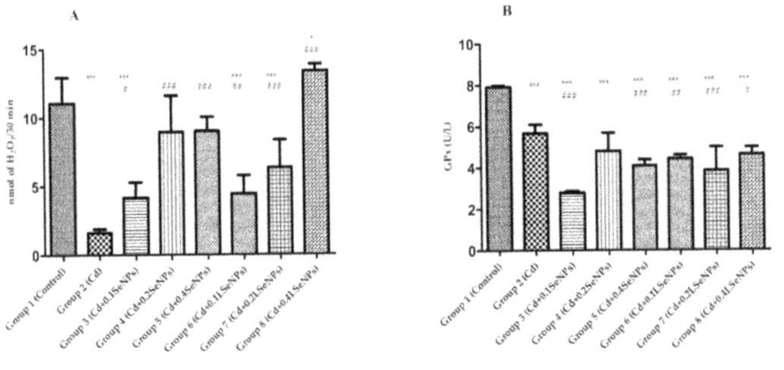

Figure 3. Effects of cadmium and selenium on catalase (**A**) and GPx (**B**) after 30 days of treatment. All values were expressed as the mean ± SD for 10 mice in each group. SeNPs+Cd and LSeNPs+Cd at different concentrations vs. control group: * $p < 0.05$; *** $p < 0.001$. SeNPs+Cd and LSeNPs+Cd at a different concentration vs. the cadmium group: # $p < 0.05$; ## $p < 0.01$; ### $p < 0.001$. Groups: 1—control; 2—$CdCl_2$; 3—$CdCl_2$ + 0.1 mg/kg SeNPs; 4—$CdCl_2$ + 0.2 mg/kg SeNPs; 5—$CdCl_2$ + 0.4 mg/kg SeNPs; 6—$CdCl_2$ + 0.1 mg/kg LSeNPs; 7—$CdCl_2$ + 0.2 mg/kg LSeNPs; 8—$CdCl_2$ + 0.4 mg/kg LSeNPs.

The significant decrease in the antioxidant enzymes after exposure to Cd was similar to that of El-Boshy et al. [48] in rats for catalase and GPx, while the sodium selenite-treated group showed significantly increased CAT and GPx activities. The combination of sodium selenite with cadmium enhanced the antioxidant activities of CAT and GPx and ameliorated the cadmium-induced liver damage by improving hepatic markers [48].

The effects of cadmium and sodium selenite on antioxidant enzymes in the liver, kidneys and testes of rats were investigated by Dzobo et al., 2013 [50]. Their results showed that in the liver, Cd treatment resulted in decreased catalase activity, while the Se treatment resulted in increased catalase activity. Instead, the co-treatment of Cd and Se resulted in an increase in GPx.

The major damage caused by toxic metals is due to the production of free radicals, which induce oxidative stress in cells, and cause damages. Oxidative stress is produced when the balance between the antioxidant system and ROS is in favor of free radicals [43].

The antioxidant capacity of selenium could be attributed to its presence in GPx or thioredoxin reductase [42,51]. The protective effect of selenium against cadmium-induced tissue damage could be attributed to its antioxidant activity through the enhancement of antioxidant enzymes from tissues.

To our knowledge, this study was one of the first which demonstrated the hepatoprotective activity of elemental selenium in the form of nanoparticles against toxicity induced by cadmium due to enhanced antioxidant enzyme activities. Several mechanisms may be involved in the protection of selenium. One hypothesis is to change the absorption of cadmium and its distribution in the body and target organs. Another hypothesis takes into account that selenium (in inorganic or organic form), is well known for its ability to eliminate ROS and improve the antioxidant system damaged caused by Cd [48].

3.4. Antioxidant Capacity of Mice Plasma

FRAP values are shown in Figure 4. The treatments with Cd and combined Cd+SeNPs did not significantly differ between groups vs. control group. The highest FRAP value was recorded for group 7 that received combination LSeNPs and Cd. This result highlights the synergistic effect between SeNPs and probiotic bacteria regarding the protective effect against Cd toxicity.

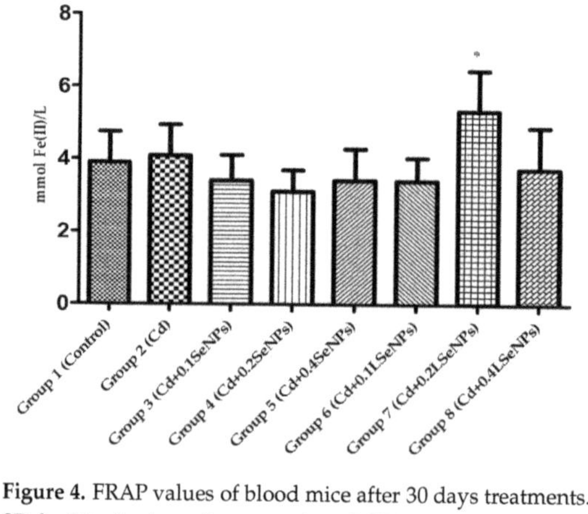

Figure 4. FRAP values of blood mice after 30 days treatments. All values were expressed as mean ± SD for 10 mice in each group. * $p < 0.05$ is statistically significant difference of sample vs. control group. Groups: 1—control; 2—$CdCl_2$; 3—$CdCl_2$ + 0.1 mg/kg SeNPs; 4—$CdCl_2$ + 0.2 mg/kg SeNPs; 5—$CdCl_2$ + 0.4 mg/kg SeNPs; 6—$CdCl_2$ + 0.1 mg/kg LSeNPs; 7—$CdCl_2$ + 0.2 mg/kg LSeNPs; 8—$CdCl_2$ + 0.4 mg/kg LSeNPs.

In one study [52], red elemental SeNPs (within a size range of 80–220 nm) were biosynthesized using the marine strain of *Bacillus* sp. and in vitro investigated in terms of antioxidant activity by DPPH and the reduction power assay. SeNPs revealed moderate antioxidant activity compared to SeO_2 and BHT due to the fact that elemental selenium is insoluble.

SeNPs have high antioxidant activity compared with other chemical forms of Se. Wang et al. [42] showed the antioxidant activity of SeNPs that demonstrated lower toxicity compared with selenomethionine.

3.5. Histopathology Analysis

The liver is an important organ for metabolism, detoxification, storage, and the excretion of xenobiotics or metabolites, which is vulnerable to injury. As the liver is an important target of cadmium [53], we assessed the structural changes by microscopic analysis.

Light microscopic examination showed a normal structure of the liver (Figure 5, Group 1) in the controls. Exposure to Cd induced degenerative changes in the liver, mainly consisting of focal hepatocyte necrosis and lysis, pycnotic nuclei with condensed chromatin, as well as sinusoidal congestion and parenchyma infiltration by mononuclear cells (Figure 5, Group 2). The co-administration of SeNPs practically prevented the changes in the liver structure in a dose-dependent manner. The best protection was obtained with LSeNPs, where we noticed the presence of rare inflammatory cells in the sinusoids (Figure 5, Groups 6–8). Moreover, histomorphometric evaluation demonstrated that the dose of 0.2 mg/kg and 0.4 mg/kg induced significantly fewer liver structural changes compared to the liver intoxicated with cadmium alone ($p < 0.001$; Figure 6).

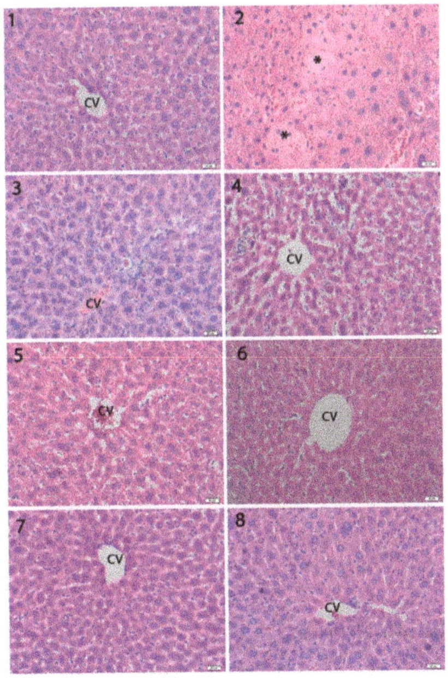

Figure 5. The histopathological sections of mice livers of the experimental groups, H&E barr 20 μm. Various degrees of histopathological changes were observed in Cd exposure groups, mainly including focal hepatocyte necrosis/lysis (*) and degeneration. The co-administration of SeNPs with Cd reduced the metal toxic histological changes in the liver in a dose-dependent manner, which was more obvious for LSeNPs. CV-centrilobular vein. Groups: 1—control; 2—$CdCl_2$; 3—$CdCl_2$ + 0.1 mg/kg SeNPs; 4—$CdCl_2$ + 0.2 mg/kg SeNPs; 5—$CdCl_2$ + 0.4 mg/kg SeNPs; 6—$CdCl_2$ + 0.1 mg/kg LSeNPs; 7—$CdCl_2$ + 0.2 mg/kg LSeNPs; 8—$CdCl_2$ + 0.4 mg/kg LSeNPs.

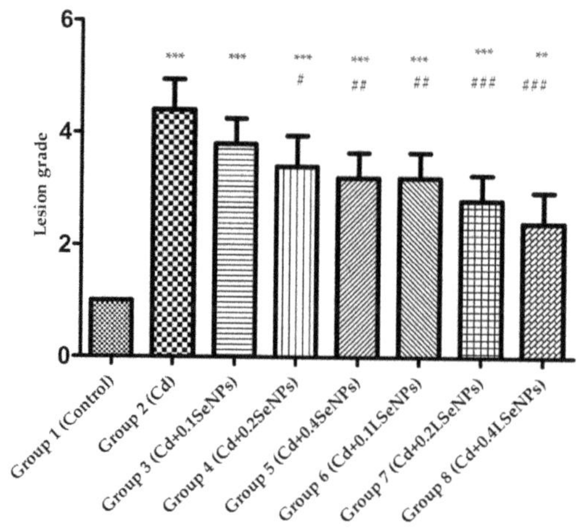

Figure 6. Semi-quantitative assessment of the severity of liver damage ranked from 1 (control status) to 5 (liver damage) induced by cadmium under the protection of SeNPs and LSeNPs. Groups: 1—control; 2—$CdCl_2$; 3—$CdCl_2$ + 0.1 mg/kg SeNPs; 4—$CdCl_2$ + 0.2 mg/kg SeNPs; 5—$CdCl_2$ + 0.4 mg/kg SeNPs; 6—$CdCl_2$ + 0.1 mg/kg LSeNPs; 7—$CdCl_2$ + 0.2 mg/kg LSeNPs; 8—$CdCl_2$ + 0.4 mg/kg LSeNPs. ** $p < 0.01$; *** $p < 0.001$ compared to control group (group 1); # $p < 0.05$; ## $p < 0.01$; ### $p < 0.001$ compared to Cd group (group2).

Regarding the antagonistic effect of SeNPs and LSeNPs, our results suggest that Se in both forms was able to reduce the hepatotoxicity of cadmium in a dose-dependent manner (Figures 5 and 6). The present study demonstrated that co-treatment with SeNPs ameliorated the histopathological damage induced by Cd in the tissues of liver by reducing the toxicity comparative to control group [54].

3.6. The SeNPs Prevent Apoptosis in Liver Parenchyma Induced by Cadmium

Exposure to Cd causes the activation of multiple death signals in parallel, including the activation of apoptosis-related mitochondrial signaling and DNA damage response [55]. Therefore, the members of the *bcl-2* protein family known to regulate the release of apoptosis-activating factors by changing ratio of *bcl-2* to *bax* which determines cell survival or cell death [56], are valuable tools to assess the hepatic protective effects of SeNPs and LSeNPs against the pro-apoptotic activity of Cd.

The immunohistochemical analysis of the pro-apoptotic *bax* and anti-apoptotic *bcl-2* markers revealed a marked expression of *bax* for the liver (Figure 7) exposed to Cd and a dose-dependent decrease in the groups receiving oral SeNPs and LSeNPs. The reduction in *bax* expression is more significant for LSeNPs. In contrast, the immunopositivity for *bcl-2* was significantly reduced in the livers of the Cd group and restored for SeNPs groups in dose-dependent manner, especially for LSeNPs.

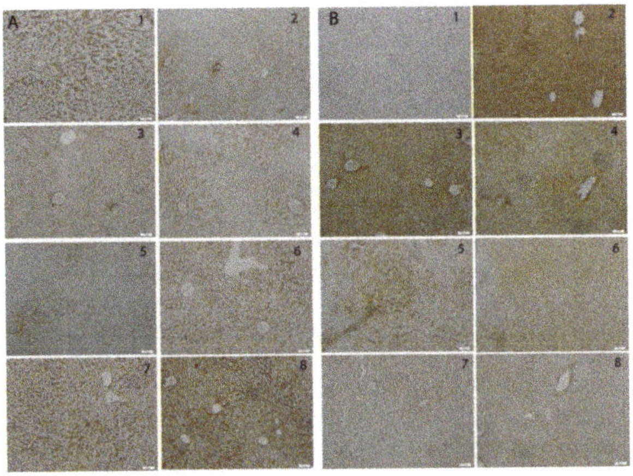

Figure 7. Hepatic immunohistochemical expression of *bcl-2* (**A**) and *bax* (**B**) in: 1—control; 2—$CdCl_2$; 3—$CdCl_2$ + 0.1 mg/kg SeNPs; 4—$CdCl_2$ + 0.2 mg/kg SeNPs; 5—$CdCl_2$ + 0.4 mg/kg SeNPs; 6—$CdCl_2$ + 0.1 mg/kg LSeNPs; 7—$CdCl_2$ + 0.2 mg/kg LSeNPs; 8—$CdCl_2$ + 0.4 mg/kg LSeNPs.

3.7. The SeNPs Prevent Inflammation in Liver Parenchyma Induced by Cadmium

To investigate the anti-inflammatory ability of SeNPs and LSeNPs, we measured the mRNA expression for the NF-kB, a key transcriptional regulator of the inflammatory response [57], and the major upregulated proinflammatory cytokines in the liver.

The level of gene expression of liver TNF-α, IL-6, NF-kB p65 was significantly increased compared to the control in the group in which Cd was administered (Figure 8), as previously described [58,59]. The co-administration of Cd with SeNPs or LSeNPs resulted in decreased gene expression, which was directly correlated with the concentration of both SeNPs forms.

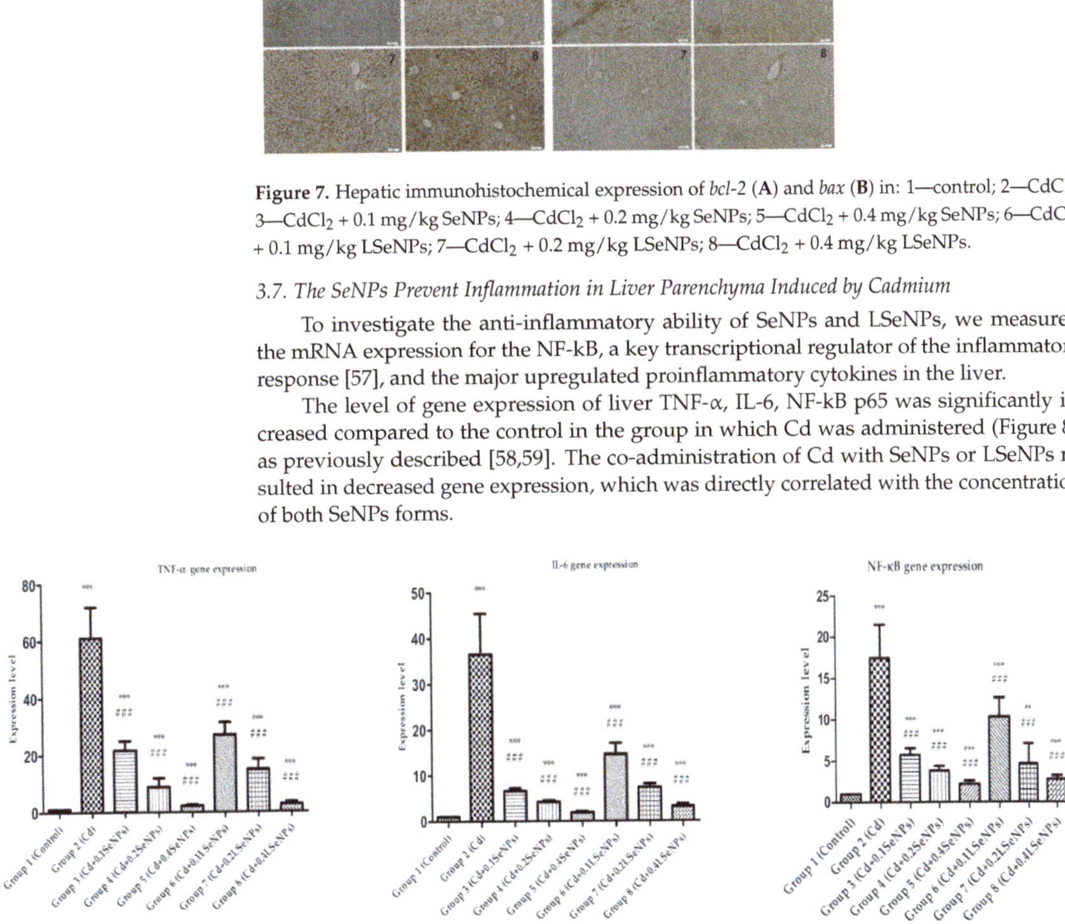

Figure 8. Hepatic TNF-α, IL-6, NF-kB p65 gene expressions. *** $p < 0.001$, ** $p < 0.01$ compared to control, ### $p < 0.001$ compared to Cd group. Groups: 1—control; 2—$CdCl_2$; 3—$CdCl_2$ + 0.1 mg/kg SeNPs; 4—$CdCl_2$ + 0.2 mg/kg SeNPs; 5—$CdCl_2$ + 0.4 mg/kg SeNPs; 6—$CdCl_2$ + 0.1 mg/kg LSeNPs; 7—$CdCl_2$ + 0.2 mg/kg LSeNPs; 8—$CdCl_2$ + 0.4 mg/kg LSeNPs.

4. Conclusions

This study proposed a new functional food based on SeNPs obtained by green synthesis using *Lactobacillus casei*, and investigated for the first time its ability to annihilate hepatic toxicity induced by cadmium, as a solution for liver injury bioremediation. Two forms of elemental SeNPs, purified SeNPs and lactic acid bacteria (*L. casei*) together with endogenous

SeNPs (called LSeNPs), were tested in Cd-induced liver toxicity to mice. Co-administration of Cd and both forms of SeNPs showed that the blood transaminases decreased significantly in a dose-dependent manner. In addition, LSeNPs at the highest dose (0.4 mg/kg b.w.) significantly increased the catalase activity comparative to the control and cadmium group. The histopathological damage induced by Cd in the mouse liver was ameliorated upon the co-treatment with both forms of SeNPs. Immunohistochemical analysis revealed a reduction in pro-apoptotic *bax* and an increase in anti-apoptotic *bcl-2* expression, especially for LSeNPs. Additionally, the co-administration of Cd with both forms of SeNPs significantly decreased the gene expression of liver inflammatory markers, with the best effects for LSeNPs. Overall, the best hepatoprotective effects were obtained for LSeNPs. A follow-up study will highlight the gender differences in the hepatoprotective effects of SeNPs on Cd-induced liver toxicity to mice, knowing that males are more susceptible to cadmium-induced hepatotoxicity than females [60,61].

A functional food that includes both probiotic bacteria and elemental SeNPs could be successfully used to annihilate Cd-induced liver toxicity, and to improve both nutritional values and health benefits. In this way, a possible new technology is provided for the food industry, the production of yogurt enriched with selenium nanoparticles produced by lactic acid bacteria with protective effects against heavy metals.

Author Contributions: Conceptualization, S.I.V., A.H., J.P.; methodology, S.I.V., V.L., A.V.T., C.B., H.H., A.C., S.G., M.R., B.M., L.C., S.C., A.H., C.C.; formal analysis, S.I.V., V.L., A.V.T., C.B., H.H., A.C., S.G., M.R., B.M., L.C., C.C.; investigation, M.P., E.M.; resources, S.I.V., A.H.; writing—original draft preparation, S.I.V., S.C., A.H.; writing—review and editing, S.I.V., S.C., A.H. All authors have read and agreed to the published version of the manuscript.

Funding: This research was funded by Romanian Ministry of Research and Innovation, project Number PN-III-P2-2.1-PED-2016-1846.

Institutional Review Board Statement: The study was conducted according to the guidelines of the Declaration of Helsinki and approved by the Ethics Committee of Vasile Goldis Western University of Arad (Certificate No. 71/07.06.2016).

Informed Consent Statement: Not applicable.

Data Availability Statement: The data presented in this study are available on request from the corresponding author. The data are not publicly available due to privacy.

Conflicts of Interest: The authors declare no conflict of interest.

References

1. Cavalu, S.; Popa, A.; Bratu, I.; Borodi, G.; Maghiar, A. New Evidences of Key Factors Involved in "Silent Stones" Etiopathogenesis and Trace Elements: Microscopic, Spectroscopic, and Biochemical Approach. *Biol. Trace Elem. Res.* **2015**, *168*, 311–320. [CrossRef]
2. Jan, A.T.; Azam, M.; Siddiqui, K.; Ali, A.; Choi, I.; Haq, Q.M.R. Heavy Metals and Human Health: Mechanistic Insight into Toxicity and Counter Defense System of Antioxidants. *Int. J. Mol. Sci.* **2015**, *16*, 29592–29630. [CrossRef]
3. Satarug, S. Dietary Cadmium Intake and Its Effects on Kidneys. *Toxics* **2018**, *6*, 15. [CrossRef] [PubMed]
4. European Food Safety Authority (EFSA). Cadmium dietary exposure in the European population. *EFSA J.* **2012**, *10*, 2551. [CrossRef]
5. European Food Safety Authority (EFSA). Statement on tolerable weekly intake for cadmium. *EFSA J.* **2011**, *9*. [CrossRef]
6. Flora, S.J.; Pachauri, V. Chelation in Metal Intoxication. *Int. J. Environ. Res. Public Health* **2010**, *7*, 2745–2788. [CrossRef]
7. Inbaraj, B.S.; Chen, B.-H. In Vitro Removal of Toxic Heavy Metals by Poly(ϒ-Glutamic Acid)-Coated Superparamagnetic Nanoparticles. *Int. J. Nanomed.* **2012**, *7*, 4419–4432.
8. Yang, J.; Hou, B.; Wang, J.; Tian, B.; Bi, J.; Wang, N.; Li, X.; Huang, X. Nanomaterials for the Removal of Heavy Metals from Wastewater. *Nanomaterials* **2019**, *9*, 424. [CrossRef] [PubMed]
9. Duan, H.; Yu, L.; Tian, F.; Zhai, Q.; Fan, L.; Chen, W. Gut microbiota: A target for heavy metal toxicity and a probiotic protective strategy. *Sci. Total Environ.* **2020**, *742*, 140429. [CrossRef] [PubMed]
10. Hardy, G. Nutraceuticals and functional foods: Introduction and meaning. *Nutrition* **2000**, *16*, 688–689. [CrossRef]
11. Martirosyan, D.M.; Singh, J. A new definition of functional food by FFC: What makes a new definition unique? *Funct. Foods Health Dis.* **2015**, *5*, 209–223. [CrossRef]
12. El-Ramady, H.R.; Domokos-Szabolcsy, É.; Abdalla, N.A.; Alshaal, T.A.; Shalaby, T.A.; Sztrik, A.; Prokisch, J.; Fári, M. Selenium and nano-selenium in agroecosystems. *Environ. Chem. Lett.* **2014**, *12*, 495–510. [CrossRef]

13. Kieliszek, M.; Błażejak, S. Current Knowledge on the Importance of Selenium in Food for Living Organisms: A Review. *Molecules* **2016**, *21*, 609. [CrossRef]
14. He, Y.; Chen, S.; Liu, Z.; Cheng, C.; Li, H.; Wang, M. Toxicity of Selenium Nanoparticles in Male Sprague–Dawley Rats at Supranutritional and Nonlethal Levels. *Life Sci.* **2014**, *115*, 44–51. [CrossRef]
15. Kieliszek, M. Selenium–Fascinating Microelement, Properties and Sources in Food. *Molecules* **2019**, *24*, 1298. [CrossRef] [PubMed]
16. Adadi, P.; Barakova, N.V.; Muravyov, K.Y.; Krivoshapkina, E.F. Designing selenium functional foods and beverages: A review. *Food Res. Int.* **2019**, *120*, 708–725. [CrossRef]
17. Zhang, J.; Wang, X.; Xu, T. Elemental Selenium at Nano Size (Nano-Se) as a Potential Chemopreventive Agent with Reduced Risk of Selenium Toxicity: Comparison with Se-Methylselenocysteine in Mice. *Toxicol. Sci.* **2007**, *101*, 22–31. [CrossRef] [PubMed]
18. Yazdi, M.H.; Mahdavi, M.; Varastehmoradi, B.; Faramarzi, M.A.; Shahverdi, A.R. The Immunostimulatory Effect of Biogenic Selenium Nanoparticles on the 4T1 Breast Cancer Model: An In Vivo Study. *Biol. Trace Elem. Res.* **2012**, *149*, 22–28. [CrossRef]
19. Hassan, C.E.; Webster, T.J. The effect of red-allotrope selenium nanoparticles on head and neck squamous cell viability and growth. *Int. J. Nanomed.* **2016**, *11*, 3641–3654. [CrossRef]
20. Chen, T.; Wong, Y.-S.; Zheng, W.; Bai, Y.; Huang, L. Selenium nanoparticles fabricated in Undaria pinnatifida polysaccharide solutions induce mitochondria-mediated apoptosis in A375 human melanoma cells. *Colloids Surf. B Biointerfaces* **2008**, *67*, 26–31. [CrossRef]
21. Vekariya, K.K.; Kaur, J.; Tikoo, K. ERα signaling imparts chemotherapeutic selectivity to selenium nanoparticles in breast cancer. *Nanomedicine* **2012**, *8*, 1125–1132. [CrossRef]
22. Piacenza, E.; Presentato, A.; Zonaro, E.; Lemire, J.A.; Demeter, M.; Vallini, G.; Turner, R.J.; Lampis, S. Antimicrobial activity of biogenically produced spherical Se-nanomaterials embedded in organic material against Pseudomonas aeruginosaand Staphylococcus aureusstrains on hydroxyapatite-coated surfaces. *Microb. Biotechnol.* **2017**, *10*, 804–818. [CrossRef]
23. Kheradmand, E.; Rafii, F.; Yazdi, M.H.; Sepahi, A.A.; Shahverdi, A.R.; Oveisi, M.R. The antimicrobial effects of selenium nanoparticle-enriched probiotics and their fermented broth against Candida albicans. *DARU J. Pharm. Sci.* **2014**, *22*, 48. [CrossRef]
24. Cavalu, S.; Kamel, E.; Laslo, V.; Fritea, L.; Costea, T.; Antoniac, I.V.; Vasile, E.; Antoniac, A.; Semenescu, A.; Mohan, A.; et al. Eco-friendly, Facile and Rapid Way for Synthesis of Selenium Nanoparticles Production, structural and morphological characterisation. *Rev. Chim.* **2018**, *68*, 2963–2966. [CrossRef]
25. Quintana, M.; Haro-Poniatowski, E.; Morales, J.; Batina, N. Synthesis of selenium nanoparticles by pulsed laser ablation. *Appl. Surf. Sci.* **2002**, *195*, 175–186. [CrossRef]
26. Ingale, A.G.; Chaudhari, A.N. Biogenic Synthesis of Nanoparticles and Potential Applications: An Eco-Friendly Approach. *J. Nanomed. Nanotechnol.* **2013**, *4*, 1–8. [CrossRef]
27. Eszenyi, P.; Sztrik, A.; Babka, B.; Prokisch, J. Elemental, Nano-Sized (100–500 Nm) Selenium Production by Probiotic Lactic Acid Bacteria. *Int. J. Biosci. Biochem. Bioinform.* **2011**, *1*, 148–152. [CrossRef]
28. Vicas, S.I.; Cavalu, S.; Laslo, V.; Tocai, M.; Costea, T.O.; Moldovan, L. Growth, Photosynthetic Pigments, Phenolic, Glucosinolates Content and Antioxidant Capacity of Broccoli Sprouts in Response to Nanoselenium Particles Supply. *Not. Bot. Horti Agrobot.* **2019**, *47*, 821–828. [CrossRef]
29. Eswayah, A.S.; Smith, T.J.; Gardiner, P.H.E. Microbial Transformations of Selenium Species of Relevance to Bioremediation. *Appl. Environ. Microbiol.* **2016**, *82*, 4848–4859. [CrossRef]
30. Vahter, M.; Åkesson, A.; Lidén, C.; Ceccatelli, S.; Berglund, M. Gender differences in the disposition and toxicity of metals. *Environ. Res.* **2007**, *104*, 85–95. [CrossRef] [PubMed]
31. Ren, X.-M.; Wang, G.-G.; Xu, N.-Q.; Luo, K.; Liu, Y.-X.; Zhong, Y.-H.; Cai, Y.-Q. The protection of selenium on cadmium-induced inhibition of spermatogenesis via activating testosterone synthesis in mice. *Food Chem. Toxicol.* **2012**, *50*, 3521–3529. [CrossRef]
32. Benzie, I.F.; Strain, J.J. The Ferric Reducing Ability of Plasma (FRAP) as a Measure of Bantioxidant PowerQ the FRAP Assay. *Anal. Biochem.* **1996**, *239*, 70–76. [CrossRef]
33. Hermenean, A.; Damache, G.; Albu, P.; Ardelean, A.; Ardelean, G.; Ardelean, D.P.; Horge, M.; Nagy, T.; Braun, M.; Zsuga, M.; et al. Histopatological alterations and oxidative stress in liver and kidney of Leuciscus cephalus following exposure to heavy metals in the Tur River, North Western Romania. *Ecotoxicol. Environ. Saf.* **2015**, *119*, 198–205. [CrossRef]
34. Livak, K.J.; Schmittgen, T.D. Analysis of Relative Gene Expression Data Using Real-Time Quantitative PCR and the 2(-Delta Delta C(T)) Method. *Methods* **2001**, *25*, 402–408. [CrossRef]
35. Cavalu, S.; Prokisch, J.; Laslo, V.; Vicas, S. Preparation, structural characterisation and release study of novel hybrid microspheres entrapping nanoselenium, produced by green synthesis. *IET Nanobiotechnol.* **2017**, *11*, 426–432. [CrossRef]
36. Gunti, L.; Dass, R.S.; Kalagatur, N.K. Phytofabrication of Selenium Nanoparticles from Emblica officinalis Fruit Extract and Exploring Its Biopotential Applications: Antioxidant, Antimicrobial, and Biocompatibility. *Front. Microbiol.* **2019**, *10*, 931. [CrossRef] [PubMed]
37. Prokisch, J.; Zommara, M. Process for Producing Elemental Selenium Nanospheres. U.S. Patent 8,003,071 B2, 23 August 2011.
38. Cavalu, S.; Antoniac, I.V.; Fritea, L.; Mates, I.M.; Milea, C.; Laslo, V.; Vicas, S.; Mohan, A. Surface modifications of the titanium mesh for cranioplasty using selenium nanoparticles coating. *J. Adhes. Sci. Technol.* **2018**, *32*, 2509–2522. [CrossRef]
39. Kannan, S.; Mohanraj, K.; Prabhu, K.; Barathan, S.; Sivakumar, G. Synthesis of selenium nanorods with assistance of biomolecule. *Bull. Mater. Sci.* **2014**, *37*, 1631–1635. [CrossRef]

40. Rayman, M.P. The use of high-selenium yeast to raise selenium status: How does it measure up? *Br. J. Nutr.* **2004**, *92*, 557–573. [CrossRef]
41. Prokisch, J.; Széles, E.; Kovács, B.; Daróczy, L.; Zommara, M. Formation of Metal Selenium Nanospheres in Bacteria: Is It a Possible Detoxification Mechanism? In Proceedings of the VII Alps-Adria Scientific Workshop, Stara Lesna, Slovakia, 28 April–2 May 2008; Volume 36, pp. 947–952. [CrossRef]
42. Wang, H.; Zhang, J.; Yu, H. Elemental selenium at nano size possesses lower toxicity without compromising the fundamental effect on selenoenzymes: Comparison with selenomethionine in mice. *Free Radic. Biol. Med.* **2007**, *42*, 1524–1533. [CrossRef] [PubMed]
43. Unsal, V.; Dalkiran, T.; Çiçek, M.; Kölükçü, E. The Role of Natural Antioxidants Against Reactive Oxygen Species Produced by Cadmium Toxicity: A Review. *Adv. Pharm. Bull.* **2020**, *10*. [CrossRef] [PubMed]
44. Jurczuk, M.; Brzóska, M.M.; Moniuszko-Jakoniuk, J.; Gałażyn-Sidorczuk, M.; Kulikowska-Karpińska, E. Antioxidant enzymes activity and lipid peroxidation in liver and kidney of rats exposed to cadmium and ethanol. *Food Chem. Toxicol.* **2004**, *42*, 429–438. [CrossRef]
45. Yuan, G.; Dai, S.; Yin, Z.; Lu, H.; Jia, R.; Xu, J.; Song, X.; Li, L.; Shu, Y.; Zhao, X. Toxicological assessment of combined lead and cadmium: Acute and sub-chronic toxicity study in rats. *Food Chem. Toxicol.* **2014**, *65*, 260–268. [CrossRef]
46. Rao, B.S.; Sreedevi, M. Cytoprotective and antigenotoxic potential of Mangiferin, a glucosylxanthone against cadmium chloride induced toxicity in HepG2 cells. *Food Chem. Toxicol.* **2009**, *47*, 592–600. [CrossRef]
47. El-Demerdash, F.M.; Yousef, M.I.; Kedwany, F.S.; Baghdadi, H.H. Cadmium-induced changes in lipid peroxidation, blood hematology, biochemical parameters and semen quality of male rats: Protective role of vitamin E and β-carotene. *Food Chem. Toxicol.* **2004**, *42*, 1563–1571. [CrossRef] [PubMed]
48. El-Boshy, M.E.; Risha, E.F.; Abdelhamid, F.M.; Mubarak, M.S.; Ben Hadda, T. Protective effects of selenium against cadmium induced hematological disturbances, immunosuppressive, oxidative stress and hepatorenal damage in rats. *J. Trace Elem. Med. Biol.* **2015**, *29*, 104–110. [CrossRef] [PubMed]
49. Alam, M.N.; Bristi, N.J.; Rafiquzzaman, M. Review on in vivo and in vitro methods evaluation of antioxidant activity. *Saudi Pharm. J.* **2013**, *21*, 143–152. [CrossRef]
50. Dzobo, K.; Naik, Y.S. Effect of Selenium on Cadmium-Induced Oxidative Stress and Esterase Activity in Rat Organs. *S. Afr. J. Sci.* **2013**, *109*, 965. [CrossRef]
51. Newairy, A.; El-Sharaky, A.; Badreldeen, M.; Eweda, S.; Sheweita, S. The hepatoprotective effects of selenium against cadmium toxicity in rats. *Toxicology* **2007**, *242*, 23–30. [CrossRef] [PubMed]
52. Forootanfar, H.; Adeli-Sardou, M.; Nikkhoo, M.; Mehrabani, M.; Amir-Heidari, B.; Shahverdi, A.R.; Shakibaie, M. Antioxidant and cytotoxic effect of biologically synthesized selenium nanoparticles in comparison to selenium dioxide. *J. Trace Elem. Med. Biol.* **2014**, *28*, 75–79. [CrossRef] [PubMed]
53. Brzoska, M.M.; Moniuszko-Jakoniuk, J.; Pilat-Marcinkiewicz, B.; Sawicki, B. Liver and Kidney Function and Histology in Rats Exposed to Cadmium and Ethanol. *Alcohol Alcohol.* **2003**, *38*, 2–10. [CrossRef]
54. Al-Waeli, A.; Pappas, A.; Zoidis, E.; Georgiou, C.; Fegeros, K.; Zervas, G. The role of selenium in cadmium toxicity: Interactions with essential and toxic elements. *Br. Poult. Sci.* **2012**, *53*, 817–827. [CrossRef]
55. Messner, B.; Türkcan, A.; Ploner, C.; Laufer, G.; Bernhard, D. Cadmium overkill: Autophagy, apoptosis and necrosis signalling in endothelial cells exposed to cadmium. *Cell. Mol. Life Sci.* **2016**, *73*, 1699–1713. [CrossRef]
56. Oltval, Z.N.; Milliman, C.L.; Korsmeyer, S.J. Bcl-2 heterodimerizes in vivo with a conserved homolog, Bax, that accelerates programed cell death. *Cell* **1993**, *74*, 609–619. [CrossRef]
57. Xiao, C.; Ghosh, S. NF-κB, an Evolutionarily Conserved Mediator of Immune and Inflammatory Responses. *Adv. Exp. Med. Biol.* **2005**, *560*, 41–45. [PubMed]
58. Bonaventura, P.; Lamboux, A.; Albarede, F.; Miossec, P. Differential effects of TNF-α and IL-1β on the control of metal metabolism and cadmium-induced cell death in chronic inflammation. *PLoS ONE* **2018**, *13*, e0196285. [CrossRef]
59. Olszowski, T.; Baranowska-Bosiacka, I.; Gutowska, I.; Chlubek, D. Pro-inflammatory properties of cadmium. *Acta Biochim. Pol.* **2012**, *59*, 475–482. [CrossRef] [PubMed]
60. Hyder, O.; Chung, M.; Cosgrove, D.; Herman, J.M.; Li, Z.; Firoozmand, A.; Gurakar, A.; Koteish, A.; Pawlik, T.M. Cadmium Exposure and Liver Disease among US Adults. *J. Gastrointest. Surg.* **2013**, *17*, 1265–1273. [CrossRef]
61. Lonardo, A.; Nascimbeni, F.; Ballestri, S.; Fairweather, D.; Win, S.; Than, T.A.; Abdelmalek, M.F.; Suzuki, A. Sex Differences in Nonalcoholic Fatty Liver Disease: State of the Art and Identification of Research Gaps. *Hepatology* **2019**, *70*, 1457–1469. [CrossRef] [PubMed]

materials

Article

Nano-Scale Modifications of Amniotic Membrane Induced by UV and Antibiotic Treatment: Histological, AFM and FTIR Spectroscopy Evidence

Simona Cavalu [1], George Roiu [1,*], Ovidiu Pop [1,*], Denisa A. Petricas Heredea [1,*], Traian Octavian Costea [2] and Claudia Florida Costea [3]

1 Faculty of Medicine and Pharmacy, University of Oradea, 10 P-ta 1 Decembrie, 410073 Oradea, Romania; scavalu@uoradea.ro
2 Advanced Materials Research Laboratory, University of Oradea, 1 University Street, 410087 Oradea, Romania; tcostea@uoradea.ro
3 Department of Ophthalmology, "G.T. Popa" University of Medicine and Pharmacy Iasi, 16 University Street, 700115 Iasi, Romania; claudia.costea@umfiasi.ro
* Correspondence: george.roiu@didactic.uoradea.ro (G.R.); popo@uoradea.ro (O.P.); heredea.amaliadenisa@student.uoradea.ro (D.A.P.H.)

Abstract: The efficiency of amniotic membrane (AM) transplantation in different types of ocular surface disorders is due to its outstanding properties such as antifibrotic, antibacterial, anti-inflammatory and antiangiogenic, working as a versatile scaffold to promote corneal tissue epithelialization. A proper preparation, preservation and clinical application are crucial for the best outcomes in the treatment of different severe ocular disorders, taking into account its fragility. In this context, by combining high-sensitivity tools such as atomic force microscopy (AFM) and Fourier transform infrared (FTIR) spectroscopy with histological and immunohistochemical examination, we aimed to investigate the ultrastructural modifications of the amniotic membrane (AM) upon UV exposure and/or antibiotic treatment, with relevance for clinical applications in ocular surface surgery. From the morphological point of view, we noticed a loss of cuboidal cells in the basal membrane, accompanied by the splitting of collagen fibers upon UV and/or gentamicin treatment, while structural alteration of proteins was evidenced by the FTIR quantitative analysis of the secondary structure. A decrease in α-helix and β-sheet content, accompanied by increased content in less ordered structures (turns, random and side chains), was noticed after all the treatments. At the nano-scale, AFM details showed modifications of collagen fibrils in terms of their thickness and network compaction upon gentamicin and/or UV treatment. The enzymatic digestion assay demonstrated that UV exposure significantly reduces the degradation rate of the AM, while gentamicin treatment promotes an accelerated enzymatic digestion upon UV exposure. In order to highlight the clinical impact of the research, a clinical case is presented showing the relevance of amniotic membrane transplantation in pterygium surgery.

Keywords: amniotic membrane; FTIR spectroscopy; immunohistochemistry; AFM; UV light; antibiotic; pterygium surgery

Citation: Cavalu, S.; Roiu, G.; Pop, O.; Heredea, D.A.P.; Costea, T.O.; Costea, C.F. Nano-Scale Modifications of Amniotic Membrane Induced by UV and Antibiotic Treatment: Histological, AFM and FTIR Spectroscopy Evidence. *Materials* **2021**, *14*, 863. https://doi.org/10.3390/ma14040863

Academic Editor: Francesco Baino
Received: 11 December 2020
Accepted: 9 February 2021
Published: 11 February 2021

1. Introduction

The structural and ultrastructural characteristics of the amniotic membrane (AM) along with its biological properties recommend this natural biomaterial as a support matrix for tissue regeneration, including corneal and conjunctiva surface reconstruction. It is not a substitute, but rather a substrate upon which the epithelial cells can easily grow, differentiate and migrate, helping tissue to regenerate [1]. The placental amniochorionic membrane comprises the inner layer, which is in direct contact with amniotic fluid (amniotic membrane), and the outer layer (chorion), which separates the amnion from the uterus, maternal blood and maternal side of the decidua (Figure 1).

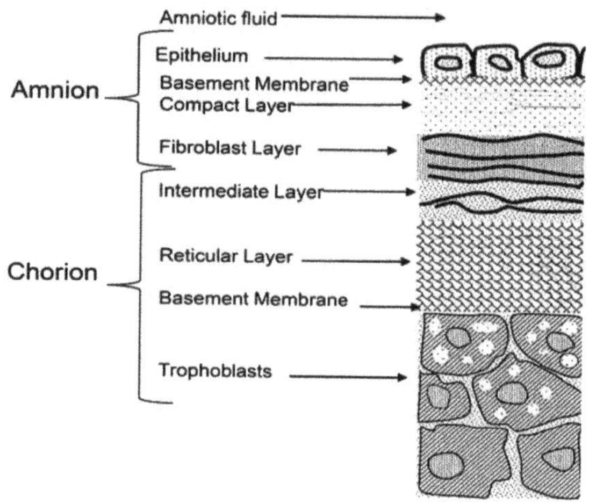

Figure 1. Schematic representation of the layered structure of the amniotic membrane (AM).

The amniotic membrane itself is a thin, transparent, tough, avascular architecture, consisting of three layers: (i) an epithelial monolayer; (ii) a basement membrane; and (iii) an avascular stroma. Histological details reveal that the epithelium consists of metabolically active cuboidal cells, uniformly arranged on the basement membrane [2,3], and is comprised mostly of collagen (type III, IV, V), fibronectin and laminin. The stroma itself is also a layered structure: (i) a compact layer—which is the main fibrous skeleton of the AM, consisting of collagen (types I, III, V, VI) and fibronectin; (ii) a fibroblast layer, composed of fibroblast cells, collagen (types I, III, IV), fibronectin, laminin and nidogen; and (iii) a spongy layer containing mainly collagen (types I, III, IV) and proteoglycans [2]. Due to its layered structure, the AM can be easily separated from the chorion by means of blunt dissection. Not only the layered structure but also the biological properties of the AM provide suitable features for potential tissue engineering applications: anti-inflammatory, antimicrobial, antiviral and antifibrotic properties and the low immunogenicity of HAM (Human Amniotic Membrane) demonstrate a favorable environment for cellular attachment and expansion through an in vivo or in vitro approach [4–8]. Both the epithelial and mesenchymal amniotic cells are pluripotent stem cell reservoirs, while the matrix of the AM is very rich in growth factors such as keratinocyte growth factors (KGFs), fibroblast growth factors (FGFs), transforming growth factor beta (TGFβ), nidogen growth factors (NGFs) and epidermal-derived growth factor (EDGF) [7].

In recent decades, a special attention was paid to applications of the AM in ophthalmic surgery, taking into account the similarities between the AM and the conjunctiva and cornea in terms of its collagen composition and other proteins such as fibronectin and laminin. Various ophthalmic disorders, such as corneal ulcer, corneal perforation, chemical burns, pterygium, infectious or vernal keratitis and bullous keratopathy, have been successfully treated and the clinical results are well documented [6–10]. Currently available commercial AM products are either cryopreserved (-80 °C) as fresh membranes or dried, γ-sterilized, denuded membranes [11]. However, the native intact AM has been found to contain higher levels of growth factors compared to the denuded AM [11,12]. The denuded AM is often used as a culture substrate for limbal epithelial cells using both allogeneic and autologous explants [13]. For this purpose, crosslinking of the AM is necessary, and hence different crosslinking strategies have been used in order to increase the thermal and mechanical stability of the AM for the culture of epithelial cells, including glutaraldehyde, carbodiimide and UV radiation crosslinking [14]. In particular, with respect to the UV crosslinking procedure, it was demonstrated that the biostability of collagenous tissue strongly depends on the number of crosslinked structures, which are closely related to the

UV exposure time [10]. Optimal crosslinking of collagen is essential for collagen binding to its receptors. The matrix permeability is drastically affected by the number of crosslinks per unit mass of the photo-crosslinked AM [15].

Although, with available modern preservation techniques, the membranes are now able to be stored for several months, allowing for scheduled surgical intervention, there is currently a debate over whether the cellular fresh, cryopreserved or dried, denuded form of the AM is considered a better substrate for promotion of epithelization in ocular surface reconstruction. While cryopreserved AMs retain the native architecture of the extracellular matrix and maintain the key biological signals, the dehydrated ones seem to be susceptible to damage in the basal lamina as a result of cell removal [11–16]. Therefore, AMs' proper preparation, preservation and clinical application are crucial for the best outcomes in the treatment of different severe ocular disorders.

Individual post-surgery regimes using topical antibiotics in the form of eye drops are usually prescribed for adjunctive therapy, in order to prevent bacterial keratitis. Commercially available topical antibiotics for this purpose include fluoroquinolones, gentamicin, ciprofloxacin and clarithromycin [17], but some of these may also affect the structural properties of the AM in terms of the hydration recovery of proteins, which has often been interpreted as the main cause of collagen maturation and aging [18]. Post-surgery exposure to UV may also influence the success of ocular surface surgery, knowing the UV influence on the structural and mechanical properties of collagen fibrils [16]. Many studies have been devoted to successful outcomes of either fresh or dehydrated membranes, performed in vivo or in vitro, but there is a lack of information related to the post-surgical conditions and potential detrimental effects of antibiotics as well as UV exposure. Hence, one of the goals of our study was to investigate to what extent the antibiotic concentration will influence the structural properties of amniotic membranes prepared for corneal reconstruction.

In the context of the above-mentioned debate, the overall aim of our work was to investigate the ultrastructural modifications of the AM upon UV exposure and/or antibiotic treatment by combining the histological/immunohistochemical examination with high-sensitivity atomic force microscopy (AFM) measurements and Fourier transform infrared (FTIR) spectroscopy. Moreover, an in vitro enzymatic assay (collagenase digestion) was performed, along with a clinical case presentation (pterygium) requiring corneal surgery and resurfacing with HA showing the most important outcome and advantages of this technique.

2. Materials and Methods

2.1. Procurement and Preparation of Biological Tissue

The research protocol was performed in agreement with the ethical standards of the Helsinki Declaration and approved by the Ethical Committee of the University of Oradea, Romania (ref. nr. 06/15.10.2020). The biological tissue was obtained under strict aseptic conditions, from a patient who had undergone cesarian section at full term, with informed consent. Under a laminar flow hood, the membrane was washed with sterile physiological saline to remove blood clots, separated from the chorion by blunt dissection and peeled. Then, the AM was cut into small-size (5 cm × 5 cm) specimens by using a sterile scalpel; each piece was again washed three times with sterile distilled water and flattened on individual Petri dishes, to which 5 mL PBS (Phosphate Buffer Saline) was added and stored at −20 °C until further treatments and preparation for histological examination, FTIR and AFM measurements.

2.2. Antibiotic and UV Treatment

The AM specimens were divided into 4 groups and the following treatments were applied:

i. Antibiotic treatment: Specimens were allowed to interact with the gentamicin injectable solution (KRKA, Novo Mesto, Slovenia) concentrations of 40 and 80 mg/mL, for 1 h, and then they were washed with PBS, rinsed with sterile distilled water, flattened on a cellulose support and stored in a refrigerator until FTIR and AFM investigations. The samples were labeled AG40 and AG80, respectively, according to each concentration.

ii. UV treatment: Specimens were exposed to UV in air using a GL4 germicidal lamp (Philips TUV 6W G6) at no more than 254 nm, for 1 h, and then kept in a refrigerator until further investigations. The samples were labeled AUV.

iii. Combined antibiotic/UV treatment: Immediately after gentamicin treatment (concentration 40 mg/mL), specimens were exposed to the UV treatment described above and then kept in a refrigerator until further investigations. The samples were labeled AGUV.

iv. The control sample was the natural amniotic membrane without any treatment, labeled AMN.

The above treatments were carried out in triplicate.

2.3. Histological and Immunohistochemical Examination

For the histological and immunohistochemical examination, the specimens were fixed in 4% formaldehyde immediately after the antibiotic or/and UV treatment. The immunohistochemical analysis was performed on 4 µm-thick sections prepared from a formalin-fixed paraffin-embedded block by using an Autostainer Link 48 (Agilent Technologies, Santa Clara, CA, USA). Immunohistochemical assays were used on paraffinized slides to perform target retrieval. The slides were developed using a DAB (3,3′-diaminobenzidine) detection kit and counterstained with hematoxylin. The sections were incubated with anti-Collagen IV (clone CIV 22), mouse monoclonal antibody (Agilent Technologies, Santa Clara, CA, USA). For each case, positive control slides were used, prepared from appendix tissue against a glandular basement membrane. Negative control was performed by omitting the primary antibody.

2.4. FTIR Spectroscopy

For the FTIR measurements, the specimens were left to dry overnight in a refrigerator, at 4 °C. A Shimadzu FT 8400 S (Shimadzu Co., Kyoto, Japan) FTIR spectrophotometer was used, equipped with an MIRacle ATR accessory (ZnSe crystal), operating in the range 400–4000 cm^{-1}, with the following spectral acquisition parameters: wavelength resolution 2.00 cm^{-1}, Happe–Genzel apodization function, 10 scans/spectrum. Data processing was performed using Origin 8 software with a Gaussian–Lorentzian function applied for fitting parameters. The amide I spectral band was baseline-corrected, and the area was normalized and deconvoluted in the interval 1600–1700 cm^{-1}. The percentage of protein secondary structures (α-helix, β-sheets, turns, unordered and side chains) was calculated based on the area under each peak, and the assignments of the components were conducted according to the literature [19–21]. Results were represented as mean ± standard deviation (SD) for three independent experiments. A probability level of $p < 0.05$ was considered statistically significant, and the analysis was conducted by Student's *t*-test.

2.5. AFM Measurement

Atomic force microscopy (Agilent 5500 AFM, Agilent Technologies, Santa Clara, CA, USA) was applied to obtain the ultrastructural details of the amniotic membrane, air-dried, after different treatments. The thickness of individual collagen fibrils was measured by recording their profile. The specimens were fixed on glass plates with a double-sided tape. The scanning was performed at room temperature and a normal humidity level (50%), in acoustic mode (taping mode), in which the AFM tip was oscillating slightly below its resonance frequency of 317.14 kHz and scanned the selected area with a speed of 5.361 m^{-6}/s at a resolution of 512 × 512 data points, in contact mode, in which the AFM tip maintained a close contact with the surface of the sample while scanning the selected area with a speed of 2.160 m^{-6}/s at a resolution of 512 × 512 data points.

2.6. Enzymatic (Collagenase) Degradation Assay

The AM specimens prepared according to Sections 2.1 and 2.2 were allowed to dry, and then they were weighted (balance model FA-G Want Balance Instr. Changzhou, China), immersed individually in a 0.1% collagenase solution (Sigma–Aldrich. St. Louis, MO, USA) with pH = 7 and incubated at 37 °C (incubator Model MCO-5 AC, Sanyo/Panasonic Biomedical, York, UK) in static conditions, under a flux of 5% CO_2. After different time intervals (12, 24, 36, 48, 60 and 72 h), the specimens were carefully removed and allowed to dry completely and weighted again. The weight of the remaining mass was expressed as a percentage (mean value ± SD) and the statistical significance was measured by an ANOVA test. A p-value of less than 0.05 was considered significant.

2.7. Clinical Case

A case of recurrent pterygium is presented, in which the AM (fresh, preserved at −20 °C) was successfully applied in order to reconstruct the corneal surface, after the surgical removal of the damaged tissue. The patient (30 years old, male, living in a rural area) was previously informed about the surgical procedure and he gave his informed consent for inclusion in the study, in accordance with the Declaration of Helsinki and the research protocol approved by the Ethical Committee of the University of Oradea, Romania (ref. nr. 06/15.10.2020). Before surgery, a complete ophthalmologic examination including measurement of intraocular pressure, visual acuity and biomicroscopy was performed.

2.8. Statistics

Results were expressed as mean ± standard deviation for three independent experiments, using Student's t-test in the case of FTIR measurements, while one way analysis of variance (ANOVA) was employed for the enzymatic digestion assay. In both cases, significance was accepted with $p < 0.05$.

3. Results

3.1. Histological Examination

The details of H&E staining reveal a uniform layer of cubic cells displayed on a basement membrane of the AMN sample (Figure 2a), while the expression of collagen IV in the basement membrane is continuous and dense (Figure 2b). After exposure to UV light, a moderate loss of cubic cells can be seen in several spots (Figure 2c), with a dense and continuous expression of collagen IV; a splitting of about 8 μm in collagen fibers was detected in a single spot (Figure 2d). The cuboidal cells slightly changed their size and shape after the treatment. The loss of cubic cells can be noticed also after gentamicin treatment (Figure 2e), accompanied by a splitting of about 3 μm in collagen fibers, over a relatively long distance (Figure 2f). By applying a double gentamicin concentration, the basement membrane maintained its integrity, but the loss of cubic cells can be noticed in a higher number of foci (Figure 2g), while the splitting of the collagen fibers is more significant, the distance between the two expressions being variable, from 8 to 20 μm (Figure 2h). The UV exposure of gentamicin-treated specimens showed extensive loss of cubic cells (Figure 2i), concomitantly with a significant splitting of collagen fibers (about 70 μm) over a long length, as shown by the expression of collagen IV (Figure 2j).

3.2. FTIR Spectroscopy

The FTIR spectra recorded in the range 400–4000 cm^{-1} are presented in Figure 3 for AM specimens belonging to each treatment group, while in Figure 4, the FTIR spectrum of pure gentamicin is presented.

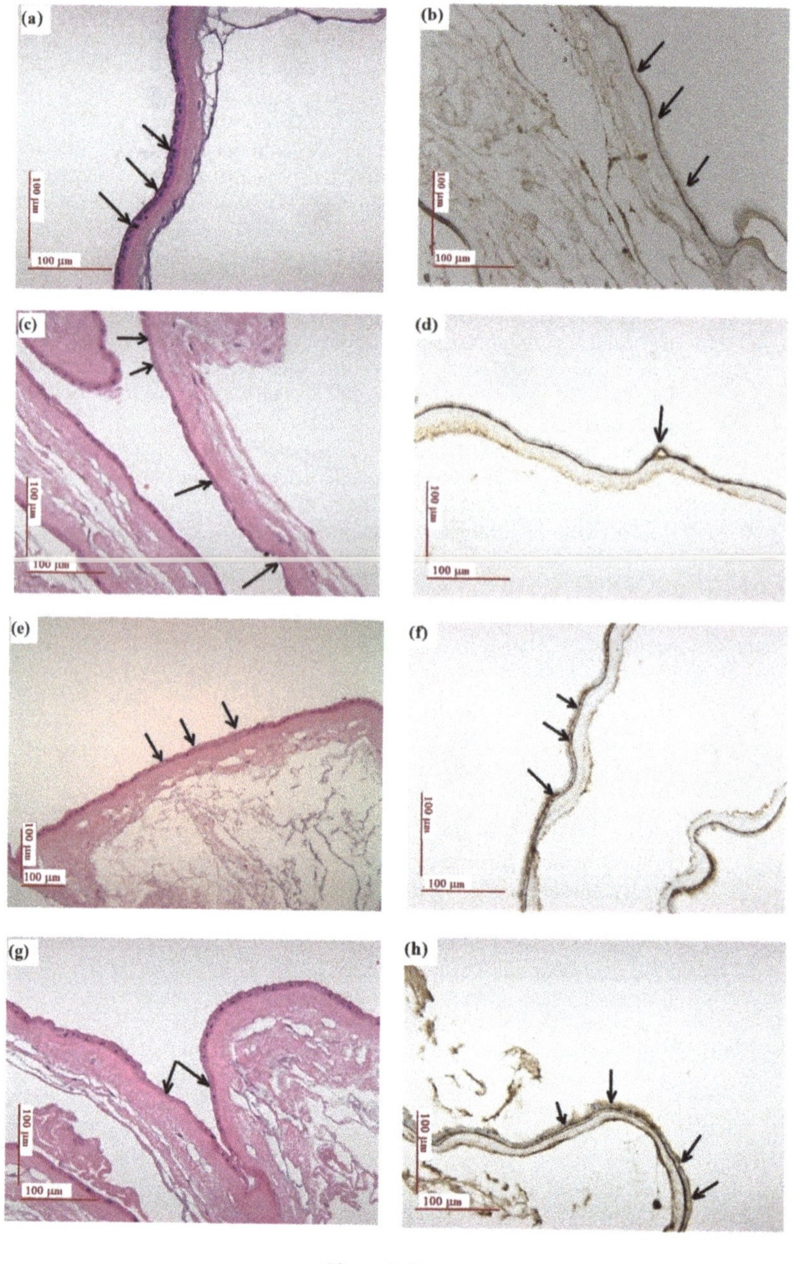

Figure 2. *Cont.*

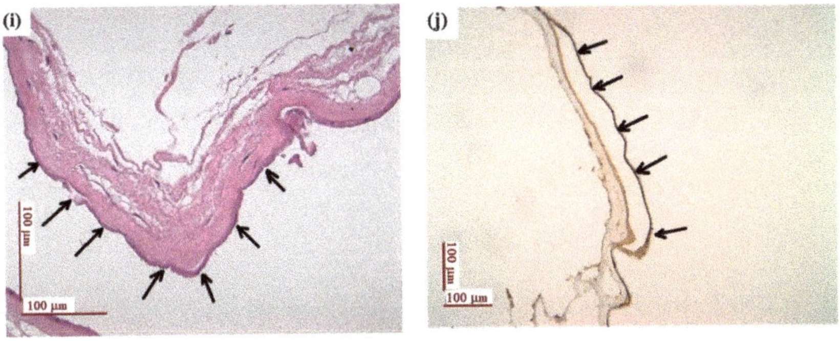

Figure 2. Histological and immunohistochemical examination: (**a**,**b**) natural amniotic membrane (AMN); (**c**,**d**) amniotic membrane exposed to UV for 1 h (AUV); (**e**,**f**) amniotic membrane treated with gentamicin (40 mg/mL) (AG40); (**g**,**h**) amniotic membrane treated with gentamicin 80 mg/mL (AG0); (**i**,**j**) amniotic membrane treated with gentamicin 40 mg/mL and exposed to UV for 1 h (AGUV). Left panel: H&E staining; right panel: immunohistochemistry staining of collagen IV (antibody clone CIV 22). Scale: 100 μm.

In the low-wavenumber region, the absorption band around 1659 cm^{-1} corresponds to the amide I protein absorption band, being assigned to the C=O stretching mode, while the absorption band around 1550 cm^{-1} corresponds to amide II absorption attributed to the N–H bending mode and C–N stretching mode. Amide III protein absorption is located between 1240 and 1300 cm^{-1}, as a result of the in-phase combination of C–N stretching and N–H in-plane bending, and also the contribution from C–C stretching and C=O bending vibrations. The band close to 1450 cm^{-1} is probably associated with C–H bending modes and the amide A band (NH stretching) observed in the high-wavenumber region, between 3190 and 3300 cm^{-1}, suggesting a low water content in the sample [21–23]. The peaks at 1399 and 630 cm^{-1} are attributable to the carboxylate ion and the C=O planar deformation. The band at 1100 cm^{-1} is attributed to the phosphodiester group of nucleic acids and glyco- and phospho-lipids, while the shoulder at 1030 cm^{-1} represents the fingerprint of gentamicin [17], as emphasized in Figure 4. In the higher-wavenumber region, the interval 3190–3300 cm^{-1} is assigned to O–H and C–H bending vibrations, while the band at 2950 cm^{-1} can be assigned to the asymmetric stretching mode of the CH_3 group. The O–H stretching region can be correlated to the hydrogen bond network around the protein, assuming that a certain hydration level still exists after the drying procedure [19,24]. However, water deprivation and restitution might have some consequences on the macromolecular structure. As a general behavior, after antibiotic and or/UV treatment, the main fingerprints of proteins shifted towards higher wavenumbers (with about 15 cm^{-1}), concomitantly with changes in the relative intensity of the amide I, II and III bands.

Computational processing of the FTIR spectra is required because FTIR spectroscopic features contain thousands of singular spectra. For this reason, a fitting procedure was applied in order to evaluate, from the quantitative point of view, the structural modifications of the protein matrix, after UV and antibiotic treatment. The most important structural changes can be evidenced through the computational analysis of the amide I band, which is commonly used to evaluate different secondary structure elements, taking into account that the amide I band is composed of superimposed absorption bands from the α-helix (1659 cm^{-1}), β-sheets (1628 cm^{-1}), turns (1678 cm^{-1}), unordered or random coil contributions (1646 cm^{-1}) and side chains (1613 cm^{-1}) [21,23–26].

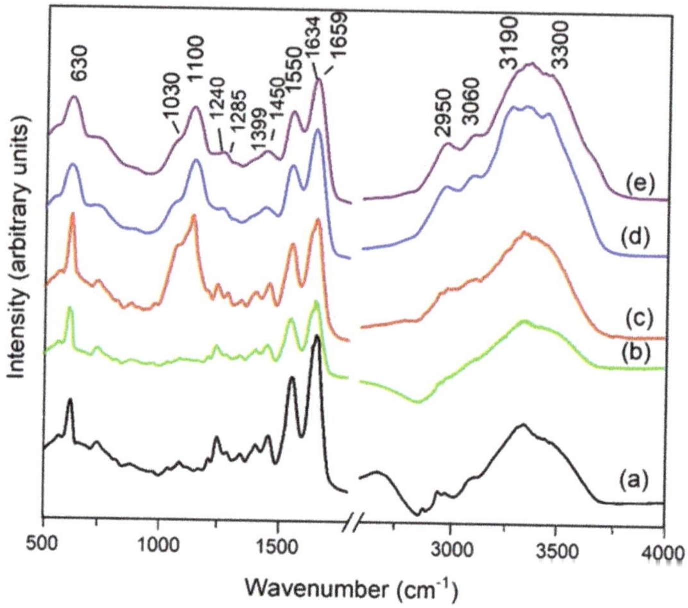

Figure 3. FTIR spectra of AM specimens after treatment with antibiotic and/or UV: (**a**) natural AM without any treatment (AMN); (**b**) AM exposed to UV for 1 h (AUV); (**c**) AM treated with gentamicin (40 mg/mL) and exposed to UV for 1 h (AGUV); (**d**) AM treated with gentamicin 40 mg/mL (AG40); (**e**) AM treated with gentamicin 80 mg/mL (AG80).

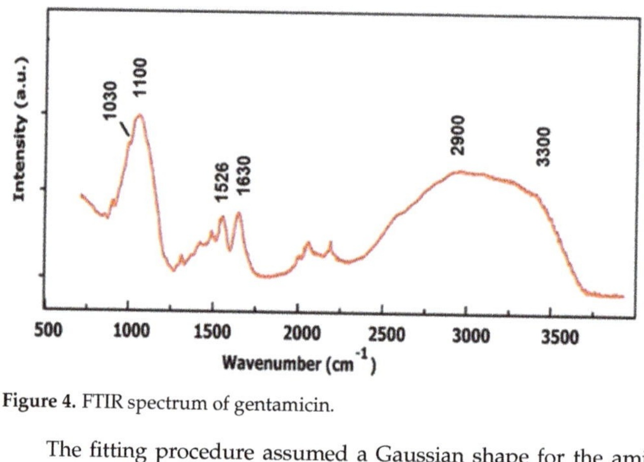

Figure 4. FTIR spectrum of gentamicin.

The fitting procedure assumed a Gaussian shape for the amide I band envelope and individual contributions. Each component of the secondary structure corresponds to different C=O stretching frequencies, resulting in different band positions. The sum of the areas under each peak represents the total amount of secondary structures in the protein. The computational fitting of the amide I vibrational band of the natural amniotic membrane and after antibiotic treatment is presented in Figure 5a–e, while the quantitative assessment of each secondary structure is presented as a diagram in Figure 6.

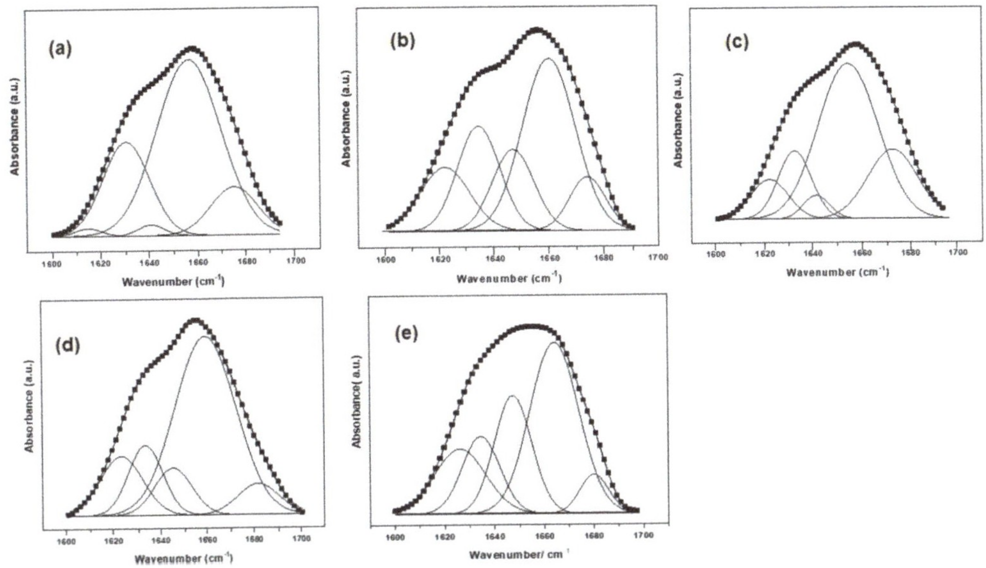

Figure 5. Computational fitting of the amide I FTIR spectroscopy absorption band after different treatments: (**a**) natural membrane (AMN), (**b**) 1 h exposure to UV (AUV), (**c**) gentamicin treatment (40 mg/mL) with 1 h exposure to UV (AGUV), (**d**) gentamicin treatment 40 mg/mL (AG40) and (**e**) gentamicin treatment 80 mg/mL (AG80).

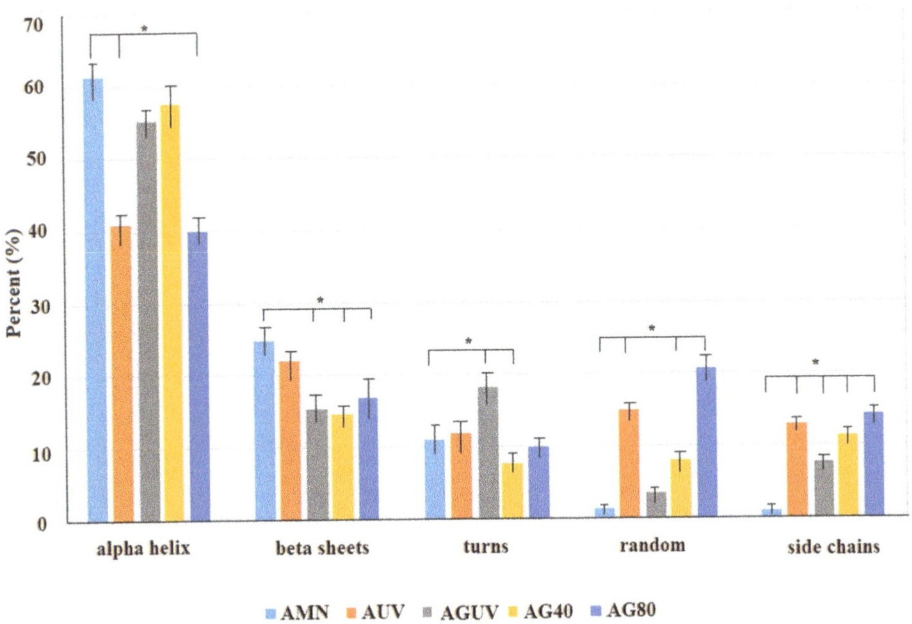

Figure 6. Quantitative analysis (percent) of the collagen secondary structure in the amniotic membrane after different treatments: natural membrane (AMN), 1 h exposure to UV (AUV), gentamicin treatment (40 mg/mL) and 1 h exposure to UV (AGUV), gentamicin treatment 40 mg/mL. * $p < 0.05$ was considered significant.

3.3. Nanotopography: AFM Examination

The 3D and 2D topographic features of the amniotic membrane after different treatments are presented in Figure 7, also indicating the details of a single collagen fibril exposed on the surface (W = width; H = height).

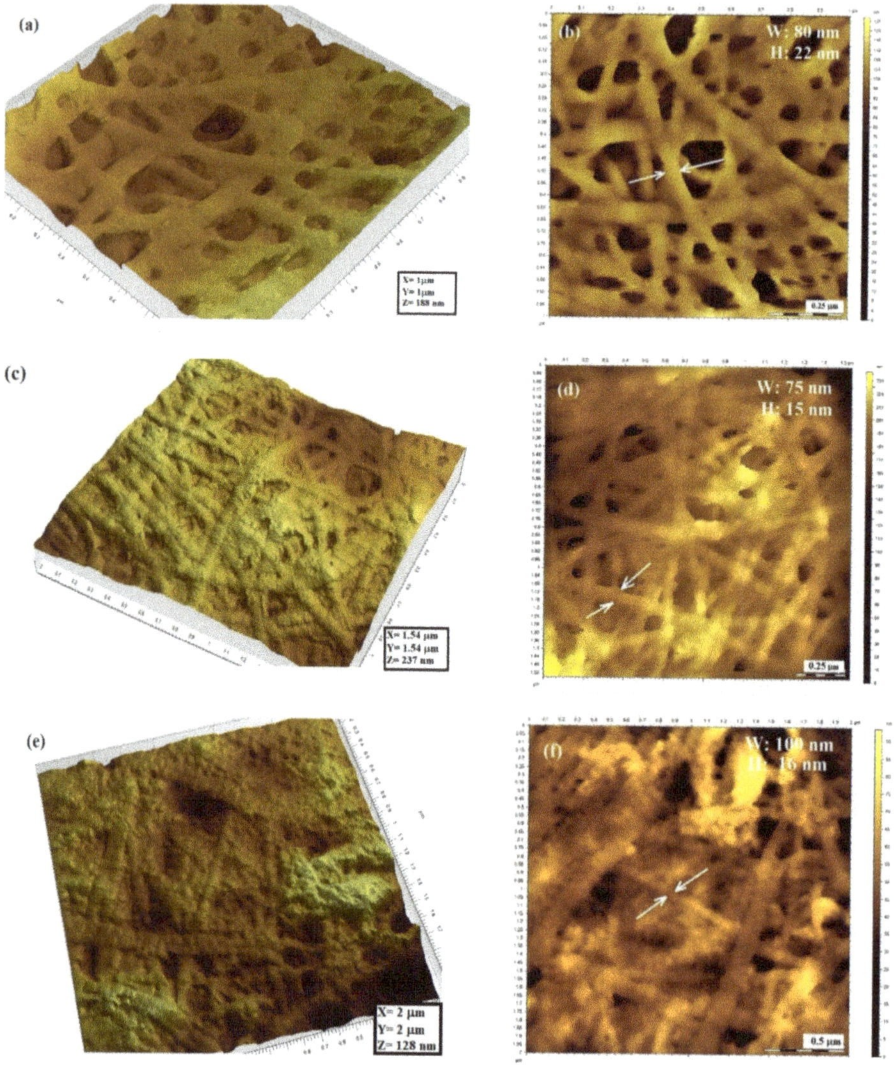

Figure 7. *Cont.*

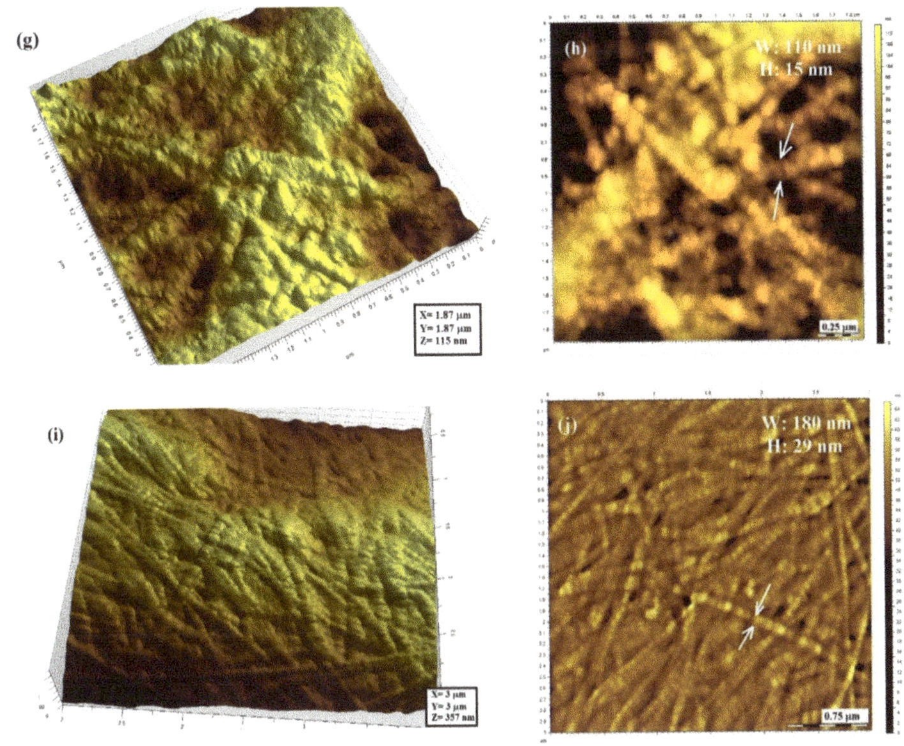

Figure 7. AFM examination of the amniotic membrane after different treatments (3D and 2D images) indicating the features of a single collagen fibril: (**a**,**b**) natural membrane (no treatment) (AMN); (**c**,**d**) membrane exposed to UV (AUV); (**e**,**f**) membrane exposed to gentamicin treatment 40 mg/mL (AG40); (**g**,**h**) membrane exposed to gentamicin treatment 80 mg/mL (AG80); (**i**,**j**) membrane exposed to gentamicin and UV treatment (AGUV). The profiles of a single collagen fibril exposed on the surface of the amniotic membrane after different treatments are presented in the Supplementary Materials (Figure S1).

As presented in Figure 7, the structural collagen fibrils comprising the AM membrane are fully exposed at the surface and randomly arranged in a 3D network. According to Figure 7a,c, the thickness of a single collagen fibril in the natural AM, untreated (AMN), is about 80 nm in width and 22 nm in height. Upon different treatments, the surface roughness was modified, and a more compact structure was noticed, accompanied by changes in the fiber thickness. The most significant change can be noticed for the AGUV sample, reaching 180 nm in width and 29 nm in height. The sole UV treatment does not show significant modifications in terms of fibril thickness. When gentamicin was applied, an increase in the fibril width was noticed, with an insignificant influence of the gentamicin concentration. Although the details of the D-bands of a single collagen fibril are not very clear in our AFM images, we can notice that the D-periodicity of collagen fibrils is not significantly altered by UV radiation or the gentamicin concentration. Further, the length of collagen chains seems to be well preserved within the 3D network.

3.4. Collagenase Digestion Assay

Figure 8 presents the results of the collagenase digestion of the amniotic membrane prior to any treatment and following the UV and/or gentamicin treatment monitored for 72 h. Within the first 12 h, a drastic degradation of AMN, AG40 and AG80 specimens was noticed, compared to AUV and AGUV specimens, and this behavior had a similar, continuous trend for the entire time interval. At each time point, insignificant modifications were noticed between gentamicin-treated samples and untreated ones, no matter the antibiotic concentration. After 72 h of digestion, less then 8% of AMN, AG40 and AG80 remained undigested, while the UV-treated specimens (AUV and AGUV) showed enzymatic resistance of 55% and 32%, respectively, in terms of the mass remaining. By comparing both UV-treated specimens, it can be noticed that gentamicin strongly influences the degradation behavior, which is reflected by the mass remaining when comparing AUV and AGUV percentages at each time point. The difference is more obvious in the time interval 36–72 h. However, for the last 12 h, a slower degradation of AUV and AGUV specimens was noticed.

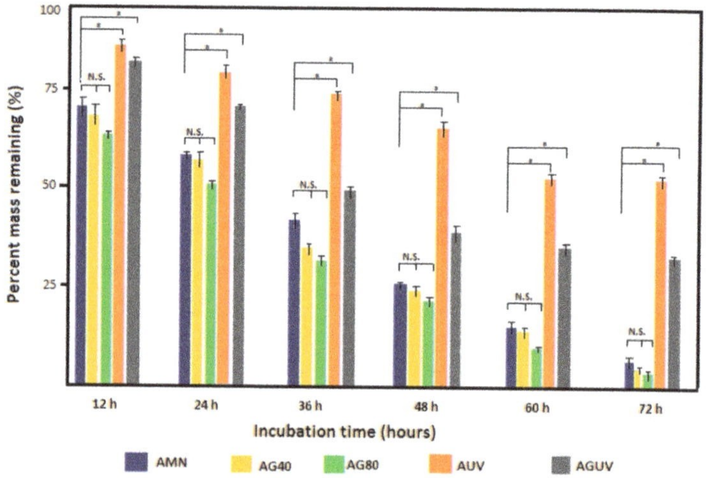

Figure 8. Collagenase digestion of the amniotic membrane after different treatments: natural membrane (AMN), gentamicin treatment with 40 mg/mL (AG40), gentamicin treatment with 80 mg/mL (AG80), 1 h exposure to UV (AUV) and 1 h exposure to UV after gentamicin treatment (AGUV). * $p < 0.05$ was considered significant; N.S.—non-significant.

3.5. Clinical Case

Figure 9a–e presents a clinical situation in which the amniotic membrane was used in ophthalmologic surgery, as a grafting biomaterial, to cover the remaining tissue defect due to the pterygium excision (or its recurrence).

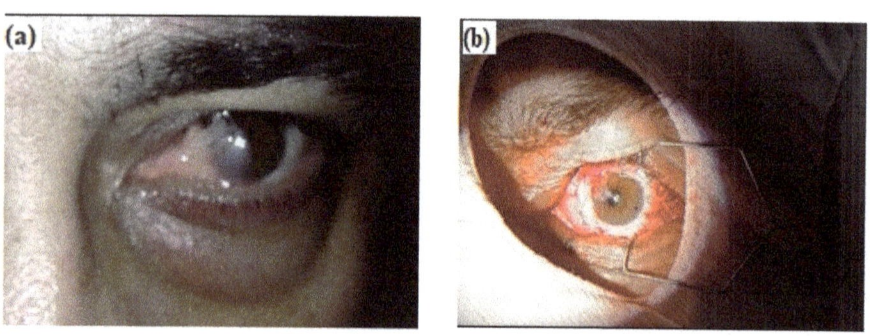

Figure 9. *Cont.*

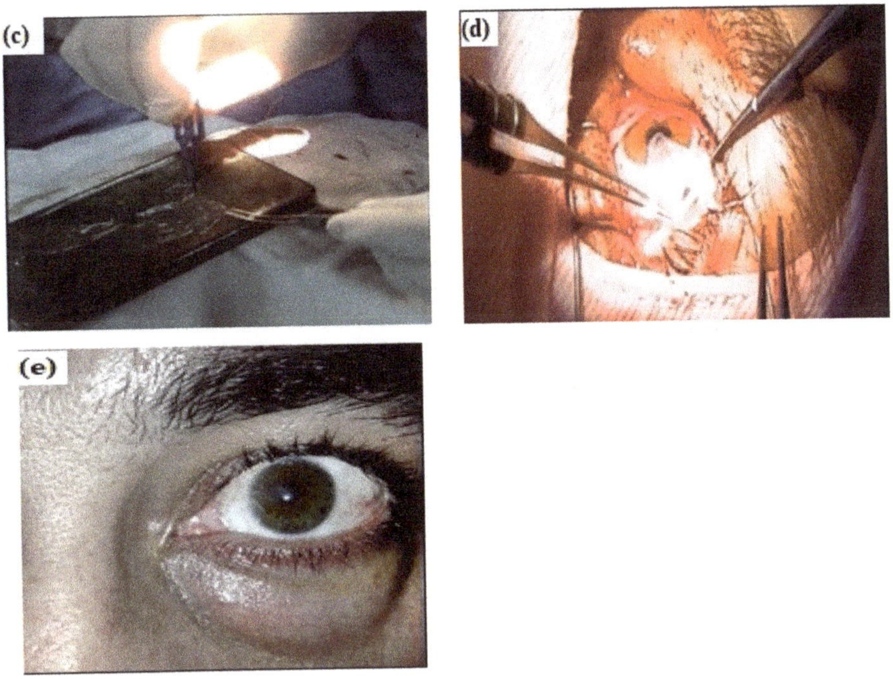

Figure 9. Amniotic membrane used in ophthalmologic surgery to cover the remaining tissue defect due to the grade 3 recurrent pterygium excision: (**a**) grade 3 recurrent pterygium, preoperative appearance; (**b**) intraoperative aspects (the same patient); (**c**) intraoperative preparation of the amniotic membrane fragment to be applied on the remaining defect following recurrent pterygium excision; (**d**) intraoperative appearance—application of the AM fragment on the remaining defect and 10.0 thread suturing to the bulbar conjunctiva of the fragment; (**e**) the aspect of the patient's cornea four weeks after surgery (from private collection of Claudia Florida Costea).

The standard protocol during pterygium surgery was applied, involving the peeling of the abnormal corneal tissue, followed by blunt dissection of the pterygium head, with caution to not affect the corneal stromal tissue. The excision of adjacent residual pterygium tissue was also performed, along with the fibrovascular tissue. The AM (preserved at −20 °C) was cut to fit over the corneal defect, with the stromal side down, in order to promote epithelial adhesion and rapid healing. Immediately after surgery, the eye was patched and shielded. The post-surgical treatment consisted of gentamicin sulfate 3%, dexamethasone and artificial tear instillations and a recommendation of total protection against UV radiation exposure. The patient was examined each day in the first week, and then once a week. After four weeks, complete healing of the corneal tissue was noticed.

4. Discussion

The clinical success of the AM in the reconstruction of ocular surfaces is related to its composition, as the AM contains and releases a remarkable mixture of growth factors and cytokines that facilitate the proliferation and differentiation of epithelial cells, concomitantly with reducing the inflammatory response by inhibiting protease activity [27–29]. It has been demonstrated that the AM could promote ocular surface tissue healing of persistent epithelial defects (corneal ulcers, pterygium, bullous keratopathy, symblepharon and eye burns), provides a substrate for cell growth, has antimicrobial effects and functions as a biological bandage [3,29,30]. As described in the literature, there are three different surgical techniques: graft or inlay, patch or overlay and combined multilayer techniques [31]. In the graft or inlay technique, the AM is administered as a permanent basement membrane substitute, being integrated in the host tissue and acting as a scaffold for epithelial cells to grow. This technique is usually applied to persistent epithelial defects, corneal ulceration or conjunctival tumors, by suture. In the patch or overlay technique, the AM is placed temporarily on the ocular surface, acting as biological bandage, but not being integrated into the host. In this case, the epithelium is expected to grow underneath rather than over the top, and the AM will disassociate from the ocular surface after a certain time [32]. Regardless the surgical technique employed, topical antibiotic application is required in order to eliminate the risk of infectious keratitis. Eye drops are preferred in most cases of multiple bacterial keratitis, consisting of cefalozin, vancomycin, tobramycin, gentamicin or fluoroquinolones, which have been approved by the Food and Drug Administration for the treatment of bacterial keratitis [33,34]. Fortified eye drops, as a recent trend in antimicrobial therapy of bacterial keratitis, relate to formulations with a very high concentration, in order to fight against increasingly resistant Gram-negative bacteria (*S. aureus, P. aeruginosa*), which are known for their ability to destroy the corneal stroma [34,35]. In vitro and in vivo experiments suggest that the antibiotic enters the cornea via direct diffusion rather than through the precorneal tear film [36]. In some animal model experiments, crystalline corneal epithelial deposits were noticed with fortified topical eye drop instillations, and a possible delay of re-epithelialization was assumed [37].

Based on these previous results, the overall goal of our study was to investigate the structural and morphological modifications of the collagen matrix in the AM, by the means of histological observation, FTIR spectroscopy and AFM images, upon treatment with gentamicin and exposure to UV-C radiation. In particular, we also aimed to investigate to what extent the antibiotic concentration will influence the structural properties of the amniotic membrane prepared for corneal reconstruction.

The native membrane (AMN) presents a typical epithelial cell layer, as a single, continuous layer of cuboidal and columnar cells on a densely eosinophilic basement membrane, as revealed in H&E staining (Figure 2a), which is in agreement with previously reported data [2,3,9]. Moreover, the immunohistochemical staining revealed that the basement membrane is very rich in collagen IV, showing a dense and continuous expression. The loss of cubic cells was noticed after gentamicin treatment, regardless of the antibiotic concentration, accompanied by the splitting of the collagen fibers while maintaining the integrity of the basement membrane. When UV treatment was applied to the natural membrane, morphological changes were noticed in terms of cell size and a tendency of collagen fibers to split in double chains. Drastic changes in the morphology of the AM were noticed when antibiotic treatment was followed by UV exposure for 1 h (specimen AGUV). The loss of cuboidal cells was observed, to a large extent, on the length of the basal membrane, while the splitting of collagen fibers was noticed over a very long length, the distance between the two expressions being larger compared to UV treatment alone, as evidenced by the immunohistochemical staining.

In order to obtain deeper details about the possible denaturation of collagen, we further performed FTIR investigations and deconvolution of the amide I absorption band and assessed structural changes in the molecules. Collagen is the principal structural component of the AM, having a unique structure characterized by three polyproline II-like helical chains, packed in a left-handed triple helix which is stabilized by glycine at every third residue. The triple helices assemble to form fibrils, which are then aligned laterally to form bundles and fibers. Under electron microscopy, the collagen molecules within the fibrils present a D-band pattern with periodic gaps and grooves, of about 67 nm, which are correlated to pathological conditions and the fibril mechanical strength [38,39]. As demonstrated by previous research [15,37,38], the native collagen triple helix is most sensitive to UV-254 nm radiation, reflected in changes in the primary and secondary structures and changes in the conformation, microstructure and material properties.

All the treatments applied in our study indicate a decrease in α-helix and β-sheet content, accompanied by increased content in less ordered structures, such as turns and random and side chains, while the position of the main FTIR fingerprints was slightly shifted toward higher wavelengths. The results are consistent with previously reported data [40], which evidenced the behavior of amide I band components in both native and denatured forms of collagen due to UV treatment, suggesting a helix–coil transformation of collagen. The triple-helical regions of collagen are known to be stabilized by hydrogen bonding and van der Waals attractions between imine residues on different chains [15]. Both chemical and UV crosslinking of collagen are associated with its denaturation due to the collapse of hydrogen bonds in polypeptides. Glutaraldehyde, chitosan, poly(2-hydroxyethyl methacrylate), riboflavin and carbodiimide were reported in the literature as efficient crosslinking agents for the AM [25,41,42], showing different degrees of denaturation.

On the other hand, the crosslinking process enhances the biostability of the AM matrices to support limbal epithelial cell growth and can significantly improve the mechanical properties of the collagen matrix [41]. Therefore, although the crosslinking may cause denaturation of the collagen matrix, in vitro experiments of cultures on modified AM samples demonstrated that epithelial progenitor cells were preserved more effectively and the benefits of crosslinking prevail in terms of AM functionality [15]. Moreover, some previous reports demonstrated that ex vivo expansion of limbal epithelial cells occurs at a faster rate on the decellularized AM (with sodium dodecyl sulfate) compared with the fresh, natural one [43]. However, according to some authors [24,42,44], a lower sheet/turn ratio after the treatment indicates inferior biocompatibility when compared to the native AM. Based on the quantitative results presented in Figure 6, the sheet/turn ratio has an average value of 2.27 for the AMN specimens, 1.83 for AUV, 1.88 for AG40, 1.70 for AG80 and 0.83 for AGUV specimens. The lowest value (0.83) indicates the lowest biocompatibility, which was noticed for AGUV specimens, the results being also supported by the histological observations.

Nanotopographic measurements on the surface of the native AM and after different treatments revealed the details of collagen fibrils in a 3D network, which is influenced by the antibiotic treatment and UV exposure. Stylianou et al. demonstrated that in the nano-scale range, the collagen topography and biological properties are affected by UV radiation, which may induce alterations in cell behavior [39]. By applying the AFM technique, they investigated the influence of UV irradiation time (1, 3 and 7 h corresponding to energy doses of 1.1–13.2 J/cm^2) on the collagen film nano-topography. The authors demonstrated that for short irradiation times (10–60 min), the film morphology did not present any significant alteration in terms of roughness, while the periodicity of the D-band pattern was preserved. After longer irradiation times (3–7 h), the roughness was gradually reduced, while after 11 h, the surface was dramatically altered, and the D-band pattern was destroyed, demonstrating an irreversible denaturation process of the triple helix. Based on these results, the authors concluded that UV irradiation can be used as a process for manipulating and controlling the surface roughness of collagen-based materials. In our work, we found modifications in terms of nano-morphology upon 1 h irradiation of the natural AM, namely, a slight reduction in the single fibril thickness from 80 in width and 22 in height to 75 in width and 15 nm in height, while a more compact structure was achieved. An increase in fibril thickness was noticed upon gentamicin treatment, depending on the concentration: 100 width/16 nm height for AG40, and 110 width/15 nm height for AG80, the length of collagen chains being well preserved within the 3D network. Notable changes were identified when UV exposure was applied immediately after gentamicin treatment (AGUV), in terms of collagen chain fragmentation, accompanied by increased thickness of collagen fibrils (180 width/29 nm height), although the D-band pattern for short collagen fragments was not altered, which is in concordance with previously reported data [39,45,46]. However, we could not find in the literature any reference related to the antibiotic influence on ultrastructural changes of the amniotic membrane, except our previous work [18], in which we demonstrated that gentamicin treatment is more favorable compared to ciprofloxacin, by assessing the denaturation process using FTIR spectroscopy. In a recent study performed by Zhang et al. [41], the evaluation of ultrastructural changes of amniotic membrane fragility was assessed after UVA/riboflavin crosslinking. The authors showed that after crosslinking treatment, the AM underwent biomechanical and ultrastructural changes: the brittleness was increased, the hardness was enhanced and the bamboo-like fiber morphology was changed. Another study was performed in order to examine ultrastructural images of the surface morphology of the air-dried and gamma-sterilized AM observed under AFM and to quantify possible changes in epithelial cell size, intercellular gap size and surface roughness [46], but with no relation to the collagen matrix. In this case, AFM results confirmed that both gamma radiation and air drying caused a reduction in cell sizes (about 32.3% reduction in the average longest diameter), but the changes did not affect the gross morphology of the membrane.

The collagenase digestion assay performed in our study demonstrated that UV exposure significantly reduces the degradation rate of AM, including the prior gentamicin treatment, while the gentamicin concentration has an insignificant influence on the dissolution time. It is well known that a fresh amnion usually dissolves within 1 week [42,47], but the crosslinking procedures (by glutaraldehyde, carbodiimide, riboflavin, UV radiation) significantly increase resistance to collagenase digestion, especially in the anterior half of the cornea [14,25,41,42,47]. Photochemical crosslinking is a rapid and efficient process with little toxicity, while chemical crosslinkers (e.g., riboflavin and glutaraldehyde) act as a photomediator, creating free radicals in the presence of UV radiation, promoting the formation of new chemical bonds bridging amino groups in collagen fibrils and subsequently increasing mechanical stiffness [48]. Our results suggest that gentamicin weakens the UV crosslinking effect of the AM, as evidenced by an accelerated degradation of AGUV compared to AUV specimens. These results might have a significant importance for in vivo application of the AM as a substrate for the growth and development of endothelial cells. According to Spoerl et al., the poor biological stability of the AM graft is a main cause of failure and early detachment after surgical transplantation in corneal tissue [42,47].

In order to highlight the clinical impact of our research, a clinical case was presented showing the relevance of pterygium surgery using the AM. Pterygium is a common benign growth of the conjunctiva from the nasal side of the sclera. According to the literature [49], the autograft surgery can be employed in these cases, but it is time-consuming and the recurrence rate is relatively high. As an alternative, the amniotic membrane is often used in order to prevent pterygium recurrence, the effectiveness of this technique being demonstrated especially for the removal of pterygium associated with severe conjunctival disease [50]. The success or failure of this procedure can be evaluated based on the purpose of surgery and by several criteria: resolution of inflammation, relief of symptoms, restoration of the corneal epithelium, recurrence. Recurrence was defined as the presence of fibrovascular proliferative tissue crossing the limbus. The selected clinical case presented in Section 3.5 was a grade 3 recurrent pterygium in which the natural AM (preserved at −20 °C) was successfully applied to cover the remaining defect following pterygium excision. In this case, by using the standard protocol, the membrane displayed scaffold-like properties, providing a healthy and intact basal membrane on which the patient's cells were able to proliferate. Further, measures were taken to prevent AM damage by UV exposure. A complete healing was achieved in 4 weeks, with excellent integration and cosmetic appearance, as well as perfect visual acuity, while gentamicin sulfate (3%) eye drop instillations every two hours and total protection against UV exposure were prescribed. The choice of antibiotics is an important step in surgical management of pterygium, either as a single drug or a combination of fortified antibiotic therapy, in terms of efficacy in the endpoints of complete re-epithelialization.

5. Conclusions

Based on histological examination and complementary AFM and FTIR spectroscopy, we investigated the ultrastructural modification of the amniotic membrane upon gentamicin treatment and/or UV exposure, in the context of its suitability to be used as a graft in corneal reconstruction. The morphological features evidenced by H&E and immunohistochemical investigations were correlated with structural and ultrastructural modifications. The loss of cuboidal cells in the basal membrane was accompanied by the splitting of collagen fibers, which was in concordance with the structural alteration of collagen molecules, as evidenced by the FTIR quantitative analysis of the protein's secondary structure. At the nano-scale, AFM details showed modifications of collagen fibrils in terms of their thickness and network compaction upon gentamicin and/or UV treatment. The collagenase digestion assay demonstrated that UV exposure significantly reduces the degradation rate of the AM, while gentamicin treatment promotes an accelerated enzymatic digestion of the AM upon UV exposure. Within the limitations of our study, the results might have importance in the context of the current debate over whether the AM should be used intact (fresh, preserved at −20 °C) or photo-crosslinked as a grafting material, taking into account the denaturation influenced by post-surgical antibiotic treatment.

Supplementary Materials: The following are available online at https://www.mdpi.com/1996-1944/14/4/863/s1: Figure S1: The AFM profiles of single collagen fibril in amniotic membrane after different treatments: (a) natural membrane (no treatment) (AMN); (b) membrane exposed to UV (AUV); (c) membrane exposed to gentamicin treatment 40 mg/mL (AG40); (d) membrane exposed to gentamicin treatment 80 mg/mL (AG80); (e) membrane exposed to gentamicin and UV treatment (AGUV).

Author Contributions: Conceptualization, S.C. and G.R.; methodology, O.P., C.F.C. and T.O.C.; software, T.O.C.; validation, S.C., C.F.C. and G.R.; formal analysis, D.A.P.H.; investigation, G.R. and D.A.P.H.; resources, G.R. and C.F.C.; data curation, T.O.C. and O.P.; writing—original draft preparation, S.C., O.P., G.R. and D.A.P.H.; writing—review and editing, S.C., O.P. and C.F.C.; visualization, T.O.C.; supervision, S.C. and C.F.C.; project administration, G.R. and C.F.C. All authors have read and agreed to the published version of the manuscript.

Funding: This research received no external funding.

Institutional Review Board Statement: The research protocol was performed in agreement with the ethical standards of the Helsinki Declaration and approved by the Ethical Committee of the University of Oradea, Romania (ref. nr. 06/15.10.2020).

Informed Consent Statement: Not applicable.

Data Availability Statement: The data presented in this study are available on request from the corresponding author.

Conflicts of Interest: The authors declare no conflict of interest.

References

1. Ibrahim, G.S.; Vitresia, H. Multilayer amniotic membrane transplantation for ocular reconstruction (Chapter 18). In *Human Amniotic Membrane: Basic Science and Clinical Application*; Abdul, A.N., Nazly, H., Norimah, Y., Eds.; World Scientific Publishing Co.: Singapore, 2017; p. 289.
2. Niknejad, H.; Peirovi, H.; Jorjani, M.; Ahmadiani, A.; Ghanavi, J.; Seifalian, A.M. Properties of amniotic membrane for potential use in tissue engineering. *Eur. Cells Mater.* **2008**, *15*, 88–99. [CrossRef] [PubMed]
3. Tighe, S.; Mead, O.G.; Lee, A.; Tseng, S.C.G. Basic science review of birth tissue uses in ophthalmology. *Taiwan J. Ophthalmol.* **2020**, *10*, 3–12. [CrossRef]
4. Lim, L.S.; Poh, R.W.Y.; Riau, A.K.; Beuerman, R.W.; Tan, D.; Mehta, J.S. Biological and Ultrastructural Properties of Acelagraft, a Freeze-Dried γ-Irradiated Human Amniotic Membrane. *Arch. Ophthalmol.* **2010**, *128*, 1303–1310. [CrossRef]
5. Inge, E.; Talmi, Y.P.; Sigler, L.; Finkelstein, Y.; Zohar, Y. Antibacterial properties of human amniotic membranes. *Placenta* **1991**, *12*, 285–288. [CrossRef]
6. Hao, Y.; Ma, D.H.-K.; Hwang, D.G.; Kim, W.-S.; Zhang, F. Identification of Antiangiogenic and Antiinflammatory Proteins in Human Amniotic Membrane. *Cornea* **2000**, *19*, 348–352. [CrossRef] [PubMed]
7. Chopra, A.; Thomas, B.S. Amniotic membrane: A novel material for regeneration and repair. *J. Biomim. Biomater. Tissue Eng.* **2013**, *18*, 1–8.
8. Tehrani, F.D.; Firouzeh, A.; Shabani, I.; Shabani, A. A Review on Modifications of Amniotic Membrane for Biomedical Applications. *Front. Bioeng. Biotechnol.* **2021**, *8*. [CrossRef]
9. Rahman, I.; Said, D.G.; Maharajan, V.; Dua, H.S. Amniotic membrane in ophthalmology: Indications and limitations. *Eye* **2009**, *23*, 1954–1961. [CrossRef] [PubMed]
10. Şapte, E.; Costea, C.F.; Cărăuleanu, A.; Dancă, C.; Dumitrescu, G.F.; Dimitriu, G.; Chihaia, M.A.; Buzdugă, C.M.; Cucu, A.; Turliuc, M.D. Histological, immunohistochemical and clinical considerations on amniotic membrane transplant for ocular surface reconstruction. *Romanian J. Morphol. Embryol. Rev. Roum. de Morphol. et Embryol.* **2017**, *58*, 363–369.

11. Utheim, T.P.; Øygunn, A.U.; Salvanos, P.; Jackson, C.J.; Schrader, S.; Geerling, G.; Sehic, A. Concise Review: Altered Versus Unaltered Amniotic Membrane as a Substrate for Limbal Epithelial Cells. *Stem. Cells Transl. Med.* **2018**, *7*, 415–427. [CrossRef]
12. Cooke, M.; Tan, E.; Mandrycky, C.; He, H.; O'Connell, J.; Tseng, S. Comparison of cryopreserved amniotic membrane and umbilical cord tissue with dehydrated amniotic membrane/chorion tissue. *J. Wound Care* **2014**, *23*, 465–476. [CrossRef]
13. Satake, Y.; Shimmura, S.; Shimazaki, J. Cultivated autologous limbal epithelial transplantation for symptomatic bullous keratopathy. *BMJ Case Rep.* **2009**, *2009*. [CrossRef]
14. Fujisato, T.; Tomihata, K.; Tabata, Y.; Iwamoto, Y.; Burczak, K.; Ikada, Y. Cross-linking of amniotic membranes. *J. Biomater. Sci. Polym. Ed.* **1999**, *10*, 1171–1181. [CrossRef]
15. Lai, J.-Y. Photo-cross-linking of amniotic membranes for limbal epithelial cell cultivation. *Mater. Sci. Eng. C* **2014**, *45*, 313–319. [CrossRef]
16. Jariashvili, K.; Madhan, B.; Brodsky, B.; Kuchava, A.; Namicheishvili, L.; Metreveli, N. Uv damage of collagen: Insights from model collagen peptides. *Biopolymers* **2012**, *97*, 189–198. [CrossRef]
17. Wang, B.; Yang, S.; Zhai, H.-L.; Zhang, Y.-Y.; Cui, C.-X.; Wang, J.-Y.; Xie, L.-X. A comparative study of risk factors for corneal infection in diabetic and non-diabetic patients. *Int. J. Ophthalmol.* **2018**, *11*, 43–47. [CrossRef] [PubMed]
18. Roiu, G.; Cavalu, S.; Teusdea, A.; Petricas-Heredea, D.A.; Fratila, O. Assessment of Antibiotic Influence on Structural Modifications of Amniotic Membrane by FTIR Spectroscopy. *Mater. Plast.* **2020**, *57*, 191–198. [CrossRef]
19. Cavalu, S.; Popa, A.; Bratu, I.; Borodi, G.; Maghiar, A. New Evidences of Key Factors Involved in "Silent Stones" Etiopathogenesis and Trace Elements: Microscopic, Spectroscopic, and Biochemical Approach. *Biol. Trace Elem. Res.* **2015**, *168*, 311–320. [CrossRef] [PubMed]
20. Bridelli, M.G. Fourier Transform Infrared Spectroscopy in the Study of Hydrated Biological Macromolecules (Chapter). In *Fourier Transforms-High-Tech Application and Current Trends;* Nikolic, G.S., Cakic, M.D., Cvetkovic, D.J., Eds.; IntechOpen: London, UK, 2017. [CrossRef]
21. Belbachir, K.; Noreen, R.; Gouspillou, G.; Petibois, C. Collagen types analysis and differentiation by FTIR spectroscopy. *Anal. Bioanal. Chem.* **2009**, *395*, 829–837. [CrossRef] [PubMed]
22. Sripriya, R.; Kumar, R. Denudation of human amniotic membrane by a novel process and its characterisations for biomedical applications. *Prog. Biomater.* **2016**, *5*, 161–172. [CrossRef]
23. Nishida, Y.; Yoshida, S.; Li, H.J.; Higuchi, Y.; Takai, N.; Miyakawa, I. FTIR spectroscopic analyses of human placental membranes. *Biopolymers* **2000**, *62*, 22–28. [CrossRef]
24. Stancanelli, R.; Ficarra, R.; Cannavà, C.; Guardo, M.; Calabrò, M.; Ficarra, P.; Ottana, R.; Maccari, R.; Crupi, V.; Majolino, D.; et al. UV–vis and FTIR-ATR characterization of 9-fluorenon-2-carboxyester/(2-hydroxypropyl)-β-cyclodextrin inclusion complex. *J. Pharm. Biomed. Anal.* **2008**, *47*, 704–709. [CrossRef]
25. Lai, J.-Y.; Lue, S.J.; Cheng, H.-Y.; Ma, D.H.-K. Effect of matrix nanostructure on the functionality of carbodiimide cross-linked amniotic membranes as limbal epithelial cell scaffolds. *J. Biomed. Nanotechnol.* **2013**, *9*, 2048–2062. [CrossRef]
26. Tunç, S.; Maitz, M.F.; Steiner, G.; Vazquez, L.; Pham, M.T.; Salzer, R. In situ conformational analysis of fibrinogen adsorbed on Si surfaces. *Colloids Surf. B Biointerfaces* **2005**, *42*, 219–225. [CrossRef] [PubMed]
27. Magyari, K.; Vanea, E.; Baia, L.; Simon, V. Attachment and conformational changes of collagen on bioactive glass surface. *Bio-Med. Mater. Eng.* **2016**, *27*, 63–74. [CrossRef]
28. Fan, J.; Wang, M.; Zhong, F. Improvement of Amniotic Membrane Method for the Treatment of Corneal Perforation. *BioMed Res. Int.* **2016**, *2016*, 1–8. [CrossRef] [PubMed]
29. Dekaris, I.; Gabrić, N.; Mravičić, I.; Karaman, Ž.; Katušić, J.; Lazić, R.; Špoljarić, N. Multilayer vs. monolayer amniotic membrane transplantation for deep corneal ulcer treatment. *Coll. Antropol.* **2001**, *25*, 23–28. [PubMed]
30. Jhanji, V.; Young, A.L.; Mehta, J.S.; Sharma, N.; Agarwal, T.; Vajpayee, R.B. Management of Corneal Perforation. *Surv. Ophthalmol.* **2011**, *56*, 522–538. [CrossRef]
31. Cauchi, P.A.; Ang, G.S.; Azuara-Blanco, A.; Burr, J.M. A Systematic Literature Review of Surgical Interventions for Limbal Stem Cell Deficiency in Humans. *Am. J. Ophthalmol.* **2008**, *146*, 251–259.e2. [CrossRef]
32. Krysik, K.; Dobrowolski, D.; Wylegala, E.; Lyssek-Boron, A. Amniotic Membrane as a Main Component in Treatments Supporting Healing and Patch Grafts in Corneal Melting and Perforations. *J. Ophthalmol.* **2020**, *2020*, 4238919-7. [CrossRef]
33. McDonald, E.M.; Ram, F.S.F.; Patel, D.V.; McGhee, C.N.J. Topical antibiotics for the management of bacterial keratitis: An evidence-based review of high quality randomized controlled trials. *Br. J. Ophthalmol.* **2014**, *98*, 1470–1477. [CrossRef] [PubMed]
34. Baum, J.; Barza, M. The Evolution of Antibiotic Therapy for Bacterial Conjunctivitis and Keratitis: 1970–2000. *Cornea* **2000**, *19*, 659–672. [CrossRef] [PubMed]
35. Nahata, M.C.; Hipple, T.F. Gentamycin sulfate fortified ophthalmic drops. *Pediatr. Drug Formul.* **2004**, *5*, 130.
36. Baum, J. Treatment of bacterial ulcers of the cornea in the rabbit: A comparison of administration by eye drops and subconjunctival injections. *Trans. Am. Ophthalmol. Soc.* **1982**, *80*, 369–390. [PubMed]
37. Kanellopoulos, A.J.; Miller, F.; Wittpenn, J.R. Deposition of Topical Ciprofloxacin to Prevent Re-epithelialization of a Corneal Defect. *Am. J. Ophthalmol.* **1994**, *117*, 258–259. [CrossRef]
38. Ortolani, F.; Giordano, M.; Marchini, M. A model for type II collagen fibrils: Distinctive D-band patterns in native and reconstituted fibrils compared with sequence data for helix and telopeptide domains. *Biopolymers* **2000**, *54*, 448–463. [CrossRef]

39. Stylianou, A.; Yova, D.; Alexandratou, E. Investigation of the influence of UV irradiation on collagen thin films by AFM imaging. *Mater. Sci. Eng. C* **2014**, *45*, 455–468. [CrossRef]
40. Rabotyagova, O.S.; Cebe, P.; Kaplan, D.L. Collagen structural hierarchy and susceptibility to degradation by ultraviolet radiation. *Mater. Sci. Eng. C* **2008**, *28*, 1420–1429. [CrossRef]
41. Zhang, C.; Du, T.; Mu, G.; Wang, J.; Gao, X.; Long, F.; Wang, Q. Evaluation and ultrastructural changes of amniotic membrane fragility after UVA/riboflavin cross-linking and its effects on biodegradation. *Medicine* **2020**, *99*, e20091. [CrossRef]
42. Spoerl, E.; Wollensak, G.; Reber, F.; Pillunat, L. Cross-Linking of Human Amniotic Membrane by Glutaraldehyde. *Ophthalmic Res.* **2004**, *36*, 71–77. [CrossRef]
43. Figueiredo, G.; Bojic, S.; Rooney, P.; Wilshaw, S.-P.; Connon, C.; Gouveia, R.; Paterson, C.; Lepert, G.; Mudhar, H.; Figueiredo, F.; et al. Gamma-irradiated human amniotic membrane decellularised with sodium dodecyl sulfate is a more efficient substrate for the ex vivo expansion of limbal stem cells. *Acta Biomater.* **2017**, *61*, 124–133. [CrossRef] [PubMed]
44. Cavalu, S.; Simon, V. Proteins adsorption to orthopedic biomaterials: Vibrational spectroscopy evidence. *J. Optoelectron. Adv. Mater.* **2007**, *9*, 3297–3302.
45. Kontomaris, S.V.; Yova, D.; Stylianou, A.; Balogiannis, G. The effects of UV irradiation on collagen D-band revealed by atomic force microscopy. *Scanning* **2014**, *37*, 101–111. [CrossRef] [PubMed]
46. Mohd, S.; Ghazali, M.I.; Yusof, N.; Sulaiman, S.; Ramalingam, S.; Kamarul, T.; Mansor, A. Quantifying the ultrastructure changes of air-dried and irradiated human amniotic membrane using atomic force microscopy: A preliminary study. *Cell Tissue Bank.* **2018**, *19*, 613–622. [CrossRef] [PubMed]
47. Spoerl, E.; Wollensak, G.; Seiler, T. Increased resistance of crosslinked cornea against enzymatic digestion. *Curr. Eye Res.* **2004**, *29*, 35–40. [CrossRef] [PubMed]
48. Chen, B.; Jones, R.R.; Mi, S.; Foster, J.; Alcock, S.G.; Hamley, I.W.; Connon, C.J. The mechanical properties of amniotic membrane influence its effect as a biomaterial for ocular surface repair. *Soft Matter* **2012**, *8*, 8379–8387. [CrossRef]
49. Lazzaro, D.R.; E Rizzuti, A.; Goldenberg, A. Amniotic membrane allografts: Development and clinical utility in ophthalmology. *Chronic Wound Care Manag. Res.* **2014**, *1*, 67. [CrossRef]
50. Röck, T.; Braumkamp, M.; Bartz-Schmidt, K.U.; Röck, D. A retrospective study to compare the recurrence rate after treatment of pterygium by conjunctival autograft, primary closure and amniotic membrane transplantation. *Med. Sci. Monit.* **2019**, *25*, 7976–7981. [CrossRef]

MDPI
St. Alban-Anlage 66
4052 Basel
Switzerland
Tel. +41 61 683 77 34
Fax +41 61 302 89 18
www.mdpi.com

Materials Editorial Office
E-mail: materials@mdpi.com
www.mdpi.com/journal/materials

www.ingramcontent.com/pod-product-compliance
Lightning Source LLC
Chambersburg PA
CBHW061812210926
51850CB00006B/686
* 9 7 8 3 0 3 6 5 4 5 3 4 9 *